DYNAMICAL SYSTEMS
Theories and Applications

Zeraoulia Elhadj

Department of Mathematics
University of Tébessa, Algeria

CRC Press
Taylor & Francis Group
Boca Raton London New York

CRC Press is an imprint of the
Taylor & Francis Group, an **informa** business

A SCIENCE PUBLISHERS BOOK

T0174833

CRC Press
Taylor & Francis Group
6000 Broken Sound Parkway NW, Suite 300
Boca Raton, FL 33487-2742

First issued in paperback 2020

ISBN-13: 978-0-367-13704-5 (hbk)
ISBN-13: 978-0-367-78039-5 (pbk)

Library of Congress Cataloging-in-Publication Data
Names: Elhadj, Zeraoulia, author.
Title: Dynamical systems : theories and applications / Zeraoulia Elhadj (Department of Mathematics, University of Tâebessa, Algeria).
Description: Boca Raton, FL : CRC Press, 2019.	"A science publishers book."	Includes bibliographical references and index.
Identifiers: LCCN 2018049243	ISBN 9780367137045 (hardback)
Subjects: LCSH: Chaotic behavior in systems.	Dynamics.
Classification: LCC Q172.5.C45 E44 2019	DDC 003/.857--dc23
LC record available at https://lccn.loc.gov/2018049243

Visit the Taylor & Francis Web site at
http://www.taylorandfrancis.com

and the CRC Press Web site at
http://www.crcpress.com

Preface

Chaos: When the present determines the future, but the approximate present does not approximately determine the future.

Edward Lorenz

Chaos theory is the study of the behavior of dynamical systems that are highly sensitive to initial conditions (The butterfly effect): *A butterfly flapping its wings in China can cause a hurricane in Texas.* In other words, small differences in initial conditions yield widely diverging solutions for such dynamical systems and the prediction of their behavior is impossible in general. However, chaos has some patterns such as: Constant feedback loops, repetition, self-similarity, fractals and self-organization. Chaos theory of dynamical systems is very complicated and contains a large number of subjects. This theory has many practical applications, such as secure communications.

The contents of this book are related to 70 different topics of discrete and continuous-time dynamical systems. This book contains more than 100 mathematical statements of various subjects and directions. Almost all of them are associated with their rigorous proofs and with many practical examples and illustrations.

The content of each chapter is as follows: In Chapter 1, we describe some techniques used in the study of chaotic dynamical systems. In particular, the Poincar map technique, the Smale horseshoe, the symbolic dynamics along different types of entropies such as the physical (or Sinai-Ruelle-Bowen) measure, Hausdorff dimension, topological entropy, Lebesgue (volume) measure, and Lyapunov exponents. Chapter 2 looks at the Human Immunodeficiency Virus and urbanization dynamics as examples of less popular topics of discrete mappings. In particular, the rural-urban interaction attractor is presented and discussed. In Chapter 3, a collection of results regarding different methods used for the rigorous proof of chaos in 2-D piecewise maps. Chapter 4 is a discussion about robust chaos in neural networks models with many illustrating examples and methods. Chapter 5 is devoted to presenting some results about the rigorous determination of the Lyapunov exponents of some 2-D discrete mappings, such as the discrete hyperchaotic double scroll mapping and the discrete butterfly. Chapter 6 is a collection of various methods of control, synchronization and chaotification of some continuous time systems and discrete time systems. In Chapter 7, deep and relevant methods, used to prove boundedness of certain forms of discrete time and continuous time systems, are presented and discussed in detail. Chapter 8 is devoted to presenting some definitions and relevant results of some forms of globally asymptotically stable discrete time mappings and continuous time systems.

Finally, Chapter 9 collects some relevant results about transforming dynamical systems to jerky and hyperjerk forms.

Readership: Advanced undergraduates and graduate students in science and engineering; researchers, instructors, mathematicians, non-linear scientists and electronic engineers interested in chaos, non-linear dynamics and dynamical systems.

This book can be considered as a guide for problems and exercises for young researchers.

At this end, we notice that the notations used in this book are independent and vary from a chapter to another. Therefore, each section has its proper notations.

Zeraoulia Elhadj
Department of Mathematics
August 2018   University of Tébessa, 12002, Algeria.

Contents

Preface   iii

1. Review of Chaotic Dynamics   1

  1.1 Introduction   1
  1.2 Poincaré map technique   1
  1.3 Smale horseshoe   3
  1.4 Symbolic dynamics   7
  1.5 Strange attractors   10
  1.6 Basins of attraction   13
  1.7 Density, robustness and persistence of chaos   16
  1.8 Entropies of chaotic attractors   23
  1.9 Period 3 implies chaos   32
  1.10 The Snap-back repeller and the Li-Chen-Marotto theorem   37
  1.11 Shilnikov criterion for the existence of chaos   38

2. Human Immunodeficiency Virus and Urbanization Dynamics   46

  2.1 Introduction   46
  2.2 Definition of Human Immunodeficiency Virus (HIV)   46
  2.3 Modelling the Human Immunodeficiency Virus (HIV)   48
  2.4 Dynamics of sexual transmission of the Human Immunodeficiency   49
       Virus
  2.5 The effects of variable infectivity on the HIV dynamics   53
  2.6 The CD4+ Lymphocyte dynamics in HIV infection   56
  2.7 The viral dynamics of a highly pathogenic Simian/Human   60
       Immunodeficiency Virus
  2.8 The effects of morphine on Simian Immunodeficiency Virus Dynamics   65
  2.9 The dynamics of the HIV therapy system   67
  2.10 Dynamics of urbanization   71

3. Chaotic Behaviors in Piecewise Linear Mappings   78

  3.1 Introduction   78
  3.2 Chaos in one-dimensional piecewise smooth maps   78
  3.3 Chaos in one-dimensional singular maps   83
  3.4 Chaos in 2-D piecewise smooth maps   86

4. Robust Chaos in Neural Networks Models   **96**

4.1   Introduction   96
4.2   Chaos in neural networks models   97
4.3   Robust chaos in discrete time neural networks   98
     4.3.1   Robust chaos in 1-D piecewise-smooth neural networks   101
     4.3.2   Fragile chaos (blocks with smooth activation function)   102
     4.3.3   Robust chaos (blocks with non-smooth activation function)   105
     4.3.4   Robust chaos in the electroencephalogram model   109
     4.3.5   Robust chaos in Diluted circulant networks   113
     4.3.6   Robust chaos in non-smooth neural networks   114
4.4   The importance of robust chaos in mathematics and some open problems   115

5. Estimating Lyapunov Exponents of 2-D Discrete Mappings   **117**

5.1   Introduction   117
5.2   Lyapunov exponents of the discrete hyperchaotic double scroll map   117
5.3   Lyapunov exponents for a class of 2-D piecewise linear mappings   121
5.4   Lyapunov exponents of a family of 2-D discrete mappings with separate variables   124
5.5   Lyapunov exponents of a discontinuous piecewise linear mapping of the plane governed by a simple switching law   128
5.6   Lyapunov exponents of a modified map-based BVP model   135

6. Control, Synchronization and Chaotification of Dynamical Systems   **142**

6.1   Introduction   142
6.2   Compound synchronization of different chaotic systems   143
6.3   Synchronization of 3-D continuous-time quadratic systems using a universal non-linear control law   151
6.4   Co-existence of certain types of synchronization and its inverse   155
6.5   Synchronization of 4-D continuous-time quadratic systems using a universal non-linear control law   164
6.6   Quasi-synchronization of systems with different dimensions   169
6.7   Chaotification of 3-D linear continuous-time systems using the signum function feedback   173
6.8   Chaos control problem of a 3-D cancer model with structured uncertainties   181
6.9   Controlling homoclinic chaotic attractor   182
6.10 Robustification of 2-D piecewise smooth mappings   186
6.11 Chaotifying stable n-D linear maps via the controller of any bounded function   192

7. Boundedness of Some Forms of Quadratic Systems — 197

7.1 Introduction — 197
7.2 Boundedness of certain forms of 3-D quadratic continuous-time systems — 197
7.3 Bounded jerky dynamics — 202
    7.3.1 Boundedness of general forms of jerky dynamics — 205
    7.3.2 Examples of bounded jerky chaos — 214
    7.3.3 Appendix A — 216
7.4 Bounded hyperjerky dynamics — 220
7.5 Boundedness of the generalized 4-D hyperchaotic model containing Lorenz-Stenflo and Lorenz-Haken systems — 224
    7.5.1 Estimating the bounds for the Lorenz-Haken system — 228
    7.5.2 Estimating the bounds for the Lorenz-Stenflo system — 229
7.6 Boundedness of 2-D Hénon-like mapping — 233
7.7 Examples of fully bounded chaotic attractors — 239

8. Some Forms of Globally Asymptotically Stable Attractors — 242

8.1 Introduction — 242
8.2 Direct Lyapunov stability for ordinary differential equations — 243
8.3 Exponential stability of non-linear time-varying — 251
8.4 Lasalle's Invariance Principle — 259
8.5 Direct Lyapunov-type stability for fractional-like systems — 261
8.6 Construction of globally asymptotically stable n-D discrete mappings — 267
8.7 Construction of superstable n-D mappings — 271
8.8 Examples of globally superstable 1-D quadratic mappings — 275
8.9 Construction of globally superstable 3-D quadratic mappings — 280
8.10 Hyperbolicity of dynamical systems — 285
8.11 Consequences of uniform hyperbolicity — 297
    8.11.1 Classification of singular-hyperbolic attracting sets — 298
8.12 Structural stability for 3-D quadratic mappings — 302
    8.12.1 The concept of structural stability — 302
    8.12.2 Conditions for structural stability — 303
    8.12.3 The Jordan normal form J_1 — 306
    8.12.4 The Jordan normal form J_2 — 309
    8.12.5 The Jordan normal form J_3 — 309
    8.12.6 The Jordan normal form J_4 — 310
    8.12.7 The Jordan normal form J_5 — 310
    8.12.8 The Jordan normal form J_6 — 311
8.13 Construction of globally asymptotically stable partial differential systems — 312
8.14 Construction of globally stable system of delayed differential equations — 322

8.15 Stabilization by the Jurdjevic-Quinn method   329
    8.15.1 The minimization problem   330
    8.15.2 The inverse optimization problem   331
    8.15.3 Input-to-state stability   332

9. Transformation of Dynamical Systems to Hyperjerky Motions   **336**

  9.1   Introduction   336
  9.2   Transformation of 3-D dynamical systems to jerk form   336
  9.3   Transformation of 3-D dynamical systems to rational and cubic   340
      jerks forms
  9.4   Transformation of 4-D dynamical systems to hyperjerk form   343
    9.4.1   The expression of the transformation between (9.45)   349
        and (9.61)-(9.62)
    9.4.2   Examples of 4-D hyperjerky dynamics   352
  9.5   Examples of crackle and top dynamics   359

References   361

Index   389

Chapter 1

Review of Chaotic Dynamics

1.1 Introduction

In this chapter, we describe the major notions and techniques used in the study of chaotic dynamical systems. In Sec. 1.2, we describe the Poincaré map technique in order to study bifurcations and chaos in dynamical systems. In Sec. 1.3, we discuss the Smale horseshoe as a fundamental tool for studying chaos in dynamical systems. The symbolic dynamics representation of a solution is presented in Sec. 1.4. The notion of chaos and strange attractors in dynamical systems is presented and discussed in Sec. 1.5 with many numerical examples. Since chaotic dynamical systems can have more than one attractor depending on the choice of initial conditions, the notion that *basin of attraction* comes from this property is discussed in Sec. 1.6. In Sec. 1.7, we discuss the possible relations between robustness and hyperbolicity discussed in Sec. 8.10. The statistical properties of chaotic attractors are presented in Sec.1.8. Different types of entropies defined for dynamical systems include the physical (or Sinai-Ruelle-Bowen) measure, Hausdorff dimension, topological entropy, Lebesgue (volume) measure and Lyapunov exponents.

Now, proving chaos in 1-D dynamics needs the so called *Period 3 implies chaos* theorem, presented in Sec. 1.9. The generalization of this theorem to high dimensions is the subject of Sec. 1.10, where the *Snap-back repeller* is defined and used instead of period 3 orbit. The Shilnikov criterion for the existence of chaos in autonomous systems is presented in Sec. 1.11. These results use the notion of homoclinic and heteroclinic orbits. The resulting chaos is called the *horseshoe type* or *Shilnikov chaos*.

1.2 Poincaré map technique

The Poincaré map technique is used to describe bifurcations and chaos in dynamical systems. The definition for the Poincaré map for a continuous-time dynamical system is given by: Let $f : \mathbb{R} \times \Omega \longrightarrow \Omega$ be a continuous flow, where $\Omega \subset \mathbb{R}^n$ is open. Consider the n-dimensional continuous-time system given by

$$x' = f(t, x) \tag{1.1}$$

Let φ_t denote the corresponding flow of the system (1.1).

Definition 1 *A map P is called a C^r-diffeomorphism if both P and its inverse P^{-1} are bijective and are r-times continuously differentiable.*

Let γ be a periodic orbit through a point p, and let U be an open and connected neighborhood of p. For any $x \in \Omega$, let $I(x) =]t_x^-, t_x^+[$ be an open interval in the real numbers. Then, one has the following definitions:

Definition 2 *(a) A positive semi-orbit through x is the set $\gamma_x^+ = \{f(t, x), t \in]0, t_x^+[\}$, and a negative semi-orbit through x is the set $\gamma_x^- = \{f(t, x), t \in]t_x^-, 0[\}$.*
(b) A Poincaré section (shown in Fig. 1.1) through a point p is a local differentiable and transversal section S of f through the point p.

Hence, the Poincaré map is defined by:

Definition 3 *A function $P : U \longrightarrow S$ is called a Poincaré map for the orbit γ on the Poincaré section S through point p if:*
(1) $P(p) = p$.
(2) $P(U)$ is a neighborhood of p and $P : U \to P(U)$ is a diffeomorphism.
(3) For every point x in U, the positive semi-orbit of x intersects S for the first time at $P(x)$.

The advantage of using the Poincaré map technique is that the Poincaré map is the intersection of a periodic orbit of the considered continuous-time system (1.1), where the transversal Poincaré section S is one dimension smaller than the original continuous dynamical system. The map P is called the *first recurrence map*, because the method of analysis is to consider a periodic orbit with initial conditions on the Poincaré section S and observe the point at which these orbits first return to the section S. The map P is used to analyze the original system (1.1) because it preserves many properties of periodic, quasiperiodic and chaotic orbits of the original continuous-time system. This

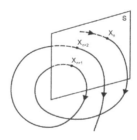

FIGURE 1.1: A Poincaré section.

can be seen via the relation between limit sets of the Poincaré map P and limit sets of the flow of the considered system (1.1) as follows:

(a) A limit cycle of φ_t is a fixed point of P, and a period-m closed orbit of P is a subharmonic solution (relative to the considered section S) of φ_t.

(b) A chaotic orbit of P corresponds to a chaotic solution of φ_t.

(c) A stable periodic point of P corresponds to a stable periodic orbit of φ_t.

(d) An unstable periodic point of P corresponds to a saddle-type periodic orbit of φ_t.

(e) There is no general method for the construction of a Poincaré map. Only numerical and rarely analytic. See [Hénon (1982), Tucker (2002b), Tsuji and Ido (2002), Fujisaka and Sato (1997), Chua, *et al.* (1986), Chua and Tichonicky (1991) and references therein].

Now, assuming that the function f given in equation (1.1) is a continuous piecewise-linear vector field, let $\Sigma_1, ..., \Sigma_p$ denote the hyperplanes separating the linear regions U_i of f, in which case their union is the set Ω. Let $\varphi(t, x)$ denote the trajectory of the system (1.1) starting at the point x.

Definition 4 *The generalized Poincaré map $H : \Omega \to \Omega$ is defined by $H(x) = \varphi(\tau(x), x)$, where $\tau(x)$ is the time needed for the trajectory $\varphi(t, x)$ to reach Ω.*

As a set of remarks concerning Definition 4:

(a) The generalized Poincaré map H has the same properties given above for the map P.

(b) The evaluation of H in regions where H is continuous is obtained by using the analytical formulas for solutions of linear systems.

(c) The evaluation of the generalized Poincaré map H on a box $X \subset \Omega$ is done by finding the return time for all points in X, i.e., the interval $\{\tau(x), x \in X\} \subset \tau(X)$, followed by the use of analytic solutions to compute $\varphi(\tau(X), X)$. $H(X)$ is enclosed in the intersection of $\varphi(\tau(X), X)$ with Ω.

(d) The Jacobian of H at X can be expressed in terms of the return time $\tau(x)$, the start box X and the image $H(X)$.

1.3 Smale horseshoe

In this section, we discuss the Smale horseshoe as a fundamental tool for studying chaos in dynamical systems. For defining the so-called *Smale's horseshoe* map, we must define the unit square as follows:

Definition 5 *A unit square D with side lengths 1 is the one with coordinates $(0,0), (1,0), (1,1), (0,1)$ in the real plane, or $0, 1, 1 + i$ in the complex plane.*

Hence, the Smale horseshoe map f described in [Smale (1967), Cvitanović, *et al.*, (1988)] consists of the following sequence of operations, as shown in Fig. 1.2 on the unit square D:

(a) Stretch in the y direction by more than a factor of two.

(b) Compress in the x direction by more than a factor of two.

(c) Fold the resulting rectangle and fit it back onto the square, overlapping at the top and bottom but not quite reaching the ends to the left and right, leaving a gap in the middle. Hence, the action of f is defined through the composition of the three geometrical transformations defined above.

(d) Repeat the above steps in order to generate the horseshoe attractor that has a Cantor set structure.

From Fig. 1.2, one sees that horseshoe maps cross the original square in a linear fashion, but in most applications horseshoe maps are rarely so regular, but very similar. Mathematically, the above actions (a) to (d) can be translated as follows [Smale (1967)]:

1. Contract the square D by a factor of λ in the vertical direction, where $0 < \lambda < \frac{1}{2}$, such that D is mapped into the set $[0, 1] \times [0, \lambda]$.

2. Expand the rectangle obtained by a factor of μ in the horizontal direction, where $2 + \epsilon < \mu$, then map the set $[0, 1] \times [0, \lambda]$ into the $[0, \mu] \times [0, \lambda]$ (the need for this ϵ factor is explained in step 3).

3. Steps 1 and 2 produce a rectangle $f(D)$ of dimensions $\mu \times \lambda$. This rectangle crosses the original square D in two sections after it has been bent, as shown in Fig. 1.2. The ϵ in step 2 indicates the extra length needed to create this bend as well as any extra on the other side of the square.

4. This process is then repeated, only using $f(D)$ rather than the unit square. The nth iteration of this process will be called $f^k(D)$, $k \in \mathbb{N}$.

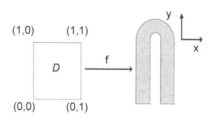

FIGURE 1.2: The Smale horseshoe map f after a single iteration.

The dynamics of the horseshoe map are considered as a tool for the reproduction of the chaotic dynamics of a flow in a small disk Δ perpendicular to
a period-T orbit Γ.

Hence, the Smale's horseshoe map consists of the iterated application of
both $f(D)$ and $f^{-1}(D)$, as shown by the following four features:

(1) When the system evolves, some orbits diverge and the points in this disk
remain close to the given periodic orbit Γ, tracing out orbits that eventually
intersect the disk once again.

(2) The intersection of the disk Δ and points in its neighborhood with the
given period-T orbit Γ come back to themselves every period T. When this
neighborhood returns, its shape is transformed.

(3) Inside the disk Δ there are some points that will leave the disk neighborhood and others that will continue to return.

(4) The set of points that never leaves the neighborhood of the given
period-T orbit Γ form a fractal.

The following result was proved in [Smale (1967)] by using the geometrical
formations of horseshoes as shown in Fig. 1.2 and its repetitions:

Theorem 1 *(a) The set $\Pi_+ = \cap_{i=0}^{i=\infty} f^i(D)$ is an interval and has an intersection with a Cantor-like set.*

*(b) The set $\Pi_- = \cap_{i=0}^{i=\infty} f^{-i}(D)$ is an interval and has an intersection with
a Cantor-like set.*

(c) The set $\Pi = \Pi_+ \cap D \cap \Pi_-$ is an intersection with a Cantor-like set.

*(d) If we consider the map $f(D)$, where f is the horseshoe map that con-
tracts by λ and expands by μ, where $0 < \lambda < \frac{1}{2}$, and $\mu > 2 + \epsilon$, the Hausdorff
dimension of Π is $\log 2 \left(\frac{1}{\log \mu} - \frac{1}{\log \lambda} \right)$.*

*(e) The limit set Π forms an uncountable, nowhere dense set (the interior
of its closure is empty) in $\mathbb{R}^2$.*

(f) $f(D)$ is equivalent to the shift map in the symbolic space.

*(g) The horseshoe map f is a diffeomorphism defined from the unit square
D of the plane into itself.*

(h) The horseshoe map is one-to-one.

(l) The domain of f^{-1} is $f(D)$.

(m) The horseshoe map f is an Axiom A diffeomorphism.

Generally, the presence of a horseshoe in a dynamical system implies the
following features:

1. There are infinitely many periodic orbits. In particular, those with arbitrarily long period.

2. The number of periodic orbits grows exponentially with the period.

3. In any small neighborhood of any point of the fractal invariant set Λ,
   there is a point on a periodic orbit.

We have the following definition:

Definition 6 *A dynamical system is chaotic in the sense of Smale if it has horseshoes of the Smale type.*

Note that Definition 6 with the so-called *Shadowing lemma* [Stoffer and Palmer (1999)] is used to prove that the Hénon map is chaotic in the sense of Smale. Generally, there are no rigorous methods for finding Smale horseshoes in a dynamical system with relatively small dimension [Smale (1967), Banks and Dragan (1994), Zgliczynski (1997)].

The most important role of the Smale horseshoe is its relationship to the homoclinic tangency discovered by Poincaré while investigating the three-body problem of celestial mechanics [Tufillaro, *et al.*, (1987)], i.e., the homoclinic tangency is the tangled intersection of such invariant manifolds with homoclinic points as shown in Fig. 1.3.

Definition 7 *(a) The invariant manifold of a map f is a set of points X such that $f(X) \subset X$.*

(b) Given a fixed point Y_0 in the invariant manifold, the manifold Y is called stable if $\forall y \in Y$, $\lim_{n \to \infty} f^n(y) \to Y_0$. Similarly, a manifold is called unstable if $\forall y \in Y$, $\lim_{n \to \infty} f^{-n}(y) \to Y_0$.

(c) A fixed point is hyperbolic if it is the intersection of one or more stable manifolds and one or more unstable manifolds.

(c) A homoclinic point is a point x different from the fixed point that lies on a stable or unstable manifold of the same fixed point Q.

Formally, the homoclinic tangency can be defined as follows:

Definition 8 *(a) (homoclinic tangency) We say that a diffeomorphism f exhibits a homoclinic tangency if there is a periodic point Q such that there is a point $x \in W^s(Q) \cap W^u(Q)$ with $T_x W^s(Q) + T_x W^u(Q) \neq T_x M$.*

(b) Given an open set V, we say that the tangency holds in V if x and Q belong to V.

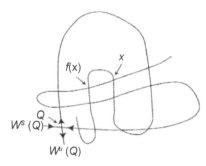

FIGURE 1.3: Part of a homoclinic tangency: Q is the hyperbolic fixed point, $W^s(Q), W^u(Q)$ are the stable and unstable manifolds, respectively, x is a homoclinic point, and thus $f(x)$, is also a homoclinic point.

Theorem 2 *(Poincaré, 1890) If there exists a single homoclinic point on a stable and an unstable invariant manifold corresponding to a particular hyperbolic fixed point, then there exists an infinite number of homoclinic points on the same invariant manifolds.*

The basic theorem for the rigorous proof of chaos for a dynamical system $f : \mathbb{R}^n \to \mathbb{R}^n$ was proved by [Smale (1967)]:

Theorem 3 *If f is a diffeomorphism and has a transversal homoclinic point, then there exists a Cantor set $\Lambda \subset \mathbb{R}^n$ in which f^m is topologically equivalent to the shift automorphism for some m.*

The existence of the shift automorphism implies the existence of a dense set of periodic orbits and an uncountably infinite collection of asymptotically aperiodic points of the map f within the set Λ. For this case, robust chaos is not possible due to the existence of the transversal homoclinic point [Weisstein (2002)].

An example can be found in [Galias (1997(b))]:

Example 1 *Let $\tilde{N}_0 = [-1,1] \times [-1,-0.5], \tilde{N}_1 = [-1,1] \times [0.5,1], P_1 = [-1,1] \times \mathbb{R}$ be the smallest vertical stripe containing $\tilde{N}_0$ and $\tilde{N}_1, M_- = [-1,1] \times (-\infty,-1), M_0 = [-1,1] \times (-0.5,0.5), M_+ = [-1,1] \times (1,\infty)$ M_-, M_0, and M_+ are subsets of P_1 lying below, between, and above $\tilde{N}_0$ and $\tilde{N}_1$. Let N_{0D}, N_{0U} be the lower and upper horizontal edges and N_{0L}, N_{0R} be the left and right vertical edges of N_0 and similarly $N_{1D}, N_{1U}, N_{1L}, N_{1R}$ be the lower upper left and right edges of $\tilde{N}_1$. Then the Smale horseshoe map is a map linear on $\tilde{N}_0$ and $\tilde{N}_1$ defined by*

$$f(x,y) = \begin{cases} \left(\frac{1}{4}x - \frac{1}{2}, \pm 5\left(y + \frac{3}{4}\right)\right), & \text{if } (x,y) \in \tilde{N}_0 \\ \\ \left(\frac{1}{4}x + \frac{1}{2}, \pm 5\left(y - \frac{3}{4}\right)\right), & \text{if } (x,y) \in \tilde{N}_1 \end{cases} \tag{1.2}$$

Note that the Jacobian of f^n is given by:

$$Df^n(x,y) = \begin{pmatrix} (0.25)^n & 0 \\ 0 & (\pm 5)^n \end{pmatrix} \tag{1.3}$$

1.4 Symbolic dynamics

Theorem 3 gives the idea that symbolic dynamics names can be given to all the orbits that remain in the neighborhood. If $\Delta = \cup_{i=1}^m D_i$ and if one knows the sequence of points $\{x_0, x_1, ...\}$ in which the orbit visits these regions D_i, then the visitation sequence $\hat{s} = \{\hat{s}_0; \hat{s}_1; ...; \hat{s}_j; ...\}$ composed of symbols $\hat{s}_j = i$ if $x_j = f^j(x_0) \in D_i$, $i = 1, .., m$ of the orbits provides a symbolic representation of the dynamics, known as *symbolic dynamics*. Indeed, let $S_m = \{0, 1, ..., m-1\}$ be the set of non-negative successive integers from 0 to $m-1$. Let Σ_m be the collection of all bi-infinite sequences of S_m with their elements, i.e., every element s of Σ_m is $s = (..., s_{-n}, ..., s_{-1}, s_0, s_1, ..., s_n, ...)$, $s_i \in \Sigma_m$. Let $\bar{s} \in \Sigma_m$ be another sequence $\bar{s} = (..., \bar{s}_{-n}, ..., \bar{s}_0, \bar{s}_1, ...)$, where $\bar{s}_i \in \Sigma_m$. Then:

Definition 9 *The distance between s and $\bar{s}$ is defined as*

$$d(s, \bar{s}) = \sum_{-\infty}^{\infty} \frac{1}{2^{|i|}} \frac{|s_i - \bar{s}_i|}{1 + |s_i - \bar{s}_i|}, \tag{1.4}$$

Hence, the set Σ_m with the distance defined as (1.18) is a metric space, and one has the following results proved in [Robinson (2004)]:

Theorem 4 *The space Σ_m is compact, totally disconnected, and perfect.*

A set having the three properties given in Theorem 1.4.1 is a Cantor set, which frequently appears in the characterization of the complex structure of invariant sets in a chaotic dynamical system. However, let an m-shift map $\sigma : \Sigma_m \to \Sigma_m$ be defined by:

$$\sigma(s)_i = s_{i+1} \tag{1.5}$$

Then the map σ defined by (1.19) has the following properties proved in [Robinson (1995)]:

Theorem 5 *(a) $\sigma(\Sigma_m) = \Sigma_m$ and σ is continuous.*
(b) The shift map σ as a dynamical system defined on Σ_m has the following properties:
(b-1) σ has a countable infinity of periodic orbits consisting of orbits of all periods,
(b-2) σ has an uncountable infinity of non-periodic orbits, and
(b-3) σ has a dense orbit.

Statements (b) of Theorem 5 imply that the dynamics generated by the shift map σ (1.5) are sensitive to initial conditions, and therefore it is chaotic in the commonly accepted sense.

Now, let X be a separable metric space and f a continuous map $f : \tilde{Q} \to X$, where $\tilde{Q} \subset X$ is locally connected and compact.

Assumption A. Suppose the following assumptions hold [Kennedy and York (2001)]:

A1 There exist two subsets of $\tilde{Q}$ denoted by $\tilde{Q}_1$ and $\tilde{Q}_2$, respectively. The sets $\tilde{Q}_1$ and $\tilde{Q}_2$ are disjoint and compact.

A2 Each connected component of $\tilde{Q}$ intersects both $\tilde{Q}_1$ and $\tilde{Q}_2$.

A3 The cross number m of $\tilde{Q}$ with respect to f is not less than 2,

where the cross number m is defined as follows:

Definition 10 *(a) A connection $\tilde{\Gamma}$ for $\tilde{Q}_1$ and $\tilde{Q}_2$ is a compact subset of $\tilde{Q}$ that intersects both $\tilde{Q}_1$ and $\tilde{Q}_2$.*
(b) A preconnection $\tilde{\gamma}$ is a compact connected subset of $\tilde{Q}$ for which $f(\tilde{\gamma})$ is a connection.
(c) The cross number m is the largest number such that every connection contains at least m mutually disjoint preconnections.

Definition 10 can be generalized to the m domain $D_1, ..., D_{m-1}$ and D_m as follows:

Definition 11 *Let $\tilde{\gamma}$ be a compact subset of $\tilde{Q}$ such that for each $1 \leq i \leq m$; $\tilde{\gamma}_i = \tilde{\gamma} \cap D_i$ is non-empty and compact. Then $\tilde{\gamma}$ is called a connection with respect to $D_1, ..., D_{m-1}$ and D_m. Let $\tilde{F}$ be a family of connections $\tilde{\gamma}'s$ with respect to $D_1, ..., D_{m-1}$ and D_m satisfying the following property:*

$$\tilde{\gamma} \in \tilde{F} \Longrightarrow f(\tilde{\gamma}_i) \subset \tilde{F} \qquad (1.6)$$

Then $\tilde{F}$ is said to be an f-connected family with respect to $D_1, ..., D_{m-1}$ and D_m.

Hence, the following fundamental result was proved in [Kennedy and York (2001)]:

Theorem 6 *Let f be the map satisfying **Assumption A**. Then there exists a compact invariant set $\tilde{Q}^I \subset \tilde{Q}$ for which $f\big|_{\tilde{Q}^I}$ is semi-conjugate to an m-shift map.*

This holds if there exists a continuous and onto map $h : \tilde{Q}^I \rightarrow \Sigma_m$ such that $h \circ f = \sigma \circ h$.

A more applicable version of the above result was given in [Yang and Tang (2004)] for piecewise-continuous maps as follows:

Theorem 7 *Let $\tilde{Q}$ be a compact subset of X and $f : \tilde{Q} \rightarrow X$ be a map satisfying the following conditions:*
(a) There exist m mutually disjoint subsets $D_1, ..., and D_m$ of $\tilde{Q}$, the restriction of f to each D_i, i.e., $f\big|_{D_i}$ is continuous.
(b)
$$\cup_{i=1}^{m} D_i \subset f(D_j), j = 1, 2, .., m \qquad (1.7)$$

then there exists a compact invariant set $K \subset \tilde{Q}$, such that $f\big|_K$ is semi-conjugate to m-shift dynamics.
(c) Suppose that there exists an f-connected family $\tilde{F}$ with respect to $D_1, ..., D_{m-1}$ and D_m. Then there exists a compact invariant set $K \subset \tilde{Q}$ such that $f\big|_K$ is semi-conjugate to m-shift dynamics.

Theorem 7 is generally used in order to estimate the topological entropy of piecewise linear systems in terms of half-Poincaré maps.

1.5   Strange attractors

In this section, we discuss the notion of chaos and strange attractors in dynamical systems. Roughly speaking, chaos is the idea that a system will produce very different long-term behaviors when the initial conditions are perturbed only slightly. Several strange attractors are studied in the current literature. The first was Lorenz [Lorenz (1963)] in 1963, where he proposed a simple mathematical model of a weather system. The model displays a complex behavior characterized by a *sensitive dependence on the initial conditions*. This implies that the prediction of a future state of the system is impossible. On one hand, the first fractal shape of the Lorenz system given by:

$$\begin{cases} x' = \sigma\left(y - x\right) \\[2mm] y' = rx - y - xz \\[2mm] z' = -bz + xy \end{cases} \tag{1.8}$$

was identified by Ruelle in 1979, and it took the form of a *butterfly* (*a butterfly in the Amazon might, in principle, ultimately alter the weather in Kansas*).

Many simple systems are now known to be chaotic, such as the logistic map modelling biological population dynamics. On the other hand, chaos is very useful and shows great potential in many technological disciplines such as in information and computer sciences, power system protection, biomedical systems analysis, flow dynamics and liquid mixing, encryption and communications, and so on [Chen and Dong (1998), Chen (1999)].

The most useful tests for characterizing chaos are:

1. The chaotic motions are more complicated than stationary, periodic or quasiperiodic, and they have very complicated shapes, called *strange attractors*.

2. There is a sensitive dependence on initial conditions, i.e., nearby solutions diverge exponentially fast.

3. There is a coexistence phenomenon, i.e., chaotic orbits coexist with a (countable) infinity of unstable periodic orbits.

Next, we define the notion of an attractor:

Definition 12 *An attractor of a dynamical system is a subset of the state space to which orbits originating from typical initial conditions tend as time increases.*

For some problems related to uniform hyperbolic dynamical systems: Convergence only for points in some *large* subset of a neighborhood of the attractor, we need the following fine definition:

Definition 13 *An attractor is a compact invariant subset A such that the trajectories of all points in a neighborhood U converge to A as time goes to infinity, and A is dynamically indecomposable (or transitive); there is some trajectory dense in A.*

Formally, Definition 13 can be reformulated as follows:

Definition 14 *A closed subset A of the non-wandering set $\Omega(f)$ is called an attractor when:*
(1) A is f-invariant, i.e., $f(A) = A$.
(2) There exists a neighborhood L of A in $\mathbb{R}^n$: $\bigcap_{n \geq 0} f^n(L) = A$.

(3) f is transitive.

and second, a definition for a strange attractor is given by:

Definition 15 *(Strange attractor) Suppose $A \subset \mathbb{R}^n$ is an attractor. Then A is called a strange attractor if it is chaotic.*

Generally, to prove that a dynamical system has a strange attractor, we might proceed as follows:

Step 1 Find in the phase space of the system a *trapping region M*.

Step 2 Show that M contains a chaotic invariant set Λ. This is done by several methods, one of which is to show that there exists a homoclinic orbit or heteroclinic cycle inside M.

Step 3 Construct the attractor A as follows:

$$\begin{cases} \bigcap_{t>0} f_t(M) = A, \text{ if } f \text{ is flow} \\ \\ \bigcap_{n>0} f^n(M) = A, \text{ if } f \text{ is a map} \end{cases} \tag{1.9}$$

So $\Lambda \subset A$, and there is sensitive dependence on initial conditions. Finally, the set A is a strange attractor if:

(a) The sensitive dependence on initial conditions on Λ extends to A.

(b) The set A is topologically transitive.

Generally, a system is said to be robust if it is capable of coping well with variations in its operating environment while experiencing minimal loss of functionality. Generally, chaotic dynamical systems display two kinds of chaotic attractors: One type has fragile chaos (the attractors disappear with perturbations of a parameter or coexist with other attractors), and the other type has robust chaos defined as follows [Banerjee, *et al.*, (1998)]:

Definition 16 *Robust chaos is defined by the absence of periodic windows and coexisting attractors in some neighborhood in the parameter space.*

The existence of these windows in some chaotic regions means that small changes of the parameters would destroy the chaos, implying the fragility of this type of chaos.

The Lorenz attractor shown in Fig. 1.4 is robust. However, the Chua attractor (the Double scroll) given by the equation:

$$
\begin{cases}
x' = \alpha \left(y - \left[m_1 x + \tfrac{1}{2} \left(m_0 - m_1 \right) \left(|x + 1| - |x - 1| \right) \right] \right) \\[2mm]
y' = x - y + z \\[2mm]
z' = -\beta y
\end{cases}
\qquad (1.10)
$$

is shown in Fig. 1.5 and it is not robust.

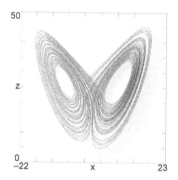

FIGURE 1.4: The Lorenz chaotic attractor obtained from equation (1.8) with $\sigma = 10, r = 28$ and $b = \tfrac{8}{3}$. [Lorenz (1963)].

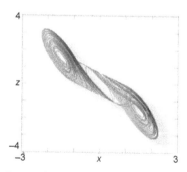

FIGURE 1.5: The classical double-scroll attractor obtained from equation (1.10) with $\alpha = 9.35, \beta = 14.79, m_0 = -\tfrac{1}{7}, m_1 = \tfrac{2}{7}$. [Chua et al. (1986)].

In many situations, robust chaos is required. For example, in the field of communications and spreading the spectrum of switch-mode power supplies to avoid electromagnetic interference [Ottino (1989), Ottino, *et al.*, (1992)] where it is necessary to obtain reliable operation in the chaotic mode. Also, in electrical engineering where robust chaos is demonstrated in [Banergee, *et al.*, (1998)] and a conjecture claiming that robust chaos cannot exist in smooth systems. However, this is not true as proved and discussed in [Andrecut and Ali (2001(a))], which includes a general theorem and a practical procedure for constructing S-unimodal maps that generate robust chaos. Many methods are available in the current literature for finding smooth and robust chaotic map [Andrecut and Ali (2001(a-b-c)), Jafarizadeh and Behnia (2002)]. In [Avrutin and Schanz (2008)], it is shown by an elementary example that the term *robust chaos* must be used carefully since chaotic attractors which are robust in the sense of [Banerjee *et al.*, (1998)] are not necessarily robust in the sense of [Milnor (1985(a))]. This confirmation was done by the study of a bifurcation phenomenon called the *bandcount adding scenario* occurring in the 2D parameter space of the one-dimensional piecewise-linear map given by:

$$x_{n+1} = \begin{cases} f_l\left(x_n\right) = ax_n + \mu + 1, \text{ if } x_n < 0 \\ \\ f_r\left(x_n\right) = ax_n + \mu - 1, \text{ if } x_n > 0 \end{cases} \tag{1.11}$$

i.e., an infinite number of interior crises bounding the regions of multi-band attractors were detected in the region of chaotic behavior.

Several methods for proving the robustness of chaos in dynamical systems in the sense that there are no coexisting attractors and no periodic windows in some neighborhood of the parameter space are as follows: Normal form analysis, unimodality, metric entropy, construction of dynamical systems using basis of the robustness or the non-robustness, geometric methods, detecting unstable periodic solutions, ergodic theory, weight-space exploration, numerical methods, and a combination of these methods. See [Zeraoulia and Sprott, (2011(g))] for more details.

1.6   Basins of attraction

Attractors of a dynamical system can be periodic, quasiperiodic, or chaotic behaviors of different types, i.e., stable, unstable, saddle, etc. Chaotic dynamical systems can have more than one attractor depending on the choice of initial conditions, thus the notion of basin of attraction comes from this property:

Definition 17 *The basin of attraction $B(A)$ for an attractor A is the set of initial conditions leading to long-time behavior that approaches that attractor.*

The basin of the Hénon map given by:

$$\begin{cases} x_{k+1} = 1 + a_1 x_k + a_4 y_k^2 \\ \\ y_{k+1} = x_k \end{cases} \tag{1.12}$$

is shown in Fig. 1.6. We remark that the basin boundaries have a very complicated fractal structure.

The following are some notes and remarks about the basin of attraction:

1. The qualitative behavior of a dynamical system depends fundamentally on which basin of attraction the initial condition resides, because the basic topological structure of such regions can vary greatly from system to system.

2. The basin boundary can be a smooth or fractal, the fractality is a result of chaotic motion of orbits on the boundary [McDonald, *et al.*, (1985)].

3. Basin boundaries can have qualitatively different types. For example, the nature of a basin can change from a simple smooth curve to a fractal (*metamorphosis*) [Grebogi, *et al.*, (1987)].

4. (Riddled basins) A smooth surface or hypersurface such that any initial condition in the surface generates an orbit that remains in the surface. Here, the system has a *bizarre* type of basin structure called a *riddled basin of attraction* [Alexander, *et al.*, (1992)]. This type of basin can exist in a dynamical system if, for example, the following conditions exist:

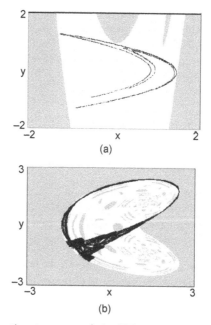

(a)

(b)

FIGURE 1.6: (a) Chaotic attractor of the Hénon map with $a_3 = -1.4$ and $a_2 = 0.3$. (b) Chaotic attractor of the map (1.12) with $a_4 = 0.59948$ and $a_1 = 1$. [Zeraoulia and Sprott (2008(a))].

(1) An invariant submanifold with chaotic dynamics exists.

(2) The Lyapunov exponent normal to the invariant submanifold is negative, i.e., the submanifold is an attractor.

(3) There exist other attractors.

The following 2-D map [Alexander, *et al.*, (1992)]:

$$
\left\{
\begin{array}{c}
z_{n+1} = z_n^2 - (1 + \lambda i)\bar{z}_n \\[2mm]
z = x + iy \\[2mm]
\bar{z} = x - iy \\[2mm]
i^2 = -1
\end{array}
\right.
\tag{1.13}
$$

or equivalently in $\mathbb{R}^2$

$$
\left\{
\begin{array}{l}
x_{n+1} = x_n^2 - y_n^2 - x_n - \lambda y_n, \\[2mm]
y_{n+1} = 2x_n y_n - \lambda x_n + y_n.
\end{array}
\right.
\tag{1.14}
$$

is an example for that situation for some values of λ. In fact, the map (1.14) has three attractors, and each internal dynamic is chaotic as in the logistic map, i.e., the basins of the three attractors are *intermingled* very complexly. Other examples of systems with riddled basins can be found in [Jing-ling Shen, *et al.*, (1996), Maistrenko, *et al.*, (1998), Cao (2004)].

Formally, it is possible to define a riddled basin as follows:

Definition 18 *(a) A closed invariant set A is said to be a weak attractor in the Milnor sense if its basin $B(A)$, i.e., the set of points whose ω-limit sets of x belongs to A has positive Lebesgue measure.*

(b) The basin of attraction $B(A)$ of an attractor A is riddled if its complement intersects every disk in a set of positive measure.

In this case, a riddled basin corresponds to an extreme form of uncertainty because it does not include any open subset and is full of holes. A set of basins which are dense in each other is called an *intermingled set*, i.e.,

Definition 19 *The basins of attraction $B(A)$ and $B(B)$ of the attractors A and B are intermingled if each disk which intersects one of the basins in a set of positive measure also intersects the other basin in a set of positive measure.*

Some illustrated examples can be found in [Djellit and Boukemara (2007), Gumowski and Mira (1980), Alexander, *et al.*, (1992), Ding and Yang (1996),

Aharonov, *et al.*, (1997), Paar and Pavin (1998), Kapitaniak, *et al.*, (2003), Taixiang and Hongjian (2007), Morales (2007)].

For quasi-attractors, coexistence of many attractors implies that the basins of attraction cannot be regular because it is well known that basin boundaries arise in dissipative dynamical systems when two or more attractors are present. In this case, each attractor has a basin of initial conditions that lead asymptotically to that attractor. The sets that separate different basins are called *basin boundaries*. In some cases, the basin boundaries can have a very complicated fractal structure and, hence, pose an additional impediment to predicting long-term behavior.

1.7   Density, robustness and persistence of chaos

In this section, we discuss the possible relations between robustness and hyperbolicity. For linear systems, one of the most interesting features of hyperbolicity is robustness, i.e., any matrix that is close to a hyperbolic one in the sense that corresponding coefficients are close, is also hyperbolic. In this case, the stable and unstable subspaces are not necessarily identical, but the dimensions remain the same. If hyperbolicity is dense, then any matrix is close to a hyperbolic one up to arbitrarily small modifications of the coefficients.

For non-linear systems, the same properties hold by replacing the word *matrix* by the word *dynamical system*. In fact, it was shown in [Zhirov, *et al.*, (2005), Zhirov (2006)] that the topological conjugacy problem is a common property for one-dimensional hyperbolic attractors of diffeomorphisms of surfaces. An algorithmic solution based on the combinatorial method for describing hyperbolic attractors of surface diffeomorphisms is given in the two following problems:

(The problem of enumeration of attractors) Let Λ_f and Λ_g be one-dimensional hyperbolic attractors of diffeomorphisms $f : M \to M$ and $g : N \to N$, where M and N are closed surfaces, either orientable or not. Does there exist a homeomorphism $h : U(\Lambda_f) \to V(\Lambda_g)$ of certain neighborhoods of attractors such that $f \circ h = h \circ g$?

(The topological conjugacy problem) Given $h > 0$, find a representative of each class of topological conjugacy of attractors with a given structure of accessible boundary (boundary type) for which the topological entropy is no greater than h.

Now, dynamical persistence means that a behavior type, i.e., equilibrium, oscillation, or chaos does not change with functional perturbation or parameter variation. Mathematically, *persistent chaos* (*p*-chaos) of degree-*p* for a dynamical system can be defined as follows:

Definition 20 *Assume a map $f_\xi : X \to X$ ($X \subset R^d$) depends on a parameter $\xi \in R^k$. The map f_ξ has chaos of degree-p on an open set $\mathcal{O} \subset X$ that is* *persistent for $\xi \in \mathcal{U} \subset R^k$ if there is a neighborhood $\mathcal{N}$ of $\mathcal{U}$ such that $\forall$ $\xi \in \mathcal{N}$, the map f_ξ retains at least $p \geq 1$ positive Lyapunov characteristic exponents (LCEs) Lebesgue a.e. in $\mathcal{O}$.*

Here, the choice of p is arbitrary. For example, the condition where p equals the number of positive LCEs is a very strict constraint; specifying a minimum p or ratio of p to the maximum number of positive exponents are weaker constraints. Flexibility allows one to analyze (say) systems with 10^6 unstable directions, in which a change in 1% of the geometry is undetectable, but a 50% change is. The notion of persistent chaos differs from that of a robust chaotic attractor in several ways: In particular, the uniqueness of the attractor is not required on the set U since there is little physical evidence indicating that such strict forms of uniqueness are present in many complex physical systems. For a low-d system, uniqueness is more difficult to establish. In [Albers, et al., (2006)] it was shown that as the dimension of a typical dissipative dynamical system is increased, the number of positive Lyapunov exponents increases monotonically and the number of parameter windows with periodic behavior decreases. The method of analysis is an extensive statistical survey of *universal approximators* (The approximation theorems of [Hornik, et al., (1990)] and time-series embedding of [Sauer, et al., (1991)] establishing an equivalence between these neural networks and general dynamical systems [Albers and Sprott (2006)].) given by single-layer recurrent neural networks of the form:

$$x'_t = \beta_0 + \sum_{i=1}^{n} \beta_i \tanh s \left(\omega_{i0} + \sum_{j=1}^{d} \omega_{ij} x_{t-j} \right) \qquad (1.15)$$

which are maps from $\mathbb{R}^d$ to $\mathbb{R}$ and denoted by $f_{s,\beta,\omega}$. Here n is the number of neurons, d is the number of time lags which determines the system's input embedding dimension and s is a scaling factor for the connection weights ω_{ij}. The initial condition is $(x_1, x_2, ..., x_d)$, and the state at time t is $(x_t, x_{t+1}, ..., x_{t+d-1})$. More details on robust chaos in neural networks can be found in Chapter 4.

A more general theory for the robustness of attractors is the subject of many non-popular works, at least for those working in the usual chaos theory. In what follows, we present some aspects and some important results in this direction. In particular, we discuss the relation between robustness and hyperbolicity of attractors in the light of the following definition about robust transitive sets:

Definition 21 *(a) An isolated transitive set Λ of a C^1 vector fields f is C^r robust transitive if there is an isolating block U of Λ so that $\Lambda(g) = \cap_{t \in R} g_t(U)$ is a transitive set for all g C^r nearby f.*

(b) Let Λ be a maximal invariant set of some diffeomorphism f, $\Lambda = \cap_{n \in \mathbb{Z}} f^n(U)$ for some neighborhood U of Λ. The set Λ is robust, or robustly transitive, if its continuation $\Lambda_g = \cap_{n \in \mathbb{Z}} g^n(U)$ is transitive for all g in a neighborhood of f.

(c) An attractor is C^r critically robust if it has a neighborhood U such that $\bigcap_{t>0} g_t(U)$ is in the closure of the closed orbits of every flow g C^r close to f.

Firstly, we present the following main facts:

1. It was shown in [Morales, et al. (1998)] that if any C^1 ($r = 1$) robust[1] singular transitive set of a C^1 vector fields is a singular hyperbolic set which is an attractor and any of its singularities is Lorenz-like.

2. There are singular hyperbolic sets[2] which are also hyperbolic, i.e., the union of a non-trivial basic set and a Lorenz-like singularity (not transitive).

3. The hyperbolic basic pieces are robust and the converse is also true when M is a surface, but not anymore if the dimension of M is at least 3. Some counter-examples in dimension 4 are given by Shub.

4. Robust sets in all dimensions always admit some dominated splitting which is volume hyperbolic. Generally, this splitting need not be partially hyperbolic, except when the ambient manifold has dimension 3.

5. Hyperbolic attractors on compact manifolds are C^r robust and C^r critically robust for all r.

6. The geometric Lorenz attractor is an example of a singular-hyperbolic attractor with only one singularity which is also C^r robust and C^r critically robust.

7. Generally, singular-hyperbolic attractors with only one singularity may be neither C^r robust nor C^r critically robust [Morales & Pujals (1997)].

8. On compact 3-manifolds all C^r critically robust singular-hyperbolic attractors with only one singularity are C^r robust, i.e., on compact 3-manifolds C^r critically robustness implies C^r robustness among singular-hyperbolic attractors with only one singularity.

More precisely, the following result is proved in [Morales & Pacifico (2004)]:

Theorem 8 *A C^r critically robust singular-hyperbolic attractor with only one singularity on compact 3-manifolds is C^r robust.*

The proof of Theorem 8 relies on the work of [Morales & Pacifico (2003(b))]. In this section, we give an outline of the proof based on several technical results given in [Morales & Pacifico (2003(b)-2004)] which we state here without proofs. Thus, let f_t be a flow induced by a C^r vector

---

[1]Robust here means that this property cannot be destroyed by small C^1-perturbations of the flow.

[2]or Lorenz-like attractor.

field f on a compact 3-manifold M equipped with the C^r topology, $r \geq 1$. The invariant manifolds $W^s_f(\sigma)$ and $W^u_f(\sigma)$ exist for a Lorenz-like hyperbolic singularity σ. This implies that the eigenspace of λ_2 is tangential to a one-dimensional invariant manifold $W^{ss}_f(\sigma)$ called the *strong stable manifold* of σ. Thus, $W^{ss}_f(\sigma)$ splits $W^s_f(\sigma)$ into two connected components $W^{s,+}_f(\sigma)$ and $W^{fs,-}_f(\sigma)$ of $W^s_f(\sigma) \backslash W^{ss}_f(\sigma)$. Let Λ be a singular-hyperbolic set with dense periodic orbits of a three-dimensional flow. Thus, it follows that every $\sigma \in Sing_f(\Lambda)$ is Lorenz-like and satisfies $\Lambda \cap W^{ss}_f(\sigma) = \{\sigma\}$ and that any compact invariant subset without singularities of Λ is hyperbolic of saddle type. From [Morales & Pacifico (2001)] it follows that if Λ is attracting, there is for every $p \in Per_f(\Lambda)$ a $\sigma \in Sing_f(\Lambda)$ such that

$$W^u_f(p) \cap W^s_f(\sigma) \neq \emptyset. \qquad (1.16)$$

For every singular-hyperbolic set Λ of a three-dimensional flow f and every Lorenz-like singularity $\sigma \in Sing_f(\Lambda)$ we define the following sets:

$$
\begin{cases}
P^+ = \left\{ p \in Per_f(\Lambda) : W^u_f(p) \cap W^{s,+}_f(\sigma) \neq \emptyset \right\} \\[2mm]
P^- = \left\{ p \in Per_f(\Lambda) : W^u_f(p) \cap W^{s,-}_f(\sigma) \neq \emptyset \right\} \\[2mm]
H^+_f = Cl(P^+) \\[2mm]
H^-_f = Cl(P^-)
\end{cases}
\qquad (1.17)
$$

and let Σ be a singular cross-section of f, i.e., it corresponds to the submanifolds $\Sigma^+ = \{x_3 = 1\}$ or $\Sigma^- = \{x_3 = -1\}$ in the stable coordinate system $(x_1, x_2, x_3) \in [-1, 1]^3$. Let l^+ and l^- be the curves obtained by intersecting $\{x_2 = 0\}$ with Σ and Σ^-, respectively; these curves are contained in $W^{s,+}_f(\sigma)$ and $W^{s,-}_f(\sigma)$ respectively.

Thus, the proof of Theorem 8 uses several lemmas, propositions and theorems:

Lemme 1.1 *Let Λ be a connected, singular-hyperbolic, attracting set with dense periodic orbits and only one singularity σ. Then $\Lambda = H^+_f \cup H^-_f$.*

Lemme 1.2 *Let Λ a singular-hyperbolic set with dense periodic orbits of a three-dimensional flow f, and fix $\sigma \in Sing_f(\Lambda)$. There are singular cross-sections Σ^+, Σ^- as above such that every orbit of Λ passing close to some point in $W^{s,+}_f(\sigma)$ (resp. $W^{s,-}_f(\sigma)$) intersects Σ^+ (resp. Σ^-). If $p \in \Lambda \cap \Sigma^+$ is close to l^+, then $W^s_f(p, \Sigma^+)$ is a vertical curve crossing Σ^+. If $p \in Per_f(\Lambda)$ and $W^u_f(p) \cap W^{fs,+}_f(\sigma) \neq \emptyset$, then $W^u_f(p)$ contains an interval $J = J_p$ intersecting l^+ transversally; and the same is true if we replace $+$ by $-$.*

The proof of transitivity is based on the two lemmas:

Lemme 1.3 *(Birkhoff's criterion). Let T be a compact, invariant set of f such that for all open sets U, V intersecting T there is $s > 0$ such that $f_s(U \cap T) \cap V \neq \emptyset$. Then T is transitive.*

Lemme 1.4 *Let Λ be a connected, singular-hyperbolic, attracting set with dense periodic orbits and only one singularity σ. Let U, V be open sets, $p \in U \cap Per_f(\Lambda)$ and $q \in V \cap Per_f(\Lambda)$. If $W_f^u(p) \cap W_f^{s,+}(\sigma) \neq \emptyset$ and $W_f^u(q) \cap W_f^{s,+}(\sigma) \neq \emptyset$, there exist $t > 0$ and $z \in W_f^u(p)$ arbitrarily close to $W_f^u(p) \cap W_f^{s,+}(\sigma)$ such that $f_t(z) \in V$. The same is true if we replace $+$ by $-$.*

Now, let Λ be a singular-hyperbolic set of $f \in C^r$ satisfying:
1. Λ is connected.
2. Λ is attracting.
3. The closed orbits contained in Λ are dense in Λ, which implies that $\Lambda = Cl(Per_f(\Lambda))$.
4. Λ has only one singularity σ.
Then, one has the following results:

Proposition 1.1 *Suppose that, for any given $p, q \in Per_f(\Lambda)$, either*
1. *$W_f^u(p) \cap W_f^{s,+}(\sigma) \neq \emptyset$ and $W_f^u(q) \cap W_f^{s,+}(\sigma) \neq \emptyset$, or*
2. *$W_f^u(p) \cap W_f^{s,-}(\sigma) \neq \emptyset$ and $W_f^u(q) \cap W_f^{s,-}(\sigma) \neq \emptyset$.*
Then Λ is transitive.

Proposition 1.2 *If there is a sequence $p_n \in PerX(\Lambda)$ converging on some point in $W_f^{s,+}(\sigma)$ such that $W_f^u(p_n) \cap W_f^{s,-}(\sigma) \neq \emptyset$ for all n, then Λ is transitive. The same is true interchanging $+$ and $-$.*

Proposition 1.3 *If there is $a \in W_f^u(\sigma) \backslash \{\sigma\}$ such that $\sigma \in \omega_f(a)$, then Λ is transitive.*

Theorem 9 *If Λ is not transitive, there is for all $a \in W_f^u(\sigma) \backslash \{\sigma\}$ a periodic orbit O of f with positive expanding eigenvalues such that $a \in W_{fs}(O)$.*

At this end, note that the flow f_t carries I^+ and I^- into I_0^+ and I_0^- respectively. These two segments play a crucial role in the following definition:

Definition 22 *We denote by $W^{u,+}$ the connected component of $W^u \backslash O$ that is accumulated (via the Inclination Lemma and the Strong λ Lemma [de Melo & Palis (1982), Deng (1989)]) by the positive orbit of I_0^+. We denote $W^{u,+}$ the connected component of $W_f^u(O) \backslash O$ that is accumulated by the positive orbit of I_0^-.*

and the three propositions given by:

Proposition 1.4 $W^{u,+} \cap W_f^{s,-}(\sigma) \neq \emptyset$ *and* $W^{u,+} \cap W_f^{s,+}(\sigma) \neq \emptyset$. *The same is true interchanging* $+$ *and* $-$.

Proposition 1.5 $H^+ = Cl(W^{u,+})$ *and* $H^- = Cl(W^{u,-})$.

Proposition 1.6 *If* $z \in Per_f(\Lambda)$ *and* $W_f^s(z) \cap W^{u,+} \neq \emptyset$, *then* $Cl(W_f^s(z) \cap W^{u,+}) = Cl(W^{u,+})$. *The same is true if we replace* $+$ *by* $-$.

Now, let $z \in Per_f(\Lambda)$ and let $H_f(z)$ be the homoclinic class associated to z.

Proposition 1.7 *If* $z \in Per_f(\Lambda)$ *is close to a point in* $W^{u,+}$, *then* $H_f(z) = Cl(W^{u,+})$. *The same is true if we replace* $+$ *by* $-$.

Thus, the following theorem was obtained:

Theorem 10 *Let* Λ *be a singular-hyperbolic set of a* C^r *flow* f *on a closed three-manifold, where* $r \geq 1$. *Suppose that the following properties hold:*
1. Λ *is connected.*
2. Λ *is attracting.*
3. *The closed orbits contained in* Λ *are dense in* Λ.
4. Λ *has a unique singularity* σ.
5. Λ *is not transitive.*
Then H^+ *and* H^- *are homoclinic classes of* f .

The proof of Theorem 10 is based on the following result:

Theorem 11 *Let* Λ *be a singular-hyperbolic set of* $f \in C^r$, $r \geq 1$. *Suppose that the following properties hold:*
1. Λ *is connected.*
2. Λ *is attracting.*
3. *The closed orbits contained in* Λ *are dense in* Λ.
4. Λ *has only one singularity.*
5. Λ *is not transitive.*
Then for every neighborhood U *of* Λ *there is a flow* g *that is* C^r *close to* f *and such that* $\Lambda(g) \not\subset \Omega(g)$, *where* $\Lambda(g)$ *is the continuation of* g.

The proof of Theorem 11 is based on Theorem 10, which imply that $\omega_f(a) = O$ for some periodic orbit with positive expanding eigenvalues of f . In particular, $W^{u,+}$ and $W^{u,-}$ are defined (recall Definition 22) and verify the following conditions given by the two lemmas:

Lemme 1.5 $Cl(W^{u,+}) \cap W^{s,-} \neq \emptyset$.

Lemme 1.6 *Let D be a fundamental domain of $W_f^{uu}(p_0)$ contained in $W^{u,+}$. There exists a neighborhood V of D and a cross-section Σ^- of f intersecting $W^{s,-}$ satisfying the following properties:*

1. Every f-orbit's sequence in Λ converging on a point in $W^{s,-}$ intersects Σ^-.

2. No positive f-orbit with initial point in V intersects Σ^-.

Now we define a perturbation (pushing) close to a point $a \in W_f^u(\sigma) \backslash \{\sigma\}$ which gives the perturbed flow g of f. This pushing in O is obtained in the standard way, as in [de Melo & Palis (1982)]. Let us define the following cross-sections:

1. Σ_a, containing a in its interior.

2. $\Sigma' = f_t(\Sigma)$.

3. Σ_0, intersecting O in a single interior point.

4. Σ^+, Σ^-, which intersect $W^{s,+}, W^{s,-}$, respectively, and point toward the side of a.

Thus, every f-orbit intersecting $\Sigma^+ \cup \Sigma^-$ intersects the set Σ and there is a well-defined neighborhood O given by $O = f_{[0,1]}(\Sigma_a)$.

Lemme 1.7 $K(g) \subsetneqq \Omega(g)$.

Proof 1 *(of Theorem 8) Let Λ be a singular-hyperbolic attractor of a C^r flow f on a compact 3-manifold M. Assume that Λ is C^r critically robust and has a unique singularity σ. Let $\Lambda(Y) = \cap_{t>0} g_t(U)$ be the continuation of Λ in a neighborhood U of Λ for g close to f. Let $C(g)$ be the union of the closed orbits of a flow g. Because Λ is C^r critically robust, then there is a neighborhood U of Λ such that $\Lambda(g) \cap C(g)$ is dense in $\Lambda(g)$ for every flow g that is C^r close to f. $\Lambda(g)$ is a singular-hyperbolic set (with a unique singularity) of f for all g close to f. Assumptions of Theorem 11 implies that Λ is an attractor. In particular, Λ and $\Lambda(g)$ are connected and so the neighborhood U above can be arranged to be connected because Λ is transitive. Thus, $\Lambda(g)$ is a singular-hyperbolic set of g satisfying conditions 1-4 of Theorem 11. If Λ is not C^r robust, one can find a g that is C^r close to f and such that $\Lambda(g)$ is not transitive and satisfies all conditions of Theorem 11, and one can find a g' that is C^r close to g and such that $\Lambda(g') \subset \Omega(g')$. This is a contradiction, because $\Lambda(g') \subsetneqq \Omega(g')$ and $\Lambda(g') \cap C(g')$ is dense in $\Lambda(g')$.*

1.8 Entropies of chaotic attractors

In this section, we discuss statistical properties of chaotic attractors. This includes the physical (or Sinai-Ruelle-Bowen) measure, Hausdorff dimension, topological entropy, Lebesgue (volume) measure, and Lyapunov exponents.

Generally, the **Lebesgue measure** is a way of assigning a length, area or volume to subsets of Euclidean space. These sets are called *Lebesgue measurable*. A measure that is supported on an attractor describes for almost every initial condition in its corresponding basin of attraction the statistical properties of the long-time behavior of the orbits of the system under consideration. The following definition gives the mathematical formulation of the Lebesgue measure of a set $A \subset \mathbb{R}^n$:

Definition 23 *(a) A box in $\mathbb{R}^n$ is a set of the form $B = \prod_{i=1}^{n} [a_i, b_i]$, where $b_i \geq a_i$.*

(b) The volume of the box B is $vol(B) = \prod_{i=1}^{n} (b_i - a_i)$.

(c) The outer measure $m^*(A)$ for a subset $A \subset \mathbb{R}^n$ is defined by:

$$m^*(A) = \inf \left\{ \sum_{j \in J} vol(B_j) : \text{with the } \textbf{Property B} \right\} \qquad (1.18)$$

$\textbf{Property } B$: the set $\{B_j : j \in J\}$ is a countable collection of boxes whose union covers A.

(d) The set $A \subset \mathbb{R}^n$ is Lebesgue measurable if

$$m^*(S) = m^*(S \cap A) + m^*(S - A) \text{ for all sets } S \subset \mathbb{R}^n. \qquad (1.19)$$

(e) The Lebesgue measure is defined by $m(A) = m^*(A)$ for any Lebesgue measurable set A.

Generally, the invariant measure for a dynamical system describes the distribution of the sequence formed by the solution for any typical initial state.

The notion of **physical (or Sinai-Ruelle-Bowen) (SRB) measure** was introduced and proved for Anosov diffeomorphisms and for hyperbolic diffeomorphisms and flows [Sinai (1972), Ruelle (1976), Ruelle and Bowen (1975)]. The common definition of SRB (Sinai-Ruelle-Bowen) measure is given here for both continuous-time discrete-time systems by:

Definition 24 *(a) Let A be an attractor for a map f and let μ be an f-invariant probability measure on A. Then μ is called an SRB (Sinai-Ruelle-Bowen) measure for (f, A) if one has for any continuous map g, that*

$$\lim_{n \to \infty} \frac{1}{n} \sum_{i} g\left(f^i(x)\right) = g d\mu \qquad (1.20)$$

where $x \in E \subset B(A)$ is in the basin of attraction for A, with $m(E) > 0$, where m is the Lebesgue measure.

(b) An invariant probability μ is a physical measure for the flow $f_t, t \in \mathbb{R}$ if the set $B(\mu)$ (the basin of μ) of points $z \in M$ satisfying

$$\lim_{T \to \infty} \frac{1}{T} \int_0^T \varphi(f_t(z))\, dt = \int \varphi d\mu \text{ for all continuous } \varphi : M \to \mathbb{R} \qquad (1.21)$$

has positive Lebesgue measure: $m(B(\mu)) > 0$.

The invariant measure is the most useful notion used in the theory of hyperbolic dynamical systems which means that the "*events*" $x \in A$ and $f(x) \in A$ are equally probable.

Definition 25 *(a) A probability measure μ in the ambient space M is invariant under a transformation f if $\mu(f^{-1}(A)) = \mu(A)$ for all measurable subsets A.*

(b) The measure μ is invariant under a flow if it is invariant under f_t for all t.

(c) An invariant probability measure μ is ergodic if every invariant set A has either zero or full measure, or equivalently, μ is ergodic if it cannot be decomposed as a convex combination of invariant probability measures. That is, one can not have $\mu = a\mu_1 + (1-a)\mu_2$ with $0 < a < 1$ and μ_1, μ_2 invariant.

A practical definition of the concept of SRB (Sinai-Ruelle-Bowen) measure is given by:

Definition 26 *(a) An f-invariant probability measure μ is called an SRB measure if and only if*
(i) μ is ergodic,
(ii) μ has a compact support,
(iii) μ has absolutely continuous conditional measures on its unstable manifolds.
(b) The set Λ carries a probability measure μ if and only if $\mu(\Lambda) = 1$.

This definition was used in [Kiriki, *et al.*, (2010)] in order to investigate the coexistence of homoclinic sets with and without SRB measures in Hénon maps. The existence of an invariant measure was proved theoretically in [Sinai (1970-1979), Ruelle (1976-1978), Bunimovich and Sinai (1980), Eckmann and Ruelle (1985), Jarvenpaa and Jarvenpaa (2001), Sanchez-Salas (2001), Jiang (2003), Bonetto, *et al.*, (2004)] for hyperbolic and nearly hyperbolic systems. For the non-existence of such a measure, one can see, for example, [Hu and Young (1995)]. In [Wolf (2006)] the notion of *generalized physical and SRB measures* was introduced, and it is naturally the generalization of classical physical and SRB measures defined for attractor measures which are supported on invariant sets that are not necessarily attractors.

Generally, if Λ is a hyperbolic attractor, then there exists a measure μ on Λ such that for any x in the basin of attraction of Λ and any observable g, one has the relation (1.20) for discrete-time systems and the relation (1.21) for continuous-time systems. In [Katok and Hasselblatt (1995)], the following conjecture was formulated:

Conjecture 12 *Let Λ be a hyperbolic set and V an open neighborhood of Λ. Does there exist a locally maximal hyperbolic set $\tilde{\Lambda}$ such that $\Lambda \subset \tilde{\Lambda} \subset V$?*

Related questions:

1. Can this set be robust?

2. Can this happen on other manifolds, in lower dimension, on all manifolds?

In [Crovisier (2001)], Crovisier answers "no" for a specific example on a four-torus. In fact, on any compact manifold M, where $\dim(M) \geq 2$, there exists a C^1 open set of diffeomorphisms, U, such that any $f \in U$ has a hyperbolic set that is not contained in a locally maximal hyperbolic set. If Λ is a hyperbolic set and V is a neighborhood of Λ, then there exists a hyperbolic set $\tilde{\Lambda}$ with a Markov partition such that $\Lambda \subset \tilde{\Lambda} \subset V$.

To define the **Hausdorff dimension**, let A be an attractor for a map or flow f. If $S \subset A$ and $d \in [0, +\infty)$, then we have the following definition:

Definition 27 *The d-dimensional Hausdorff content of S is defined by:*

$$C_H^d(S) = \inf \left\{ \sum_i r_i^d : \boldsymbol{P} \text{ is verified} \right\}. \tag{1.22}$$

$\boldsymbol{P}$: *There is a cover of S with balls of radii $r_i > 0$. The Hausdorff dimension of A is defined by*

$$D_H(A) = \inf \left\{ d \geq 0, C_H^d(S) = 0 \right\}. \tag{1.23}$$

By Definition 27, the Hausdorff dimension of the Euclidean space $\mathbb{R}^n$ is n, for a circle $\mathbb{S}^1$ it is 1, and for a countable set it is 0. The Hausdorff dimension of a triadic Cantor set is $\frac{\ln 2}{\ln 3} = 0.630\,93$, and for the triadic Sierpinski triangle it is $\frac{\ln 3}{\ln 2} = 1.585\,0$. Generally, if the Hausdorff dimension $D_H(A)$ strictly exceeds the topological dimension of A, then the set A is a fractal. Also, $D_H(A) \geq D_{ind}(A)$, where $D_{ind}(A)$ is the inductive dimension for A which is defined recursively as an integer or $+\infty$, $D_{ind}(A) = \inf_Y D_H(Y)$, where Y ranges over metric spaces homeomorphic to A [Szpilrajn (1937)] and if $A = \cup_{i \in I} A_i$ is a finite or countable union, then $D_H(A) = \sup_{i \in I} D_H(A_i)$, and if $A = A_1 \times A_2$ then $D_H(A) \geq D_H(A_1) + D_H(A_2)$.

The **topological entropy** of a dynamical system is a non-negative real number that measures the complexity of the system. This entropy can be defined in various equivalent ways.

Let A be a compact Hausdorff topological space. For any finite cover C of A, let the real number $H(C)$ is given by

$$H(C) = \log_2 \inf_j \{j = \text{The number of elements of } C \text{ that cover } A\}. \quad (1.24)$$

For two covers C and D, let $C \vee D$ be their (minimal) common refinement, which consists of all the non-empty intersections of a set from C with a set from D, and similarly for multiple covers. For any continuous map $f : A \to A$, the following limit exists:

$$H(C, f) = \lim_{n \to +\infty} \frac{1}{n} H \left(C \vee f^{-1} \vee ... \vee f^{-n+1} C \right). \quad (1.25)$$

Then one has the following definition:

Definition 28 *The topological entropy $h(f)$ of f is the supremum of $H(C, f)$ over all possible finite covers C, i.e.,*

$$h(f) = \sup_C \{H(C, f), \ C \text{ is a finite cover of } A\}. \quad (1.26)$$

A useful definition of the topological dimension of a space E in the case of singular-hyperbolic attractors was given in [Hurewicz and Wallman (1984)]:

Definition 29 *The topological dimension of a space E is either -1 (if $E = \emptyset$) or the last integer k for which every point has arbitrarily small neighborhoods whose boundaries have dimension less than k.*

For practical applications, we give a more appropriate definition for the topological entropy using the notion of (n, ϵ)-separated sets and interval arithmetic, namely the method of calculation using the number of periodic orbits of the system under consideration in a specific range.

Let $f : X \to X$ be a map. Then one has the following definition:

Definition 30 *A set $E \subset X$ is called (n, ϵ)-separated if for every two different points $x, y \in E$, there exists $0 \leq j < n$ such that the distance between $f^j(x)$ and $f^j(y)$ is greater than ϵ. Let us define the number $s_n(\epsilon)$ as the cardinality of a maximum (n, ϵ)-separated set:*

$$s_n(\epsilon) = \max \{ \text{card } E : \ E \text{ is } (n, \epsilon)\text{-separated} \} \quad (1.27)$$

The number

$$H(f) = \lim_{\epsilon \to 0} \lim_{n \to \infty} \sup \frac{1}{n} \log s_n(\epsilon) \quad (1.28)$$

is called the topological entropy of f.

Under certain assumptions [Bowen (1971)], the topological entropy can be expressed in terms of the number of periodic orbits as follows:

Definition 31 *If f is an Axiom A diffeomorphism, then*

$$H\left(f\right) = \lim_{n\to\infty} \sup \frac{\log C\left(f^n\right)}{n}, \tag{1.29}$$

where $C\left(f^n\right)$ is the number of fixed points of f^n.

Formula (1.29) can be used as the lower bound for the topological entropy $H\left(f\right)$ when the distance between periodic orbits of length n is uniformly separated from zero, i.e., the formula

$$H\left(f\right) = \frac{\log C\left(f^n\right)}{n}, \tag{1.30}$$

for sufficiently large n.

In a topological sense, we have the following definition:

Definition 32 *A dynamical system is called chaotic if its topological entropy is positive.*

For the Hénon map, the upper and lower bounds for the topological entropy were estimated based on the method given in [Misiurewicz and Szewc (1980), Zgliczynski (1997), Galias (1998)] using the interval technique and the following result:

Theorem 13 *The topological entropy $H\left(h\right)$ of the Hénon map is located in the interval*

$$0.3381 < H\left(h\right) \le \log 2 < 0.6932. \tag{1.31}$$

The **Lyapunov exponent** of a dynamical system with evolution equation f_t in a n-dimensional phase space is a quantity that characterizes the rate of separation of infinitesimally close trajectories. If δx_0 is an initial separation of two trajectories in phase space, then the λ is defined by:

$$\lambda = \lim_{t\to+\infty} \frac{1}{t} \ln \frac{\left|\delta x\left(t\right)\right|}{\left|\delta x_0\right|}. \tag{1.32}$$

Generally, the spectrum of Lyapunov exponents $\{\lambda_1, \lambda_2, ..., \lambda_n\}$ depends on the starting point x_0. The Lyapunov exponents can be defined from the Jacobian matrix $J^t\left(x_0\right) = \frac{df_t}{dx}\big|_{x=x_0}$, which describes how a small change at the point x_n propagates to the final point $f_t(x_0)$.

Definition 33 *(Lyapunov exponent for a continuous-time dynamical system) The matrix $L\left(x_0\right)$ (when the limit exists[3]) defined by*

$$L(x_0) = \lim_{t\to+\infty} \left(J^t\left(x_0\right)\left(J^t\left(x_0\right)\right)^T\right)^{\frac{1}{2t}} \tag{1.33}$$

---

[3]The conditions for the existence of the limit are given by the Oseledec theorem.

has n eigenvalues $\Lambda_i(x_0), i = 1, 2, ..., n$, and the Lyapunov exponents $\lambda_i, i = 1, 2, ..., n$ are defined by

$$\lambda_i(x_0) = \log \Lambda_i(x_0), i = 1, 2, ..., n \qquad (1.34)$$

For the zero vector, we define $\lambda_i(x_0) = -\infty$.

In the next definition, we present the standard definition of the Lyapunov exponents for a discrete n-dimensional mapping.

Definition 34 *(Lyapunov exponent for a discrete dynamical system): Consider the following n-dimensional discrete dynamical system:*

$$x_{k+1} = f(x_k), x_k \in \mathbb{R}^n, k = 0, 1, 2, ... \qquad (1.35)$$

where $f : \mathbb{R}^n \longrightarrow \mathbb{R}^n$ is the vector field associated with system (1.35). Let $J(x)$ be its Jacobian evaluated at x, and consider the matrix:

$$T_r(x_0) = J(x_{r-1}) J(x_{r-2}) ... J(x_1) J(x_0). \qquad (1.36)$$

Moreover, let $J_i(x_0, l)$ be the module of the ith eigenvalue of the lth matrix $T_r(x_0)$, where $i = 1, 2, ..., n$ and $r = 0, 1, 2, ...$
Now, the Lyapunov exponents of an n-dimensional discrete-time systems are defined by:

$$\lambda_i(x_0) = \ln \left(\lim_{r \longrightarrow +\infty} J_i(x_0, r)^{\frac{1}{r}} \right), i = 1, 2, ..., n. \qquad (1.37)$$

Generally, the analytic calculation of Lyapunov exponents using the matrix $L(x_0)$ is not possible. In fact, there are many numerical methods and algorithms that deal with some estimates [Benettin, *et al.*, (1980), Shimada and Nagashima (1979), Sprott (2003)].

Example 2 *Examples of maps characterized by a positive Lyapunov exponent for the critical value are the Collet-Eckmann maps:*
(a) A unimodal map f is called Collet-Eckmann (CE) if there exists constants $C > 0, \lambda > 1$ such that for every $n > 0$,

$$|Df^n(f(0))| > C\lambda^n$$

(b) A unimodal map f is called backwards Collet-Eckmann (BCE) if there exists $C > 0, \lambda > 1$ such that for any $n > 0$ and any x with $f^n(x) = 0$, we have

$$|Df^n(x)| > C\lambda^n$$

The unimodal Collet-Eckmann and the backwards Collet-Eckmann maps are strongly hyperbolic along the critical orbit. In fact, all S-unimodal maps and Collet-Eckmann maps are also backwards Collet-Eckmann maps.

A general procedure for constructing S-unimodal maps that generate robust chaos was done in [Andrecut and Ali (2001)] as follows:

Let $f(x) : I \to I$ and $g(x) : I \to I$ be S-unimodal maps on the interval $I = [0, 1]$, with $c_f \in (0, 1)$ the unique maximum of f and $c_g \in (0, 1)$ the unique maximum of g, then the map

$$F_\nu(x) = \frac{f\left(v \frac{g(x)}{g(c_g)}\right)}{f(v)}, x \in I \tag{1.38}$$

is S-unimodal and it generates robust chaos for $\nu \in (0, c_f)$.

As an example, let $f(x) = \pi^{-1} \sin(\pi x)$ and $g(x) = \sqrt{x}(1 - \sqrt{x})$ the function $F_\nu(x)$ given by (1.38) generates robust chaos for $x \in [0, 1]$ and $\nu \in (0, \frac{1}{2})$ and numerical computation of the Lyapunov exponent is $\lambda(\nu) = \ln(2) > 0, \forall \nu \in (0, \frac{1}{2})$.

some possible relations between the geometry of chaotic attractors of typical analytic unimodal maps and the behavior of the critical orbit, i.e. an explicit formula relating the combinatorics of the critical orbit with the exponents of periodic orbits are given by:

Theorem 14 *(a) Let f_λ be a non-trivial analytic family of unimodal maps. Then, for almost every non-regular parameter λ, and for every periodic orbit p in the non-trivial attractor A_{f_λ}, the exponent of p is determined by an explicit combinatorial formula involving the kneading sequence of f and the itinerary of p.*

(b) Let f_λ be a non-trivial analytic family of quasiquadratic maps. Then, for almost every non-regular parameter λ, the critical point belongs to the basin of μ_{f_λ} (the absolutely continuous invariant measure of f_λ).

In the setting of Theorem 14(b), one also has equality between the Lyapunov exponent of the critical value and the Lyapunov exponent of μ_{f_λ}.

The inverse of the largest Lyapunov exponent is called the *Lyapunov time* which is finite for chaotic orbits and infinite for non-chaotic orbits. Notice that the Lyapunov exponents are asymptotic quantities. Thus, some conditions for the existence of the solution are needed. For almost systems these conditions are: The phase space is a compact, boundaryless manifold and the initial condition lies in a positively invariant region. On the other hand, Lyapunov exponents are not continuous functions of the orbits, but the spectrum of Lyapunov exponents $\{\lambda_1, \lambda_2, ..., \lambda_n\}$ depends on the starting point x_0. However, these exponents are the same for almost all starting points x_0 for an ergodic component of the dynamical system. Also, we have $\sum_{i=1}^{n} \lambda_i = 0$ if the system is conservative and if the system is dissipative, then $\sum_{i=1}^{n} \lambda_i < 0$. If the dynamical system is a smooth flow, then one exponent is always zero (Haken's theorem). For non-smooth systems, this condition is not typical. The Lyapunov spectrum can be used to estimate some topological dimensions, such as the so-called *Kaplan-Yorke dimension* D_{KY} that is defined as follows:

$$D_{KY} = k + \sum_{i=1}^{k} \frac{\lambda_i}{|\lambda_{k+1}|}$$

where k is the maximum integer such that the sum of the k largest exponents is still non-negative, and in this case D_{KY} is an upper bound for the information dimension of the system [Kaplan and Yorke (1987)]. Pesin's theorem [Pesin (1992)] implies that the sum of all the positive Lyapunov exponents gives an estimate of the *Kolmogorov-Sinai entropy*. For any continuous-time vector fields $f, g \in \mathbb{R}^n$, and non-zero constant $c \in \mathbb{R}$, one has

$$\begin{cases} \lambda_i(f+g) \leq \max\{\lambda_i(f), \lambda_i(g)\} \\ \\ \lambda_i(cg) = \lambda_i(g) \end{cases} , i = 1, 2, ..., n.$$

In some situations, it is important to determine the upper and lower bounds for all the Lyapunov exponents of a given n-dimensional system. A recent result was given in [Li and Chen (2004)] for discrete mappings as follows:

Theorem 15 *If a system $x_{k+1} = f(x_k), x_k \in \Omega \subset \mathbb{R}^n$, verify*

$$\|Df(x)\| = \|J\| = \sqrt{\lambda_{\max}(J^T J)} \leq N < +\infty, \tag{1.39}$$

with a smallest eigenvalue of $J^T J$ that satisfies

$$\lambda_{\min}(J^T J) \geq \theta > 0, \tag{1.40}$$

where $N^2 \geq \theta$, then, for any $x_0 \in \Omega$, all the Lyapunov exponents at x_0 are located inside $\left[\frac{\ln \theta}{2}, \ln N\right]$. That is,

$$\frac{\ln \theta}{2} \leq l_i(x_0) \leq \ln N, i = 1, 2, ..., n, \tag{1.41}$$

where $l_i(x_0)$ are the Lyapunov exponents for the map f.

The notion of entropies defined above can be generalized to singular-hyperbolic attractors discussed in Sec. 8.11.1. The relation between dynamics and topological dimension was considered in [Hirsch (1970), Bowen (1970), Bowen (1973), Przytycki (1980)]: Any Anosov diffeomorphism on n-manifolds have no compact invariant $(n-1)$-dimensional sets [Hirsch (1970)], and any hyperbolic automorphisms on n-tori have compact invariant k-dimensional sets for all $0 \leq k \leq n-2$, namely, we have the two following theorems proved in [Przytycki (1980)]:

Theorem 16 *For any Anosov diffeomorphism f of an n-dimensional torus $\mathbb{T}^n$ and for any integer k such that $0 \leq k \leq n$ and $k \neq n - 1$ there exists a compact f-invariant subset N^K of topological dimension k.*

Theorem 17 *For any Anosov diffeomorphism f of an n-dimensional manifold $\mathbb{M}^n$ and for any integer k such that $0 \leq k \leq n - 2$ there exists a compact f-invariant subset N^K such that*

$$k \leq \dim N^K \leq \min\left(k + s - 1, k + u - 1, \frac{k + n - 1}{2}\right)$$

where s and u denote dimensions of stable and unstable manifolds, respectively.

Also, minimal systems for Axiom A diffeomorphisms are always 0-dimensional and minimal systems for Axiom A flows are 1-dimensional or 0-dimensional, i.e., a single point [Bowen (1970-1973)]. For singular-hyperbolic systems the following result was proved in [Morales (2003)] where the question is whether the non-singular minimal subsets of such sets are one dimensional:

Theorem 18 *All compact non-singular invariant subsets of a transitive singular set of a C^1 vector fields f are one-dimensional.*

Theorem 18 generalizes the minimal set's results for Axiom A flows [Bowen (1973)] to a class of non-expansive flows systems with some hyperbolicity [Morales, et al. (1999)].

The following result was proved in [Morales (2008)]:

Theorem 19 *Singular-hyperbolic attractors on compact 3-manifolds have topological dimension ≥ 2.*

The existence of unstable manifolds for a subset of the attractor which is visited infinitely many times by a residual subset and the densness of the set of periodic orbits was proved in [Arroyo & Pujals (2007)]:

Theorem 20 *If Λ is a singular-hyperbolic transitive attractor for a flow Φ_t, then the set of periodic orbits in Λ is dense in Λ, that is, $Cl\left(Per(\Lambda)\right) = \Lambda$.*

Also, a complete description of the spectral decomposition of a singular-hyperbolic transitive attractor is given by the following result [Arroyo & Pujals (2007)]:

Theorem 21 *If Λ is a singular-hyperbolic transitive attractor for a flow Φ_t, then there exists a periodic orbit p such that $H(p) = \Lambda$.*

and that there is an SRB-measure supported on the attractor [Arroyo & Pujals (2007)]:

If Λ is a singular-hyperbolic transitive attractor of a $C^{1+\alpha}$ flow, $\alpha > 0$, then it has an SRB measure supported on it.

The proof for Theorem 20 and Theorem 21 is based on the construction of a family of return maps on a certain collection (finite) of small transversal sections with markovian properties.

Examples of singular-hyperbolic sets which do not have any periodic orbit at all are shown in [Morales (2004)]. These sets are neither attractors nor robust. Also, any transitive singular-hyperbolic attractor has at least one periodic orbit [Bautista & Morales (2006)]. Also, a general singular-hyperbolic invariant set does not allow a decomposition into a finite number of disjoint homoclinic classes [Bautista, *et al.* (2004)]. In [Morales & Pujals (1997)], some examples are given of singular-hyperbolic transitive attractors with a periodic orbit in the closure of a homoclinic class to which it is not homoclinically related. These systems are examples of singular hyperbolic transitive attractors that are not C^1-robustly transitive sets.

1.9   Period 3 implies chaos

The following definition collects the most used notions in the dynamics of 1-D mappings:

Definition 35 *(a) Let S be a set and f a function mapping the set S into itself. i.e., $f : S \to S$. The functions*

$$f^1 = f, f^2 = f \circ f, ..., f^n = f^{n-1} \circ f, ...$$

are said to be an iterate of f.

(b) Let $f : S \to S$. If $x_0 \in S$ then $x_0, f^1(x_0), ..., f^n(x_0), ...$ is called the orbit of x_0. x_0 is called the seed of the orbit.

(c) Let $f : S \to S$. A point $a \in S$ is said to be a fixed point of f if $f(a) = a$.

(d) Let $f : S \to S$. A point $x \in S$ is said to be periodic if there exists a positive integer p such that $f^p(x) = x$. If m is the least $n \in \mathbb{N}$ such that $f^n(x) = x$ then m is called the period of x.

(e) Let $f : S \to S$. The point $x_0 \in S$ is said to be eventually periodic if x_0 is not periodic, but some point in the orbit of x_0 is periodic.

Example 3 *As an example of Definition 35(d). Let $f = x^2 - 1$. If $x_0 = 0$ then $f(x_0) = -1$ and $f^2(x_0) = f(-1) = 0$. Therefore x_0 is periodic with period 2. As an example of Definition 35(e). Let $f = x^2 - 1$ again. If we look at $x_0 = 1$ then we have $f(x_0) = 0$ and we have already seen than $x = 0$ is periodic with period 2 for this function.*

The following definition collects the most used notions in the nature of fixed points of a 1-D mappings:

Definition 36 *(a) Let a be a fixed point of $f : \mathbb{R} \to \mathbb{R}$. The point a is said to be an attracting fixed point if there is an open interval I containing a such that if $x \in I$, then $f^n(x) \to a$ as $n \to \infty$.*

(b) Let a be a fixed point of $f : \mathbb{R} \to \mathbb{R}$. The point a is said to be a repelling fixed point if there is an open interval I containing a such that if $x \in I$ and $x \neq a$, then there exists an integer n such that $f^n(x) \notin I$.

(c) A fixed point that is neither attracting nor repelling is called a neutral fixed point.

The existence of fixed points is established by the following theorem:

Theorem 22 *(The Weierstrass Intermediate Value Theorem): Let f be a continuous mapping from $[0,1]$ into $[0,1]$. Then there exists a point $z \in [0,1]$ such that $f(z) = z$.*

Proof 2 *If $f(0) = 0$ or $f(1) = 1$ then the theorem is evident. Assuming that $f(0) > 0$ and $f(1) < 1$. Let $g : [0,1] \to \mathbb{R}$ be defined by $g(x) = x - f(x)$. g is continuous, $g(0) = -f(0) < 0$ and $g(1) = 1 - f(1) > 0$. Then there exists $z \in [0,1]$ such that $g(z) = 0$, i.e., $f(z) = z$.*

The following theorems explain the stability of fixed points:

Theorem 23 *Let S be an interval in $\mathbb{R}$ and a be a point in the interior of S. Furthermore, let a be a fixed point of a function $f : S \to \mathbb{R}$. If f is differentiable at the point a and $|f'(a)| < 1$, then a is an attracting fixed point of f.*

Proof 3 *As $|f'(a)| < 1$, we have $|f'(a)| < k < 1$ where $k = \frac{|f'(a)|+1}{2}$. By definition, $f'(a) = \lim_{x \to a} \frac{f(x)-f(a)}{x-a}$. There exists $\delta > 0$ and interval $I = [a - \delta, a + \delta]$ such that $|\frac{f(x)-f(a)}{x-a}| \leq k$ for all $x \in I$ with $x \neq a$. Since a is a fixed point $f(a) = a$. So we get $|f(x) - a| \leq k|x - a|, \forall x \in I$. noting that $k < 1$ the last result implies that $f(x)$ is closer to a than x. Therefore, we get that $f(x) \in I$. So we can repeat the argument using $f(x)$ instead of x. We get:*

$$|f^2(x) - a| \leq k|f(x) - a|, \forall x \in I$$

Combining this two results we get:

$$|f^2(x) - a| \leq k^2|x - a|, \forall x \in I$$

Since $|k| < 1$ then $k^2 < 1$. We can repeat the argument again and we can obtain by induction

$$|f^n(x) - a| \leq k^n|x - a|, \forall x \in I, \text{ and } \forall n \in \mathbb{N}$$

As $|k| < 1$, $\lim_{n \to \infty} k^n = 0$. This implies that $f^n(x) \to a$ as $n \to \infty$. Finally, a is an attracting point.

Theorem 24 *Let S be an interval in $\mathbb{R}$ and a be a point in the interior of S. Let a be a fixed point of a function $f : S \rightarrow \mathbb{R}$. If f is differentiable at a and $|f'(a)| > 1$, than a is a repelling fixed point of f.*

One of the most known mappings in the theory of 1-D dynamics is the quadratic map: $Q_c : \mathbb{R} \rightarrow \mathbb{R}$ defined by

$$Q_c(x) = x^2 + c \tag{1.42}$$

where $c \in \mathbb{R}$. We have the following result:

Theorem 25 *(The first bifurcation theorem): Let Q_c be the quadratic function (1.42) and $c \in \mathbb{R}$.*
(1) If $c > \frac{1}{4}$, then all orbits tend to infinity. That is, $\forall x \in \mathbb{R}$, $(Q_c)^n(x) \rightarrow \infty$ as $n \rightarrow \infty$.
(2) If $c = \frac{1}{4}$, then Q_c has exactly one fixed point at $x = \frac{1}{2}$ and this is a neutral fixed point.
(3) If $c < \frac{1}{4}$, then Q_c has two fixed points $a_+ = \frac{1}{2}(1 + \sqrt{1 - 4c})$ and $a_- = \frac{1}{2}(1 - \sqrt{1 - 4c})$.
(3-1) The point a_+ is always repelling.
(3-2) If $-\frac{3}{4} < c < \frac{1}{4}$, than a_- is attracting.
(3-4) If $c < -\frac{3}{4}$, then a_- is repelling.

Before presenting the so called second bifurcation theorem, we need the following definition:

Definition 37 *(a) Let f be a function mapping the set S into itself. If the point $x \in S$ has period m, then the orbit of x is $\{x, f(x), ..., f^{m-1}(x)\}$ and the orbit is called an **m**-cycle.*
(b) Let a be a periodic point of a function $f : S \rightarrow S$ of period m, for some $m \in \mathbb{N}$. a is clearly a fixed point of $f^m : S \rightarrow S$. a is said to be an attracting period point of f if it is an attracting fixed point of f^m. Similarly a is said to be a repelling period point of f if it is a repelling fixed point of f^m.

Theorem 26 *(The second bifurcation theorem): Let Q_c be the quadratic function (1.42) and $c \in \mathbb{R}$.*
(1) If $-\frac{3}{4} \leq c < \frac{1}{4}$, then Q_c has no 2-cycles.
(2) If $-\frac{5}{4} < c < -\frac{3}{4}$, then Q_c has an attracting 2-cycle, $\{q_-, q_+\}$, where $q_+ = \frac{1}{2}(-1 + \sqrt{-4c - 3})$ and $q_- = \frac{1}{2}(-1 - \sqrt{-4c - 3})$.
(3) If $c < -\frac{5}{4}$, then Q_c has a repelling 2-cycle $\{q_-, q_+\}$.

The period 3 theorem was proved in 1975 by Yorke and Li [Yorke and Li (1975)]. In order to present its proof, we need some helpful lemmas.

Lemme 1.8 *If $f : \mathbb{R} \rightarrow \mathbb{R}$ then for every interval $I \subset \mathbb{R}$, then $f(I)$ is also an interval.*

Proof 4 *Indeed, a continuous mapping maps connected sets to connected sets, and all connected sets in $\mathbb{R}$ are intervals.*

Lemme 1.9 *Let $a, b \in \mathbb{R}$ with $a < b$ and $f : I = [a, b] \to \mathbb{R}$ a continuous function. If $f(I) \supseteq I$, then f has a fixed point in I.*

Proof 5 *Define $f[I] = [c, d]$ where $c < a$ and $d > b$, then there exists two points $s, t \in [a, b]$ such that $f(s) = c \leq a \leq s$ and $f(t) = d \geq b \geq t$ otherwise $f(I)$ will not cover I. We define $g(x) : I \to \mathbb{R}$ as follows: $g(x) = f(x) - x$. Thus, $g(x)$ is continuous and $g(s) \leq 0$, $g(t) \geq 0$. By Weirstrass Intermediate Value Theorem 22, there exists a point $a \in I$ such that $g(a) = 0 \to f(a) = a$ and a is a fixed point of f in I.*

Lemme 1.10 *Let $a, b \in \mathbb{R}$ with $a < b$. Let $f : [a, b] \to \mathbb{R}$ be a continuous function and $f([a, b]) \supseteq J = [c, d]$, for $c, d \in \mathbb{R}$ with $c < d$. Then there exists a subinterval $I' = [s, t]$ of $I = [a, b]$ such that $f(I') = J$.*

Proof 6 *We first observe that $\{c\}$ and $\{d\}$ are closed sets in the usual topology and since $f([a, b]) \supseteq J = [c, d]$ and f is continuous we get that $f^{-1}\{c\}$ and $f^{-1}\{d\}$ are non-empty closed sets. Then there is a largest number s such that $f(s) = c$.*
Firstly, assuming that there is $x > s$ such that $f(x) = d$. Then there is a smallest number t such that $t > s$ and $f(t) = d$. Now we suppose in contradiction that there is a $y \in [s, t]$ such that $f(y) < c$. If so, we can look at the interval $[y, t]$ and by the Weirstrass intermediate value theorem 22, there is $z > y > s$ such that $f(z) = c$ in contradiction with the maximality of s. Similarly, there is no $y \in [s, t]$ such that $f(y) > d$. Thus, $f([s, t])$ contains no points outside the interval $[c, d]$ and by the Weirstrass intermediate value theorem 22, the image $f[s, t]$ covers $[c, d]$, therefore $f[s, t] = [c, d] = J$ as required.
Secondly, assuming that there is no $x > s$ such that $f(x) = d$. Let s' be the largest number that $f(s') = d$. Clearly $s' < s$. Let t' be the smallest number such that $t' > s'$ and $f(t') = c$. As before, we get $f([s', t']) = [c, d] = J$ as required.

Theorem 27 *(The period 3 theorem): Let $f : \mathbb{R} \to \mathbb{R}$ be a continuous function. If f has a point of period 3, then for each $n \in \mathbb{N}$ it has a periodic point of period n.*

Proof 7 *There exists a point $a \in \mathbb{R}$ such that $f(a) = b$, $f(b) = c$ and $f(c) = a$ where $a \neq b \neq c$. We shall consider the case where $a < b < c$, the other cases are similar.*
Define $I_0 = [a, b]$ and $I_1 = [b, c]$. Using Lemma 1.8 we get $f(I_0) \supseteq I_1$. Using Lemma 1.8 again we can get $f(I_1) \supseteq I_0 \cup I_1$. If we apply Lemma 1.10 on this result, then there exists an interval $A_1 \subseteq I_1$ such that $f(A_1) = I_1$. Note that $f(A_1) = I_1 \supseteq A_1$, using Lemma 1.10 again will give us $A_2 \subseteq A_1$ such that $f(A_2) = A_1$. So far we have $A_2 \subseteq A_1 \subseteq I_1$ and $f^2(A_2) = I_1$.

At this stage, the induction is used to extend this result. Assuming that for $n - 3$ *there exists a series* $A_{n-3} \subseteq A_{n-4}... \subseteq A_2 \subseteq A_1 \subseteq I_1$ *and* $f(A_i) = A_{i-1}$ $\forall 2 \leq i \leq n - 2$ *and* $f(A_1) = I_1$. *Now using Lemma 1.10 again we can get* $A_{n-2} \subseteq A_{n-3}$ *such that* $f(A_{n-2}) = A_{n-3}$. *Using this result we get that* $f^{n-2}(A_{n-2}) = I_1$ *and* $A_{n-2} \subseteq I_1$.

Noting that $f(I_0) \supseteq I_1 \supseteq A_{n-2}$ *we can have a closed interval* $A_{n-1} \subseteq I_0$ *such that* $f(A_{n-1}) = A_{n-2}$. *Finally as* $f(I_1) \supset I_0 \supseteq A_{n-1}$ *there is a closed interval* $A_n \subseteq I_1$ *such that* $f(A_n) = A_{n-1}$.

Putting the above parts together we get; $\forall 2 \leq i \leq n$ $f(A_i) = A_{i-1}$ *and* $f^n(A_n) = I_1$. *Now using the fact that* $A_n \subset I_1$ *and using Lemma 1.9 we get that there is a point* $x_0 \in A_n$ *such that* $f^n(x_0) = x_0$, *i.e.,* x_0 *is a periodic point of* f *of period* n. *We now have to show that* x_0 *period is* n.

To do this, assuming by contradiction that x_0 *is a periodic point of period* k *where* $k < n$. *We notice that* $f(x_0) \in A_{n-1} \subseteq I_0$ *and* $\forall 2 < i < n$ $f^i(x_0) \in I_1$. *Now* $I_0 \cap I_1 = \{b\}$. *We first assume that* $f(x_0) \neq b$. *If so, we get that* $f^k(x_0) = x_0$ *by assumption and therefore* $f^{k+1}(x_0) = f(x_0) \in I_0$ *in contradiction to the fact that* $f^i(x_0) \in I_1, \forall 2 < i < n$. *Now we show that* $f(x_0) \neq b$. *Assuming by contradiction that* $f(x_0) = b$. *We have* $f^2(x_0) = f(b) = a \notin I_1$. *Hence, for any* $n \geq 3$ *there exists a point* x_0 *such that* x_0 *has a period* n *of* f. *For the case* $n = 1$, *we use the fact that* $f(I_1) \supseteq I_1$ *and by Lemma 1.9, there is a fixed point of* f *in* I_1. *For the case* $n = 2$, *we have* $f(I_0) \supseteq I_1$ *and* $f(I_1) \supseteq I_0$. *Using Lemma 1.10, there is a closed interval* $B \subseteq I_0$ *such that* $f(B) = I_1$. *Now* $f^2(B) \supseteq I_0$, *and by using Lemma 1.9, there exists a point* $x_1 \in B$ *such that* $f^2(x_1) = x_1$. *Now* $x_1 \in B \subseteq I_0 = [a, b]$ *and* $f(x_1) \in f(B) \subseteq I_1$ $[b, c]$. *Moreover* $x_1 \neq b$ *as* $f^2(b) = f(c) = a \neq b$. *Finally,* x_1 *is of period 2.*

The more general theorem is called the *Sarkovskii theorem* proved in [Sarkovskii (1964)]:

Theorem 28 *(Sarkovskii's Theorem): Let* $f : \mathbb{R} \to \mathbb{R}$ *be a continuous function. If* f *has a point of period* n *and* n *precedes* k *in Sarkovskii's ordering of the natural numbers, then* f *has a periodic point of period* k.

The method of the proof of this result is based essentially on the Sarkovskii's ordering ($\triangleright$) of the natural numbers given by:

$$3 \triangleright 5 \triangleright 7 \triangleright 9 \triangleright ... \tag{1.43}$$

$$2 \cdot 3 \triangleright 2 \cdot 5 \triangleright 2 \cdot 7 \triangleright ...$$

$$2^2 \cdot 3 \triangleright 2^2 \cdot 5 \triangleright 2^2 \cdot 7, ...$$

$$2^3 \cdot 3 \triangleright 2^3 \cdot 5 \triangleright 2^3 \cdot 7, ...$$

$$........................$$

$$..., 2^n \triangleright 2^{n-1} \triangleright ...., 2^2 \triangleright 2^1 \triangleright 1$$

Since the number 3 is the first number in the ordering (1.42), then Theorem 28 implies Theorem 27.

The converse of Sarkovskii's theorem 28 is also true and given by:

Theorem 29 *(Converse of Sarkovskii's Theorem): Let $n \in \mathbb{N}$ and l precedes n in Sarkovskii's ordering of the natural numbers given by (1.43). Then there exists a continuous function $f : \mathbb{R} \to \mathbb{R}$ which has a periodic point of period n, but no periodic point of period l.*

The generalization of Theorem 28 to hight dimensions is the subject of the next section, where the *Snap-back* repeller is defined and used instead of period 3 orbit.

1.10   The Snap-back repeller and the Li-Chen-Marotto theorem

In this section, we present a very powerful tool used for the rigorous proof of chaos in the sense of Li-Yorke [Li and Yorke (1975)] using the so-called Marotto theorem [Marotto (1978)] along some other versions of it. The Marotto theorem is the best one in predicting and analyzing discrete chaos in higher-dimensional difference equations to date. It is well known that there exists an error in the condition of the original, and several authors had tried to correct it in different ways [Chen, *et al.* (1998), Lin, *et al.* (2002), Li and Chen (2003)]. A corrected and improved version of the Marotto theorem was derived very recently by Li and Chen [Li and Chen (2004)].

Let us consider the following n-dimensional dynamical system:

$$X_{k+1} = f(X_k), X_k \in \mathbb{R}^n, k = 0, 1, 2, . \qquad (1.44)$$

where the map $f : \mathbb{R}^n \longrightarrow \mathbb{R}^n$ is continuous. Denote by $B_r(P)$ the closed ball in $\mathbb{R}^n$ of radius r centered at a point $P \in \mathbb{R}^n$. Let f be differentiable in $B_r(P)$. The point $P \in \mathbb{R}^n$ is an expanding fixed point of f in $B_r(P)$ if $f(P) = P$ and all eigenvalues of $Df(X)$ exceed 1 in absolute value for all $X \in B_r(P)$, then one has the following definition:

Definition 38 *(Snap-back repeller): Assume that P is an expanding fixed point of f in $B_r(P)$ for some $r > 0$. P is then said to be a snap-back repeller of f if there exists a point $P_0 \in B_r(p)$ with $P_0 \neq P$ such that $f^m(P_0) = P$ and the determinant $|Df^m(P_0)| \neq 0$ for an integer $m > 0$.*

Theorem 30 *(Li-Chen-Marotto Theorem) Suppose that in system (1.44), f is a map from $\mathbb{R}^n$ to itself and P is its fixed point. Assume also that*
  (i) $f(X)$ is continuously differential in $B_r(P)$ for some $r > 0$;
  (ii) all eigenvalues of $Df(P)^T Df(P)$ are greater than 1;

(iii) there exists a point

$$P_0 \in B_{r_1}(P) = \{X \in \mathbb{R}^n / \|X - P\| \le r_1 \le r\}$$

and all eigenvalues of $Df(X)^T Df(X) > 1$, with $P_0 \ne P$ such that $f^m(P_0) = P$, where $f^i(P_0) \in B_r(P), i = 0, 1, 2, ..., m$; and $\det Df^m(P_0) \ne 0$, for some positive integer m.

Then, system (1.44) is chaotic in the sense of Li-Yorke and we have:

(a) there is a positive integer N such that for each integer $p \ge N$, f has a point of period p;

(b) there is a "scrambled set" of f, i.e., an uncountable set S containing no periodic points of f such that:

(b_1) $f(S) \subset S$,

(b_2) for every $X_S; Y_S \in S$ with $X_S \ne Y_S$,

$$\limsup_{k \to \infty} \|f^k(X_S) - f^k(Y_S)\| > 0,$$

(b_3) for every $X_S \in S$ and any periodic point Y_{per} of f,

$$\limsup_{k \to \infty} \|f^k(X_S) - f^k(Y_{per})\| > 0,$$

(c) there is an uncountable subset S_0 of S such that for every $X_{S_0}; Y_{S_0} \in S_0$ we have

$$\limsup_{k \to \infty} \|f^k(X_{S_0}) - f^k(Y_{S_0})\| = 0.$$

1.11   Shilnikov criterion for the existence of chaos

In this section, we present some Shilnikov theorems about the existence of chaos in dynamical systems. These results are the commonly agreed-upon analytic criteria for proving chaos in autonomous systems. These results use the notion of homoclinic and heteroclinic orbits. The resulting chaos is called the *horseshoe type* or *Shilnikov chaos*. These horseshoes give the extremely complicated behavior typically observed in chaotic systems [Guckenheimer and Holmes (1983)].

Let $f : \mathbb{R}^n \longrightarrow \mathbb{R}^n$ be a real function defining a discrete mapping and let $Df(x)$ be its Jacobian matrix. To simplify the notion of *homoclinic bifurcations*, we need the following definitions:

Definition 39 *(a) A saddle point q of f is a point where $Df(q)$ has some eigenvalues λ such that $|\lambda| < 1$, and the remaining satisfy $|\lambda| > 1$.*

(b) A homoclinic point p of a map $f : \mathbb{R}^n \longrightarrow \mathbb{R}^n$ lies inside the intersection of its stable and unstable separatrix (invariant manifold), i.e., $\lim_{n \longrightarrow +\infty} f(x) = \lim_{n \longrightarrow -\infty} f(x) = P$.

(c) The map $f : \mathbb{R}^n \longrightarrow \mathbb{R}^n$ has a hyperbolic period-k orbit γ if there exists $q \in \mathbb{R}^n$ such that $f^k(q) = q$, and q is a saddle point with its stable and unstable manifold $W^s(q), W^u(q)$ intersecting transversely in a point p.

(d) The point p in the Definition 39(b) is called a transverse homoclinic point.

The most important results for the existence of a transverse homoclinic point are the existence of infinitely many periodic and homoclinic points in a small neighborhood of this point, as shown in the *Smale-Moser Theorem* [Smale (1965)]:

Theorem 31 *In the neighborhood of a transverse homoclinic point, there exists an invariant Cantor set on which the dynamics are topologically conjugate to a full shift on N symbols.*

Theorem 31 means that there is a positively and a negatively invariant Cantor set containing infinitely many saddle-type (unstable) periodic orbits of arbitrarily long periods, uncountably many bounded non-periodic orbits, and a dense orbit. Hence, the horseshoe persists under perturbations. Indeed, if q is a transversal homoclinic point to a hyperbolic fixed point p of a diffeomorphism f, then, for any neighborhood U of $\{p, q\}$, there is a positive integer k such that f^k has a hyperbolic invariant set $\{p, q\} \in \Gamma \subset U$ on which f^k is topologically conjugate to the two-sided shift map on two symbols. In addition, in any neighborhood of a transversal homoclinic point, there are infinitely many periodic and non-periodic points.

In this case, the definition of homoclinic bifurcations is given by:

Definition 40 *A homoclinic bifurcation occurs when periodic orbits appear from homoclinic orbits to a saddle, saddle-focus, or focus-focus equilibrium.*

FIGURE 1.7: Homoclinic bifurcation.

A good reference for homoclinic bifurcations in dynamical system is [Kuznetsov (2004)].

Now, consider the third-order autonomous system

$$x' = f(x) \tag{1.45}$$

where the vector field $f : \mathbb{R}^3 \longrightarrow \mathbb{R}^3$ belongs to class $C^r(r \geq 1)$, $x \in \mathbb{R}^3$ is the state variable of the system and $t \in \mathbb{R}$ is the time. Suppose that f has at least an equilibrium point P.

Definition 41 *(a) The point P is called a hyperbolic saddle focus for system (1.45), if the eigenvalues of the Jacobian $A = Df(P)$ are γ and $\rho + i\omega$, where $\rho\gamma < 0$ and $\omega \neq 0$.*

(b) A homoclinic orbit $\gamma(t)$ refers to a bounded trajectory of system (1.45) that is doubly asymptotic to an equilibrium point P of the system, i.e., $\lim_{t\longrightarrow+\infty} \gamma(t) = \lim_{t\longrightarrow-\infty} \gamma(t) = P$.

(c) A heteroclinic orbit $\delta(t)$ is similarly defined, except that there are two distinct saddle foci, P_1 and P_2, being connected by the orbit, one corresponding to the forward asymptotic time, the other to the reverse asymptotic time limit, i.e., $\lim_{t\longrightarrow+\infty} \delta(t) = P_1$, and $\lim_{t\longrightarrow-\infty} \gamma(t) = P_2$.

The homoclinic and heteroclinic Shilnikov methods are summarized in the following theorem [Shilnikov (1965-1970)]:

Theorem 32 *Assume the following:*
(i) The equilibrium point P is a saddle focus and $|\gamma| > |\rho|$.
(ii) There exists a homoclinic orbit based at P.
Then,
(1) The Shilnikov map, defined in a neighborhood of the homoclinic orbit of the system, possesses a countable number of Smale horseshoes in its discrete dynamics.

(2) For any sufficiently small C^1-perturbation g of f, the perturbed system

$$x' = g(x) \tag{1.46}$$

has at least a finite number of Smale horseshoes in the discrete dynamics of the Shilnikov map defined near the homoclinic orbit.

(3) Both the original system (1.45) and the perturbed system (1.46) exhibit horseshoe chaos.

Similarly, there is also a heteroclinic Shilnikov theorem [Shilnikov (1965-1970)] given by:

Theorem 33 *Suppose that two distinct equilibrium points, denoted by P_1 and P_2, respectively, of system $\dot{x} = f(x)$ are saddle foci whose characteristic values γ_k and $\rho_k + i\omega_k$ $(k = 1, 2)$ satisfy the following Shilnikov inequalities: $\rho_1\rho_2 > 0$ or $\gamma_1\gamma_2 > 0$. Suppose also that there exists a heteroclinic orbit joining P_1 and P_2. Then the system $x' = f(x)$ has both Smale horseshoes and the horseshoe type of chaos.*

As an example of the application of Theorem 33. Let us consider the 3D polynomial differential system (called the T system) studied in [Tigan and Opriş (2008)] and given by:

$$\begin{cases} x' = a(y - x) \\ y' = (c - a)x - axz \\ z' = -bz + xy, \end{cases} \tag{1.47}$$

with a, b, c real parameters and $a \neq 0$. If $\frac{b}{a}(c-a) > 0$, the system T possesses three equilibrium isolated points:

$$
\begin{cases}
O = (0,0,0) \\
\\
E_1 = \left(\sqrt{\frac{b}{a}(c-a)}, \sqrt{\frac{b}{a}(c-a)}, \frac{c-a}{a} \right) \\
\\
E_2 = \left(-\sqrt{\frac{b}{a}(c-a)}, -\sqrt{\frac{b}{a}(c-a)}, \frac{c-a}{a} \right)
\end{cases}
\tag{1.48}
$$

and if $b \neq 0$, $\frac{b}{a}(c-a) \leq 0$, it has only one isolated equilibrium point, $O(0,0,0)$. If $b \neq 0$ and $a > 0, b > 0, c \leq a$, then $O(0,0,0)$ is asymptotically stable. If $b \neq 0$ and $b < 0$ or $a < 0$ or $a > 0, c > a$, then $O(0,0,0)$ is unstable. If $b \neq 0$ and $a + b > 0, ab(c - a) > 0, b(2a^2 + bc - ac) > 0$, then the equilibrium points $E_{1,2}$ are asymptotically stable.

The characteristic polynomial associated to the Jacobian matrix of the system T in E_1 and E_2 is:

$$
f(\lambda) = \lambda^3 + \lambda^2(a + b) + bc\lambda + 2ab(c - a) = 0
\tag{1.49}
$$

Let $\lambda = \mu - (a + b)/3$, then

$$
\mu^3 + p\mu + q = 0
\tag{1.50}
$$

where

$$
\begin{cases}
p = bc - \frac{1}{3}a^2 - \frac{2}{3}ab - \frac{1}{3}b^2 \\
\\
q = \frac{2}{27}a^3 - \frac{16}{9}a^2b + \frac{2}{9}ab^2 + \frac{2}{27}b^3 + \frac{5}{3}bca - \frac{1}{3}b^2c
\end{cases}
\tag{1.51}
$$

Let $\Delta = \left(\frac{q}{2}\right)^2 + \left(\frac{p}{3}\right)^3$. If $\Delta > 0$, then Eq. (1.50) has a negative solution, α_1, together with a pair of complex solutions, $\alpha_2 \pm i\alpha_3$, where

$$
\begin{cases}
\alpha_1 = \sqrt[3]{-\frac{q}{2} + \sqrt{\Delta}} + \sqrt[3]{-\frac{q}{2} - \sqrt{\Delta}} \\
\\
\alpha_2 = \frac{-1}{2}\left(\sqrt[3]{-\frac{q}{2} + \sqrt{\Delta}} + \sqrt[3]{-\frac{q}{2} - \sqrt{\Delta}} \right) \\
\\
\alpha_3 = \frac{\sqrt{3}}{2}\left(\sqrt[3]{-\frac{q}{2} + \sqrt{\Delta}} - \sqrt[3]{-\frac{q}{2} - \sqrt{\Delta}} \right)
\end{cases}
$$

and the three roots of (1.49) are given by:

$$
\begin{cases}
\lambda_1 = -\frac{a+b}{3} + \alpha_1 < 0 \\
\\
\lambda_2 = -\frac{a+b}{3} + \alpha_2 + i\alpha_3 \\
\\
\lambda_3 = -\frac{a+b}{3} + \alpha_2 - i\alpha_3,
\end{cases}
\tag{1.52}
$$

The Lyapunov exponents of system (1.47) for $(a, b, c) = (2.1, 0.6, 30)$ and the initial conditions $(0.1, -0.3, 0.2)$ are $\lambda_1 = 0.37 > 0$, $\lambda_2 = 0.00$ and $\lambda_3 = -3.07$. So the system T is chaotic. Indeed, this was proved by using the Shilnikov heteroclinic method [Silva (1993)]:

Theorem 34 *If a 3D given system $x' = F(x)$ has two equilibrium points E_1, E_2, of type saddle-focus, i.e., the eigenvalues of the Jacobian matrix associated to the system in these points are $\gamma_k \in \mathbb{R}$ and $\alpha_k \pm i\beta_k \in \mathbb{C}$, $k = 1, 2$, such that*

$$\alpha_1 \alpha_2 > 0 \quad or \quad \gamma_1 \gamma_2 > 0 \tag{1.53}$$

and (the Shilnikov inequality)

$$|\gamma_k| > |\alpha_k|, k = 1, 2, \tag{1.54}$$

and if the system has a heteroclinic orbit connecting the equilibrium points E_1 and E_2, then the Poincaré map defined on a transversal section of the flow in a neighborhood of the heteroclinic orbit presents chaos of horseshoe type.

The following result was proved in [Tigan and Opriş (2008)]:

Theorem 35 *If $\Delta > 0$ and $\alpha_1 + \alpha_2 < -\frac{2(a+b)}{3}$, then the system T has a heteroclinic orbit given by*

$$\phi(t) = \begin{cases} -x_0 + \sum\limits_{n=1}^{\infty} a_n e^{n\alpha t} & for \quad t > 0 \\ \\ 0 & for \quad t = 0 \\ \\ x_0 - \sum\limits_{n=1}^{\infty} a_n e^{-n\alpha t} & for \quad t < 0 \end{cases} \tag{1.55}$$

which connects the equilibrium points E_1, E_2, so the chaos is of horseshoe type.

Proof 8 *From the first two Eqs. of (1.47) one gets*

$$\begin{cases} y = x + \frac{\dot{x}}{a} \\ \\ z = -\frac{\dot{y} - (c-a)x}{ax} = -\frac{\ddot{x} + a\dot{x}}{a^2 x} + \frac{c-a}{a} \end{cases}$$

and using the last Eq. of (1.47) we get

$$\frac{d}{dt}\left(\frac{x'' + ax'}{x}\right) + b\frac{x'' + ax'}{x} + a^2 x^2 + axx' - ab(c-a) = 0$$

or, equivalently

$$xx''' + (a+b)xx'' - x'x'' - a(x')^2 + abxx' + a^2 x^4 + ax^3 x' - ab(c-a)x^2 = 0. \tag{1.56}$$

If $x(t)$ is determined, then we can find $y(t)$, $z(t)$. Thus, we have to find a function $\phi(t)$ such that $x(t) = \phi(t)$ to satisfy Eq. (1.56) and $\phi(t) \to -\sqrt{\frac{b}{a}(c-a)}$ for $t \to +\infty$, $\phi(t) \to \sqrt{\frac{b}{a}(c-a)}$ for $t \to -\infty$ or, inversely $\phi(t) \to \sqrt{\frac{b}{a}(c-a)}$ for $t \to +\infty$, $\phi(t) \to -\sqrt{\frac{b}{a}(c-a)}$ for $t \to -\infty$.

Without loosing the generality, we can assume that the direction from E_1 to E_2 corresponds to $t \to \infty$, and from E_2 to E_1 corresponds to $t \to -\infty$. Assuming that $\phi(t)$ has the following form $\phi(t) = -x_0 + \sum\limits_{n=1}^{\infty} a_n e^{n\alpha t}$, with α real number and $x_0 = \sqrt{\frac{b}{a}(c-a)}$. By identifying the coefficients of $e^{n\alpha t}$ in (1.56) we get:

$$x_0 \left[\alpha^3 + \alpha^2(a+b) + bc\alpha + 2ab(c-a)\right] a_1 = 0, \tag{1.57}$$

$$a_2 = \frac{\left(\alpha^2 b + ab\alpha - ab(c-a) + 6a^2 x_0^2 + 3\alpha a x_0^2\right) a_1^2}{f(2\alpha) x_0} \tag{1.58}$$

$$
\begin{aligned}
a_3 &= -\frac{\left(4a^2 x_0 a_1^3 + 3a\alpha x_0 a_1^3\right)}{f(3\alpha) x_0} \\
&\quad + \frac{\left[3\alpha^3 + \alpha^2 a + 5\alpha^2 b + 3ab\alpha - 2abc + 2a^2 b + 12a^2 x_0^2 + 9\alpha a x_0^2\right] a_1 a_2}{f(3\alpha) x_0} \\
&\quad + \frac{\left[3\alpha^3 + \alpha^2 a + 5\alpha^2 b + 3ab\alpha - 2abc + 2a^2 b + 12a^2 x_0^2 + 9\alpha a x_0^2\right] a_1 a_2}{f(3\alpha) x_0}
\end{aligned}
\tag{1.59}
$$

and for $n \geq 4$

$$
\begin{aligned}
a_n &= \frac{a^2 a_{ijpq} + ab_{ijpq}}{f(n\alpha) x_0} + \frac{1}{f(n\alpha) x_0} \sum_{i+j=n} \alpha^3 \left(j^3 - ij^2\right) a_i a_j \\
&\quad + \frac{\sum\limits_{i+j=n} \left[\alpha^2 \left((a+b) j^2 - ija\right) + ab\alpha j - ab(c-a) + 6a^2 x_0^2 + 3\alpha a j x_0^2\right] a_i a_j}{f(n\alpha) x_0} \\
&\quad + \frac{\sum\limits_{i+j=n} \left[\alpha^2 \left((a+b) j^2 - ija\right) + ab\alpha j - ab(c-a) + 6a^2 x_0^2 + 3\alpha a j x_0^2\right] a_i a_j}{f(n\alpha) x_0}
\end{aligned}
$$

where

$$f(n\alpha) = (n\alpha)^3 + (n\alpha)^2 (a+b) + bcn\alpha + 2ab(c-a) \tag{1.60}$$

$$a_{ijpq} = -4x_0 \sum_{i+j+p=n} a_i a_j a_p + \sum_{i+j+p+q=n} a_i a_j a_p a_q \tag{1.61}$$

$$b_{ijpq} = -3\alpha x_0 \sum_{i+j+p=n} p a_i a_j a_p + \alpha \sum_{i+j+p+q=n} p a_i a_j a_p a_q \tag{1.62}$$

with $i, j, p, q \geq 1$.

We put the sums (1.62) and (1.61) in the following forms:

$$a_{ijpq} = -4x_0 \sum_{i=1}^{n-2} a_i \left(\sum_{j=1}^{n-i-1} a_j a_{n-i-j} \right) + \sum_{i=1}^{n-3} a_i \left(\sum_{j=1}^{n-i-1} a_j \left(\sum_{p=1}^{n-i-j-1} a_p a_{n-i-j-p} \right) \right)$$

(1.63)

$$b_{ijpq} = -3\alpha x_0 \sum_{i=1}^{n-2} a_i \left(\sum_{j=1}^{n-i-1} (n-i-j) a_j a_{n-i-j} \right) + $$

(1.64)

$$\alpha \sum_{i=1}^{n-3} a_i \left(\sum_{j=1}^{n-i-1} a_j \left(\sum_{p=1}^{n-i-j-1} p a_p a_{n-i-j-p} \right) \right) + $$

$$\alpha \sum_{i=1}^{n-3} a_i \left(\sum_{j=1}^{n-i-1} a_j \left(\sum_{p=1}^{n-i-j-1} p a_p a_{n-i-j-p} \right) \right)$$

Assume $a_1 = 0$, then all coefficients $a_n, n \geq 2$, are zero. Thus, assuming that $a_1 \neq 0$ we get

$$\alpha^3 + \alpha^2(a+b) + bc\alpha + 2ab(c-a) = 0$$

(1.65)

i.e., α is the negative root of characteristic polynomial (1.49). Because Eq. (1.49) has a single negative solution for $\Delta > 0$, we get

$$f(n\alpha) = (n\alpha)^3 + (n\alpha)^2(a+b) + bcn\alpha + 2ab(c-a) \neq 0, n > 1.$$

(1.66)

Hence, the coefficients a_n are completely determined by a, b, c, α and a_1, and they are of the following form:

$$a_n = g(n)a_1^n, n > 1$$

(1.67)

where the terms $g(n)$ are known functions and the corresponding branch of the heteroclinic orbit for $t > 0$ is determined. For the case $t < 0$, define $\phi(t) = x_0 + \sum_{n=1}^{\infty} a_n e^{-n\beta t}$, with β real, and find $\beta = \alpha$ and $b_n = -a_n, n > 0$. Finally, the heteroclinic orbit $\phi(t)$ is given by (1.55) and it is possible to show that $\phi(t)$ is uniformly convergent.

Example 4 *As a numerical example, let us consider the values: $a = 2.1, b = 0.6, c = 30$. Then*

$$\begin{cases} \lambda_1 = \alpha = -3.429 \\ \\ \lambda_2 = 0.364 - 4.513i \\ \\ \lambda_3 = 0.364 + 4.513i \\ \\ \Delta = 911.69 > 0 \\ \\ x_0 = 2.8234. \end{cases}$$

and the equilibrium points E_1, E_2 are saddle-focus. Assuming that $\phi(t)$ is at least continuous, then from $\phi(0_-) = \phi(0_+) = 0$, we find the equation

$$-x_0 + a_1 + a_2 + ... + a_n = 0, \ \ n > 1. \tag{1.68}$$

Thus, the first coefficient is $a_1 = 3.051$ for any $n > 10$. Also, numerical calculations demonstrate that the series (1.55) which describes the heteroclinic orbit between E_1 and E_2, is rapidly convergent. For example, considering the first ten terms of the series, for $t = 10$ we get $\phi(t) = -x_0 + \sum\limits_{n=1}^{10} a_n e^{nat} =$

-2.8234, for $t = -10$, $\phi(t) = x_0 - \sum\limits_{n=1}^{10} a_n e^{-nat} = 2.8234$ and for $t = 0$,

$\phi(t) = -x_0 + \sum\limits_{n=1}^{10} a_n e^{nat} = 0.0000$. So the equilibrium points E_1, E_2 and O are

found with an approximation of four exact decimals because $x_0 = \sqrt{\frac{b}{a}(c-a)} =$

2.8234.

Chapter 2

Human Immunodeficiency Virus and Urbanization Dynamics

2.1 Introduction

This chapter is devoted to presenting some results regarding the dynamics of the Human Immunodeficiency Virus (HIV). In Sec. 2.2, we give an overview of the definition of Human Immunodeficiency Virus (HIV) as a retrovirus. In particular, some statistics are also given and discussed. In Sec. 2.3, we present the standard systems modelling HIV. The dynamics of sexual transmission is presented in Sec. 2.4 via a mathematical distributed delay equation. The effects of variable infectivity on the HIV dynamics are presented in Sec. 2.5. The CD4$^+$ Lymphocyte dynamics in HIV infection are discussed in Sec. 2.6 where a model of CD4$^+$ lymphocyte dynamics in HIV-infected persons is presented. A mathematical model that tracks both infectious and total viral load was presented in Sec. 2.7 in order to describe the *in vitro* kinetics of the SHIV-KS661 viral infection. In Sec. 2.8, the effects of morphine on Simian Immunodeficiency Virus Dynamics is presented via a mathematical model that incorporates experimentally observed effects of morphine on inducing HIV-1 co-receptor expression. Finally, the dynamics of an HIV therapy system are presented in Sec. 2.9. We show that the concentrations of the CD4 lymphocyte population converge to a fixed quantity and the CD8 lymphocyte population converges to another fixed quantity and the HIV-1 viral load converges also to another fixed quantity. Thus, the system never returns to the normal unperturbed concentrations of the CD4 and CD8 lymphocyte population.

In Sec. 2.10, we present some properties of dynamics of urbanization. In particular, the rural-urban interaction attractor is presented and discussed.

2.2 Definition of Human Immunodeficiency Virus (HIV)

The Human Immunodeficiency Virus (HIV) is defined as a virus spread through certain body fluids that attacks the body's immune system.

The image of the virus is shown in Fig. 2.1.

The virus HIV is a retrovirus defined by:

Definition 42 *(Retrovirus): A type of virus that uses RNA as its genetic material. After infecting a cell, a retrovirus uses an enzyme called reverse transcriptase in order to convert its RNA into DNA. The retrovirus then integrates its viral DNA into the DNA of the host cell, which allows the retrovirus to replicate.*

The procedure of infection is shown in Fig. 2.2. In particular, the CD4 T cells (these cells help the immune system fight off infections) where it has the possibility to destroy them over time. The body is unable to fight off infections and disease since HIV reduces the number of T cells in the body. The most dangerous stages are the opportunistic infections or cancers that take advantage of a very weak immune system and signal that the person has AIDS (the damage of the immune system characterized by an increasing number of severe illnesses, known as *opportunistic infections*). Unfortunately, no effective cure currently exists and the human body can't get rid of HIV completely, even with treatment. The good news is that HIV can be controlled by the so called *antiretroviral therapy* or ART (introduced in the mid-1990s). In the

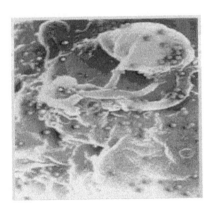

FIGURE 2.1: Images of the Human Immunodeficiency Virus.

FIGURE 2.2: Viral DNA integrates into the host cell genome.

optimal conditions, this medicine can prolong the lives of many people infected with HIV. The HIV has the following three stages of development: (1) acute HIV infection, (2) clinical latency, and (3) AIDS (Acquired Immunodeficiency Syndrome). Depending on many factors, such as genetic makeup, people living with HIV may progress through these stages at different rates. Unfortunately, no effective cure currently exists for HIV and the only solution is the different types of preventions.

The number of people infected with HIV has increased from a few persons in 1984 to 3.5 million per year in 1997. After this year, the number of new diagnoses began to decrease and in 2015 it was reduced to 2.1 million per year. The number of AIDS-related deaths increased throughout the 1990s and about 2 million people died in 2004 and 2005. Since then, the number of deaths has fallen to 1.1 million people in 2015. However, in the same year, about 36.7 million people were living with HIV globally.

2.3   Modelling the Human Immunodeficiency Virus (HIV)

Modeling of HIV infection dynamics with adaptive immune responses contains many factors, such as: uninfected cells, infected cells in the intracellular eclipse phase before the production of virus begins, productively infected cells, free virus, $CD8^+$ effector cells, and antibodies.

Generally, standard models [Perelson and Ribeiro (2013), Wodarz (2014)] of HIV dynamics consider three populations: Virus-free, uninfected (or target) cells, and infected cells. The resulting systems are simple and explain early viral load data in individuals infected with HIV. However, these systems do not explicitly take into consideration the immune response which has an effect on the overall viral dynamics. This is a result of the experimental studies showing that $CD8^+$ T cell depletion in Simian Immunodeficiency Virus (SIV)-infected macaques results in an increase in viral load [Jin, et al., (1999), Schmitz, et al., (1999)]. Also, these systems indicate a role for immune control [Regoes, et al., (2004), Kouyos, et al., (2010)]. However, the $CD8^+$ T cells play a limited role in HIV control. In particular in elite controllers.

Several modifications to the standard models of virus dynamics are reported in this chapter. In particular, the inclusion of the immune responses to keep the modified model tractable. Several models contain multiple subsets of infected cells: (a) infected non-virus producing cells and (b) productively infected cells. Other models include multiple populations of viral strains, with their corresponding infected cells and the effector responses levied against them. Other models include immune escape, or the process of viral evolution to evade the established immune responses.

In fact, the HIV models can be divided into the following classes depending on the problem or the question related to experimental data:

1. Models of intracellular eclipse phase for a better fit to HIV data, improve escape rate estimates, estimate the timing of MHC downregulation, and resolve the mechanism of CD8$^+$ activity.

2. Models that include CD8$^+$ T cell responses delineate controllers from viral rebounders or establish thresholds between control and AIDS.

3. Models estimate the timing and the rate of escape, for single and multiple escape.

4. Models showing the triggers, characteristics and mechanisms of escape.

5. Models explore the conditions and consequences of CTL escape.

6. Models predict how neutralizing antibodies may determine virus control or persistence.

7. Models estimate antibody escape rates.

8. Models predict the roles for antibodies in vaccine strategies.

Detailed explanations on these types of modelling can be found in [Schwartz, *et al.*, (2016)].

2.4 Dynamics of sexual transmission of the Human Immunodeficiency Virus

In [Castillo-Chavez, *et al.*, (1989)] a mathematical distributed delay model was utilized in order to present an analysis of single and multiple group models for the spread of the Human Immunodeficiency Virus (HIV), which is the etiological agent for the Acquired Immunodeficiency Syndrome (AIDS):

$$\frac{dS(t)}{dt} = \Lambda - \lambda C(T(t)) S(t) \frac{I(t)}{T(t)} - \mu S(t) \tag{2.1}$$

$$I(t) = I_0(t) + \int_0^t \lambda C(T(x)) S(x) \frac{I(x)}{T(x)} \exp(-\mu(t-x)) P(t-x) dx, \tag{2.2}$$

$$
\begin{aligned}
A\left(t\right) \;=\; & A_0\left(t\right)+A_1 \exp\left(-\left(\mu+d\right)t\right) && \text{(2.3)} \\
& +\int\limits_0^t \left\{
\begin{array}{l}
\int\limits_0^\tau \lambda C\left(T\left(x\right)\right) S\left(x\right) \frac{I(x)}{T(x)} \exp\left(-\mu\left(\tau-x\right)\right) \\[4pt]
\times\left[-P'\left(\tau-x\right)\exp\left(-\left(\mu+d\right)\left(t-\tau\right)\right)\right] dx
\end{array}
\right\} d\tau
\end{aligned}
$$

The system (2.1)-(2.3) describes a single sexually active homosexual population. This population is subdivided into three groups of active individuals: S (uninfected), I (HIV-infectious), and A (AIDS-infectious) with the assumption that the A-individuals (individuals with *full-blown* AIDS) are sexually inactive, i.e., they do not contribute to the dynamics of AIDS.

The parameters in system (2.1)-(2.3) have the following meaning:

- The functions (with compact support) $I_0\left(t\right)$, $A_0\left(t\right)$, and the constant A_1, are introduced in order to take care of the initial conditions.

- A is the *recruitment* rate into S

- μ is the natural mortality rate.

- d is the AIDS induced mortality rate.

- $\lambda = i\phi$ is the transmission rate per infectious partner.

- i is the probability of transmission per contact with an infectious individual.

- ϕ is the average number of contacts per sexual partner.

- $C(T)$ is the mean number of sexual partners an average individual has per unit time. Generally, $C(T)$ increases linearly for small T and saturate for large T. Thus, we assume that $C(T)$ is a non-decreasing function of T.

- $T = S + I$ is the sexually active population.

- The factor $\frac{I}{T}$ is the probability that a randomly selected individual will be infectious.

- $P(s)$ is the proportion of the individuals that become infectious at time t and that, if alive, are still infectious at time $t + s$ (they survive as infectious).

- $-P'\left(x\right)$ is the rate of removal of individuals from group I into group A, x time units after infection.

These definitions imply that for a homogeneously mixed sexually active population, the incidence rate (the number of new cases per unit time) is given by $\lambda C(T)\frac{SI}{T}$. The proportion $P(s)$ is non-negative and non-decreasing and $P(0) = 1$ and $\int_0^\infty P(s)ds < +\infty$.

Let $P(s) = \exp(\alpha s)$, thus we have the following result proved in ([Castillo-Chavez, *et al.*, (1989(a))] about the dynamics of a single group model:

Theorem 36 *(A) The disease-free state $\left(\frac{\Lambda}{\mu},0\right)$ is a globally asymptotically stable equilibrium, if and only if, the reproductive number*

$$R = \lambda C\left(\frac{\Lambda}{\mu}\right)\frac{1}{\mu+\alpha} \leq 1 \tag{2.4}$$

(B) If $R > 1$, then there is a unique endemic state, which is a global attractor for all positive solutions.
(C) The infection-free state is a global attractor whenever the reproductive number

$$R = \lambda C\left(\frac{\Lambda}{\mu}\right)\int_0^\infty P(s)\exp(-\mu s)\,ds \leq 1$$

(D) If $R > 1$, then the limiting system

$$\frac{dS(t)}{dt} = \Lambda - \lambda C(T(t))S(t)\frac{I(t)}{T(t)} - \mu S(t) \tag{2.5}$$

$$I(t) = \int_{-\infty}^t \lambda C(T(t))S(x)\frac{I(x)}{T(x)}\exp(-\mu(t-x))P(t-x)\,dx, \tag{2.6}$$

has a unique endemic state. It is locally asymptotically stable, provided

$$0 \leq C'(t) \leq \frac{C(T)}{T}\ \left(or\ \frac{d}{dT}\left(\frac{C(T)}{T}\right) \leq 0\right)$$

Notice that R in (2.4) controls whether or not the disease can be maintained.

At this end, for the single group models (2.1)-(2.3) are robust, in the sense that only *simple* dynamics are possible. Also, the reproductive number is not significantly affected by the shape of the survivorship function if one assumes that the function is biologically reasonable.

Now, we consider n sexually active subpopulation, each divided into three classes S_i, I_i, and A_i. Assuming that the mixing between groups is proportionate to their sexual activity (proportionate mixing). Hence, the ith-incidence rate is given by the following expression:

$$B_i(t) = S_i(t)C(T(t))\sum_{j=1}^n \lambda_{ij}p_{ij}(t)\frac{I_j(t)}{T_j(t)}\ \text{where}\ p_{ij}(t) = \frac{C_j(T(t))T_j(t)}{\sum_{k=1}^n C_k(T(t))T_k(t)}$$

where $T_k(t) = S_k(t) + l_k(t)$, and where the functions C_i are non-decreasing. Let $\sigma_i = \frac{1}{\alpha_i}$ and the exponential removal $P_i(s) = \exp(-\alpha_i s)$, then the dynamics of transmission can be rescaled as follow:

$$\frac{dS_i(t)}{dt} = \Lambda_i - B_i(t) - \mu S_i(t) \tag{2.7}$$

$$\frac{dI_i(t)}{dt} = B_i(t) - \mu(\sigma_i + 1) I_i(t) \tag{2.8}$$

$$\frac{dA_i(t)}{dt} = \alpha_i I_i(t) - (d + \mu) A_i(t), i = 1, 2, ..., n \tag{2.9}$$

Define the probability $\kappa(i, T^*)$ by:

$$\kappa(i, T^*) = \frac{C_i(T^*) \frac{\Lambda_i}{\mu}}{\sum\limits_{k=1}^{n} C_k(T^*) \frac{\Lambda_k}{\mu}}, \quad \text{where } T^* = \sum_{k=1}^{n} \frac{\Lambda_k}{\mu}, i = 1, 2, ..., n$$

and define the matrix L given by $L = \left[\frac{C_j(T^*)\lambda_{ij}\kappa(i)}{\sigma_i + 1}\right]_{n \times n}$ and the function $H(\mu) = L - \mu E$, where E is the $n \times n$ identity matrix. Let

$$M(H(\mu)) = \sup\{Re\rho : \det(\rho E - H(\mu)) = 0\}$$

Then there is a unique μ_0 such that

$$M(H(\mu)) = \begin{cases} < 0 \text{ if } \mu > \mu_0 \\ \\ 0 \text{ if } \mu = \mu_0 \\ \\ > 0 \text{ if } \mu < \mu_0 \end{cases}$$

and the infection-free state $\bar{S} = \left(\frac{\Lambda_1}{\mu}, ..., \frac{\Lambda_n}{\mu}, 0, ..., 0\right)$ is locally asymptotically stable if $M(H(\mu)) < 0$.

Assuming that $C_i(T) = c_i$ (a constant), for $i = 1, 2, ..., n$; L is irreducible; and ,μ_0 is such that $M(H(\mu_0)) = 0$, define $h(\mu_0)$ by:

$$h(\mu_0) = \sum_{i,j=1}^{n} \bar{I}_i I_i (c_i \lambda_{ij} - \mu_0 \sigma_j) c_j I_j$$

where $I = (I_1, ..., I_n)$ and $\bar{I} = (\bar{I}_1, ..., \bar{I}_n)$ are positive eigenvectors of $H(\mu_0)$ and $H^T(\mu_0)$ corresponding to the zero eigenvalue, respectively. The existence of these positive eigenvectors (all entries are positive) is guaranteed by M-matrix theory.

Thus, we have the following result proved in [Castillo-Chavez, *et al.*, (1989(a))] about the bifurcations of the n-group model (2.1)-(2.3):

Theorem 37 *(A) If $h(\mu_0) \neq 0$, then μ_0 is a bifurcation point. More specifically, if $h(\mu_0) > 0$ ($h(\mu_0) < 0$) then there is an $\epsilon > 0$ and unique continuously differentiable functions S and I mapping $(\mu_0 - \epsilon, \mu_0] \rightarrow \mathbb{R}_+^n$ ($(\mu_0, \mu_0 + \epsilon] \rightarrow \mathbb{R}_+^n$) such that $(S(\mu_0), I(\mu_0)) = \left(\frac{\Lambda_1}{\mu}, ..., \frac{\Lambda_n}{\mu}, 0, ..., 0\right)$ and $(S(\mu), I(\mu))$ is a positive endemic equilibrium of (2.7)-(2.9) and this endemic equilibrium is locally asymptotically stable for each μ in $(\mu_0 - \epsilon, \mu_0)$ (unstable for each μ in $(\mu_0, \mu_0 + \epsilon)$).*

(B) For each μ in $(0, \mu_0)$, the system (2.7)-(2.9) has a positive endemic equilibrium.

(C) If $h(\mu_0) < 0$, then there is an $\epsilon > 0$ such that the system (2.7)-(2.9) has at least two positive equilibria for each μ in $(\mu_0, \mu_0 + \epsilon)$.

2.5   The effects of variable infectivity on the HIV dynamics

In [Thieme and Castillo-Chavez (1989)] a dynamical model was formulated as follows:

$$\frac{dS(t)}{dt} = \Lambda - B(t) - S(t) \tag{2.10}$$

$$\left(\frac{\partial}{\partial t} + \frac{\partial}{\partial \tau}\right) i(t, \tau) = -(1 + \alpha(\tau)) i(t, \tau) \tag{2.11}$$

$$i(t, 0) = B(t) = C(T(t)) S(t) \frac{W(t)}{T(t)} \tag{2.12}$$

$$T = I + S \tag{2.13}$$

$$I(t) = \int_0^\infty i(t, \tau) \, d\tau \tag{2.14}$$

$$W(t) = \int_0^\infty \lambda(\tau) i(t, \tau) \, d\tau \tag{2.15}$$

$$\frac{dA(t)}{dt} = \int_0^\infty \alpha(\tau) i(t, \tau) \, d\tau - (1 + v) A(t)$$

The population in system (2.10)-(2.15) is divided into three groups: S (uninfected, but susceptible), I (HIV infected), and A (fully developed AIDS symptoms). Here the A-individuals are assumed to be sexually inactive and sexually active individuals (S and I) are supposed to choose their partners at random.

The parameters in system (2.10)-(2.15) have the following meaning:

- A is the number of individuals with fully developed AIDS symptoms. This parameter does not play any further role in the dynamics of the epidemic.

- v is the rate at which an individual with fully developed AIDS symptoms dies from the disease.

- t is the time

- τ is the time since the moment of being infected, i.e., infection-age.

- Λ is the constant rate for which the individuals are recruited into the sexually active population.

- μ is the constant rate at which the length of the sexually active period is exponentially distributed such that healthy individuals become sexually inactive. It was assumed that the average length $\frac{1}{\mu}$ of the activity period is 1, $\mu = 1$.

- $\alpha(\tau)$ is the rate at which the infected individuals with infection-age τ stop being sexually active by force of the disease. So the chance of an individual still being sexually active if he has been infected τ time units ago is given by $\exp\left(-\tau - \int_0^\tau \alpha(\rho)\, d\rho\right)$. Here model (2.10)-(2.15) does not assume that at the moment of infection, an individual follows a severe or a mild course of the disease, i.e., by assuming that

$$\int_0^\infty \alpha(\tau)\, d\tau < +\infty$$

- $I(t) = \int_0^\infty i(t, \tau)\, d\tau$ is the infected part of the population according to age of infection.

- $i(t, \tau)$ is the infection-age density. Here, the chance that a randomly chosen partner is infected and has infection-age τ is $\frac{i(t,\tau)}{T(t)}$.

- $T + S + I$ is the size of the sexually active population.

- $\lambda(\tau)$ is the mean risk for which an average susceptible contracts the disease from an infected partner with age of infection τ. Thus, the chance of an average susceptible individual being infected at time t (under the

condition that he has had a sexual contact at that time) is given by $\frac{W(t)}{T(t)}$
where $T = S + I$ and $W(t) = \int_0^\infty \lambda(\tau) i(t, \tau) d\tau$.

- $C(T)$ is the number of sexual contacts $C(T)$ that an average individual has per unit of time. Assume that this number is a function of the size of the sexually active population $T = S + I$.

- $B(t) = C(T(t)) S(t) \frac{W(t)}{T(t)}$ is the incidence rate (number of new cases of infection per unit time).

For model (2.10)-(2.15) the following ingredients are assumed:

1. A non-linear functional relationship between mean sexual per capita activity and the size of the sexually active population.

2. A stratification of the infected part of the sexually active population according to infection age, i.e., time since the moment of infection.

3. An infection-age-dependent rate of leaving the sexually active population due to disease progression.

4. An infection-age-dependent infectivity.

The above four facts mimic the HIV dynamics in a homogeneously mixing male homosexual population.

For the functions defining model (2.10)-(2.15) we assume the following:

1. $\alpha(\tau)$ is a non-negative measurable function.

2. $\lambda(\tau)$ is a non-negative integrable function of infection age.

3. $C(T)$ is a non-decreasing function of T, where $C(T) > 0$ whenever $T > 0$.

4. Also, $M(t) = \frac{C(T)}{T}$ is a non-increasing function of T, i.e., C increases in a sublinear way, reflecting some kind of saturation effect.

A basic reproductive number R_0 was used in order to study the effects of variable infectivity in combination with a variable incubation period on the dynamics of HIV in the homogeneously mixing population (2.10)-(2.15). In fact, R_0 gives the average number of secondary infections that a typical infectious individual can produce if it is introduced into the disease-free population.

The following result was proved in [Thieme and Castillo-Chavez (1989)]:

Theorem 38 *If $R_0 \leq 1$, there exists only the disease-free equilibrium. If $R_0 > 1$, there is a unique endemic equilibrium.*

Theorem 38 provides information regarding the existence of a state in which the disease persists. The next result connects partially the basic reproductive number and the actual disease dynamics:

Theorem 39 *Let $R_0 > 1$. Then the disease-free equilibrium is globally attractive. In particular we have*

$$B(t), I(t), W(t) \to 0, S(t) - \Lambda, \; for \; t \to \infty$$

The global convergence result if $R_0 > 1$ is not possible in general. However, if a trajectory is not attracted to the endemic equilibrium, it has to oscillate around it. The following result was proved in [Thieme and Castillo-Chavez (1989)] (Analogous statements can be derived for $S, I,$ and W):

Theorem 40 *Let $R_0 > 1$. The following holds:*
(a)

$$\limsup_{t \to \infty} B(t) \le B^*$$

Let $\lambda(\tau) \ne 0$ and τ_0 be the smallest $\bar{\tau}$ such that $\lambda(\tau) \ne 0$ for almost all $\bar{\tau} \ge \tau$. Let

$$\int_0^{\tau_0} i(0, \tau)\, d\tau > 0$$

Then

$$\limsup_{t \to \infty} B(t) \le B^*$$

The boundary flow, i.e., $i = 0$, is attracted to the infection-free state as shown in [Hale & Waltman (1989), Theorem 4.2]:

Theorem 41 *Let $R_0 > 1$ and $\lambda(\tau) \ne 0$ and τ_0 be the smallest $\bar{\tau}$ such that $\lambda(\tau) \ne 0$ for almost all $\bar{\tau} \ge \tau$. Let*

$$\int_0^{\tau_0} i(0, \tau)\, d\tau > 0$$

Then

$$\liminf_{t \to \infty} I(t) > \epsilon > 0$$

with ϵ not depending on the initial conditions.

However, the dynamical systems persistence theory does not give an answer whether or not B and W are bounded away from zero.

2.6   The CD4$^+$ Lymphocyte dynamics in HIV infection

In [Hraba and Dolezal (1996)] a model of CD4$^+$ lymphocyte dynamics in HIV-infected persons was presented as follows:

$$\begin{cases} \frac{d\bar{P}(t)}{dt} = \frac{I_p + f[(P_0 - P(t)) + (R_0 - R(t))]}{d(t)} - \bar{\tau}_p \bar{P}(t) - \bar{c}_p a(t) C(t) \bar{P}(t) \\ \bar{P}(0) = \bar{P}_0 \end{cases} \tag{2.16}$$

$$\begin{cases} \frac{dP(t)}{dt} = \bar{\tau}_p \bar{P}(t) - \tau_p P(t) - c_p a(t) C(t) P(t) \\ P(0) = P_0 \end{cases} \tag{2.17}$$

$$\begin{cases} \frac{d\bar{R}(t)}{dt} = \frac{2}{3} \frac{I_p + f[(P_0 - P(t)) + (R_0 - R(t))]}{d(t)} - \bar{\tau}_p \bar{R}(t) \\ \bar{R}(0) = \frac{2}{3} \bar{P}_0 \end{cases} \tag{2.18}$$

$$\begin{cases} \frac{dR(t)}{dt} = \bar{\tau}_p \bar{R}(t) - (\tau_R - \rho_R) R(t) \\ R(0) = \frac{2}{3} P_0 \end{cases} \tag{2.19}$$

$$\begin{cases} \frac{da(t)}{dt} = a(t) [\theta - \zeta - \gamma C(t)] \\ a(0) = a_0 \end{cases} \tag{2.20}$$

$$\begin{cases} \frac{dC(t)}{dt} = \lambda a(t) [\varepsilon I_C + \alpha C(t)] \left(\frac{P(t)}{P_0}\right)^\upsilon - (\tau_C - \rho_C) C(t) \\ C(0) = C_0 \end{cases} \tag{2.21}$$

where the influx-constraining function was

$$d(t) = \begin{cases} 1, & \text{if } \ln \frac{a(t)}{a_0} < L \\ h \ln \frac{a(t)}{a_0}, & \text{if } \ln \frac{a(t)}{a_0} \geq L \end{cases} \tag{2.22}$$

The parameters in system (2.17)-(2.22) have the following meaning:

- I_P is the influx of P cells, i.e., the rate (all rates are in $days^{-1}$) of differentiation of P cells from stem cells.

- $\bar{\tau}_p$ is the rate of maturation of P cells into P cells.

- τ_P is the rate of natural death of P cells.

- The quantities $\bar{\tau}_R$ and τ_R are defined in a fully analogical way.

- f is the amplifying coefficient of the linear feedback effect of P and/or R cell decrease on the influx of $\bar{P}$ and $\bar{R}$ cells at time t.

- The quantity $\bar{c}_p a\left(t\right) C\left(t\right)$ is the rate of elimination of P cells due to the amount of HIV products $a(t)$ and the number of cytotoxic T cells $C(t)$ at time t.

- The quantity $c_p a\left(t\right) C\left(t\right)$ is the rate of elimination of P cells.

- The value a_0 is the function of the infectious dose of HIV.

- θ characterizes the growth rate of HIV.

- γ is the rate of inactivation of HIV products mediated by cytotoxic C cells. The maturation of the cells C from their precursors is assumed to be dependent on the encounter with HIV products and the effect of HIV specific helper T cells.

- I_C is the influx of C cell precursors.

- ε is their maturation rate.

- α is the proliferation rate of C cells under the antigenic stimulation by HIV products and helper T cell influence.

- τ_C is their natural death rate.

- The helper T cell effect on maturation and proliferation of C cells is expressed by the ratio $\frac{P(t)}{P_0}$.

- The coefficient v is introduced in order to characterize the intensity of this helper effect.

- The value h characterizes the HIV-constraining intensity on the $\bar{P}$ and $\bar{R}$ cell influx.

- The value L given in (2.22) defines the level, where such constraining (limiting) effect of $d(t)$ starts.

- ζ is the HIV elimination rate by AZT or passive immunization.

- λ is the immune response-enhancing factor.

- ρ_R and ρ_C are the elimination rates of CD8$^+$ and C cells, respectively, by anti-CD8 antibodies.

The model (2.17)-(2.22) considers immature and mature CD4$^+$ ($\bar{P}$ and P cells) and CD8$^+$ lymphocytes ($\bar{R}$ and R cells). Here, $R = \frac{2}{3}P$ and $\bar{R} = \frac{2}{3}\bar{P}$. Eqs. (2.17)-(2.20) describe the sizes of these cell compartments at time t. Eq. (2.21) describe the amount of HIV products at time t and Eq. (2.22) gives the number of cytotoxic T cells specific for HIV (C cells) at time t. Also, these cells both limit proliferation of HIV, as described by Eq. (2.21), and effect destruction of CD4$^+$ cells presenting HIV products according to Eqs. (2.17)-(2.18).

The model parameters in simulation are given by:

$$\begin{cases} \bar{\tau}_P = 0.2, \tau_P = 0.01, \bar{\tau}_R = 0.2, \tau_R = 0.01, \tau_C = 0.01 \\\\ I_P = 1.0, I_C = 0.2, \bar{P}_0 = 5.0, P_0 = 100.0, \bar{R}_0 = .33, R_0 = 66.7 \\\\ C_0 = 0.0, a_0 = 0.0005, f = 0.01, \alpha = 0.7, \varepsilon = 0.512, \gamma = 0.3 \\\\ \theta = 0.02, v = 1.6, h = 3.5, L = 3.0 \end{cases}$$

The main assumption here is that only mature CD4$^+$ lymphocytes are susceptible to HIV products, i.e., $\bar{c}_P = 0.0$, $c_P = 20.0$. The parameter ε was used for final adjustment of the simulation. If no therapeutic interventions are assumed, i.e., $\lambda = 1.0, \zeta = 0.0, \rho_R = 0.0, \rho_C = 0.0$, the resulting CD4$^+$ standard curve characterizes best fit of the observed clinical data.

The model (2.17)-(2.22) incorporates a feedback mechanism regulating the production of T lymphocytes and simulates the dynamics of CD8$^+$ lymphocytes where its production is closely linked to that of CD4$^+$ cells. Since CD4$^+$ lymphocyte counts are considered as indicator of HIV infection, then the model (2.17)-(2.22) can be used to simulate the therapeutic interventions as chemotherapy and active and passive immunization and the therapeutic administration of anti-CD8 antibodies.

Notice that the CD4$^+$ cell observed values are depicted as circles and the CD8$^+$ lymphocytes as squares. These values are depicted in Fig. 2.3 to Fig. 2.8 as a percentage of normal CD4$^+$ lymphocyte numbers (the normal value of CD8$^+$ lymphocytes is thus 66.7%).

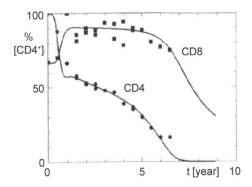

FIGURE 2.3: Simulated CD4$^+$ and CD8$^+$ lymphocyte dynamics of system (2.17)-(2.22) in HIV infection compared with observed mean T-cell values for CD4$^+$ lymphocytes (circles) and CD8+ lymphocytes (squares). Reused with permission from [Hraba and Dolezal (1996)].

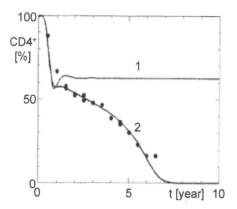

FIGURE 2.4: The linear curve 1 (v = 1.0) decreases with the non-linear curve 2 (v = 1.6) of system (2.17)–(2.22). Reused with permission from [Hraba and Dolezal (1996)].

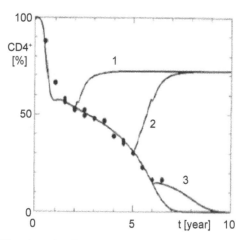

FIGURE 2.5: Simulated effect of permanent AZT treatment of system (2.17)-(2.22) ($\zeta = 0.005$) started $2m$ $5m$ or 6 years after the acquisition of HIV infection-curves 1, 2, and 3, respectively. Reused with permission from [Hraba and Dolezal (1996)].

2.7 The viral dynamics of a highly pathogenic Simian/ Human Immunodeficiency Virus

In [Iwami, *et al.*, (2012)] a mathematical model that tracks both infectious and total viral load was utilized in order to describe the *in vitro* kinetics of the SHIV-KS661 viral infection:

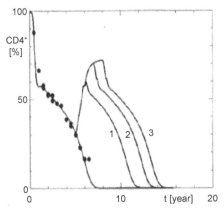

FIGURE 2.6: Simulated effect of temporary AZT treatment of system (2.17)-(2.22) ($\zeta = 0.005$) started 5 years after the acquisition of HIV infection and lasted 1, 2, or 3 years-curves 1, 2, and 3, respectively. Reused with permission from [Hraba and Dolezal (1996)].

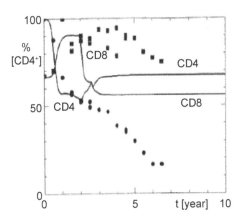

FIGURE 2.7: Simulated CD4$^+$ and CD8$^+$ lymphocyte dynamics of system (2.17)-(2.22) after permanent treatment with anti-CD8 antibodies started 2 years after the acquisition of the HIV infection. Cells mediating the protective anti-HIV immune reaction are not affected by this treatment ($\rho_R = 0.007, \rho_C = 0.0$). Reused with permission from [Hraba and Dolezal (1996)].

$$\frac{dx}{dt} = -\beta x v_I - dx \tag{2.23}$$

$$\frac{dy}{dt} = \beta x v_I - ay \tag{2.24}$$

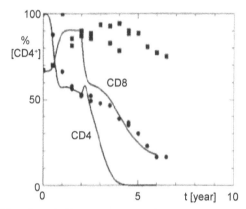

FIGURE 2.8: Simulated CD4+ and CD8$^+$ lymphocyte dynamics of system (2.17)-(2.22) after permanent treatment with anti-CD8 antibodies started 2 years after the acquisition of the HIV infection. Cells mediating the protective anti-HIV immune reaction are also affected by this treatment ($\rho_R = 0.007, \rho_C = 0.007$). Reused with permission from [Hraba and Dolezal (1996)].

$$\frac{dv_I}{dt} = pky - r_I v_I - r_{RNA} v_I \tag{2.25}$$

$$\frac{dv_{NI}}{dt} = (1 - p)\, ky + r_I v_I - r_{RNA} v_{NI} \tag{2.26}$$

where the parameters in the equations (2.23)-(2.26) have the following meanings:

- x and y are the number of target (susceptible) and infected (virus-producing) cells per ml of medium.

- v_I and v_{NI} are the number of RNA copies of infectious and non-infectious virus per ml of medium, respectively.

- d, a, r_{RNA}, and β represent the death rate of target cells, the death rate of infected cells, the degradation rate of viral RNA, and the rate constant for infection of target cells by virus, respectively.

- k is the viral production rate of an infected cell.

The model (2.23)-(2.26) was obtained by assuming the following facts:

- Each infected cell releases k virus particles per day, i.e., a fraction p are infectious and $1 - p$ are non-infectious.

- Infectious virions lose infectivity at rate r_I, becoming non-infectious.

- Once a cell is infected by an infectious virus it immediately begins producing progeny virus.

A schematic of model (2.23)-(2.26) is shown in Fig. 2.9. Due to some causes, such as fitting the observed viral load data and to account the partial removal of cells and virus due to sampling, the system (2.23)-(2.26) was transformed into the following scaled model:

$$\frac{dx}{dt} = -\beta_{50}xv_{50} - dx - \delta x \tag{2.27}$$

$$\frac{dy}{dt} = \beta_{50}xv_{50} - ay - \delta y \tag{2.28}$$

$$\frac{dv_{RNA}}{dt} = ky - r_{RNA}v_{RNA} - r_c v_{RNA} \tag{2.29}$$

$$\frac{dv_{50}}{dt} = k_{50}y + r_I v_{50} - r_{RNA}v_{50} - r_C v_{50} \tag{2.30}$$

where the parameters in the equations (2.27)-(2.30) have the following meanings:

- $v_{RNA} = v_I + v_{NI}$ is the total concentration of viral RNA copies.

- $v_{50} = \alpha v_I$ is the infectious viral load expressed in $TCID_{50}/ml$.

- $0 < \alpha \le 1.47TCID_{50}$ is the conversion factor from infectious viral RNA copies to $TCID_{50}$. The bounds in the interval are due to the fact that the measure of $1TCID_{50}$ corresponds to an average of 0.68 infection events by Poisson statistics, we have $0 < \alpha \le 1.47TCID50$ per RNA copies of infectious virus.

- $\beta_{50} = \frac{\beta}{\alpha}$ and $k_{50} = \alpha pk$ are the converted infection rate constant and production rate of infectious virus, respectively.

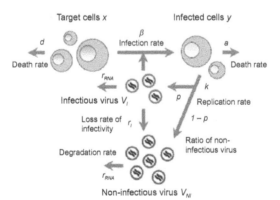

FIGURE 2.9: A schematic representation of model (2.23)-(2.26). Reused with permission from [Iwami, *et al.* (2012)].

Since at each sampling time the concentration of Nef-negative and Nef-positive HSC-F cells must be reduced in model (2.27)-(2.30) by 5.5% and the viral loads (RNA copies and TCID50) by 99.93% to account for the experimental harvesting of cells and virus, these losses are modeled in system (2.27)-(2.30) by approximating the sampling of cells and virus as a continuous exponential decay; $\delta = 0.057$ per day for cell harvest and $r_c = 7.31$ per day for virus harvest. The rates of RNA degradation and loss of infectivity for SHIV-KS661 are shown in Fig. 2.10.

Notice that d, r_I and r_{RNA} are determined by direct measurements in separate experiments. The parameters $\beta_{50}, \alpha, k,$ and k_{50} and the 16 initial $(t = 0)$ values for the variables are determined by fitting the model to some data.

The mathematical model (2.27)-(2.30) for SHIV-KS661 infection on HSC-F cells gives the following results regarding the viral dynamics of a highly pathogenic Simian/Human Immunodeficiency Virus:

1. Reliable estimation of the parameters characterizing cell-virus interactions *in vitro*.

2. A consistent quantitative description of SHIV-KS661 kinetics in HSC-F cell cultures.

3. Reliable estimation of the minimum fraction of infectious virus produced by an infected cell.

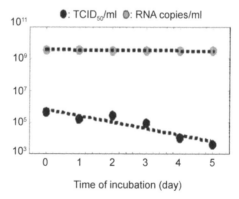

FIGURE 2.10: Rates of RNA degradation and loss of infectivity for SHIV-KS661. Stock virus was incubated under the same conditions as the infection experiments, but in the absence of cells, then sampled every day and stored at -80°C. After the sampling, the RNA copy number (gray circles) and 50% tissue culture infectious dose (black circles) of the samples were measured. Linear regressions yielded an RNA degradation rate of $r_{RNA} = 0.039$ per day and a loss of viral infectivity rate of $r_I = 0.93$ per day. Reused with permission from [Iwami, *et al.* (2012)].

4. The improved method for quantifying viral kinetics *in vitro* can be applied to other viral infections because it improves the understanding of the differences in replication across different strains, between complete and protein-deficient viruses, the differences in viral pathogenesis, and the effects of anti-viral therapies, etc...

2.8 The effects of morphine on Simian Immunodeficiency Virus Dynamics

In [Vaidya, *et al.*, (2016)] a mathematical model that incorporates experimentally observed effects of morphine on inducing HIV-1 co-receptor expression is given by:

$$\frac{dT_I}{dt} = \lambda + qT_h - dT_I - rT_I - \beta_I VT_I, \quad T_I(0) = T_{I0} \tag{2.31}$$

$$\frac{dT_h}{dt} = rT_I - dT_h - \beta_h VT_h - qT_h, \quad T_h(0) = T_{h0} \tag{2.32}$$

$$\frac{dI}{dt} = \beta_I VT_I + \beta_h VT_h - \delta I, \quad I(0) = I_0 \tag{2.33}$$

$$\frac{dV}{dt} = pI - cV, \quad V(0) = V_0 \tag{2.34}$$

The system (2.31)-(2.34) is based on *in vitro* observations regarding the effects of morphine conditioning on CCR5 levels. The system contained two subpopulations of target cells (CD4$^+$ T cells): One with lower susceptibility to infection (lower infection rate) due to a low level of co-receptor expression and another with higher susceptibility (higher infection rate). We notice that the morphine enhances the rate of cells transferring from the T_I-group to the T_h-group resulting in a higher proportion of T_h cells in the target cell population. A schematic diagram of the model (2.31)-(2.34) is presented in Fig. 2.11.

The parameters in system (2.31)-(2.34) have the following meanings:

- T_I is the low level of co-receptor expression.

- T_h is the high level of co-receptor expression

- λ is the constant rate for which the target cells are generated.

- d is the per capita net loss rate of the target cells, i.e., the difference between the rate of loss from cell death and rate of gain due to cell division.

- T_I and T_h are the target cells.

- T_I-group is the set of the newly generated target cells.

- r is the rate of transition from T_I to T_h.

- q is the rate of transition from T_h to T_I. In this case T_I and T_h, become productively infected cells, I, upon contact with free virus, V, at rates β_I and β_h, respectively.

- δ, p and c are the rate constants of infected cell loss, virus production by infected cells, and virus clearance, respectively.

In this model, the total CD4 count is given by $T = T_I + T_h + I$, with $T(0) = T_0$.

Many experimental studies showed that the levels of CCR5 and CXCR4 expression in various cell types are increased [Li, *et al.*, (2003)] in morphine-dependent animals and a high correlation between co-receptors expression and the pathogenesis of HIV and SIV infection. Therefore, system (2.31)-(2.34) confirms the hypothesis that morphine increases the susceptibility of target cells to HIV/SIV infection.

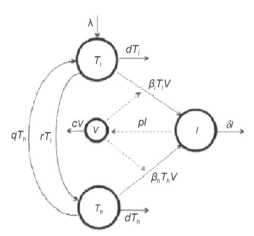

FIGURE 2.11: Schematic diagram of the model (2.31)-(2.34). The model contains two subpopulations, T_l and T_h of target cells, with low and high susceptibilities to infection. Cells within these populations can switch susceptibilities with rates r and q, respectively. The target cells are infected, upon contact with virus, V, at rates β_I and β_h, respectively, and become productively infected cells, I. Reused with permission from [Vaidya, *et al.* (2016)].

2.9 The dynamics of the HIV therapy system

In this section, we give some aspects of the dynamics of the Human Immunodeficiency Virus (HIV). Today, HIV is the most dangerous and lethal disease and there are millions of infected population. The dynamics of the HIV therapy system is given by the following first order non-linear differential equation [Filter *et al.*, (2005), Xia & Moog (2003), Hammond (1993), Wein *et al.*, (1997), de Boer & Perelson (1998), Wein *et al.*, (1998), Dixit & Perelson (2004), Strafford *et al.*, (2000), Fishman & Perelson (1994), Essunger & Perelson (1994), Berry & Nowak (1994), Lipsitch & Nowak (1995), Nowak *et al.*, (1995), Courchamp *et al.*, (1995), Iwasa *et al.*, (2005), Nowak *et al.*, (1997), Wodarz *et al.*, (1999), Gilchrist *et al.*, (2004), Jafelice *et al.*, (2004)]:

$$\begin{cases} x' = a\,(x_0 - x) - bxz \\ \\ y' = c\,(y_0 - y) + dyz \\ \\ z' = z\,(ex - fy) \end{cases} \qquad (2.35)$$

where $(a, b, c, d, e, f) \in \mathbb{R}^6$ are positive bifurcation parameters. The variables $x\,(t)$, $y\,(t)$ and $z\,(t)$ are the concentrations of the CD4 lymphocyte population, the CD8 lymphocyte population and the HIV-1 viral load, respectively. The positive qunatities x_0 and y_0 are the normal unperturbed concentrations of the CD4 and CD8 lymphocyte population, respectively. The system (2.35) has two equilibria

$$\begin{cases} P_1 = (x_0, y_0, 0) \\ \\ P_2 = \left(\frac{adx_0e + bcfy_0}{e(ad+bc)}, \frac{eadx_0 + bcfy_0}{adf + bcf}, \frac{ac(x_0e - fy_0)}{adx_0e + bcfy_0} \right) = (u, v, w) \end{cases} \qquad (2.36)$$

in which P_1 is stable when $ex_0 - fy_0 < 0$ and unstable when $ex_0 - fy_0 > 0$. From our numerical calculations, we remark that system (2.35) converges always to the equilibrium point P_2. The stability of P_2 can be studied using the Jacobian matrix given by:

$$J\,(P_2) = \begin{pmatrix} -a - bw & 0 & -bu \\ 0 & -c + dw & dv \\ ew & -fw & eu - fv \end{pmatrix} \qquad (2.37)$$

The characteristic polynomial at P_2 is given by $\lambda^3 + A\lambda^2 + B\lambda + C = 0$, where

$$\left\{ \begin{array}{l} A = a + c - ue + bw - dw + fv \\[2mm] B = ac - bdw^2 - aue - cue - adw + bcw + afv + cfv + duwe + bfvw \\[2mm] C = -acue + acfv + aduwe + bcfvw \end{array} \right.$$

$$(2.38)$$

The exact value of the eigenvalues is obtained by using the Cardan's method for solving a cubic characteristic equation, but due to the complicated formulas in this case, we use the Routh-Hurwitz conditions, leading to the conclusion that the real parts of the roots λ are negative if and only if $A > 0$, $C > 0$ and $AB - C > 0$. Thus, P_2 is stable if and only if:

$$\left\{ \begin{array}{l} a + c - ue + bw - dw + fv > 0 \\[2mm] acfv - acue + aduwe + bcfvw > 0 \\[2mm] \mu_1 + \mu_2 + \mu_3 + \mu_4 + \mu_5 + \mu_6 + \mu_7 > 0 \end{array} \right. \qquad (2.39)$$

where

$$\left\{ \begin{array}{l} \mu_1 = a^2 c - a^2 dw + a^2 fv - ea^2 u + 2abcw - 2abdw^2 \\[2mm] \mu_2 = 2abfvw - eabuw + ac^2 - 2acdw + 2acfv - 2eacu + ad^2 w^2 \\[2mm] \mu_3 = -2adfvw + 2eaduw + af^2 v^2 - 2eafuve^3 au^2 + b^2 cw^2 \\[2mm] \mu_4 = -b^2 dw^3 + b^2 fvw^2 + bc^2 w - 2bcdw^2 + 2bcfvw - 2ebcuw \\[2mm] \mu_5 = bd^2 w^3 - 2bdfvw^2 + 2ebduw^2 + bf^2 v^2 w - ebfuvw \\[2mm] \mu_6 = c^2 fv - ec^2 u - cdfvw + 2ecduw + cf^2 v^2 \\[2mm] \mu_7 = -2ecfuv + e^3 cu^2 - ed^2 uw^2 + edfuvw - e^3 du^2 w \end{array} \right. \qquad (2.40)$$

We notice that it is not easy to solve a such inequalities. The simple way is to consider some real values of the 8 parameters and vary one of them as shown by the next section.

In this section, we show numerically that the so-called *HIV therapy system* is regular and converges to a stable equilibrium point for the almost real values (in the medical sense) of its bifurcation parameters. Since there is no rigorous proof of that convergence property, it is conjectured that the system is not chaotic for all positive bifurcation parameters.

It was claimed in [Charlotte & Bingo (2010)] that the dynamics of the HIV system (2.35) are chaotic by calculating the Lyapunov exponents of this system. Apparently, this solution is only transiently chaotic. In fact, we show numerically that system (2.35) is not chaotic for $a = 0.25, b = 50, c = 0.25$, $e = 0.01$, $f = 0.006$, $x_0 = 1000$, $y_0 = 550$, $x(0) = x_0, y(0) = y_0$, $z(0) = 0.03$

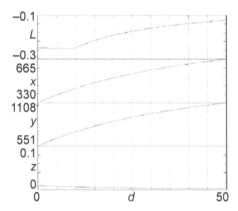

FIGURE 2.12: Bifurcation diagram of the variables x, y and z (with the variation of the largest Lyapunov exponent L) of system (2.35) plotted versus $d \in [0, 50]$ with $a = 0.25, b = 50, c = 0.25, e = 0.01, f = 0.006, x_0 = 1000, y_0 = 550, x(0) = x_0, y(0) = y_0$, and $z(0) = 0.03$.

and $0 < d < 50$ as shown in Fig. 2.12. In this case, there is only a single stable equilibrium with a negative L and no indication of multistability. Numerical calculations show that this stable equilibrium is

$$P_2 = \left(\frac{1000.0d + 16500.0}{d + 50.0}, \frac{2.5d + 41.25}{0.0015d + 0.075}, \frac{0.41875}{2.5d + 41.25} \right)$$

It is easy to verify that the first and the second components of P_2 are increasing and the third one is decreasing with respect to the variations of d, that is,

$$\begin{cases} \frac{d}{d(d)} \left(\frac{1000.0d + 16500.0}{d + 50.0} \right) = \frac{33500.0}{d^2 + 100.0d + 2500.0} > 0 \\\\ \frac{d}{d(d)} \left(\frac{2.5d + 41.25}{0.0015d + 0.075} \right) = \frac{1.675 \times 10^5}{3.0d^2 + 300.0d + 7500.0} > 0 \\\\ \frac{d}{d(d)} \left(\frac{0.41875}{2.5d + 41.25} \right) = -\frac{1.0469}{(2.5d + 41.25)^2} < 0 \end{cases}$$

The graphs of these functions (in d) correlate with bifurcation diagrams shown in Fig. 2.13. This method is used also for all other bifurcation parameters and shows the convergence of system (2.35) on its equilibrium point P_2. To prove this result, we have from the above analysis

$$\begin{cases} A = \frac{3325.0d + 1.6625 \times 10^5}{5000(2.0d + 33.0)} \\\\ B = \frac{6.749 \times 10^9 d + 2.8424 \times 10^8 d^2 + 3.3913 \times 10^6 d^3 + 5.0759 \times 10^{10}}{500\,000(d + 50.0)(2.0d + 33.0)^2} \\\\ C = \frac{1675.0d + 83752.}{4000(d + 50.0)} \end{cases} \tag{2.41}$$

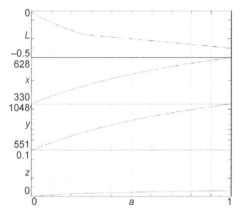

FIGURE 2.13: Bifurcation diagram of the variables x, y and z (with the variation of the largest Lyapunov exponent L) of system (2.35) plotted versus $a \in [0, 1]$ with $b = 50, c = 0.25, d = 10, e = 0.01, f = 0.006, x_0 = 1000, y_0 = 550, x(0) = x_0, y(0) = y_0,$ and $z(0) = 0.03$.

then P_2 is stable if and only if

$$
\begin{cases}
\dfrac{3325.0d + 1.662\,5 \times 10^5}{2.0d + 33.0} > 0 \\[12pt]
\dfrac{1675.0d + 83752.}{d + 50.0} > 0 \\[12pt]
\dfrac{3.644\,6 \times 10^{11}d + 1.685\,1 \times 10^{10}d^2 + 2.702\,3 \times 10^8 d^3 + 1.160\,4 \times 10^6 d^4 + 2.623 \times 10^{12}}{1000\,000(d + 50.0)(2.0d + 33.0)^3} > 0
\end{cases}
$$

$$(2.42)$$

which proves that P_2 is stable for all $d > 0$. The same analysis can be done for the other parameters and the behavior of system (2.35) can be seen in Fig. 2.14 to Fig. 2.19.

So, from all these figures, we conclude that the HIV therapy system (2.35) is regular and converges to its stable equilibrium point P_2 for all a, b, c, d, e, f, x_0 and y_0 in the indicated ranges. Thus, we can formulate the following conjecture:

Conjecture 42 *The HIV therapy system (2.35) converges to its stable equilibrium point P_2 for all a, b, c, d, e, f, x_0 and y_0.*

As a result of the previous conjecture, we can conclude that the concentrations of the CD4 lymphocyte population converges to the fixed quantity $\frac{adx_0e + bcfy_0}{e(ad + bc)}$ and the CD8 lymphocyte population converges to $\frac{eadx_0 + bcfy_0}{adf + bcf}$ and the HIV-1 viral load converges to $\frac{ac(x_0e - fy_0)}{adx_0e + bcfy_0}$ with the condition $x_0e - fy_0 > 0$ since it is a positive quantity. In this case, the system (2.35) never converges to P_1 because it is always unstable. This means that the system (2.35) never returns to the normal unperturbed concentrations of the CD4 and CD8 lymphocyte population denoted by x_0 and y_0.

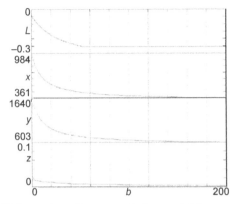

FIGURE 2.14: Bifurcation diagram of the variables x, y and z (with the variation of the largest Lyapunov exponent L) of system (2.35) plotted versus $b \in [0, 200]$ with $a = 0.25$, $c = 0.25$, $d = 10$, $e = 0.01$, $f = 0.006$, $x_0 = 1000$, $y_0 = 550$, $x(0) = x_0$, $y(0) = y_0$, and $z(0) = 0.03$.

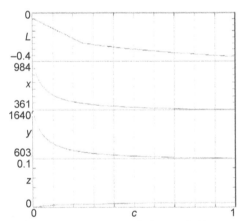

FIGURE 2.15: Bifurcation diagram of the variables x, y and z (with the variation of the largest Lyapunov exponent L) of system (2.35) plotted versus $c \in [0, 1]$ with $a = 0.25, b = 50$, $d = 10$, $e = 0.01$, $f = 0.006$, $x_0 = 1000$, $y_0 = 550$, $x(0) = x_0$, $y(0) = y_0$, and $z(0) = 0.03$.

2.10   Dynamics of urbanization

By assuming the pre-existence of a Central Business District (CBD), classical urban models explain the spatial structure of an ideal mono-centric city. These models are defined by geographic locations in proximity to the CBD. This fact implies monotonically decreasing land rents and density as the distance

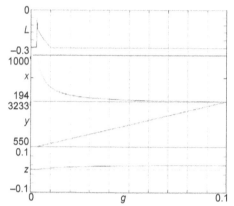

FIGURE 2.16: Bifurcation diagram of the variables x, y and z (with the variation of the largest Lyapunov exponent L) of system (2.35) plotted versus $e \in [0, 0.1]$ with $a = 0.25, b = 50, c = 0.25, d = 10, f = 0.006, x_0 = 1000, y_0 = 550, x(0) = x_0, y(0) = y_0,$ and $z(0) = 0.03$.

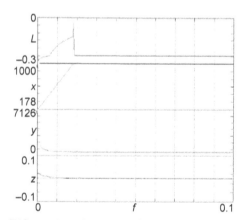

FIGURE 2.17: Bifurcation diagram of the variables x, y and z (with the variation of the largest Lyapunov exponent L) of system (2.35) plotted versus $f \in [0, 0.1]$ with $a = 0.25, b = 50, c = 0.25, d = 10, e = 0.01, x_0 = 1000, y_0 = 550, x(0) = x_0, y(0) = y_0,$ and $z(0) = 0.03$.

from the CBD increases. Due to several factors, the classical urban economics models are unable to explain the formation of modern cities with polycentric structures and the dynamic processes that characterize them. Only ideal polycentric cities and their associated dynamics can be explained by these models. Most methods of analysis are given by the means of *agent-based modeling approach*. The result is that the urban dynamics are characterized by a sudden emergence of sub-centers and the complex dynamics that emerge in the mod-

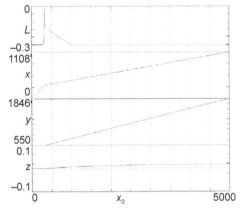

FIGURE 2.18: Bifurcation diagram of the variables x, y and z (with the variation of the largest Lyapunov exponent L) of system (2.35) plotted versus $x_0 \in [0, 5000]$ with $a = 0.25, b = 50, c = 0.25, d = 10, e = 0.01, f = 0.006, y_0 = 550, x(0) = x_0, y(0) = y_0$, and $z(0) = 0.03$.

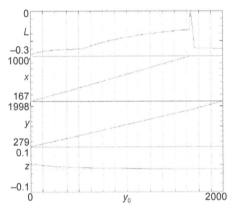

FIGURE 2.19: Bifurcation diagram of the variables x (with the variation of the largest Lyapunov exponent L) of system (2.35) plotted versus $y_0 \in [0, 2000]$ with $a = 0.25, b = 50, c = 0.25, d = 10, e = 0.01, f = 0.006, x_0 = 1000, x(0) = x_0, y(0) = y_0$, and $z(0) = 0.03$.

els cause patterns similar to those created by *self organizing criticality* models (fractals).

By using an iteration method to find the numerical solution of rural-urban interaction models of population presented in [Chen (2009)] a 2-dimensional map was obtained as follows:

$$
f(r, u) = \begin{pmatrix} (1 + a)\, r - eru \\ dr + (1 + c)\, u + fru \end{pmatrix}
\tag{2.43}
$$

where r and u respectively represent the rural population and urban population in a region at time t $(r(t) > 0, u(t) > 0)$, and a, c, d, e, f are all parameters. The map (2.43) displays very complex periodic oscillation and chaotic behavior. It was shown in [Chen (2009)] that map (2.43) is chaotic for $r(0) = 1, u(0) = 0.2, a = 1.05, c = -2, d = 0.9, e = 0.8$, and $f = 0.2$. This strange attractor was called *rural-urban interaction attractor*, whose box-counting dimension is about 1.5, and the correlation dimension is around 0.75. See some of these chaotic attractors in Figs. 2.20-2.21

The map (2.43) has two fixed points $P_1 = (0,0)$ for all bifurcation parameters and $P_2 = \left(\frac{a}{e}, \frac{-ac}{de+af} \right)$ if $de + af \neq 0$. The Jacobian matrix of the map (2.43) is given by:

$$J(r, u) = \begin{pmatrix} a - ue + 1 & -re \\ d + fu & c + fr + 1 \end{pmatrix}$$

and its characteristic polynomial for a fixed point (x, x) is given by:

$$\lambda^2 + (ue - c - a - fr - 2)\lambda + ((a - ue + 1)(c + fr + 1) + re(d + fu))$$

Thus, after some calculations, one can ascertain that P_1 is unstable if $|a + 1| > 1, |c + 1| > 1$, and P_1 is stable if $|a + 1| < 1, |c + 1| < 1$. P_1 is a saddle point if $|a + 1| > 1, |c + 1| < 1$, or $|a + 1| < 1, |c + 1| > 1$. The stability of P_2 can be determined by finding the roots of the corresponding characteristic polynomial $\lambda^2 + \alpha_1 \lambda + \alpha_2$ but with some tedious calculations, where

$$\begin{cases} \alpha_1 = -\frac{(e + ce + af)}{e} - \frac{(a^2 f + de + af + ace + ade)}{de + af} \\ \\ \alpha_2 = \frac{a(ed^2 + afd - acf)}{de + af} + \frac{e^{-1}(e + ce + af)(a^2 f + de + af + ace + ade)}{de + af} \end{cases}$$

First of all, we note that r and u stand for the plots of every r and u iterates. A stands for the regions in $a \in [0, 1.2]$ with multiple attractors. D stands for dissipation, i.e., the sum of Lyupanov exponents is negative. The line in D marks the location of maximum dissipation. See Fig. 2.23 for dissipative and bounded orbits in the intervals $(a, d) \in [-5, 5] \times [-5, 5]$.

The symbol H stands for homoclinic orbit for the map (2.43). The symbol L stands for Lyapunov exponent. The symbol P stands for periodic orbits. The symbol S stands for the size of the attractor.

It appears there are coexisting attractors throughout the range $0 < a < 1.161$. The route to chaos is fairly complicated with fixed-point, quasiperiodicity, period-5, quasiperiodicity, chaos, period-6, quasiperiodicity, chaos, and then a boundary crisis at $a = 1.161$. See Fig. 2.22.

In the corresponding interval, the map (2.43) where homoclinic orbits appear, the resulting chaotic attractor is a *quasi-attractor*. This means that the chaotic attractor presented by map (2.43) is the limit set enclosing periodic

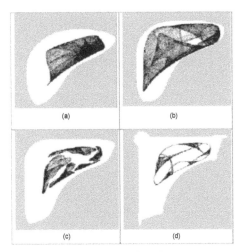

FIGURE 2.20: Attractors for the map (2.43) with their basins of attraction when $r_0 = 1, u_0 = 0.2, a = 1.05, d = 0.9, e = 0.8$ (a) $c = -2.3, f = 0.2$, (b) $c = -2.2, f = 0.3$, (c) $c = -2.1, f = 0.2$,(d) $c = -2.0, f = 0.2$.

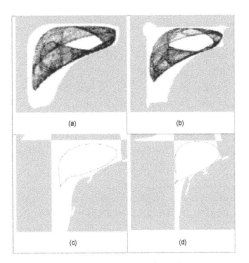

FIGURE 2.21: Attractors for the system (2.43) with their basins of attraction when $r_0 = 1, u_0 = 0.2, a = 1.05, d = 0.9, e = 0.8$ (a) $c = -2.1, f = 0.3$, (b) $c = -2, f = 0.3$, (c) $c = -1.5, f = 0.2$,(d) $c = -1.1, f = 0.2$.

orbits of different topological types (e.g., stable and saddle periodic orbits) and structurally unstable orbits.

Generally, the most known observed chaotic attractors are quasiattractors [Anishchenko (1990-1995)]. Thus, the chaotic attractor presented by map (2.43) has the following important properties:

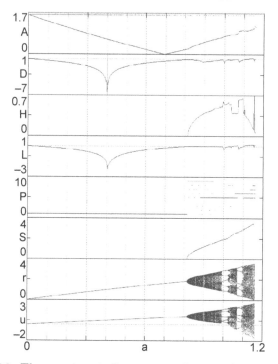

FIGURE 2.22: The quasi-periodic route to chaos and variation of the Lyapunov exponents for the map (2.43) obtained for $r_0 = 1, u_0 = 0.2, a = 1.05$, $d = 0.9, e = 0.8, f = 0.2$ and $0 < a \leq 1.2$. The symbol A stands for the regions with multiple attractors; D stands for dissipation. The line in D marks the location of maximum dissipation. The symbol H stands for homoclinic orbit for the map (2.43). The symbol L stands for Lyapunov exponent. The symbol P stands for periodic orbits. The symbol S stands for the size of the attractor.

1. It is the separatrix loops of saddle-focuses or homoclinic orbits of saddle cycles in the moment of tangency of their stable and unstable manifolds, because they enclose non-robust singular trajectories that are *dangerous*.

2. In the neighborhood of its trajectories, there exists a map of *Smale's horseshoe-type*. This map contains both a non-trivial hyperbolic subset of trajectories and a denumerable subset of stable periodic orbits [Shilnikov (1965-1970)].

3. This attractor is the unified limit set of the whole attracting set of trajectories, including a subset of both chaotic and stable periodic trajectories which have long periods and weak and narrow basins of attraction and stability regions.

4. The basins of attraction of stable cycles are very narrow.

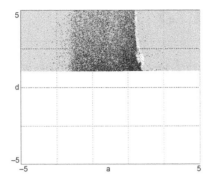

FIGURE 2.23: The sign of the average of $\log |(0.2 + 0.8d + 0.2a)\, r - 0.8\,(-2.0 + 1)\, u + (a - 2.0 - 2.0a + 1)|$ over the orbit on the attractors of the system (2.43) for $(a, d) \in [-5, 5] \times [-5, 5]$ defines the regions of net expansion and contraction.

5. Some orbits do not reveal themselves ordinarily in numeric simulations except some quite large stability windows where they are clearly visible.

From the above properties, it is easy to conclude that the chaotic attractor presented by map (2.43) has a very complex structure of embedded basins of attraction in terms of initial conditions and a set of bifurcation parameters of non-zero measure. The homoclinic tangency of stable and unstable manifolds of saddle points in the Poincaré section is the principal cause of this complexity as shown in [Gavrilov and Shilnikov (1973), Afraimovich and Shilnikov (1983)].

Chapter 3

Chaotic Behaviors in Piecewise Linear Mappings

3.1 Introduction

In this chapter, we discuss the different methods used for rigorous proof of chaos in piecewise maps. The importance of these maps can be seen from the fact that the dynamics of a number of switching circuits can be represented by one-dimensional piecewise smooth maps under discrete modeling, in particular in power electronic circuits [Wolf, *et al.*, (1994), Banerjee and Chakrabarty (1998), Banerjee, *et al.*, (2000), Robert and Robert (2002)]. The study of chaos in this type of system is based on the affinity of the corresponding normal forms for fixed points on borders, and the behavior of fixed points and periodic points depend on the bifurcation parameter for the scenarios associated with the various cases. A detailed overview of these maps is presented in [Zeraoulia and Sprott (2011(g))].

In Sec. 3.1, we present the essential results on the chaotic dynamics and bifurcations in 1-D piecewise smooth maps. In Sec. 3.2, the dynamical properties of a 1-D singular map are given and discussed along some rare types of bifurcations. Robust chaos and several types of border collision bifurcations are discussed in Sec. 3.3.

3.2 Chaos in one-dimensional piecewise smooth maps

Consider the one-dimensional piecewise smooth system:

$$x_{k+1} = f(x_k, \mu) = \begin{cases} g(x, \mu), x \leq x_b \\ \\ h(x, \mu), x \geq x_b \end{cases} \qquad (3.1)$$

where μ is the bifurcation parameter and x_b is the border. The map $f : \mathbb{R} \times \mathbb{R} \to \mathbb{R}$ is assumed to be piecewise smooth, i.e., f depends smoothly on x everywhere except at x_b, where it is continuous in x. It is also assumed that f depends smoothly on μ everywhere. Let R_L and R_R be the two regions in state space separated by the border:

$$\begin{cases} R_L = \{x : x \le x_b\} \\ \\ R_R = \{x : x \ge x_b\} \end{cases} \tag{3.2}$$

Let $x_0(\mu)$ be a possible path of fixed points of f, this path depends continuously on μ. Suppose also that the fixed point hits the boundary at a critical parameter value μ_b: $x_0(\mu_b) = x_b$.

Theorem 43 *The normal form of the piecewise smooth one-dimensional map (3.1) is given by [Yuan (1997), Banerjee, et al., (2000)]:*

$$G_1(x, \mu) = \begin{cases} ax + \mu, x \le 0 \\ \\ bx + \mu, x \ge 0 \end{cases} \tag{3.3}$$

Proof 9 *The normal form (3.3) at a fixed point on the border is a piecewise affine approximation of the map in the neighborhood of the border point x_b. The method of derivation of such a form is as follows: Let $\bar{x} = x - x_b$ and $\bar{\mu} = \mu - \mu_b$, Eq. (3.1) becomes*

$$\bar{f}(\bar{x}, \bar{\mu}) = \begin{cases} g(\bar{x} + x_b, \bar{\mu} + \mu_b), \bar{x} \le 0 \\ \\ h(\bar{x} + x_b, \bar{\mu} + \mu_b), \bar{x} \ge 0 \end{cases} \tag{3.4}$$

Hence, for map (3.4), the border is at $\bar{x} = 0$, and the state space is divided into two halves, $\mathbb{R}_- = (-\infty, 0]$ and $\mathbb{R}_+ = [0, \infty)$ and the fixed point of (3.4) is at the border for the parameter value $\bar{\mu} = 0$. Expanding $\bar{f}$ to first order about $(0, 0)$ gives

$$\begin{cases} \bar{f}(\bar{x}, \bar{\mu}) = \begin{cases} a\bar{x} + \bar{\mu}v + O\left(\bar{x}, \bar{\mu}\right), \bar{x} \le 0 \\ \\ b\bar{x} + \bar{\mu}v + O\left(\bar{x}, \bar{\mu}\right), \bar{x} \ge 0 \end{cases} \\ \\ a = \lim_{x \to 0^-} \frac{\partial}{\partial x}\bar{f}(\bar{x}, 0) \qquad (3.5) \\ \\ b = \lim_{x \to 0^+} \frac{\partial}{\partial x}\bar{f}(\bar{x}, 0) \\ \\ v = \lim_{x \to 0} \frac{\partial}{\partial \mu}\bar{f}(\bar{x}, 0) \end{cases}$$

Note that the last limit in (3.5) doesn't depend on the direction of approach of 0 by x, due to the smoothness of f in μ. Assume that $v \ne 0$, $|a| \ne 1$ and $|b| \ne 1$, then the non-linear terms are negligible close to the border. Let us define a new parameter $\mu'' = \bar{\mu}v$ and dropping the higher order terms as in [Yuan (1997), Banerjee, et al., (2000)], then the 1-D normal form is given by

$$G_1(\bar{x}, \bar{\mu}) = \begin{cases} a\bar{x} + \mu'', \bar{x} \leq 0 \\ \\ b\bar{x} + \mu'', \bar{x} \geq 0 \end{cases} \tag{3.6}$$

which has the same form of (3.3).

Next, we discuss some border collision bifurcation scenarios from x_b for μ near μ_b: Let x_R^* and x_L^* be the possible fixed points of the system near the border to the right $(x > x_b)$ and left $(x < x_b)$ of the border, respectively. Then in the normal form (3.3), one has $x_R^* = \frac{\mu}{1-b} \geq 0$, if $b < 1$ and $x_L^* = \frac{\mu}{1-a} \leq 0$, if $a < 1$.

Scenario A. Persistent fixed point (non-bifurcation)

Border collision bifurcation scenarios can be obtained by various combinations of the parameters $a \geq b$ as μ is varied, because the normal form (3.3) is invariant under the transformation $x \to -x$, $\mu \to -\mu$, $a \rightleftarrows b$. Two situations lead to this scenario:

Scenario A1: (Persistence of stable fixed point) or Period-1 $\to$ Period-1 in [Yuan (1997), Banerjee, *et al.*, (2000)]. If

$$-1 < b \leq a < 1 \tag{3.7}$$

then a stable fixed point for $\mu < 0$ persists and remains stable for $\mu > 0$. Here, the fixed point changes continuously as a function of the bifurcation parameter and the eigenvalue associated with the system linearization at the fixed point changes discontinuously from a to b at $\mu = 0$.

Scenario A2: (Persistence of unstable fixed point) or No Attractor $\to$ No Attractor in [Yuan (1997), Banerjee *et al.*, (2000)]. If

$$1 < b \leq a \tag{3.8}$$

or

$$b \leq a < -1 \tag{3.9}$$

then an unstable fixed point for $\mu < 0$ persists and remains unstable for $\mu > 0$. Here, there is one unstable fixed point (that depends continuously on μ) for both positive and negative values of μ and no local attractors exist and the system trajectory diverges for all initial conditions.

Scenario B. Border collision pair bifurcation

For other values of the parameters a and b, given in (3.7), (3.8) and (3.9), there are two main kinds of border collision bifurcation: *Border collision pair bifurcation*, which is similar to saddle node bifurcation (or tangent bifurcation) in smooth systems, and *border crossing bifurcation*, which has some similarities with period doubling bifurcation in smooth maps (supercritical period doubling bifurcation in smooth maps with one distinction). In border collision pair bifurcation, the map (3.3) has two fixed points (one side of the border and the other fixed point is on the opposite side) for positive (respectively, negative) values of μ, and no fixed points for negative (respectively, positive) values of μ. Hence, the border collision pair bifurcation occurs if

$$b < 1 < a. \tag{3.10}$$

The three situations that lead to this scenario are as follows:

Scenario B1: (Merging and annihilation of stable and unstable fixed points) or No Fixed Point $\rightarrow$ Period-1 in [Yuan (1997), Banerjee, *et al.*, (2000)]. If

$$-1 < b < 1 < a \tag{3.11}$$

then there is a bifurcation from no fixed point to two period-1 fixed points. Here, there is no fixed point for $\mu < 0$ while there are two fixed points x_L^* (unstable) and x_R^* (stable) for $\mu \geq 0$.

Scenario B2: (Merging and annihilation of two unstable fixed points, plus chaos) or No Fixed Point $\rightarrow$ Chaos in [Yuan (1997), Banerjee, *et al.*, (2000)]. If

$$\begin{cases} a > 1 \\ \frac{-a}{a-1} < b < -1 \end{cases} \tag{3.12}$$

then there is a bifurcation from no fixed point to two unstable fixed points, plus a growing chaotic attractor as μ is increased through zero.

Scenario B3: (Merging and annihilation of two unstable fixed points) or No fixed point $\rightarrow$ No attractor in [Yuan (1997), Banerjee, *et al.*, (2000)]. If

$$\begin{cases} a > 1 \\ b < \frac{-a}{a-1} \end{cases} \tag{3.13}$$

then there is a bifurcation from no fixed point to two unstable fixed points as μ is increased through zero and the system trajectory diverges for all initial conditions.

Scenario C. border crossing bifurcation

In border crossing bifurcation, the fixed point persists and crosses the border as μ is varied through zero and other attractors or repellers appear or disappear as a result of the bifurcation. Indeed, border crossing bifurcation occurs if

$$\begin{cases} a > -1 \\ b < -1 \end{cases} \tag{3.14}$$

There are three situations that lead to this scenario.

Scenario C1: (Supercritical border collision period doubling) or Period-1 $\rightarrow$ Period-2 in [Yuan (1997), Banerjee, *et al.*, (2000)]. If

$$\begin{cases} b < -1 < a < 0 \\ ab < 1 \end{cases} \tag{3.15}$$

then there is a bifurcation from a stable fixed point to an unstable fixed point, plus a stable period-2 orbit. The first condition of (3.15) implies that there is a bifurcation from a stable period-1 fixed point to an unstable period-1 fixed point, the second implies the emergence of a stable period-2 solution for $\mu > 0$.

Scenario C2: (Subcritical border collision period doubling) or Period-1 $\rightarrow$ No Attractor in [Yuan (1997), Banerjee, *et al.*, (2000)]. If

$$\begin{cases} b < -1 < a < 0 \\ \\ ab > 1 \end{cases} \tag{3.16}$$

then there is a bifurcation from a stable fixed point to an unstable fixed point. In this case, there is a period-1 attractor and an unstable period-2 orbit for $\mu < 0$ and an unstable fixed point for $\mu > 0$ and the system trajectory diverges to infinity for $\mu > 0$.

Scenario C3: (Emergence of periodic or chaotic attractor from stable fixed point) or Period-1 $\rightarrow$ Periodic or Chaotic Attractor in [Yuan (1997), Banerjee, *et al.*, (2000)]. If

$$\begin{cases} 0 < a < 1 \\ \\ b < -1 \end{cases} \tag{3.17}$$

then there is a bifurcation from a stable fixed point to an unstable fixed point, plus a period-n attractor, $n \geq 2$ or a chaotic attractor as μ is increased through zero. This scenario depends on the pair (a, b). For more details, the reader can see [Nusse and Yorke (1995)].

To show the bifurcation of a subcritical period-2 orbit for $\mu < 0$, in **Scenario C2**, let us consider the first and second return maps (for $\mu < 0$), given by:

$$x_{k+1} = \begin{cases} ax_k + \mu, x_k \leq 0 \\ \\ bx_k + \mu, x_k \geq 0 \end{cases} \tag{3.18}$$

$$x_{k+2} = \begin{cases} abx_k + \mu(1+b), x_k \leq -\frac{\mu}{a} \\ \\ a^2 x_k + \mu(1+a), -\frac{\mu}{a} \leq x_k \leq 0 \\ \\ abx_k + \mu(1+a), x_k \geq 0 \end{cases} \tag{3.19}$$

The first return map (3.18) has a stable fixed point, $x_L^* = \frac{\mu}{1-a}$, and the second return map (3.19) has three fixed points, one of which coincides and two unstable ones $x_1^* = \frac{\mu(1+b)}{1-ab}$, $x_2^* = \frac{\mu(1+a)}{1-ab}$ which form a period-2 orbit for the first return map (3.18). The above analysis implies that it is possible to determine regions in the $a-b$ plane in which robust chaos can occur. Namely, Scenario B2 or Scenario C3,. i.e., (3.12) or (3.17) that gives the following interval for robust chaos denoted by:

$$RC = (1, \infty) \times \left(\frac{-a}{a-1}, -1 \right) \cup (0, 1) \times (-\infty, -1) \qquad (3.20)$$

In fact, for $\mu > 0$, the behavior is robust chaotic for the whole region of (3.20), i.e., the chaotic attractor of the 1-D normal form (3.3) is robust in the continuous region of the parameter space where no periodic windows exist [Banerjee, *et al.*, (2000)].

Via an elementary example [Avrutin and Schanz (2008)], it was shown that chaotic attractors which are robust in the sense of [Banerjee *et al.*, (1998)] are not necessarily robust in the sense of [Milnor (1985(a))]. Indeed, a bifurcation phenomenon called the *bandcount adding scenario* occurring in the 2D parameter space of the one-dimensional piecewise-linear map given by:

$$x_{n+1} = \begin{cases} f_l(x_n) = ax_n + \mu + 1, & \text{if } x_n < 0 \\ \\ f_r(x_n) = ax_n + \mu - 1, & \text{if } x_n > 0 \end{cases} \qquad (3.21)$$

i.e., an infinite number of interior crises bounding the regions of multi-band attractors were detected in the region of chaotic behavior. This fact leads to a self-similar structure of the chaotic region in parameter space.

3.3   Chaos in one-dimensional singular maps

These maps play a key role in the theory of *up-embedability of graphs* [Chen and Liu (2007)]. In [Alvarez-Llamoza, *et al.*, (2008)], the critical behavior of the Lyapunov exponent of a 1-D singular map (which has only one face on a surface) near the transition to robust chaos via type-III intermittency was determined for a family of one-dimensional singular maps. The calculation of critical boundaries separating the region of robust chaos from the region of stable fixed points was given and discussed.

We need the following definitions:

Definition 43 *Intermittent chaos is characterized by the irregular switchings between long periodic-like signals, called laminar phase, and comparatively short chaotic bursts.*

Definition 44 *The statistical signature of intermittency is given by the scaling relations describing the dependence of the average length (l) of the laminar phases with a control parameter ϵ that measures the distance from the bifurcation point.*

Different types of intermittencies was given in [Pomeau and Manneville (1980)] according to the local Poincaré map associated to the system under consideration:

Definition 45 *(a) Type-I intermittency occurs by a tangent bifurcation when the Floquet's multiplier for the Poincaré map crosses the circle of unitary norm in the complex plane through +1.*

(b) Type-II intermittency is due to a Hopf's bifurcation which appears as two complex eigenvalues of the Floquet's matrix cross the unitary circle off the real axis.

(c) Type-III intermittency is associated to an inverse period doubling bifurcation whose Floquet's multiplier is −1.

- For chaotic systems, the statistical signature is the positive largest Lyapunov exponent λ.

- For type-III intermittency, one has $(l) \sim \epsilon^{-1}$, and $\lambda \sim \epsilon^{\frac{1}{2}}$, both when $\epsilon \to 0$ [Pomeau and Manneville (1980)].

- We have $(l) \sim \epsilon^v$, with $\frac{1}{2} < v < 1$ [Khan, *et al.*, (1992), Fukushima and Yazaki (1995)].

In [Alvarez-Llamoza, *et al.*, (2008)] the following singular map:

$$\begin{cases} x_{n+1} = f(x_n) = \mu - |x_n|^z \\ \frac{d}{dx}(\mu - |x|^z) = -z|x|^{z-1} \end{cases} \tag{3.22}$$

where $n \in \mathbb{Z}$, and μ is a real parameter, $|z| < 1$ describes the order of the singularity at the origin.

The map (3.22) has the following properties:

1. The map is unbounded, displaying robust chaos on a single interval of the parameter μ and associated to type-III intermittency.

2. The Schwarzian derivative is given by

$$S(f, x) = \frac{1 - z^2}{2x^2} > 0, \text{ for all } |z| < 1 \tag{3.23}$$

   and it is always positive.

3. The map is not a unimodal map and it does not exhibit a sequence of period doubling bifurcations.

4. If a stable fixed point loses its stability at some critical value of the parameter to yield robust chaos, then condition (3.23) implies the occurrence of an inverse period doubling bifurcation.

5. The map has two stable fixed points that exist for each value of z : $x_-^* < 0$ and $x_+^* \geq 0$, satisfying $f(x^*) = x^*$ and $|f'(x^*)| < 1$

6. The fixed point x_-^* becomes unstable at the parameter value

$$\mu_- (z) = |z|^{\frac{z}{1-z}} - |z|^{\frac{1}{1-z}} \tag{3.24}$$

through an inverse period doubling bifurcation that gives robust chaos via type-III intermittency.

7. the fixed point x_+^* originates from a tangent bifurcation at the value

$$\mu_+ (z) = |z|^{\frac{z}{1-z}} + |z|^{\frac{1}{1-z}} \tag{3.25}$$

with type-I intermittency transition to chaos.

8. The boundaries $\mu_+ (z)$ and $\mu_- (z)$ correspond to the Lyapunov exponent values $\lambda = 0$.

9. For $z \in (0, 1)$ the behaviors of the fixed points are interchanged:

(a) The fixed point x_-^* displays a tangent bifurcation at (3.24) and a type-I intermittent transition to chaos occurs.

(b) The fixed point x_+^* undergoes an inverse period doubling bifurcation at (3.25), giving the scenario for a type-III intermittent transition to chaos.

10. The transition to chaos via type-III intermittency occurs at the critical parameter values $\mu_m (z) = \mu_- (z)$ for $z \in (-1, 0)$, and $\mu_m (z) = \mu_+ (z)$ for $z \in (0, 1)$.

11. Near the critical boundary $\mu_m (z)$ within the chaotic region, the map (3.23) shows type-III intermittency.

12. Type-I or type-III intermittencies appear at the boundaries of the robust chaos intervals, .i.e. $z = -0.5$ or $z = 0.5$.

13. The width of the interval for robust chaos on the parameter μ for a given $|z| < 1$ is given by:

$$\Delta\mu (z) = \mu_+ (z) - \mu_- (z) = 2 |z|^{\frac{1}{1-z}} \tag{3.26}$$

14. The Lyapunov exponent is positive on the robust chaos interval $\Delta\mu (z)$.

15. The transition to chaos through type-I intermittency is smooth.

16. The transition to chaos via type-III intermittency has a discontinuity (due to the sudden loss of stability of the fixed point associated to the inverse period doubling bifurcation) of the derivative of the Lyapunov exponent at the parameter values corresponding to the critical curve $\mu_m (z)$.

17. The behavior of the Lyapunov exponent near the boundary $\mu_m(z)$ within the chaotic region is characterized by a scaling relation

$$\lambda \sim \epsilon^{\beta(z)} \qquad (3.27)$$

where $\epsilon = |\mu - \mu_m|$ and $\beta(z)$ is a critical exponent expressing the order of the transition.

18. The exponent $\beta(z)$ is calculated from the slopes of each curve in the $\log - \log$ plot of the Lyapunov exponent versus ϵ.

19. The scaling behavior of the Lyapunov exponent at the transition to chaos is characterized by the values of the critical exponent β in the interval $0 \leq \beta < 1/2$ as the singularity z of the map varies in $(-1, 1)$.

3.4   Chaos in 2-D piecewise smooth maps

Power electronics is an area concerned with the problem of the efficient conversion of electrical power from one form to another [Robert and Robert (2002), Banerjee and Grebogi (1999), Banergee, *et al.*, (1998), Hassouneh, *et al.*, (2002), Banerjee and Verghese (2000), Kowalczyk (2005), Banerjee, *et al.*, (1999)]. *Power converters* [Banerjee, *et al.*, (1999)] exhibit several non-linear phenomena, such as border-collision bifurcations, coexisting attractors (alternative stable operating modes or fragile chaos), and chaos (apparently random behavior). These phenomena are created by switching elements [Banerjee, *et al.*, (1999)]. Border-collision bifurcations in piecewise-smooth systems are studied in [Robert and Robert (2002), Banerjee and Grebogi (1999), Banergee, *et al.*, (1998), Hassouneh, *et al.*, (2002), Banerjee and Verghese (2000), Kowalczyk (2005), Banerjee, *et al.*, (1999)]. Piecewise-smooth systems can exhibit classical smooth bifurcations, but if the bifurcation occurs when the fixed point is on the border, there is a discontinuous change in the elements of the Jacobian matrix as the bifurcation parameter is varied. A variety of these bifurcations have been reported in [Robert and Robert (2002), Banerjee and Grebogi (1999)].

Let us consider the following 2-D piecewise smooth system given by:

$$g(x,y;\rho) = \left(\begin{array}{c} g_1 = \left(\begin{array}{c} f_1(x,y;\rho) \\ f_2(x,y;\rho) \end{array} \right), \text{ if } x < S(y,\rho) \\ \\ g_2 = \left(\begin{array}{c} f_1(x,y;\rho) \\ f_2(x,y;\rho) \end{array} \right), \text{ if } x \geq S(y,\rho) \end{array} \right) \qquad (3.28)$$

where the smooth curve $x = S(y, \rho)$ divides the phase plane into two regions R_1 and R_2, given by:

$$R_1 = \{(x, y) \in \mathbb{R}^2, x < S(y, \rho)\} \tag{3.29}$$

$$R_2 = \{(x, y) \in \mathbb{R}^2, x \geq S(y, \rho)\}. \tag{3.30}$$

and the boundary between them as

$$\Sigma = \{(x, y) \in \mathbb{R}^2, x = S(y, \rho)\}. \tag{3.31}$$

It is assumed that the functions g_1 and g_2 are both continuous and have continuous derivatives. Then the map g is continuous, but its derivative is discontinuous at the borderline $x = S(y, \rho)$. It is further assumed that the one-sided partial derivatives at the border are finite and in each subregion R_1 and R_2 and the map (3.28) has one fixed point in R_1 and one fixed point in R_2 for a value ρ_* of the parameter ρ.

It is shown in [Banerjee and Grebogi (1999)] that the normal form of the map (3.28) is given by:

$$N(x, y) = \begin{cases} \begin{pmatrix} \tau_L & 1 \\ -\delta_L & 0 \end{pmatrix} \begin{pmatrix} x \\ y \end{pmatrix} + \begin{pmatrix} 1 \\ 0 \end{pmatrix} \mu, & \text{if } x < 0 \\[3mm] \begin{pmatrix} \tau_R & 1 \\ -\delta_R & 0 \end{pmatrix} \begin{pmatrix} x \\ y \end{pmatrix} + \begin{pmatrix} 1 \\ 0 \end{pmatrix} \mu, & \text{if } x > 0, \end{cases} \tag{3.32}$$

where μ is a parameter and $\tau_{L,R}, \delta_{L,R}$ are the traces and determinants of the corresponding matrices of the linearized map in the two subregions R_L and R_R given by:

$$\begin{cases} R_L = \{(x, y) \in \mathbb{R}^2, x \leq 0, y \in \mathbb{R}\} \\[2mm] R_R = \{(x, y) \in \mathbb{R}^2, x > 0, y \in \mathbb{R}\} \end{cases} \tag{3.33}$$

evaluated at the fixed point

$$P_L = \left(\frac{\mu}{1 - \tau_L + \delta_L}, \frac{-\delta_L \mu}{1 - \tau_L + \delta_L} \right) \in R_L \tag{3.34}$$

(with eigenvalues $\lambda_{L1,2}$) and the fixed point

$$P_R = \left(\frac{\mu}{1 - \tau_R + \delta_R}, \frac{-\delta_R \mu}{1 - \tau_R + \delta_R} \right) \in R_R \tag{3.35}$$

(with eigenvalues $\lambda_{R1,2}$), respectively. The stability of the fixed points is determined by the eigenvalues of the corresponding Jacobian matrix, i.e.,

$$\lambda = \frac{1}{2} \left(\tau \pm \sqrt{\tau^2 - 4\delta} \right) \tag{3.36}$$

in the regions R_L and R_R the map (3.32) is smooth and the boundary between them is given by:

$$\Sigma = \left\{ (x,y) \in \mathbb{R}^2, x = 0, y \in \mathbb{R} \right\}. \tag{3.37}$$

Next, we summarize the known sufficient conditions for the possible bifurcation phenomena in the normal form (3.32). In fact, the dynamics of piecewise smooth maps is the first reason for some investigations into border collision bifurcations phenomena [Feigin (1970), Nusse and Yorke (1992)]. Later the development is based on the observation that most power electronic circuits yield piecewise smooth maps under discrete-time modeling, and non-smooth bifurcations are quite common in them, as demonstrated in [Yuan, *et al.*, (1998)]. In [Feigin (1970), di Bernardo, *et al.*, (1999)] it was shown that for a two-dimensional PWS map, the stability of its fixed points is determined by the eigenvalues of the corresponding Jacobian matrices. Also, the existence of period-1 and period-2 orbits before and after border collision was studied with a classification of border-collision bifurcations in n-dimensional piecewise smooth systems based on the various cases depending on the number of real eigenvalues greater than 1 or less than -1. Other methods of classifications can be found for example in [Banerjee and Grebogi (1999), Banerjee, *et al.*, (2000)] for one and two-dimensional maps. Their methods are based on the consideration of the asymptotically stable orbits (including chaotic orbits) before and after border collision and the assumption that $|\delta_L| < 1$ and $|\delta_R| < 1$. However, it was proved that attractors can exist if the determinant in one side is greater than unity in magnitude, if that determinant in the other side is smaller than unity. The case where $|\delta_L| > 1$ and $\delta_R = 0$ (which occurs in some classes in power electronic systems) has been studied in [Parui and Banerjee (2002)].

Types of fixed points of the normal form (3.32)

The possible types of fixed points of the normal form map (3.32) are given by:

(1) **For positive determinant**:

(1-a) For $2\sqrt{\delta} < \tau < (1 + \delta)$, then the Jacobian matrix has two real eigenvalues $0 < \lambda_{1L}, \lambda_{2L} < 1$, and the fixed point is a regular attractor.

(1-b) For $\tau > (1 + \delta)$, then the Jacobian matrix has two real eigenvalues $0 < \lambda_{1L} < 1, \lambda_{2L} > 1$ and the fixed point is a regular saddle.

(1-c) For $-(1 + \delta) < \tau < -2\sqrt{\delta}$, then the Jacobian matrix has two real eigenvalues $-1 < \lambda_{1L} < 0, -1 < \lambda_{2L} < 0$ and the fixed point is a flip attractor.

(1-d) For $\tau < -(1+\delta)$, then the Jacobian matrix has two real eigenvalues $-1 < \lambda_{1L} < 0, \lambda_{2L} < -1$ and the fixed point is a flip saddle.

(1-e) For $0 < \tau < 2\sqrt{\delta}$, then the Jacobian matrix has two complex eigenvalues $|\lambda_{1L}|, |\lambda_{2L}| < 1$ and the fixed point is a clockwise spiral.

(1-g) For $-2\sqrt{\delta} < \tau < 0$, then the Jacobian matrix has two complex eigenvalues $|\lambda_{1L}|, |\lambda_{2L}| < 1$ and the fixed point is a counter-clockwise spiral.

(2) **For negative determinant**:

(2-a) For $-(1+\delta) < \tau < (1+\delta)$, then the Jacobian matrix has two real eigenvalues $-1 < \lambda_{1L} < 0, 0 < \lambda_{2L} < 1$ and the fixed point is a flip attractor.

(2-b) For $\tau > 1 + \delta$, then the Jacobian matrix has two real eigenvalues $\lambda_{1L} > 1, -1 < \lambda_{2L} < 0$ and the fixed point is a flip saddle.

(2-c) For $\tau < -(1+\delta)$, then the Jacobian matrix has two real eigenvalues $0 < \lambda_{1L} < 1, \lambda_{2L} < -1$ and the fixed point is a flip saddle.

Regions for non-chaos

The known cases in which a locally unique fixed point before the border yields, after the border, either a new locally unique fixed point or a locally unique period-2 attractor, i.e., no chaos are as follows depending on the sign of the system determinants on both sides of the normal form (3.32) are described in the following cases:

The case of positive determinants on both sides of the border

Scenario A: (*Locally unique stable fixed point on both sides of the border*).

If

$$
\left\{
\begin{array}{l}
\delta_L > 0, \delta_R > 0 \\[2mm]
-(1+\delta_L) < \tau_L < (1+\delta_L) \\[2mm]
-(1+\delta_R) < \tau_R < (1+\delta_R)
\end{array}
\right.
\tag{3.38}
$$

then a stable fixed point persists as the bifurcation parameter μ is increased (or decreased) through zero.

For the parameter range given by (3.38), a stable fixed point yields a stable fixed point after the border crossing with or without extraneous periodic orbits emerging from the critical point. These cases are as follows:

1. If $2\sqrt{\delta_L} < \tau_L < (1+\delta L), -(1+\delta R) < \tau R < -2\sqrt{\delta_R}$, then a regular attractor yields a flip attractor.

2. If $2\sqrt{\delta_L} < \tau_L < (1+\delta L), 2\sqrt{\delta_R} < \tau R < (1+\delta R)$ then a regular attractor yields a regular attractor.

3. If $-(1+\delta L) < \tau_L < -2\sqrt{\delta_L}, -(1+\delta L) < \tau_L < -2\sqrt{\delta_L}$, then a flip attractor yields a regular attractor.

4. If $-(1+\delta L) < \tau_L < -2\sqrt{\delta_L}, -(1+\delta R) < \tau R < -2\sqrt{\delta_R}$, then a flip attractor yields a flip attractor.

5. If $2\sqrt{\delta L} < \tau L < (1+\delta L), -2\sqrt{\delta_R} < \tau R < 2\sqrt{\delta_R}$, then a regular attractor yields a spiral attractor.

6. If $-2\sqrt{\delta_L} < \tau_L < 2\sqrt{\delta_L}, 2\sqrt{\delta_R} < \tau R < (1+\delta R)$, then a spiral attractor yields a regular attractor.

7. If $-(1+\delta L) < \tau L < -2\sqrt{\delta_L}, -2\sqrt{\delta_R} < \tau R < 2\sqrt{\delta_R}$, then a flip attractor yields a spiral attractor.

8. If $-2\sqrt{\delta_R} < \tau R < 2\sqrt{\delta_R}, -(1+\delta R) < \tau R < -2\sqrt{\delta_R}$, then a spiral attractor yields a flip attractor.

9. If $0 < \tau_L < 2\sqrt{\delta_L}, -2\sqrt{\delta_R} < \tau_R < 0$, then a clockwise spiral attractor yields an anit-clockwise spiral attractor.

10. If $-2\sqrt{\delta_L} < \tau_L < 0, 0 < \tau_R < 2\sqrt{\delta_R}$, then an anit-clockwise spiral attractor yields a clockwise spiral attractor.

11. If $0 < \tau_L < 2\sqrt{\delta_L}, 0 < \tau_R < 2\sqrt{\delta_R}$, then a clockwise spiral attractor yields a clockwise spiral attractor.

12. If $-2\sqrt{\delta_L} < \tau_L < 0, -2\sqrt{\delta_R} < \tau_R < 0$, then an anit-clockwise spiral attractor yields an anit-clockwise spiral attractor.

Note that extraneous periodic orbits occur (either before, after, or on both sides of the border) at bifurcation in the cases (9) and (10). For the cases (5),(6), (11) and (12) there are no known examples, but also no proof that extraneous periodic orbits do not occur. For the cases (1) to (4) the extraneous periodic orbits cannot appear, i.e., the fixed points on both sides of the border are locally unique and stable. This happens when the fixed point changes from:

(1) regular attractor to flip attractor,

(2) regular attractor to regular attractor,

(3) flip attractor to regular attractor and

(4) flip attractor to flip attractor, as μ is varied through its critical value.

Scenario B: (*Supercritical period doubling border collision bifurcation*): There are two regions in the parameter space where period doubling border collision bifurcation occurs, i.e., a locally unique stable fixed point leads to an unstable fixed point, plus a locally unique attracting period two orbit. This scenario is divided into **Scenario B1** and **Scenario B2** as follows:

Scenario B1: If

$$\begin{cases} \delta_L > 0, \delta_R > 0 \\ -(1+\delta_L) < \tau_L < -2\delta_L \\ \tau_R < -(1+\delta_R) \\ \tau_R\tau_L < (1+\delta_R)(1+\delta_L) \end{cases} \tag{3.39}$$

Scenario B2: If

$$\begin{cases} \delta_L > 0, \delta_R > 0 \\ 2\delta_L < \tau_L < (1+\delta_L) \\ \tau_R < -(1+\delta_R) \\ \tau_R\tau_L > -(1-\delta_R)(1-\delta_L) \end{cases} \tag{3.40}$$

The case of negative determinants on both sides of the border

We know that if the determinants on both sides of the border are negative, then if the fixed point is stable, it is locally unique. The condition for a locally unique stable fixed point on both sides of the border is given in **Scenario C** below.

Scenario C: (*Locally unique stable fixed point on both sides of the border*). If

$$
\begin{cases}
\delta_L < 0, \delta_R < 0 \\
-(1 + \delta_L) < \tau_L < (1 + \delta_L) \\
-(1 + \delta_R) < \tau_R < (1 + \delta_R)
\end{cases}
\tag{3.41}
$$

then a locally unique stable fixed point leads to a locally unique stable fixed point as μ is increased through zero.

Scenario D: (*Supercritical border collision period doubling*). If

$$
\begin{cases}
\delta_L < 0, \delta_R < 0 \\
-(1 + \delta_L) < \tau_L < (1 + \delta_L) \\
\tau_R < -(1 + \delta_R) \\
\tau_R \tau_L < (1 + \delta_R)(1 + \delta_L) \\
\tau_R \tau_L > -(1 - \delta_R)(1 - \delta_L)
\end{cases}
\tag{3.42}
$$

then a locally unique stable fixed point to the left of the border for $\mu < 0$ crosses the border and becomes unstable and a locally unique period two orbit is born as μ is increased through zero, i.e., this is a condition for supercritical period doubling border collision with no extraneous periodic orbits.

The case of negative determinant to the left of the border and positive determinant to the right of the border

If the determinant is negative to the left of the border, i.e., $\delta_L < 0$, then the eigenvalues are real. If the determinant is positive to the right of the border, i.e., $\delta_R > 0$, then the eigenvalues are real if $\tau_R^2 > 4\delta_R$. Then, a sufficient condition for having a locally unique fixed point leading to a locally unique fixed point as μ is varied through the critical value is given as follows:

Scenario E: (*Locally unique stable fixed point on both sides of the border*). If

$$
\begin{cases}
\delta_L < 0, \delta_R > 0 \\
-(1 + \delta_L) < \tau_L < (1 + \delta_L) \\
-(1 + \delta_R) < \tau_R < (1 + \delta_R) \\
\tau_R^2 > 4\delta_R
\end{cases}
\tag{3.43}
$$

The conditions (3.43) can be divided into two cases:

Scenario E1: This occurs if

$$\begin{cases} \delta_L < 0, \delta_R > 0 \\ -(1+\delta_L) < \tau_L < 1+\delta_L \\ -(1+\delta_R) < \tau_R < -2\delta_R \end{cases} \tag{3.44}$$

Scenario E2: This occurs if

$$\begin{cases} \delta_L < 0, \delta_R > 0 \\ -(1+\delta_L) < \tau_L < 1+\delta_L \\ 2\delta_R < \tau_R < 1+\delta_R \end{cases} \tag{3.45}$$

The case of positive determinant to the left of the border and negative determinant to the right of the border

If the determinant is negative to the right of the border, i.e., $\delta_R < 0$, then the eigenvalues are real. If the determinant is positive to the left of the border, i.e., $\delta_L > 0$, then the eigenvalues are real if $\tau_L^2 > 4\delta_L$. Hence, a sufficient condition for having a locally unique fixed point leading to a locally unique fixed point as μ is varied through the critical value is given as follows:

Scenario F: (*Locally unique stable fixed point on both sides of the border*). If

$$\begin{cases} \delta_L > 0, \delta_R < 0 \\ -(1+\delta_L) < \tau_L < 1+\delta_L \\ \tau_L^2 > 4\delta_L \\ -(1+\delta_R) < \tau_R < 1+\delta_R \end{cases} \tag{3.46}$$

The conditions (3.46) can be divided into two cases:

Scenario F1: This occurs if

$$\begin{cases} \delta_L > 0, \delta_R < 0 \\ -(1+\delta_L) < \tau_L < -2\delta_L \\ -(1+\delta_R) < \tau_R < 1+\delta_R \end{cases} \tag{3.47}$$

Scenario F2: This occurs if

$$\begin{cases} \delta_L > 0, \delta_R < 0 \\ 2\delta_L < \tau_L < 1+\delta_L \\ -(1+\delta_R) < \tau_R < 1+\delta_R \end{cases} \tag{3.48}$$

In fact, there are no known conditions for the supercritical border collision period doubling that occurs without EBOs when the determinants on both sides of the border are of opposite signs.

Stable fixed point leading to stable fixed point plus extraneous periodic orbits

There are certain border collision bifurcations that, while not causing a *catastrophic collapse* of the system, may lead to undesirable system behavior [Banerjee and Grebogi (1999), Banerjee, *et al.*, (2000)], i.e., the case of stable fixed point leading to stable fixed point plus extraneous periodic orbits, that can display multiple attractor bifurcation on either side of the border or both sides of the border in addition to the stable fixed points. Typical conditions for this case are not available in the current literature. The first example is the case of stable fixed point plus period-4 attractor to stable fixed point, plus period-3 attractor as μ is increased through zero:

Example 5 *This situation is described by the following map:*

$$
l(x, y) = \begin{cases} \begin{pmatrix} 0.50 & 1 \\ -0.90 & 0 \end{pmatrix} \begin{pmatrix} x \\ y \end{pmatrix} + \begin{pmatrix} 1 \\ 0 \end{pmatrix} \mu, & \text{if } x < 0 \\[2mm] \begin{pmatrix} -1.22 & 1 \\ -0.36 & 0 \end{pmatrix} \begin{pmatrix} x \\ y \end{pmatrix} + \begin{pmatrix} 1 \\ 0 \end{pmatrix} \mu, & \text{if } x > 0, \end{cases}
\tag{3.49}
$$

Here, the fixed point for $\mu < 0$ is spirally attracting and for $\mu > 0$ is a flip attractor. The second example is given by the case of stable fixed point to stable fixed point plus period-7 attractor displayed by the following map:

$$
m(x, y) = \begin{cases} \begin{pmatrix} 1.6 & 1 \\ -0.8 & 0 \end{pmatrix} \begin{pmatrix} x \\ y \end{pmatrix} + \begin{pmatrix} 1 \\ 0 \end{pmatrix} \mu, & \text{if } x < 0 \\[2mm] \begin{pmatrix} -1.4 & 1 \\ -0.6 & 0 \end{pmatrix} \begin{pmatrix} x \\ y \end{pmatrix} + \begin{pmatrix} 1 \\ 0 \end{pmatrix} \mu, & \text{if } x > 0, \end{cases}
\tag{3.50}
$$

Here, the fixed point for $\mu < 0$ is spirally attracting and for $\mu > 0$ is also spirally attracting with opposite sense of rotation.

Regions for robust chaos: undesirable and dangerous bifurcations

Definition 46 *The dangerous bifurcations considered begin with a system operating at a stable fixed point on one side of the border, say the left side.*

The main dangerous bifurcations that can result from border collision are:
Border collision pair bifurcation: This occurs if

$$
\begin{cases} -(1 + \delta_L) < \tau_L < 1 + \delta_L \\[2mm] \tau_R > (1 + \delta_R) \end{cases}
\tag{3.51}
$$

where a stable fixed point and an unstable fixed point merge and disappear as μ is increased through zero. This is analogous to saddle node bifurcations in smooth maps. In this case, the system trajectory diverges for positive values of μ since no local attractors exist.

Subcritical border collision period doubling: This occurs if

$$\begin{cases} -(1+\delta_L) < \tau_L < 1+\delta_L \\ \\ \tau_R < -(1+\delta_R) \\ \\ \tau_R\tau_L > (1+\delta_R)(1+\delta_L). \end{cases} \tag{3.52}$$

where a bifurcation from a stable fixed point and an unstable period-2 orbit to the left of border to an unstable fixed point to the right of the border occurs as μ is increased through zero.

Supercritical border collision period doubling

This is not a *dangerous bifurcation*, but it is undesirable in some applications such as the one called *cardiac conduction model* proposed in [Sun, et al., (1996)] as a two-dimensional PWS map in which the atrial His interval A is that between cardiac impulse excitation of the lower interatrial septum to the Bundle of His.

Example 6 *This example is given by:*

$$\begin{pmatrix} A_{n+1} \\ B_{n+1} \end{pmatrix} = f\left(A_n, R_n, H_n\right) \tag{3.53}$$

where

$$f\left(A_n, R_n, H_n\right) = \begin{cases} \begin{cases} A_{\min} + D_n, \text{ if } A_n < 130 \\ \\ A_{\min} + C_n, \text{ if } A_n > 130 \end{cases} \\ \\ R_n = \exp\left(-\frac{(H_n+A_n)}{\tau_{fat}}\right) + \gamma\exp\left(-\frac{H_n}{\tau_{fat}}\right) \\ \\ D_n = R_{n+1} + (201 - 0.7A_n)\exp\left(-\frac{H_n}{\tau_{rec}}\right) \\ \\ C_n = R_{n+1} + (500 - 3.0A_n)\exp\left(-\frac{H_n}{\tau_{rec}}\right) \end{cases} \tag{3.54}$$

where $R_0 = \gamma\exp\left(-\frac{H_n}{\tau_{fat}}\right)$, H_0 is the initial H interval, the parameters $A_{\min}$, τ_{fat}, γ and τ_{rec} are positive constants. The variable H_n (usually is the bifurcation parameter) is the interval between bundle of His activation and the subsequent activation (the AV nodal recovery time). The variable R_n (usually is the bifurcation parameter) is a drift in the nodal conduction time. For $\tau_{rec} = 70ms$, $\tau_{fat} = 30000ms$, $A_{\min} = 33ms$, $\gamma = 0.3ms$ a stable period-2 orbit is born after the border collision.

Stable fixed point leading to chaos: Or *instant chaos* where chaotic behaviors are developed following border collision [Ohnishi and Inaba (1994)], i.e., the behavior of this system changes directly to chaos (without period doubling bifurcation or an intermittency) when a limit cycle loses its stability.

Regions for robust chaos: It was shown in [Banergee, *et al.*, (1998-1999)] that the robust chaos occurs in the map (3.3) if the following cases holds true:

(1)

$$\begin{cases} \tau_L > 1 + \delta_L, \text{ and } \tau_R < -(1 + \delta_R) \\ \\ 0 < \delta_L < 1, \text{ and } 0 < \delta_R < 1, \end{cases} \tag{3.55}$$

and the parameter range for boundary crisis is given by:

$$\delta_L \tau_L \lambda_{1L} - \delta_L \lambda_{1L} \lambda_{2L} + \delta_R \lambda_{2L} - \delta_L \tau_R + \delta_L \tau_L - \delta_L - \lambda_{1L} \delta_L > 0 \tag{3.56}$$

where the inequality (3.56) determines the condition for stability of the chaotic attractor. The robust chaotic orbit continues to exist as τ_L is reduced below $1 + \delta_L$.

(2)

$$\begin{cases} \tau_L > 1 + \delta_L, \text{ and } \tau_R < -(1 + \delta_R) \\ \\ \delta_L < 0, \text{ and } -1 < \delta_R < 0 \\ \\ \frac{\lambda_{1L} - 1}{\tau_L - 1 - \delta_L} > \frac{\lambda_{2R} - 1}{\tau_R - 1 - \delta_R}, \end{cases} \tag{3.57}$$

The condition for stability of the chaotic attractor is also determined by (3.56). However, if the third condition of (3.57) is not satisfied, then the condition for existence of the chaotic attractor changes to:

$$\frac{\lambda_{2R} - 1}{\tau_R - 1 - \delta_L} < \frac{(\tau_L - \delta_L - \lambda_{2L})}{(\tau_L - 1 - \delta_L)(\lambda_{2L} - \tau_R)} \tag{3.58}$$

(3) The remaining ranges for the quantity $\tau_{L,R}, \delta_{L,R}$ can be determined in some cases using the same logic as in the above two cases, or there is no analytic condition for a boundary crisis, and it has to be determined numerically.

Chapter 4

Robust Chaos in Neural Networks Models

4.1   Introduction

Chaotic dynamical systems display two kinds of chaotic attractors: One type has *fragile chaos* (the attractors disappear with perturbations of a parameter or coexist with other attractors), and the other type has *robust chaos* defined in [Banerjee, *et al.*, (1998)]: *Robust chaos is defined by the absence of periodic windows and coexisting attractors in some neighborhood in the parameter space.* The existence of these windows in some chaotic regions means that small changes of the parameters would destroy the chaos, implying the fragility of this type of chaos. In [Caroppo, *et al.*, (1999)] the dynamic behavior of attractor neural networks with non-monotonic transfer function was analyzed using mean-field equations whose macroscopic dynamics can be analytically calculated in which the time evolution of the macroscopic parameters describing the system is determined by a chaotic two-dimensional (2-D) map. Structure and hyperbolicity of the resulting strange attractor are studied using Hausdorff dimension and Lyapunov exponents. The fragility of chaos in this case, can be seen from the fact that the present model behaves in agreement with the conjecture of Barreto-Hunt-Grebogi-Yorke claiming that robust chaos cannot appear in smooth systems [Barreto, *et al.*, (1997)]. Also, fully connected networks with non-monotonic activation function and non-monotonic transfer functions displays robust chaos in time series generated by feed-forward neural networks, as shown in [Laughton & Coolen (1994)] and [Priel & Kanter (2000)] respectively.

This chapter is a continuation of the work given in [Zeraoulia & Sprott (2011(g))] where the details of this particular topic are not discussed enough. In this chapter, we discuss robust chaos in neural networks with many illustrating examples and methods. Indeed, in Sec. 4.2, we give an overview about chaos in the field of neural networks with some suggested applications. Sec. 4.3 deals with the method of weight-space exploration that can be used to predict robust chaos in neural networks. In Sec. 4.3.1, we give an example of robust chaos generated by a neural network with a smooth activation function. While in Sec. 4.3.2, we give an example of fragile chaos generated by a neural network with a smooth activation function. In Sec. 4.3.3 we present and discuss robust chaos in the case of blocks with non-smooth activation function. In Sec. 4.3.4 we present and discuss robust chaos in the *electroencephalogram model* by using numerical calculations. As a very important result, we present in Sec. 4.3.5 an example of robust chaos in *Diluted circulant networks* by using the largest Lyapunov exponent as a signature for chaos. In Sec. 4.3.6, we discuss robust chaos in non-smooth neural networks. Finally, in Sec. 4.4 a collection of open problems is given concerning some aspects relating mathematics to the notion of robust chaos.

4.2 Chaos in neural networks models

Chaos in neural networks models has a long history. The pioneering work in [Sompolinsky, *et al.*, (1988)] shows that a continuous-time dynamic model of a network of N non-linear elements interacting via random asymmetric couplings displays chaotic phase occurring at a *critical value* of the gain parameter. Following this work, the occurrence of chaos in asymmetric synapses was examined in a number of subsequent papers, for example in [Tirozzi & Tsokysk (1991)], the critical capacity and the transition point to chaos were found using the theory of chaotic regimes for asymmetric networks to the case of highly diluted neural networks with Hebb learning rule. In [Cessac, *et al.*, (1994)] an investigation of the dynamical behavior of neural networks with asymmetric synaptic weights, in the presence of random thresholds show chaos obtained via a *quasi-periodicity route to chaos*. It was shown in [Molgedey, *et al.* (1992)] by using Lyapunov exponents calculations that chaos can occur in discrete parallel dynamics of a fully connected network of non-linear elements interacting via long-range random asymmetric couplings under the influence of external noise. In [Laughton & Coolen (1994)] bifurcation phenomena were studied for a separable, stochastic neural network, which becomes deterministic in the thermodynamic limit. In particular, quasi-periodic solutions displaying mode locking in the discrete time case. The occurrence of oscillations and chaos in the frame of neural networks were studied in [Amit (1989), Hertz, *et al.*, (1990)]. Neural networks with symmetric synaptic connections were studied in [Fischer & Hertz (1991)] using methods closely related to those used in the theoretical description of the *spin glasses* because they admit an *energy function*.

Neural networks have several applications in the real world. The brain, for example, is a highly dynamic system with a rich temporal structure of neural processes [Eckhorn, *et al.*, (1988), Charles, *et al.*, (1989)., Steriade, *et al.*, (1993)]. Chaotic behaviour can also be found in the nervous system, which implies that chaos is useful in the comprehension of *cognitive processes* in the brain [Skarda & Freeman (1987)]. Also, real life applications include: *function approximation*, including time series prediction and modeling, *clas-*

sification, including pattern and sequence recognition, novelty detection and sequential decision making and *data processing*, including filtering, clustering, blind signal separation and compression.

Application areas of ANNs include the following domains:

- system identification and control (vehicle control, process control).

- game-playing and decision making (backgammon, chess, racing).

- pattern recognition (radar systems, face identification, object recognition, etc.)

- sequence recognition (gesture, speech, handwritten text recognition).

- medical diagnosis, financial applications, data mining (or knowledge discovery in databases).

- visualization and e-mail spam filtering.

4.3   Robust chaos in discrete time neural networks

In [Dogaru, *et al.*, (1996)] the method of weight-space exploration is essentially based on two concepts. The first is the concept of *running*, which is defined as a mapping from the high-dimensional space of the neural network structure (defined by the interconnection weights, the initial conditions, and the non-linear activation function) to one or two scalar values (called *dynamic descriptors*) giving the essential information about the dynamic behavior of the network, as obtained by running it on a manageable enough large time period. The second concept is called *descriptor map*, in which some of the interconnection weights are selected as variable parameters by splitting them into two groups corresponding with two parameters p_0 and p_1 in a specified domain, using a finite specified resolution (res), i.e., $res = 32$ which corresponds to a population of $res^2 = 1024$ neural networks with the property that all individuals share the same *generic structure* with some differences in view point of values of the parameter weights.

The method of analysis in [Dogaru, *et al.*, (1996)] is based on the computation of a descriptor map for each population, i.e., either a colored pixel (neural network individual) in the descriptor map is assigned with the spatial position given by the coordinates $x, y \in \{0, res - 1\}$, where the 0 values corresponding with the upper left corner of the map, or brightness proportional with the value of the dynamic descriptor obtained by applying to it the running procedure. The structure of each pixel is established by taking

$$p_0 = p_0^{\min} + \frac{y \left(p_0^{\max} - p_0^{\min} \right)}{res} \qquad (4.1)$$

We note that it is possible to define a similar relation between the horizontal position x and the other parameter just like relation (4.1) that indicates that the descriptor map have no significant structure differences for neighboring individuals. This fact implies robustness of dynamic descriptors, i.e., when large compact zones having the same brightness (color) or bifurcations as crisp borders between compact zones in the descriptors maps. In this case, a calibration pattern is down displayed for each map, having the left sided color or gray level corresponding to the minimum value of the dynamic descriptor and the right side color to the maximum value of the same dynamic descriptor described by the following equation

$$X(t) \longleftarrow W.Y(t) + I, t = T_i, ..., T_i + T_a - 1 \qquad (4.2)$$

where N is the number of neurons and X, Y and I are N-dimensional vectors indicating respectively the state, the output and the input of the network. The component "s" of the vector Y is the *activation function* of the "source" "s" neuron applied to it's state at the previous discrete time moment, i.e.,

$$y_s(t) = f(x_s(t-1)) \qquad (4.3)$$

in which the dynamics of (4.2) are performed in the interval $[0, T_i]$ without computing the dynamic descriptors (who are evaluated only during the analysis period T_a during a running procedure) in order to avoid transitory effects. The $N \times N$ matrix W is composed by the weights (some of them are variable parameters) which connect the source and target neurons, i.e.,

$$W = \{w_{st}\}, s, t \in \{0, N-1\} \qquad (4.4)$$

Now, the *entropic descriptor* (which gives information about the disorder in the neural network) used in [Dogaru, et al., (1996)] is essentially based on the work of Fraser in 1989. While the number of distinct states (*nrds*) of the system during the analysis period was considered in [Dogaru & Murgan (1994)] as a entropic descriptor. In this case, fixed points are characterized by $nrds = 1$ and large $nrds$ values characterize large period limit cycles. On the other hand, the evaluation of the *degree of chaos* is done using the sensitivity descriptor which is based on evaluating the Lyapunov exponents $\lambda_i, i \in \{0, N-1\}$ of the dynamic system (4.2) defined in [Kanou & Horio (1995)] by:

$$\lambda_i = \frac{1}{delt} \lim_{T_a \to \infty, \Delta x \to 0} \frac{1}{T_a} \sum_{t=T_i}^{T_i+T_a-1} \ln \left| \frac{x_i(t+delt) - xp_i(t+delt)}{\Delta x} \right| \qquad (4.5)$$

where $x_i(t+delt)$ is the not perturbed state of the neuron i and $xp_i(t+delt)$ is the perturbed state at the moment $t+delt$, when the status at the moment t was perturbed with the small amount Δx. From [Kanou & Horio (1995)], it is known that system (4.2) is chaotic if $\lambda_i > 0$ and can display different chaotic attractors with a very small amount Δx on direction i, after a period *delt* or it being structurally stable and, thus, system (4.2) evolving on the same attractor as in the not perturbed case.

The sensitivity for all directions is characterized by the following sensitivity descriptor (sd) given by:

$$sd = \sum_{i=0}^{N-1} \sigma\left(\lambda_i\right), \text{ where } \sigma\left(x\right) = \begin{cases} 0, \text{ if } x \le 0 \\ \\ 1, \text{ if } x > 0 \end{cases} \qquad (4.6)$$

in which $sd > 0$ for chaotic dynamics with the remark that, generally, networks with large sd values have wider attractors distributed in the whole state space.

The strategy used in [Dogaru, *et al.*, (1996)] for finding robust chaotic neural networks is based on the six following steps:

Step 1: Choose the generic structure, i.e., neural network dimension, nonlinear activation function and the groups of parameter weights.

Step 2: Specify for each parameter the resolution (the size of the population of neural networks) and the variation domains.

Step 3: Evaluate the entropic and the sensitivity maps.

Step 4: Select a specific network as a pixel in the descriptor maps, according with the desired behavior, i.e., the network must be *hyperchaotic* (two positive Lyapunov exponents).

Step 5: (Synthesis) Compute the parameter weights based on the coordinates of the pixel associated with the network.

Step 6: Evaluate the dynamics or/and use the chaotic neural network individual synthesized in **Step 5**.

Note that steps 1 to 5 can be repeated (this strategy was successfully applied in [Dogaru, *et al.*, (1996)]) after *freezing* the last group of parameter weights to those values, giving the best fitness with the desired behavior. The use of different non-linear activation functions with different numbers of neurons generates many descriptor maps displaying robust chaos, while monotone increasing functions are not suitable for generating robust chaos.

Example 7 *As an example, the saturated modulus (sat-mod) or saturated rectifier function used in electronic instrumentation and given by:*

$$f\left(x\right) = \begin{cases} |x|, \text{ if } |x| < 1 \\ \\ 1, \text{ if } |x| \ge 1 \end{cases} \qquad (4.7)$$

displays robust chaos in system (4.2), i.e., the sd values are almost the same for the same non-linearity instead of choosing different groups of parameter weights for 9 neurons with sigmoid function $f(x) = \tanh 2x$ (with very few individuals with chaotic behavior) and 3 neurons with the function (4.7) (without chaotic behavior).

The interconnection matrix W given by

$$\begin{cases} W = \begin{pmatrix} 1 & p_0 & 0 \\ p_1 & 1 & p_0 \\ 0 & p_1 & 1 \end{pmatrix}, I = \begin{pmatrix} -0.5 \\ -0.5 \\ -0.5 \end{pmatrix} \\ p_0^{\min} = p_1^{\min} = -2, p_0^{\max} = p_1^{\max} = 2, res = 32 \end{cases} \quad (4.8)$$

corresponds with a 2-D cellular neural network structure, where each neuron is connected with its four topologic neighbors (left, right, up and down) by the same set of synaptic weights in which for most of the pairs (p_0, p_1) periodic signals with high entropy are generated. In this case, the sensitivity to initial conditions is very low, which indicates the small values of the *sd* defined by (4.6) representing respectively the entropic and sensitivity of the resulted maps for the non-monotonic saturated *rectifier* obtained for a three neurons generic structure characterized by high entropy, i.e., more than 70% of chaotic individuals having a high sensitivity, that is, $sd = 3$. On the other hand, for $p_0 = -1, 75$ and $p_1 = 0.25$, a well-defined robust chaotic neural network shows the *chaoticity* of all state variables in which the associated signals have a power spectral density which is exponentially proportional to the frequency.

4.3.1 Robust chaos in 1-D piecewise-smooth neural networks

As an example of the occurrence of robust chaos in 1-D piecewise-smooth maps, the following model of networks of neurons with the activation function:

$$f(x) = |\tanh s(x - c)| \quad (4.9)$$

is studied in [Potapov & Ali (2000)], where it was shown that in a certain range of s and c the dynamical system

$$x_{k+1} = |\tanh s(x_k - c)| \quad (4.10)$$

cannot have stable periodic solutions, which prove the robustness of chaos. In fact, robust chaos can occur in a class of artificial neural networks (ANN) that are the discrete mappings $f : \mathbb{R}^n \to \mathbb{R}^n$ given by

$$\hat{x}_i = f\left(\sum_{j=1}^{n} w_{ij} x_j + b_i\right), i = 1, ..., n. \quad (4.11)$$

where f is the activation function and $x_i = x_i(t)$, $\hat{x}_i = x_i(t+1)$ with *recurrent* networks (otherwise, the dynamics are *trivial*), i.e., the variables x_i cannot be renumbered such that the matrix w becomes lower triangle, that is, $w_{ij} = 0$ for $j > i$. The properties of map (4.11) essentially depend on f. In most cases, the activation function f has the so called *sigmoid* form[1], for

---

[1]Monotonous function that tends to a constant value as $x \to \pm\infty$.

example $f(x) = \tanh x$ or $f(x) = (1 + e^{-x})^{-1}$. In some case *non-monotonous* functions have been used in [Morita (1993), Tan, et al., (1998)] and they represent a block of *sigmoid* neurons that can generate robust chaos. In other cases, a sum of several activation functions can be used in which *complex neurons* are obtained with two or more simple non-linear elements acting simultaneously as follows:

$$\hat{x}_i = f_1\left(\sum_{j=1}^{n} w_{ij}x_j + b_i\right) + f_2\left(\sum_{j=1}^{n} w_{ij}x_j + b_i\right), i = 1, ..., n. \qquad (4.12)$$

If map (4.12) (with standard sigmoid neurons) displays robust chaos, then *chaotic blocks* can be obtained along a robust network of required complexity. In fact, numerical calculations show the fragility of chaos in them, which correlates with the Barreto-Hunt-Grebogi-Yorke conjecture [Barreto, et al., (1997)]. However, robust chaos can occur with the non-monotonous and non-smooth functions of the form

$$\begin{cases} f(x) = |\tanh s(x - c)| \\[2mm] f(x) = \frac{s|x-c|}{(1+s|x-c|)} \end{cases} \qquad (4.13)$$

as shown in [Potapov & Ali (2000)], in which both functions can be obtained with the help of equation (4.13) or block building technique from the non-smooth *gated* sigmoid functions presented in [Carpenter & Grossberg (1991)], i.e., $f^+(x) = \max(0, \tanh(x))$.

For a certain range of s and c values (probably in a wider interval of parameter values) in (4.13) chaos is robust, because the resulting chaotic attractor satisfies $\left|\frac{df}{dx}\right| > 1$. Also, for s not too large, even when $\left|\frac{df}{dx}\right| < 1$ (a part of the attractor expands to the domain), the Lyapunov exponent decreases gradually to zero, without *abrupt falls*, which confirms that chaos in a network of *weakly* connected robust chaotic neurons is also robust.

4.3.2   Fragile chaos (blocks with smooth activation function)

On one hand, combinations of sigmoid-type neurons are equivalent to a *complex* neuron with non-monotonous function, in which robust chaos occurs. On the other hand, chaos in (4.12) is not very typical and it is more common (robust) if f is non-monotonous as shown in [Shuai, et al., (1997)]. In [Potapov & Ali (2000)] fragile chaos in network of type (4.12) can be obtained using the usual sigmoid f, i.e., in this case, it is easy to leave the domain of chaos by changing only one of w_{ij}.

The first example of such a situation is when $f(x) = \tanh(x)$ in which chaos can arise in a network of 3 sigmoid neurons [Barton (1995)]. This case is possible with the following function:

$$g(x) = \tanh s(x - c_1) - \tanh s(x - c_2) \qquad (4.14)$$

that resembles the well known logistic map as an example of a smooth chaotic system. The same construction can arise in a usual 2-neuron network with special choice of connection matrix w, i.e., a network equations of the form

$$\hat{x} = f(wx + b), \ \hat{x}, x, b \in \mathbb{R}^2. \tag{4.15}$$

where $\hat{x}$ stands for x at the next time step. If T is a 2×2 matrix, the change of variable $y = Tx$, transform system (4.12) to the following map

$$\hat{y} = T\hat{x} = Tf(wx + b) = Tf(wT^{-1}y + b) \tag{4.16}$$

For example, if

$$T = \frac{1}{2}\begin{pmatrix} 1 & -1 \\ 1 & 1 \end{pmatrix}, wT^{-1} = \begin{pmatrix} s & 0 \\ s & 0 \end{pmatrix}, w = \frac{1}{2}\begin{pmatrix} s & -s \\ s & -s \end{pmatrix}, \tag{4.17}$$

then (4.12) becomes

$$\hat{y}_1 = \frac{1}{2}(f(sy_1 + b_1) - f(sy_1 + b_2)) \tag{4.18}$$

$$\hat{y}_2 = \frac{1}{2}(f(sy_1 + b_1) + f(sy_1 + b_2)) \tag{4.19}$$

or, in original variables,

$$\hat{x}_1 = f\left(\frac{s}{2}x_1 - \frac{s}{2}x_2 + b_1\right) \tag{4.20}$$

$$\hat{x}_2 = f\left(\frac{s}{2}x_1 - \frac{s}{2}x_2 + b_2\right) \tag{4.21}$$

In this case, a one dimensional dynamical system (in the variable y_1 while y_2 is its function) is obtained because the matrix w is singular. If the matrix w is non-singular, then the same dynamics can be also obtained if the second eigenvalue is a small non-zero number ϵ. Thus, the block can be written as a one dimensional map of the form

$$\hat{x} = \tanh s(x - c + d) - \tanh s(x - c - d) \tag{4.22}$$

where

$$\begin{cases} c = \frac{-(b_1 + b_2)}{2s} \\ \\ d = \frac{b_1 - b_2}{2s} \end{cases} \tag{4.23}$$

Numerical simulations given in [Potapov & Ali (2000)] show the fragility of chaos, because for example if $s = 6$, $d = 0.1$ and c is the bifurcation parameter, then some periodic windows occur in the ranges of c where chaos seems persistent, i.e., numerical experiments indicate the presence of stable periodic solutions among chaotic ones. The method of analysis is done by plotting a bifurcation diagram and Lyapunov exponent for this map. Similar dynamics are obtained with the variation of s or d.

The second example of a such situation is a block of coupled oscillatory neurons which is not related directly to the logistic map. This model is obtained using the *oscillating neurons*. That is,

$$\begin{cases} \hat{x} = f\left(s\left(x - c\right)\right), s > 1 \\ \\ f\left(x\right) = \tanh x \end{cases} \tag{4.24}$$

Note that system (4.24) always has a fixed point, the solution of equation

$$x_* = \tanh\left(s_*\left(x - c\right)\right) \tag{4.25}$$

The derivative at this point is given by

$$\frac{df}{dx}|_{x=x_*} = \frac{s}{\cosh^2 s\left(c - x_*\right)} = -s\left(1 - \tanh^2 s\left(c - x_*\right)\right) = -s\left(1 - x_*^2\right) \tag{4.26}$$

It was proved in [Pasemann (1997)] that when $s\left(1 - x_*^2\right) > 1$, this fixed point of map (4.24) becomes unstable and a stable period-2 cycle arises. To achieve maximum *versatility of coupling*, the value of s was chosen (using the analysis of bifurcation points) for which oscillations in a single neuron are observed in the widest range of c considered as a bifurcation parameter when s is being fixed. Thus, the bifurcation points c_b can be obtained from the equation $|f'\left(x_*\right)| = 1$ or

$$\cosh s\left(c_b - x_*\right) = \sqrt{s} \tag{4.27}$$

i.e.,

$$x_* = \tanh s\left(c_b - x_*\right) = \pm\sqrt{\frac{s-1}{s}} \tag{4.28}$$

From (4.27) it follows that

$$\begin{cases} s\left(c_b - x_*\right) = \ln\left(\sqrt{s} \pm \sqrt{s-1}\right) = \pm sr \\ \\ r = s^{-1}\ln\left(\sqrt{s} + \sqrt{s-1}\right) \\ \\ \Delta c = c_{b+} - c_{b-} = 2\left(\sqrt{\frac{s-1}{s}} + \frac{1}{s}\ln\left(\sqrt{s} + \sqrt{s-1}\right)\right) \end{cases} \tag{4.29}$$

In this case, there is s value corresponding to the largest Δc, which satisfy $\frac{d\Delta c}{ds} = 0$ or

$$\sqrt{\frac{s-1}{s}} = \ln\left(\sqrt{s} + \sqrt{s-1}\right) \tag{4.30}$$

Numerical solution of (4.13) gives $s_0 = s \approx 3.2767$ with $\Delta c \approx \left(s\left(s-1\right)\right)^{-\frac{1}{2}} \approx 2.3994$. The value s_0 was used in the coupled system given by

$$\hat{x}_1 = f\left(c_1 - sx_1 + ax_2\right) \tag{4.31}$$

$$\hat{x}_2 = f\left(c_2 - ax_1 - sx_2\right) \tag{4.32}$$

Similar analysis as before confirms the fragility of chaos for some choices of c_i and a.

4.3.3 Robust chaos (blocks with non-smooth activation function)

For higher-dimensional networks, it is easy to find chaos if the matrix w is chosen randomly for large network size n. However, it is hard to find a high-dimensional (robust) attractor with random choice, because the largest dimension of randomly found attractor was about $\frac{n}{3}$ for $n > 5$ and the number of positive Lyapunov exponents also did not exceed $\sim \frac{n}{3}$, while the number of parameters was $\sim n^2$ which implies the fragility of chaos due to the Barreto-Hunt-Grebogi-Yorke conjecture [Barreto, *et al.*, (1997)]. In this case, the use of block with effective non-monotonous activation function (to diminish the number of parameters) is a good step for finding robust chaos. But, it seems that the non-robustness of chaos is a result of the smoothness (at least differentiable) of the function f. In [Potapov & Ali (2000)] the robustness of chaos was proved in neurons with the properties close to the *tent map* where the function f is non-differentiable in one point. The method of analysis is based on the fact that the tent map satisfies $\left| \frac{df}{dx} \right| > 1$, and therefore there is no stabilizing spine locus just like in the logistic map or *logistic network*. Also, the tent map is non-differentiable at its vertex, where it behaves like $|x|$, and the Barreto-Hunt-Grebogi-Yorke conjecture [Barreto, *et al.*, (1997)] does not hold for it.

The first example of this situation is obtained using an activation function of the form (4.13) which displays two *wings* which meet at $x = c$, where $f = 0$, i.e., *non-smooth neuron* and the *pincers map*. Two kinds of such tent neurons displays robust chaos for a range of s and c, i.e., chaos can exist only for $s > 1$ and $c > 0$.

Like in the case of the logistic block (4.20)–(4.21), the transfer function can be implemented as a two-neuron network with singular connection matrix w and the non-smooth sigmoid activation function given by

$$f^+ (x) = \begin{cases} \tanh (x), x \geq 0 \\ \\ 0, x < 0 \end{cases} \tag{4.33}$$

These types of functions have been used in [Carpenter & Grossberg (1991)], that is,

$$f^+ (x) = f^+ (s (x - c)) + f^+ (s (c - x)) \tag{4.34}$$

Using the activation function (4.33), the so called *pincers map* [Potapov & Ali (2000)] can be written as follow:

$$\hat{x} = |\tanh s (x - c)| \tag{4.35}$$

Numerical experiment confirms the robustness of chaos due to the *visibly continuous* dependence of the Lyapunov exponent $\lambda (c)$ in the domain where $\lambda > 0$ within a certain c interval and the fact that $\left| \frac{df}{dx} \right| > 1$ everywhere on the attractor and stable periodic solutions cannot exist, i.e., the domain $x_{r_1} = c - r < x < x_{r_2} = c + r$, where r is defined in (4.29) and the points $x = x_{r_i}$, $i = 1, 2$ are obtained from the equation $\left| \frac{df}{dx} (x_{ri}) \right| = 1$, that is, $\cosh s (x_r - c) = \sqrt{s}$ coincides with (4.27).

For

$$
\begin{cases}
s < 2 \\[2mm]
c \leq r \\[2mm]
c > \sqrt{\frac{s-1}{s}} - r
\end{cases}
\tag{4.36}
$$

the chaotic attractor of map (4.35) is strictly robust. Indeed, the map (4.35) maps the interval $[0, 1]$ into itself, but there exists a smaller subinterval, which is also mapped into itself. The map (4.35) takes its smallest value $f = 0$ at $x = c$ which mapped to $f(0) = \tanh sc$. If

$$
f(f(0)) \leq f(0)
\tag{4.37}
$$

then this interval maps onto itself since $f(x) \leq f(0)$ for all $x \in [0, f(0)]$. The symmetry of (4.35) with respect to the point $x = c$ implies that the inequality in (4.37) become equality when $f(0) = 2c$, then $f(2c) = f(0) = 2c$ and the condition for strict robustness is given by

$$
\tanh sc \leq 2c
\tag{4.38}
$$

Hence, inequality (4.38) always holds if (4.36) is verified. For larger s, the first two terms in the expansion of $\tanh x$, are used to obtain

$$
\begin{cases}
sc - \frac{1}{3} (sc)^3 \leq \tanh sc \leq 2c, \\[2mm]
c \geq c_{*1} = \frac{1}{s} \sqrt{\frac{3(s-2)}{s}}
\end{cases}
\tag{4.39}
$$

in which for $s \leq 2$ chaos begins at $c = 0$, while for $s > 2$ it begins at $c > 0$. The second condition of (4.13) is obtained from the inequality $|f'| > 1$ everywhere within the attractor. Indeed, on the interval $[0, f(0)]$, the quantity $|f'|$ takes its lowest value at $x = 0$, i.e., the point $x = 0$ must falls within the interval $[x_{r_1}, x_{r_2}]$, that is,

$$
c \leq r
\tag{4.40}
$$

Note that conditions (4.39) and the second one of (4.13) do not contradict each other because $\ln \left(\sqrt{s} + \sqrt{s - 1} \right) \geq \sqrt{\frac{3(s-2)}{s}}$.

The third condition of (4.36) is obtained by avoiding the problem that for small c values, the chaotic attractor can coexist with a stable fixed point. Indeed, this fixed point disappears when the *right wing* of the map is tangential to the line $f = x$. Thus, $x = x_{r_2}$. This gives the relations

$$
x_{r_2} = c + r = f(c + r) = \tanh sr = \frac{\sinh sr}{\cosh sr} = \sqrt{\frac{s-1}{s}}
\tag{4.41}
$$

The length of c interval, where chaos is robust is given by:

$$\Delta c = 2r - \sqrt{\frac{s-1}{s}} \qquad (4.42)$$

with its largest value $s \approx 1.30$, that is $\Delta c \approx 1.32$ ($c_{min} = 0.077, c_{max} = 0.402$), i.e., when $\frac{d(\Delta c)}{ds} = 0$. Relation (4.13) shows that if c grows, $r \to 0$, and at some $s = s_u$ ($s_u \approx 2.73$) chaos loses its robustness. On the other hand, numerical calculations show that chaos is robust on the larger c interval and the Lyapunov exponent gradually tends to 0 as c increases. The reason for this is that the attractor gradually extends to the domain, where $|f'| < 1$, but the fact that $|f'(x)| > |f'(0)| = s \cosh^2(sc)$ implies that the mean value $(\ln |f'(x)|)$ still remains positive. An estimate of the largest possible c value[2] is given by

$$c_{max} = s^{-1} \ln\left(\sqrt{s} + \sqrt{s^2 - 1}\right) \qquad (4.43)$$

based on the assumption that chaos disappears close to the c value where $|f'(c)| |f'(0)| \approx 1$ which gives the condition $\cosh(sc) = s$ or (4.13). Another estimate of c_{max} can be obtained from the fact that chaos cannot exist when the image of 0 is less than c, i.e., the system behaves like a neuron with monotonous activation function with a period-2 oscillation as the most complex behavior when $\tanh(sc) \leq c$.

The second example of this situation is obtained using *hyperbolic activation function* instead of hyperbolic tangent

$$\begin{cases} \hat{x} = f(x) \\ f(x) = \frac{s|x-c|}{(1+s|x-c|)} \end{cases} \qquad (4.44)$$

Qualitatively, the map (4.44) behaves like (4.13), though all estimates are different (the bifurcations of chaos disappearance are simpler) because the use of this function strongly reduces the amount of computations, with a wider maximal interval of guaranteed chaos robustness as $\Delta c = 0.125$ at $s = \frac{4}{3}$.

In the rest of this section, we will discuss robust chaotic networks. Indeed, robust chaotic neuron permits the organization of a robust chaotic network of the form

$$\hat{x}_i = \left| \tanh s \left(x_i + a \sum_{j=1}^{n} w_{ij} x_i - c \right) \right|, w_{ii} = 0, |w_{ij}| < 1, a \geq 0, i = 1, ..., n \qquad (4.45)$$

---

[2]This value is very close to the point of chaos disappearance, or, for greater $s > 2.5$, to the point of appearing of windows of periodicity.

or as a vector system

$$\begin{cases} \hat{X} = F(X, a) \\ X = (x_1, x_2, ..., x_n) \\ \hat{X} = (\hat{x}_1, \hat{x}_2, ..., \hat{x}_n) \\ F(X, a) = \left| \tanh s \left(x_i + a \sum_{j=1}^{n} w_{ij} x_i - c \right) \right| \end{cases} \qquad (4.46)$$

where $c = r - \frac{1}{2} \Delta c$ is in the middle of the guaranteed robustness interval. For $a = 0$, equation (4.13) represents n independent chaotic maps, and its matrix of derivatives $DF = diag(DF_{ii})$ with all $|DF_{ii}| > \alpha(c) > 1$ where $\alpha(c)$ is given by

$$\alpha(c) = |f'(0)| = \frac{s}{\cosh^2(sc)} \qquad (4.47)$$

If we take into account that $|x_i| \le 1$, let $a > 0$ but small enough, such that inequality

$$g = \left| \sum_{j=1}^{n} w_{ij} x_i \right| \le a \sum_{j=1}^{n} |w_{ij}| \le k \qquad (4.48)$$

holds. Obviously, the following relations for the matrix of derivatives DF holds true

$$\begin{cases} |DF_{ij}| = \left| (\tanh s (x - c))' \right| . a \, |w_{ij}| \le sa \, |w_{ij}|, i \ne j \\ |DF_{ii}| = \left| ((\tanh s (x - c)))' \left(x + a \sum_{j=1}^{n} w_{ij} x_j \right) \right| \ge \alpha(c + g) \ge \alpha(c + k) \end{cases} \qquad (4.49)$$

(the greater k, the smaller is $\left| (\tanh s (x - c))'(0) \right|$). Thus, all eigenvalues μ_k of DF belong to the circles on the complex plane with the centers at $P_i = DF_{ii}$ and radii $R_i = \sum_{j \ne i} |DF_{ij}|$. This fact is a result of the Gershgorin circles theorem[3]. Hence, one has

$$|\mu_k| \ge \min_i \left\{ |DF_{ii}| - \sum_{j \ne i} |DF_{ij}| \right\} \ge \alpha(c + k) - sk \qquad (4.50)$$

and if $\alpha(c + k) > 1 + sk$ then $|\mu_k| > 1$. Using the expression for $\alpha(c)$ given by (4.47), we can write it as follows:

$$\cosh s(c + k) < \sqrt{\frac{s}{1 + sk}} \qquad (4.51)$$

---

[3] This theorem is used to bound the spectrum of a square matrix.

For $k = 0$, inequality (4.51) holds because of the choice of c. For $k < 0$, we have the exponentially growing function of k, while for $k > 0$ this function is decreasing. Hence, at some $k = k_{max} > 0$ inequality (4.51) holds. On the other hand, the inequality $\left| (\tanh s\,(x - c))'(0) \right| > 1$ gives the rough upper bound $k < k'_{max} = r - c$. Linear approximation of (4.51) gives the lower bound for k_{max}, i.e., To avoid overestimation, we linearize the right hand side at $k = 0$, and take the estimates of the derivative at $k = k_{max}$ for the left hand side, and for the latter we take its upper bound k'_{max}. Thus,

$$\left\{ \begin{array}{c} \cosh sc + sk_{max} \sinh s\,(c + k'_{max}) = \sqrt{s}\left(1 - \frac{sk_{max}}{2}\right) \\[2mm] c + k'_{max} = r \\[2mm] \sinh s = \sqrt{s - 1} \end{array} \right. \qquad (4.52)$$

implies that

$$k_{max} = \frac{1}{s}\frac{\sqrt{s} - \cosh sc}{\sqrt{s - 1} + \frac{1}{2}\sqrt{s}} \qquad (4.53)$$

Therefore, for $k < k_{max}$ all eigenvalues of DF satisfy the following inequality

$$|\mu_k| \geq |\mu_{min}| = \alpha\,(c + k) - sk > 1 \qquad (4.54)$$

which confirms that for any solution of linearized equation (4.45), $\hat{u} = DF(X)\,\mathbf{u}$ grows at least as $|\mu_{min}|^t$, and hence all n Lyapunov exponents satisfy the following inequalities

$$\lambda_k > \ln|\mu_{min}| > 0 \qquad (4.55)$$

even with small variations of w_{ij} which gives sufficient, but not necessary conditions for the robustness of chaos for small enough a in the network (4.45). Numerical calculations show that, as a varies, the interval of robustness is wider, and practical requirement is that

$$c' = c + \left| a \sum_{j \neq i} w_{ij} x_i \right| \qquad (4.56)$$

belongs to the interval of chaos robustness in a single map.

4.3.4 Robust chaos in the electroencephalogram model

Firstly, the *electroencephalogram* (EEG) is a signal recorded by scalp electrodes reflecting the synchronous activity of many millions of neurons. Secondly, looking to the history of the experimental analysis of the EEG, it is possible to say that these analyses have failed to show clear evidence of chaotic activity [Nunez (1981), Liley, et al., (1999), Christiansen & Rugh (1997), Freeman (1995)]. But in [Dafilis, *et al.*, (2001)] and evidence for the existence of robust chaotic dynamics (in a theoretical viewpoint) in a neurophysiologically plausible continuum theory of electrocortical activity (electroencephalogram (EEG)) was described and it was shown that the set of parameter values supporting robust chaos within parameter space has positive measure and exhibits fat fractal scaling.

Before describing the results given in [Dafilis, et al., (2001)] we state here some important results reported on the topic of the occurrence of chaos in the electroencephalogram (EEG) model:

1. The electroencephalogram (EEG) is directly proportional to the local field potential recorded by electrodes on the brain's surface [Nunez (1981)].

2. One single EEG electrode placed on the scalp records the aggregate electrical activity from up to 6 cm^2 of brain surface, i.e., many millions of neurons [Cooper, *et al.*, (1965)]. In this case, the modulation of such a system via discrete enumeration of these neurons is infeasible.

3. It was shown in [Nunez (2000)] that continuum models of neocortex to date fall into two broad classes: Those which describe the dynamics of a neocortical macrocolumn, where in a small volume of neocortex[4], the number of neurons is between 40000 and 100000, and those which describe the activity of the whole neocortical mantle[5].

4. The model considered in [Dafilis, *et al.*, (2001)] is a local model derived from the more general global theory presented in [Liley, et al., (1999)] and it comes from the simplest physiologically and anatomically consistent theory of electrocortical dynamics, i.e., This model considers the behavior of the mean soma membrane potential of two functionally distinct neural populations.

With these considerations, the model of electroencephalogram (EEG) is composed of a set of coupled first and second order non-linear ordinary differential equations (ODEs) given by:

$$\tau_e \frac{dh_e}{dt} = (h_{er} - h_e) + \frac{h_{eeq} - h_e}{|h_{eeq} - h_{er}|} I_{ee} + \frac{h_{ieq} - h_e}{|h_{ieq} - h_{er}|} I_{ie} \qquad (4.57)$$

$$\tau_i \frac{dh_i}{dt} = (h_{ir} - h_i) + \frac{h_{eeq} - h_i}{|h_{eeq} - h_{ir}|} I_{ei} + \frac{h_{ieq} - h_i}{|h_{ieq} - h_{ir}|} I_{ii} \qquad (4.58)$$

$$\frac{d^2 I_{ee}}{dt^2} + 2a\frac{dI_{ee}}{dt} + a^2 I_{ee} = Aae \left\{ N_{ee} S_e \left(h_e \right) + p_{ee} \right\} \qquad (4.59)$$

---

[4]called *local models.*
[5]Called *global models.*

$$\frac{d^2 I_{ie}}{dt^2} + 2b\frac{dI_{ie}}{dt} + b^2 I_{ie} = Bbe\left\{N_{ie}S_i\left(h_i\right) + p_{ie}\right\} \tag{4.60}$$

$$\frac{d^2 I_{ei}}{dt^2} + 2a\frac{dI_{ei}}{dt} + a^2 I_{ei} = Aae\left\{N_{ei}S_e\left(h_e\right) + p_{ei}\right\} \tag{4.61}$$

$$\frac{d^2 I_{ii}}{dt^2} + 2b\frac{dI_{ii}}{dt} + b^2 I_{ii} = Bbe\left\{N_{ii}S_i\left(h_i\right) + p_{ii}\right\} \tag{4.62}$$

where

$$S_q\left(h_q\right) = \frac{q_{\max}}{1 + \exp\left(\frac{-\sqrt{2}(h_q - \theta_q)}{s_q}\right)}, q = e, i. \tag{4.63}$$

Firstly, the system (4.57)-(4.62) can be written as a set of ten non-linear first order ODEs and this fact was used for numerical simulations. Secondly, equations (4.57) and (4.58) describe the temporal evolution of the mean soma membrane potentials of the excitatory denoted by h_e and the inhibitory populations denoted by h_i. Thirdly, equations (4.59)-(4.62) represent a spatially homogeneous form of a more complete model of spatio-temporal electrocortical dynamics studied in [Liley, *et al.*, (1999)] and they describe the temporal evolution of the *synaptic* activity. The functions S converting the mean soma membrane potential h_e of the respective population into an equivalent mean firing rate. This means that firing rate acts as a drive to the second order *synapse* described by the left-hand-sides of equations (4.59)-(4.62).

Parameters A and B are the excitatory and inhibitory population postsynaptic potential peak amplitudes, with a and b are the respective synaptic rate constants, with the multiplier e (the base of natural logarithms). Parameters τ_e, τ_i are population membrane time constants, and h_{er}, h_{ir}, and h_{eeq} and h_{ieq} are the resting and equilibrium potentials. Parameters p_{ee} and p_{ei} are excitatory inputs to the respective populations, with p_{ie} and p_{ii} as inhibitory inputs. Parameters N_{ee} and N_{ie} are respectively the number of synapses received by each excitatory neuron from nearby excitatory and inhibitory neurons, with N_{ei} and N_{ii} as the number of synapses received by inhibitory neurons. Parameters θ_e, θ_i are the excitatory and inhibitory population firing thresholds and s_e and s_i represents the standard deviations for these thresholds, with mean maximal firing rates for each population given by $e_{\max}$ and $i_{\max}$.

To solve equations (4.57)-(4.62), the authors in [Dafilis, *et al.*, (2001)] uses the backward differentiation formula (BDF) method implemented by a software library written in C called CVODE [Cohen & Hindmarsh (1996)] with a user-supplied analytic Jacobian matrix, and CVODE's dense linear solver option. Random initial conditions were generated using the random number generation library SPRNG, given in [Mascagni (1999)] with 10^{-9} as a value of the scalar absolute and relative tolerances within 10^5 milliseconds (100 seconds) which is a sufficient amount of time for the system (4.57)-(4.62) to converge to a stable estimate of the largest Lyapunov exponent (LLE).

The Lyapunov exponents of the system (4.57)-(4.62) were determined using the continuous Gram–Schmidt orthonormalization algorithm of Christiansen and Rugh presented in [Christiansen & Rugh (1997)]. For the attractor dimensions, the authors uses the Kaplan–Yorke formula which relates the Lyapunov exponents [Kaplan & Yorke (1979), Farmer, *et al.*, (1983)]. Regarding the attractors' reconstruction, they are created by using CVODE option of XPPAUT [G. B. Ermentrout, XPPAUT, http://www.pitt.edu/;phase] from time series 100 seconds long after removing an initial transient of 5 seconds. For the noise-driven attractor, a fourth order Runge–Kutta integrator method was used with a fixed timestep of 0.1 milliseconds. For $N = 25$, $SD0.08$, the attractor has an LLE of 5.51 s^{-1} base e, compared to the attractor in the case where $N = 25$, $SD0.4$ which has $LLE = 42.9s^{-1}$ base e.

For the calculation of the autocorrelations and delay-time embeddings, the authors used a package software called TISEAN [Hegger, et al., (1999)]. To determine the *fat fractal* nature[6] of the set supporting chaotic dynamics, the authors selected all points in which their $LLE \geq 0.1s^{-1}$ base e and this selection gave a set of 578227 points which they covered with squares of variable side length (ε) and they calculated the number of squares $[N(\varepsilon)]$ occupied as a function of ε. The area of the covering set is given by

$$\mu(\varepsilon) = N(\varepsilon)\varepsilon^2 \qquad (4.64)$$

and in [Umberger & Farmer (1985)] Umberger and Farmer assume a scaling relation for small ε of the form

$$\mu(\varepsilon) = \mu_0 + K\varepsilon^\gamma \qquad (4.65)$$

where K is a constant and μ_0 is the limiting measure of the set, and γ is the fat fractal exponent. Finally, they calculate the parameters μ_0, K, and γ by performing a Levenberg–Marquardt non-linear fit of model (4.57)-(4.62) to the data set over the scaling region $0.02 < \varepsilon < 1.0$.

Using the above methods, the main results are:

1. The stability of chaos in a particular parameter set with respect to perturbations in p_{ee} and p_{ei} is guaranteed because they are consistent with physiology while the remaining parameters are fixed.

2. The effects of varying p_{ee} and p_{ei} over the range 0 to 15 afferent pulses per neuron per millisecond [Freeman (1995)] clearly reveals three distinct dynamical regimes: Point attractor with a negative LLE, limit cycle with a zero LLE, and chaotic dynamics with a positive LLE.

3. Simple fractals like the *Cantor set* and the *Sierpinski gasket*, can be seen

---

[6]Fat fractal is an object which show structure at all scales and which have positive measure.

in the limit of smaller and smaller spatial scales by using an Umberger–Farmer box-counting method [Umberger & Farmer (1985)] in order obtain a convergence over a two decade scaling region and some sets has a fat fractal exponent of approximately 0.5, which indicate an object of some fat fractal structure in a set of evident positive measure.

4.3.5 Robust chaos in Diluted circulant networks

In [Sprott (2008)] the author examines (using the largest Lyapunov exponent as the signature chaos) the prevalence and degree of chaos on large unweighted recurrent networks of ordinary differential equations with *sigmoidal non-linearities* and unit coupling given by

$$x_i' = -b_i x_i + \tanh \left(\sum_{j=1, j \neq i}^{N} a_{ij} x_j \right) \qquad (4.66)$$

where N is the dimension of the system. Equation (4.66) is an example of a general system of coupled ordinary differential equations with a sigmoidal non-linearity, such as the hyperbolic tangent, which define a complex network [Bornholdt & Schuster (2003)].

Some important remarks about equation (4.66) can be summarized in the following:

1. The non-linearity tanh of equation (4.66) is appropriate because it models the common situation in nature.

2. With all $x_i = 0$, equation (4.66) has a *static equilibrium* which represents the state of the system in the absence of external resources and the damping coefficient b describes the rate at which the system decays to its equilibrium state in the absence of external resources. More details on discrete-time neural networks iterated maps[7] can be found in [Cessac, et al., (1994), Albers, et al., (1998)].

3. Appropriate choice of the vector b_i and the matrix a_{ij}, system (4.66) can exhibit a wide range of dynamics, in particular, chaos for $N \leq 4$ with a sufficiently large N. This can be seen from the fact that system (4.66) can approximate to arbitrary accuracy any dynamical system as shown in [Funahashi & Nakamura (1993)].

4. If $b_i = b$ for all i is the bifurcation parameter, then system (4.66) is equivalent to an alternate form given in [Funahashi & Nakamura (1993), Sompolinsky, et al., (1988)].

---

[7]These models display also chaos, but they are less common in the physical sciences, i.e., they are not be physically *realizable*.

5. Equation (4.66) can be considered as an artificial neural network in the sense that the neurons accept weighted inputs from all the other neurons and non-linearly *squash* their sum, as shown in [Haykin (1999)]. Also, Equation (4.66) can be considered as a model of the brain, of a financial market and of a *political system* in which the signals represent voters' position along some political spectrum such as Democrat/ Republican, and the connections are the flow of information between individuals that determine their views. Generally, the network can represent any collection of non-linearly interacting agents such as people, firms, animals, cells, molecules, or any number of other entities, with the connections representing the flow of energy, data, goods, or information among them.

The interesting case here is the so-called *Diluted circulant networks* studied in [Bauer & Martienssen (1989)] using both discrete-time and continuous-time models in [Tirozzi & Tsokysk (1991)] without circulant matrices. Generally, chaos is rare in diluted circulant networks and it is absent in *maximally diluted circulant networks*, where each neuron receives an input from only one other. However, there are some highly diluted cases of system (4.66) with $N = 317$ given in [Sprott (2008)] that exhibit chaos and robust chaos, one example of which is given by:

$$x_i' = -b_i x_i + \tanh(x_{i+42} - x_{i+126} + x_{i+254}) \tag{4.67}$$

Example 8 *As an elementary example: For $b = 0.36$, the largest Lyapunov exponent is $\lambda = 0.0381$. An interesting unanswered question is: Why do those three connections give chaos while most others do not? In fact, changing any of the connections one place to the right or left replacing 126 with 125 or 127 in Eq. (4.67), for example, or even changing N from 317 to 316 or 318 destroys the chaos for $b = 0.36$. However, the case in Eq. (4.67) is robustly chaotic in the sense that no periodic windows are evident over a wide range of b. We recall the fact that 126 is 3×42, while 254 is approximately 6×42.*

4.3.6   Robust chaos in non-smooth neural networks

An example of robust chaos in non-smooth maps occurs with the perceptron [Widrow & Lehr (1990)], which is the simplest kind of *feedforward neural network* and is calculated as

$$f(x) = \begin{cases} 1, & \text{if } \omega \cdot x + b > 0 \\ 0, & \text{else,} \end{cases} \tag{4.68}$$

where ω is a vector of real-valued weights, $\omega \cdot x$ is the dot product (which computes a weighted sum), b is the *bias*, and a is a constant term that does not depend on any input value. The value of $f(x)$ (0 or 1) is used in order to classify x as either positive or negative in the case of a binary classification problem. The properties of the time series generated by a perceptron with monotonic and non-monotonic transfer functions were examined in [Priel & Kanter (2000)]. The results show that a perceptron with a monotonic function can produce fragile chaos only, whereas a non-monotonic function can generate robust chaos.

4.4 The importance of robust chaos in mathematics and some open problems

First of all, robust chaos is generated by discrete maps and differential equations. So, it is an important field of applied mathematics. Secondly, it was shown in [Albers & Sprott (2006)] that the use of approximation theorems of [Hornik, *et al.* (1990)] and time-series embedding of [Sauer, *et al.*, (1991)] establish an equivalence between single-layer recurrent neural networks of the form [Albers, *et al.*, (2006)]

$$F_{d,n,\beta} : x_t = \beta_0 + \sum_{i=1}^{n} \beta_i \tanh s \left(\omega_{i0} + \sum_{j=1}^{d} \omega_{ij} x_{t-j} \right) \qquad (4.69)$$

and general dynamical systems. Models of the form (4.69) are maps from $\mathbb{R}^d$ to $\mathbb{R}$. Here n is the number of neurons, d is the number of time lags, which determines the system's input embedding dimension, and s is a scaling factor for the connection weights ω_{ij}. The initial condition is $(x_1, x_2, ..., x_d)$, and the state at time t is $(x_t, x_{t+1}, ..., x_{t+d-1})$. In some references such as [Albers, *et al.*, (2006)] the $k = n(d+2) + 1$-dimensional parameter space was taken as follows:

1. $\beta_i \in [0, 1]$ is uniformly distributed and rescaled to satisfy $\sum_{i=1}^{n} \beta_i^2 = n$.

2. ω_{ij} is normally distributed with zero mean and unit variance.

3. The initial condition $x_j \in [-1, 1]$ is uniform.

Equation (4.69) remains very mysterious and requires additional researches from mathematicians and physicists. One of the major problems related to robust chaos in dynamical systems are as follows:

1. What are the main and new behaviors in the dynamics of the sum of n maps of the form $F_{d,n,\beta}$?

2. Is it possible to generate homoclinic chaos *via* the composition of two continuous mappings $F_{d,n,\beta}$?

3. What is the behavior of maps (or systems) resulting from compositions of several maps (or systems) of the form $F_{d,n,\beta}$?

4. Find a new method for constructing robust chaotic attractors by using the properties of the map $F_{d,n,\beta}$.

5. Find sufficient conditions for the non-existence of periodic orbits in $F_{d,n,\beta}$. The importance of this idea is to distinguish between the type of chaos obtained *via* the usual routes to chaos and chaos coming from nothing (without a transient from periodic orbits).

6. Find a method for chaotifying $F_{d,n,\beta}$ if it is not chaotic.

Chapter 5

Estimating Lyapunov Exponents of 2-D Discrete Mappings

5.1 Introduction

This chapter is devoted to presenting some results about the rigorous determination of the Lyapunov exponents of some 2-D discrete mappings. In Sec. 5.1, the Lyapunov exponents of the *discrete hyperchaotic double scroll* mapping are calculated and the bifurcation parameters, in which the result attractor is hyperchaotic, are estimated. In Sec. 5.2, the Lyapunov exponents of a class of 2-D piecewise linear mappings are calculated. this case includes the so called: *The discrete butterfly*. In Sec. 5.3, the Lyapunov exponents of a family of 2-D discrete mappings with separate variables are presented under some comparable conditions on the functions defining the system. Finally, the Lyapunov exponents of a discontinuous piecewise linear mapping of the plane governed by a simple switching law are calculated and discussed in Sec. 5.4. Some interesting dynamical phenomena are also present for an elementary example.

5.2 Lyapunov exponents of the discrete hyperchaotic double scroll map

The so called *Discrete hyperchaotic double scroll map* is presented and studied in [Zeraoulia and Sprott (2009(a))]:

$$
f\left(x, y\right) = \begin{pmatrix} x - a\left(\frac{2m_1 y + (m_0 - m_1)(|y+1| - |y-1|)}{2}\right) \\ bx \end{pmatrix} \tag{5.1}
$$

where a and b are the bifurcation parameters. The function

$$
h(x) = \frac{2m_1 x + (m_0 - m_1)\left(|x+1| - |x-1|\right)}{2}
$$

is the characteristic function of the so-called *double scroll attractor* [Chua, et al., 1986], and m_0 and m_1 are respectively the slopes of the inner and outer sets of the original Chua circuit.

The map (5.1) is extremely simple and minimal in view of the number of terms. Also, it has some of the classical double scroll properties.

1. The associated function $f(x, y)$ is continuous in $\mathbb{R}^2$, but it is not differentiable at the points $(x, -1)$ and $(x, 1)$ for all $x \in \mathbb{R}$.

2. The map (5.1) is a diffeomorphism except at points $(x, -1)$ and $(x, 1)$ when $abm_1m_0 \neq 0$, since the determinant of its Jacobian is non-zero if and only if $abm_1 \neq 0$ or $abm_0 \neq 0$.

3. The map (5.1) does not preserve area and it is not a reversing twist map for all values of the system parameters.

4. The map (5.1) is symmetric under the coordinate transformation $(x, y) \longrightarrow (-x, -y)$, and this transformation persists for all values of the system parameters.

5. The chaotic attractor of map (5.1) is symmetric just like the classic double scroll [Chua, et al., 1986].

6. Due to the shape of the vector field f of the map (5.1), the plane can be divided into three linear regions denoted by:

$$\begin{cases} R_1 = \{(x, y) \in \mathbb{R}^2 / \ y \geq 1\} \\ R_2 = \{(x, y) \in \mathbb{R}^2 / \ |y| \leq 1\} \\ R_3 = \{(x, y) \in \mathbb{R}^2 / \ y \leq -1\} \end{cases}$$

where in each of these regions the map (5.1) is linear.

7. For all values of the parameters m_0, m_1 such that $m_0m_1 > 0$, the map (5.1) has a single fixed point $(0, 0)$.

8. If $m_0m_1 < 0$, then the map (5.1) has three fixed points:

$$\begin{cases} P_1 = \left(\frac{m_1-m_0}{bm_1}, \frac{m_1-m_0}{m_1}\right) \\ P_2 = (0, 0) \\ P_3 = \left(\frac{m_0-m_1}{bm_1}, \frac{m_0-m_1}{m_1}\right) \end{cases}$$

9. The Jacobian matrix of the map (5.1) evaluated at the fixed points P_1 and P_3 is the same and is given by $J_{1,3} = \begin{pmatrix} 1 & -abm_1 \\ 1 & 0 \end{pmatrix}$.

10. The two equilibrium points P_1 and P_3 have the same stability type.

11. The Jacobian matrix of the map (5.1) evaluated at the fixed point P_2 is given by $J_2 = \begin{pmatrix} 1 & -abm_0 \\ 1 & 0 \end{pmatrix}$, and the characteristic polynomials for $J_{1,3}$ and J_2 are given respectively by $\lambda^2 - \lambda + abm_1 = 0$ and $\lambda^2 - \lambda + abm_0 = 0$. The local stability of these equilibria can be studied by evaluating the eigenvalues of the corresponding Jacobian matrices given by the solution of their characteristic polynomials.

Now, by using Theorem 15, it is possible to prove rigorously that the attractor of map (5.1) is hyperchaotic. Indeed, we have

$$\|f'(x,y)\| = \begin{cases} \sqrt{\dfrac{b^2+a^2m_1^2+\sqrt{2b^2+b^4+2a^2m_1^2+a^4m_1^4-2a^2b^2m_1^2+1}+1}{2}} < +\infty, \\[2em] \qquad \text{if } |y| \geq 1 \\[1em] \sqrt{\dfrac{b^2+a^2m_0^2+\sqrt{2b^2+b^4+2a^2m_0^2+a^4m_0^4-2a^2b^2m_0^2+1}+1}{2}} < +\infty, \\[2em] \qquad \text{if } |y| \leq 1 \end{cases}$$

(5.2)

and

$$\lambda_{\min}\left(f'(x)^T (f'(x))\right) = \begin{cases} \dfrac{b^2+a^2m_1^2-\sqrt{2b^2+b^4+2a^2m_1^2+a^4m_1^4-2a^2b^2m_1^2+1}+1}{2}, \\[2em] \qquad \text{if } |y| \geq 1 \\[1em] \dfrac{b^2+a^2m_0^2-\sqrt{2b^2+b^4+2a^2m_0^2+a^4m_0^4-2a^2b^2m_0^2+1}+1}{2}, \\[2em] \qquad \text{if } |y| \leq 1. \end{cases}$$

(5.3)

Now, if

$$\begin{cases} |a| > \max\left(\dfrac{1}{|m_1|}, \dfrac{1}{|m_0|}\right) \\[1.5em] |b| > \max\left(\dfrac{|am_1|}{\sqrt{a^2m_1^2-1}}, \dfrac{|am_0|}{\sqrt{a^2m_0^2-1}}\right) \end{cases}$$

(5.4)

then both Lyapunov exponents of the map (5.1) are positive for all initial conditions $(x_0, y_0) \in \mathbb{R}^2$. Therefore, the attractor is hyperchaotic.

Example 9 *As a numerical example, let $m_0 = -0.43$ and $m_1 = 0.41$, then $|a| > 2.439$, and for $b = 1.4$, one has that $|a| > 3.323$. Fig. 5.1 shows the Lyapunov exponent spectrum for the map (5.1) for $m_0 = -0.43$, $m_1 = 0.41, b = 1.4$, and $-3.365 \leq a \leq 3.365$. The regions of hyperchaos are*

$-3.365 \leq a \leq -3.323$ *and* $3.323 \leq a \leq 3.365$. *The discrete hyperchaotic dou-ble scroll shown in Fig. 5.2 results from a stable period-3 orbit transitioning to a fully developed chaotic regime. This is the border-collision bifurcation shown in Fig. 5.3. If* $b = 1.4, m_0 = -0.43$, *and* $m_1 = 0.41$ *and vary* $a \in \mathbb{R}$, *then the map (5.1) exhibits the following dynamical behaviors as shown in Fig. 5.3.*

(1) In the interval $a < -3.365$, *the map (5.1) does not converge.*

(2) For $-3.365 \leq a \leq 3.365$, *the map (5.1) begins with a reverse border-collision bifurcation, leading to a stable period-3 orbit, and then collapses to a point that is reborn as a stable period-3 orbit leading to fully developed chaos.*

(3) For $a > 3.365$, *the map (5.1) does not converge.*

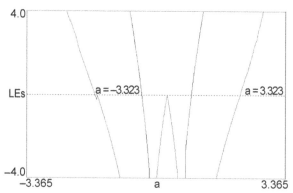

FIGURE 5.1: Variation of the Lyapunov exponents of map (5.1) versus the parameter $-3.365 \leq a \leq 3.365$ with $b = 1.4, m_0 = -0.43$, and $m_1 = 0.41$.

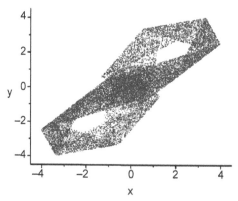

FIGURE 5.2: The discrete hyperchaotic double scroll attractor obtained from the map (5.1) for $a = 3.36, b = 1.4, m_0 = -0.43$, and $m_1 = 0.41$ with initial conditions $x = y = 0.1$.

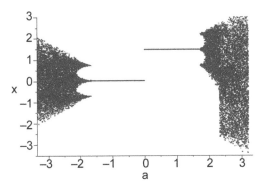

FIGURE 5.3: The border collision bifurcation route to chaos of map (5.1) versus the parameter $-3.365 \leq a \leq 3.365$ with $b = 1.4$, $m_0 = -0.43$, and $m_1 = 0.41$.

5.3 Lyapunov exponents for a class of 2-D piecewise linear mappings

Let us consider an arbitrary piecewise-linear map $f : D \to D$, where $D = D_1 \cup D_2 \subset \mathbb{R}^2$, defined by:

$$
f(X_k) = \begin{cases} AX_k + b + \begin{pmatrix} (b_{11} + b_{11} - a_{11} - a_{22})\, x_k \\ \left(\dfrac{-a_{12}a_{21} + b_{11}a_{22} - b_{11}b_{22} + b_{12}b_{21} + a_{22}b_{22} - a_{22}^2}{a_{12}} \right) x_k \end{pmatrix}, \\ \qquad\qquad \text{if } X_k \in D_1 \\ \\ BX_k + c, \text{ if } X_k \in D_2 \end{cases}
$$

(5.5)

where $A = \begin{pmatrix} a_{11} & a_{12} \\ a_{21} & a_{22} \end{pmatrix}$ and $B = \begin{pmatrix} b_{11} & b_{12} \\ b_{21} & b_{22} \end{pmatrix}$ are 2×2 real matrices, and $b = \begin{pmatrix} b_1 \\ b_2 \end{pmatrix}$ and $c = \begin{pmatrix} c_1 \\ c_2 \end{pmatrix}$ are 2×1 real vectors, and $X_k = \begin{pmatrix} x_k \\ y_k \end{pmatrix} \in \mathbb{R}^2$ is the state variable.

The following theorem was proved in [Zeraoulia and Sprott (2010(b))]:

Theorem 44 *The piecewise linear map (5.5) is chaotic in the following case:*

$$
\begin{cases} 2 < b_{11}^2 + b_{12}^2 + b_{21}^2 + b_{22}^2 < (b_{11}b_{22} - b_{12}b_{21})^2 + 1 \\ \\ |b_{11}b_{22} - b_{12}b_{21}| > 1. \end{cases}
$$

(5.6)

Proof 10 *The system (5.5) can be written as follows:*

$$g\left(X_k\right) = \begin{cases} QX_k + b \text{ if } X_k \in D_1 \\ \\ BX_k + c \text{ if } X_k \in D_2, \end{cases} \tag{5.7}$$

where the matrix Q is given by:

$$Q = \begin{pmatrix} b_{11} + b_{22} - a_{22} & a_{12} \\ \frac{b_{11}a_{22} - b_{11}b_{22} + b_{12}b_{21} + a_{22}b_{22} - a_{22}^2}{a_{12}} & a_{22} \end{pmatrix} \tag{5.8}$$

The Jacobian matrix of the map (5.5) is given by:

$$J\left(X_k\right) = \begin{cases} Q \text{ if } X_k \in D_1 \\ \\ B \text{ if } X_k \in D_2. \end{cases} \tag{5.9}$$

We remark that $J\left(X_k\right)$ given in (5.9) is not well-defined due to the discontinuity, but, since B and Q have the same eigenvalues, then one has that

$$\|Q\| = \|B\| = \sqrt{\lambda_{\max}\left(B^T B\right)}$$

because

$$B^T B = \begin{pmatrix} b_{11}^2 + b_{21}^2 & b_{11}b_{12} + b_{21}b_{22} \\ b_{11}b_{12} + b_{21}b_{22} & b_{12}^2 + b_{22}^2 \end{pmatrix}$$

is at least a positive semi-definite matrix, then all its eigenvalues are real and positive, i.e.,

$$\lambda_{\max}\left(B^T B\right) \geq \lambda_{\min}\left(B^T B\right) \geq 0$$

Hence the eigenvalues of $B^T B$ are given by:

$$\begin{cases} \lambda_{\max}\left(B^T B\right) = \frac{1}{2}b_{11}^2 + \frac{1}{2}b_{12}^2 + \frac{1}{2}b_{21}^2 + \frac{1}{2}b_{22}^2 + \frac{1}{2}\sqrt{d} \\ \\ \lambda_{\min}\left(B^T B\right) = \frac{1}{2}b_{11}^2 + \frac{1}{2}b_{12}^2 + \frac{1}{2}b_{21}^2 + \frac{1}{2}b_{22}^2 - \frac{1}{2}\sqrt{d}, \end{cases} \tag{5.10}$$

where

$$d = \left(\left(b_{11} + b_{22}\right)^2 + \left(b_{12} - b_{21}\right)^2\right)\left(\left(b_{12} + b_{21}\right)^2 + \left(b_{11} - b_{22}\right)^2\right) > 0 \tag{5.11}$$

for all b_{11}, b_{12}, b_{21}, and b_{22}. Condition (1.39) gives

$$\|f'(x)\| = \|B\| = \|Q\| = \sqrt{\lambda_{\max}\left(B^T B\right)} = N < +\infty$$

because B and Q have the same eigenvalues. Condition (1.40) gives the following inequality:

$$\theta^2 - \left(b_{11}^2 + b_{12}^2 + b_{21}^2 + b_{22}^2\right)\theta + \left(b_{11}b_{22} - b_{12}b_{21}\right)^2 \geq 0 \tag{5.12}$$

with the condition

$$\theta < \frac{b_{11}^2 + b_{12}^2 + b_{21}^2 + b_{22}^2}{2}. \tag{5.13}$$

Since the discriminant of (5.12) is equal to $d > 0$, then (5.12) holds when:

$$\theta \geq \lambda_{\max}\left(B^T B\right), \ \text{or} \ \theta \leq \lambda_{\min}\left(B^T B\right). \tag{5.14}$$

The condition $\theta \geq \lambda_{\max}\left(B^T B\right)$ is impossible because of condition (5.13), so that θ must satisfy the condition:

$$\theta < \lambda_{\min}\left(B^T B\right) = \frac{b_{11}^2 + b_{12}^2 + b_{21}^2 + b_{22}^2 - \sqrt{d}}{2}. \tag{5.15}$$

Now, if (5.6) is verified, then $\lambda_{\min}\left(B^T B\right) > 1$, i.e. $\theta = 1$, and one has:

$$0 < l_i\left(x_0\right) \leq \ln N, i = 1, 2, \tag{5.16}$$

i.e., the map (5.5) converges to a hyperchaotic attractor for all parameters b_{11}, b_{12}, b_{21}, and b_{22} satisfying (5.6).

Example 10 *As a numerical example, let us consider the following map:*

$$g\left(x_k, y_k\right) = \begin{cases} \left(\begin{array}{cc} 1 & -\alpha \\ \beta & 0 \end{array}\right)\left(\begin{array}{c} x_k \\ y_k \end{array}\right) + \left(\begin{array}{c} 1 \\ 0 \end{array}\right), & \text{if } y_k \geq 0 \\ \left(\begin{array}{cc} 1 & \alpha \\ -\beta & 0 \end{array}\right)\left(\begin{array}{c} x_k \\ y_k \end{array}\right) + \left(\begin{array}{c} 1 \\ 0 \end{array}\right), & \text{if } y_k < 0 \end{cases} \tag{5.17}$$

$$= \left(\begin{array}{c} 1 - \alpha\left|y_k\right| + x_k \\ \beta x_k sgn(y_k) \end{array}\right)$$

where α, β, and γ are bifurcation parameters. The chaotic attractor of the map (5.17) is called discrete butterfly [Zeraoulia (2011(c))], where sgn(.) is the standard signum function that gives ± 1 depending on the sign of its argument. Applying Theorem 15, we get:

$$\begin{cases} |\alpha| > \max\left(\frac{|\beta|}{\sqrt{\beta^2 - 1}}, \frac{1}{|\beta|}\right) \\ |\beta| > 1 \end{cases} \tag{5.18}$$

Assuming that $\beta < -1$. Then we remark that

$$\frac{|\beta|}{\sqrt{\beta^2 - 1}} = \frac{-\beta}{\sqrt{\beta^2 - 1}} < \frac{1}{|\beta|} = \frac{-1}{\beta}$$

and, thus, we get

$$|\alpha| > \frac{-1}{\beta}, \beta < -1. \tag{5.19}$$

Fig. 5.4 shows that for $-2 \le \beta < -1$, and $-0.1 - \frac{1}{\beta} < \alpha < \frac{-1}{\beta} + 0.4$, the map (5.17) converges to bounded hyperchaotic attractors or unbounded orbits for $\alpha > -\frac{1}{\beta}$. A chaotic attractor for the case with $\alpha = 0.6$ and $\beta = -2$ is shown in Fig. 5.5. It is necessary to verify the hyperchaoticity of the attractors by calculating both Lyapunov exponents using the formula:

$$l_1(x_0) + l_2(x_0) = \ln|\det(J)| = \ln|\alpha\beta|$$

averaged along the orbit where $\det(J)$ is the determinant of the Jacobian matrix. The result is shown in Fig. 5.6 for $0.4 \le \alpha \le 0.9$, with $\beta = -2$.

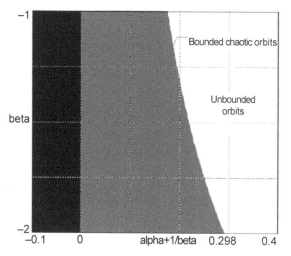

FIGURE 5.4: Regions of dynamical behaviors in the $\alpha\beta$-plane for the map (5.17) with $-2 \le \beta < -1$ and $-0.1 - \frac{1}{\beta} < \alpha < \frac{-1}{\beta} + 0.4$.

5.4 Lyapunov exponents of a family of 2-D discrete mappings with separate variables

There are many works that focus on the topic of chaotic behaviors of a discrete mapping. For example it has been studied with control and anti-control (chaotification) schemes or using Lyapunov exponents [Baillieul, *et al.*, (1980), Chen and Lai (1997-2003), Chen and Lai (1998), Li, *et al.*, (2002(a-b-c)), Wang and Chen (1999-2000), Zhang and Chen (2002)], or using several modified versions of the Marotto Theorem (Sec. 1.10) [Marotto (1978), Li and Yorke (1975),

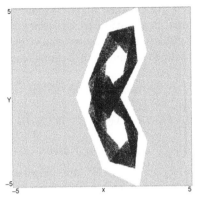

FIGURE 5.5: The piecewise linear chaotic attractor obtained from the map (5.17) with its basin of attraction in white for $\alpha = 0.6$, $\beta = -2$, and the initial condition $x_0 = y_0 = 0.01$.

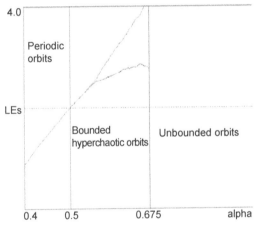

FIGURE 5.6: Variation of the Lyapunov exponents of the map (5.17) versus the parameter $0.4 \leq \alpha \leq 0.9$, with $\beta = -2$.

Lin, *et al.*, (2002), Chen, *et al.*, (1998), Li and Chen (2003)] in order to prove the existence of chaos in n-dimensional dynamical discrete system, where the results are in some way making an originally non-chaotic dynamical system chaotic, or enhancing the existing chaos of a chaotic system.

In this section, we prove rigorously the existence of chaotic behavior in a family of 2-dimensional discrete mappings, using the standard definition of Lyapunov exponents given in Sec. 1.8. The shape of the map under consideration permits to impose sufficient conditions on its components to allow this rigorous proof of chaos. The choice of dimension two is robust since almost real systems are in three dimensions, and therefore the associated Poincaré map is of two dimensions, in addition to the effect that these maps has an important variety of applications in sciences and engineering.

Let us consider the following family of 2-dimensional discrete mappings:

$$\begin{cases} x_{k+1} = \alpha x_k + g_1\left(y_k, \mu\right) \\ \\ y_{k+1} = h\left(x_k, \mu\right) + g_2\left(y_k, \mu\right) \\ \\ X_0 = \left(x_0, y_0\right)\text{-given} \end{cases} \tag{5.20}$$

where $\alpha \in \mathbb{R}, \mu = (\mu_1, \mu_2, ..., \mu_j) \in \mathbb{R}^j$ are bifurcation parameters, $X_k = (x_k, y_k) \in \mathbb{R}^2$ is the state space and h, g_1, g_2 are real functions defined as follows: $h : \Omega \subset \mathbb{R} \times \mathbb{R}^j \longrightarrow \mathbb{R}, g_1 : \Omega_1 \subset \mathbb{R} \times \mathbb{R}^j \longrightarrow \mathbb{R}$, and $g_2 : \Omega_2 \subset \mathbb{R} \times \mathbb{R}^j \longrightarrow \mathbb{R}$ and are assumed to be first-order differentiable, at least locally in a region of interest.

The presence of chaos in a dynamical system is proved by measuring the largest Lyapunov exponent (Sec. 1.8) which quantifies the exponential divergence of initially close state-space trajectories and estimates the amount of chaos in a system.

Theorem 45 *Let us consider the family of 2-d discrete mappings given by equation (5.20) and suppose the following conditions:*

(i) There is at least one fixed point for map (5.20), i.e., the equation:

$$h\left(\left(\frac{g_1\left(y, \mu\right)}{1 - \alpha}\right), \mu\right) + g_2\left(y, \mu\right) - y = 0 \tag{5.21}$$

has at least one real solution.

(ii) The functions h, g_1, g_2 are first-order differentiable, at least locally in a region of interest.

(iii) The function h verifies:

$$h\left(x, \mu\right) \neq 0, \text{ and } \frac{\partial h\left(x, \mu\right)}{\partial x} = 0 \text{ for all } (x, \mu) \in \Omega. \tag{5.22}$$

(iv) The parameter α verifies $\alpha > 1$.
(v) There exists a real constant $M > 0$, such that:

$$\left| \frac{\partial g_2\left(y, \mu\right)}{\partial y} \right| < M, \text{ for all } (y, \mu) \in \Omega_2. \tag{5.23}$$

uniformly with respect to y.

Hence, the family of maps (5.20) is chaotic in the sense that the largest Lyapunov exponent is positive.

Proof 11 *The Jacobian matrix J of map (5.20) is given by:*

$$J\left(x_k, y_k\right) = \begin{pmatrix} \alpha & \frac{\partial g_1(y_k, \mu)}{\partial y} \\ \frac{\partial h(x, \mu)}{\partial x} & \frac{\partial g_2(y_k, \mu)}{\partial y} \end{pmatrix} = \begin{pmatrix} \alpha & \frac{\partial g_1(y_k, \mu)}{\partial y} \\ 0 & \frac{\partial g_2(y_k, \mu)}{\partial y} \end{pmatrix} \tag{5.24}$$

Since $\frac{\partial h(x,\mu)}{\partial x} = 0$ *for all* $(x, \mu) \in \Omega$. *The matrix* $T_m(X_0)$ *defined by (1.36) is given by:*

$$T_m(X_0) = \begin{pmatrix} \alpha & \frac{\partial g_1(y_m,\mu)}{\partial y} \\ 0 & \frac{\partial g_2(y_m,\mu)}{\partial y} \end{pmatrix} \begin{pmatrix} \alpha & \frac{\partial g_1(y_{m-1},\mu)}{\partial y} \\ 0 & \frac{\partial g_2(y_{m-1},\mu)}{\partial y} \end{pmatrix} \cdots \begin{pmatrix} \alpha & \frac{\partial g_1(y_0,\mu)}{\partial y} \\ 0 & \frac{\partial g_2(y_0,\mu)}{\partial y} \end{pmatrix}$$
$$(5.25)$$

We set: $\beta_k = \frac{\partial g_1(y_k,\mu)}{\partial y}$ *and* $\gamma_k = \frac{\partial g_2(y_k,\mu)}{\partial y}, k = 0, 1, ..., m$, *then the matrix* $T_m(X_0)$ *can be rewritten as follows:*

$$T_m(X_0) = \begin{pmatrix} \alpha & \beta_m \\ 0 & \gamma_m \end{pmatrix} \begin{pmatrix} \alpha & \beta_{m-1} \\ 0 & \gamma_{m-1} \end{pmatrix} \cdots \begin{pmatrix} \alpha & \beta_1 \\ 0 & \gamma_1 \end{pmatrix} \begin{pmatrix} \alpha & \beta_0 \\ 0 & \gamma_0 \end{pmatrix} \quad (5.26)$$

Now, some algebra leads to the following simplified form of the matrix $T_m(X_0)$:

$$T_m(X_0) = \begin{pmatrix} \alpha^{m+1} & * \\ 0 & \prod_{k=0}^{k=m} \gamma_k \end{pmatrix} \quad (5.27)$$

Where the entry $*$ *in the above upper triangular matrix is not important for computing the eigenvalues of* $T_m(X_0)$, *so it is not displayed here. The eigenvalues of* $T_m(X_0)$ *are given by:*

$$\delta_1 = \alpha^{m+1} \quad (5.28)$$

$$\delta_2 = \prod_{k=0}^{k=m} \gamma_k \quad (5.29)$$

and therefore one has $J_1(X_0, m) = |\alpha^{m+1}|$ *and* $J_2(X_0, m) = \left| \prod_{k=0}^{k=m} \gamma_k \right|$ *and it follows from the definition of the Lyapunov exponent (1.37) that:*

$$\begin{cases} \lambda_1(X_0) = \lim_{m \longrightarrow +\infty} \ln\left((\alpha^{m+1})^{\frac{1}{m}} \right) = \ln(\alpha) \\ \\ \lambda_2(X_0) = \lim_{m \longrightarrow +\infty} \ln\left(\left(\prod_{k=0}^{k=m} \gamma_k \right)^{\frac{1}{m}} \right) \end{cases} \quad (5.30)$$

Finally, the first Lyapunov exponent $\lambda_1(X_0)$ *given by (5.30) is positive if and only if*

$$\alpha > 1 \quad (5.31)$$

then the family of maps (5.20) is chaotic in the sense that the largest Lyapunov exponent is positive.

We notice the following important remarks:

(a) The assumption (i) about the existence of a fixed point for map (5.20) is essential here because, in most cases, the non-existence of a fixed point implies that the dynamics of the map goes to infinity.

(b) The assumption (ii) is also essential for the computing of the Jacobian matrix of map (5.20).

(c) The set of functions h that verify:

$$h(x, \mu) \neq 0, \frac{\partial h(x, \mu)}{\partial x} = 0 \text{ for all } (x, \mu) \in \Omega$$

(assumption (iii)) is not empty because the functions

$$h(x, \mu) = sgn(x) \text{ and } h(x, \mu) = \begin{cases} a, & \text{if } (x, \mu) \in \Gamma_1 \\ \\ b, & \text{if } (x, \mu) \in \Gamma_2 \end{cases}$$

are examples of a such situation, where: Γ_1 and Γ_2 are subsets of Ω. The same remake about the set of functions g_2 (assumption (iv)), for example:

$$g_2(y, \mu) = \sin(y) \text{ and } g_2(y, \mu) = \cos(y)$$

(d) The assumption (v) is used here for practical reasons, i.e., to guarantee the boundedness of the solution of the map under consideration.

5.5   Lyapunov exponents of a discontinuous piecewise linear mapping of the plane governed by a simple switching law

There are many works that focus on the topic of the rigorous mathematical proof of chaos in a discrete mapping (continuous or not). For example, it has been studied rigorously with control and anti-control schemes or using Lyapunov exponents in order to prove the existence of chaos in n-dimensional dynamical discrete system, since a large number of physical and engineering systems have been found to exhibit a class of continuous or discontinuous piecewise maps [Zeraoulia & Sprott (2009), Feely & Chua (1991), Ogorzalek (1992), Banerjee & Verghese (2001), Yuan, et al., (1998), Feely & Chua (1992), Sharkovsky & Chua (1993)] where the discrete-time state space is divided into two or more compartments with different functional forms of the map separated by borderlines [Yuan, et al., (1998), Jain & Banerjee (2003)]. The theory for discontinuous maps has some progress reported for 1-D and n-D discontinuous maps in [Chua & Lin (1988), Feely & Chua (1991), Ogorzalek (1992)], these results are restrictive and cannot be obtained in the general n-dimensional context. Discontinuous switching systems are a special type of discontinuous piecewise map that occur in many industrial control systems such as multi-body systems with unilateral constraints, cooperative and autonomous systems, intelligent systems, robots and social systems [Liberzon

(2003), Zhang, *et al.*, (2006)], thus making the analysis of this type of applications fundamental and extremely important.

In this section, we rigorously determine the dynamics of a discontinuous piecewise linear switching map of the plane governed by a simple switching law, using single input parameter. The proof required the use of the standard definition of the Lyapunov exponents as the usual test for chaos. Some discussions on the problems related to the lack of continuity in the map are also given here.

Let us consider the following 2-dimensional discrete switching mapping:

$$\begin{pmatrix} x_{n+1} \\ y_{n+1} \end{pmatrix} = f(x_n, y_n) = \begin{pmatrix} -ay_n sgn(x_n) + x_n \\ b - |x_n| \end{pmatrix} \tag{5.32}$$

where a, b, are positive bifurcation parameters, and $sgn(.)$ is the standard signum function that gives the sign of its argument. Hence, one can rewrite the map (5.32) as follows:

$$\begin{pmatrix} x_{n+1} \\ y_{n+1} \end{pmatrix} = \begin{cases} \begin{pmatrix} 1 & a \\ 1 & 0 \end{pmatrix} \begin{pmatrix} x_n \\ y_n \end{pmatrix} + \begin{pmatrix} 0 \\ b \end{pmatrix}, & \text{if } x_n < 0 \\ \begin{pmatrix} 1 & -a \\ -1 & 0 \end{pmatrix} \begin{pmatrix} x_n \\ y_n \end{pmatrix} + \begin{pmatrix} 0 \\ b \end{pmatrix}, & \text{if } x_n \geq 0 \end{cases} \tag{5.33}$$

Note that the switching map (5.32) is a local diffeomorphism when $a \neq 0$, and its associated function $f(x_n, y_n)$ is discontinuous in the points $(0, y_n)$. Due to the shape of the vector field f of the switching map (5.32), the plane can be divided into two linear regions denoted by:

$$D_1 = \{(x_n, y_n) \in \mathbb{R}^2 : x_n < 0\} \tag{5.34}$$

$$D_2 = \{(x_n, y_n) \in \mathbb{R}^2 : x_n \geq 0\} \tag{5.35}$$

where in each of these regions the switching map (5.32) is linear and it is possible to determine the exact solution in terms of the initial condition and the bifurcation parameters a, b. On the other hand, there is really no physical basis for the system equations (5.32), or at least none is present. Then, the study is reduced to some computations and numerical simulations.

First, we begin by studying the existence of the fixed point of f mapping and determine their local stability type. Hence, the fixed points of the switching map (5.32) are the real solutions of the system:

$$\begin{cases} -ay sgn(x) + x = x \\ y = b - |x| \end{cases}$$

Thus, one may easily obtain for all $a \neq 0$ and $b > 0$ that: $y = 0$, and $x = \pm b$.

Thus, the fixed points of the switching map (5.32) are $P_1 = (-b, 0) \in D_1$, and $P_2 = (b, 0) \in D_2$. While, if $a = 0$, then there are infinite non-isolated fixed points of the form $(x, b - |x|)$. Obviously, the Jacobian matrix of the map (5.32) evaluated at the fixed points P_1 and P_2 is given respectively by

$$\begin{cases} J_1 = \begin{pmatrix} 1 & a \\ 1 & 0 \end{pmatrix} \\ \\ J_2 = \begin{pmatrix} 1 & -a \\ -1 & 0 \end{pmatrix} \end{cases}$$

and the characteristic polynomials for J_1 and J_2 are the same and are given by:

$$\lambda^2 - \lambda - a = 0 \tag{5.36}$$

Thus, the two fixed points P_1 and P_2 have the same local stability type. Hence, it is easy to verify that:

1. If $a < -1$, then both P_1 and P_2 are sources.

2. If $-1 < a < 0$, then both P_1 and P_2 are sinks.

3. If $a = -1$, or $a = 0$, or $a = 2$, then both P_1 and P_2 are non-hyperbolic fixed points.

4. If $0 < a < 2$, then both P_1 and P_2 are saddle fixed points.

5. If $a > 2$, then both P_1 and P_2 are sources.

The idea of the proof is to find the roots λ_1 and λ_2 of equation (5.36) which can be found according to the signum of its discriminant $\Delta = 4a + 1$, then there exist three different cases, where in each case it is easy to give the expression of the corresponding eigenvalues and the local stability type of a fixed point. The expressions of the eigenvalues are as follows:

1. If $a \geq \frac{-1}{4}$, then the eigenvalues λ_1 and λ_2 are real and they are given by:

$$\begin{cases} \lambda_1 = \frac{1}{2}\sqrt{4a + 1} + \frac{1}{2} \\ \\ \lambda_2 = \frac{1}{2} - \frac{1}{2}\sqrt{4a + 1} \end{cases} \tag{5.37}$$

2. If $a < \frac{-1}{4}$, then the eigenvalues λ_1 and λ_2 are complex and they are given by:

$$\begin{cases} \lambda_1 = \frac{1 + i\sqrt{-(4a+1)}}{2} \\ \\ \lambda_2 = \frac{1 - i\sqrt{-(4a+1)}}{2} \end{cases} \tag{5.38}$$

From the cases (1), (2) one may conclude that: if $-1 < a < 0$, then both fixed points P_1 and P_2 are asymptotically stable, if $a = -1$, or $a = 0$, or $a = 2$, then both P_1 and P_2 are non-hyperbolic, if $0 < a < 2$, then both P_1 and P_2 are saddle fixed points. Finally, if $a > 2$, or $a < -1$, then both P_1 and P_2 are unstable.

In the following, we will analytically compute the largest Lyapunov exponent of the switching map (5.32) in each linear region D_1 and D_2 separately, and we will show that these exponents are identical in each linear region D_1 and D_2 defined above, then we have the following result.

Theorem 46 *For all positive values of the parameter b, the dynamics of the switching map (5.32) are given in terms of the parameter a rigorously as follows:*

(1) The map (5.32) converges to a hyperchaotic attractor when $a < -1$.

(2) The behavior of the map (5.32) is not clear when $a = -1$.

(3) The map (5.32) converges to stable fixed point when $-1 < a < 0$.

(4) The map (5.32) converges to a non-isolated fixed point $(x_0, b - |x_0|)$ for all $x_0 \in \mathbb{R}$ when $a = 0$.

(5) The map (5.32) converges to a chaotic attractor when $a > 0$.

Proof 12 *The essential idea of this proof is that the matrices J_1 and J_2 are equivalent, then one can consider the Jacobian matrix $Df(x, y)$ of the switching map (5.32) as J_1 or J_2, and in this case we denoted it by $A = \begin{pmatrix} 1 & a \\ 1 & 0 \end{pmatrix}$, then, the matrix $T_n(X_0)$ is given by $T_n(X_0) = A^n$. Hence, if we compute the Lyapunov exponents in each linear region, we found that are the same since J_1^n and J_2^n have the same eigenvalues for all integers n. Finally, one can obtain the following results:*

(I) If $a \geq \frac{-1}{4}$, then the matrix A has two real and distinct eigenvalues λ_1 and λ_2, given by (5.37), hence, if we consider the Jordan normal form for the matrix A, i.e., $A = PJP^{-1}$, where P is the matrix whose columns consist of the two eigenvectors of the matrix A, and $J = \begin{pmatrix} \lambda_1 & 0 \\ 0 & \lambda_2 \end{pmatrix}$, then the formula $A = PJP^{-1}$ implies that $A^n = PJ^nP^{-1}$, then the eigenvalues of A^n are the same as the eigenvalues of the matrix J^n, thus, the eigenvalues of $T_n(X_0)$ are $\delta_1 = \left(\frac{1+\sqrt{4a+1}}{2}\right)^n$, and $\delta_2 = \left(\frac{1-\sqrt{4a+1}}{2}\right)^n$, and therefore one has:

$$\begin{cases} J_1(X_0, n) = |\delta_1|^{\frac{1}{n}} = \left|\frac{1}{2}\sqrt{4a+1} + \frac{1}{2}\right| \\ \\ J_2(X_0, n) = |\delta_2|^{\frac{1}{n}} = \left|-\frac{1}{2}\sqrt{4a+1} + \frac{1}{2}\right| \end{cases} \tag{5.39}$$

One remarks that $J_1(X_0, n) \geq J_2(X_0, n)$, then the largest Lyapunov exponent is given by:

$$\omega_1(X_0) = \ln \left|\frac{1}{2}\sqrt{4a+1} + \frac{1}{2}\right| \tag{5.40}$$

in each linear region D_1 and D_2 defined above in (5.34) and (5.35).

Hence, we obtain the following:

(i) The switching map (5.32) converges to a stable fixed point when $\omega_1(X_0) < 0$, i.e., $\frac{-1}{4} \leq a < 0$, and converges to a chaotic attractor when $\omega_1(X_0) > 0$, i.e., $a > 0$. The case where $\omega_1(X_0) = 0$, i.e., $a = 0$, in this case the switching map (5.32) converges to the point $(x_0, b - |x_0|)$ for all $x_0 \in \mathbb{R}$. Note that if $a > 2$, then both $\omega_1(X_0)$, and $\omega_2(X_0)$ are positive, then the switching map (5.32) has hyperchaotic attractors.

(ii) If $a < \frac{-1}{4}$, then the matrix A has two conjugate complex eigenvalues λ_1 and λ_2, given by (5.37), hence, if we consider again the Jordan normal form of the matrix A, i.e., $A = PJP^{-1}$, where P is the matrix whose columns consist of the two eigenvectors of the matrix A, and $J = \begin{pmatrix} \alpha & \beta \\ -\beta & \alpha \end{pmatrix}$, where $\alpha = \frac{1}{2}$, and $\beta = \frac{\sqrt{4a+1}}{2}$, then the formula $A = PJP^{-1}$ implies that $A^n = PJ^nP^{-1}$, then the eigenvalues of A^n are the same as the eigenvalues of the matrix J^n, thus, the eigenvalues of $T_n(X_0)$ are:

$$
\begin{cases}
\delta_1 = \left(\frac{1 + \sqrt{-(4a+1)}i}{2} \right)^n \\[3mm]
\delta_2 = \left(\frac{1 - \sqrt{-(4a+1)}i}{2} \right)^n
\end{cases}
\tag{5.41}
$$

and therefore one has:

$$
J_1(X_0, n) = J_2(X_0, n) = |\delta_1|^{\frac{1}{n}} = |\delta_2|^{\frac{1}{n}} = \sqrt{-a}
\tag{5.42}
$$

and it follows from the definition of the Lyapunov exponents (1.37) that:

$$
\omega_1(X_0) = \omega_2(X_0) = \frac{1}{2} \ln(-a)
\tag{5.43}
$$

in each linear region D_1 and D_2 defined above in (5.34) and (5.35). Finally, one obtains that the switching map (5.32) converges to a stable fixed point when $-1 < a < \frac{-1}{4}$, and the behavior is not clear if $a = -1$. This case is discussed below and converges to a super chaotic attractor if $a < -1$. Finally, one can obtain the desired results using the two cases (I) and (ii).

In the sequel, we will illustrate some observed chaotic attractors as shown in Fig. 5.8 to Fig. 5.10, along with some interesting dynamical phenomena. The dynamical behaviors of the switching map (5.32) are investigated numerically, toward this aim, fix the parameter $b = 0.3$, and the initial condition $x_0 = y_0 = 0.01$, and let $a \in \mathbb{R}$ vary. The switching map (5.32) exhibits the dynamical behaviors shown in Fig. 5.7 as follows:

1. Unbounded orbits for $a < -1.6$.

2. Chaos for $-1.6 \leq a < -1$.

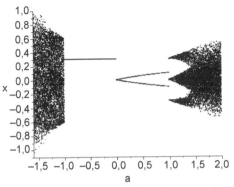

FIGURE 5.7: The bifurcation diagram for the map (5.32) for: $-1.6 \leq a \leq$ 2.0, and $b = 0.3$, and the initial condition $x_0 = y_0 = 0.01$.

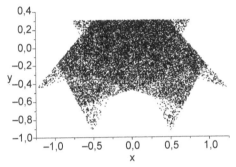

FIGURE 5.8: The piecewise linear chaotic attractor of map (5.32) obtained for: $a = -1.6$, and $b = 0.3$, and the initial condition $x_0 = y_0 = 0.01$.

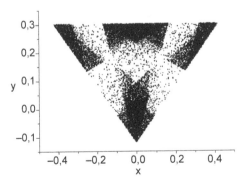

FIGURE 5.9: The piecewise linear chaotic attractor of map (5.32) obtained for: $a = 1.4$, and $b = 0.3$, and the initial condition $x_0 = y_0 = 0.01$.

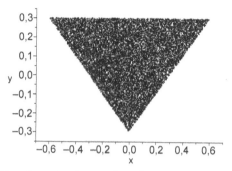

FIGURE 5.10: The piecewise linear chaotic attractor of map (5.32) obtained for: $a = 2$, and $b = 0.3$, and the initial condition $x_0 = y_0 = 0.01$,

3. Period-6 orbit for $a = -1$.

4. Stable fixed point for $-1 < a < 0$.

5. The point $(0.01, 0.29)$ when $a = 0$ (it is a non-isolated fixed point, that depends on the initial condition (x_0, y_0)).

6. Stable period-2 orbit collapses to a point and is reborn as a stable period-3 orbit for $0 < a \leq 1$.

7. Chaos for $1 < a < 2$.

8. Unbounded orbits for $a > 2$.

This scenario mainly depends on the choice of the initial conditions.

Generally, for a continuous map, positive Lyapunov exponents indicate chaos, negative exponents indicate fixed points, and if the Lyapunov exponent is equal to 0, then the dynamics are periodic, while for a discontinuous map a zero Lyapunov exponent does not still indicate periodic behavior, for this reason and for the case where $a = -1$, the switching map (5.32) generates a symbolic sequence $\hat{s} = \{\hat{s}_0; \hat{s}_1; ...; \hat{s}_j; ...\}$ composed of symbols $\hat{s}_j = i$ if $X_j = f^j(X_0) \in D_1$ or D_2. Each of those symbolic sequences is called "*admissible*" and its symbols describe the order in which trajectories, starting from any initial condition $X_0 \in \mathbb{R}^2$, visit the various sub regions D_1 and D_2. Finally, we conclude that there are some cases (depending on the position of the initial conditions) where the behavior of the switching map (5.32) is not periodic in spite of its Lyapunov exponent being zero. On one hand, we observe via numerical simulations that a positive Lyapunov exponent does not still indicate chaotic behaviors, for example when $0 < a \leq 1$, one can observe a stable period-2 orbit collapses to a point and is reborn as a stable period-3 orbit, this is like *border collision bifurcations* including "*period two to period three*" for piecewise smooth systems studied in [Nusse & Yorke (1992)]. On the other hand, one can observe that unbounded orbits also occur for $a > 2$, or $a < -1.6$. So, we conclude that one has to be very careful when dealing with this type of system because of its bewildering complexity. Note that if $a = 0$, then the map (5.32) takes the following form:

$$x_{n+1} = x_n, y_{n+1} = b - |x_n|$$

Thus, it converges to the non-isolated fixed point $(x_0, b - |x_0|)$, for all $x_0 \in \mathbb{R}$.

5.6 Lyapunov exponents of a modified map-based BVP model

In [Cao & Wang (2010)], the continuous Bonhoeffer–van der Pol (BVP) oscillator was transformed into a map-based BVP model by using the forward Euler method. This map is given by

$$h(x, y) = \begin{pmatrix} x + \delta \left(y - \frac{1}{3}x^3 + x + \mu \right) \\ y + \delta\rho \left(a - x - by \right) \end{pmatrix} \tag{5.44}$$

where

$$0 < \rho \leq 1, 0 < a < 1, 0 < b < 1, 0 < \delta < 1, \tag{5.45}$$

μ is a stimulus intensity, and δ is the step size. Here the variable x represents the electric potential across the cell membrane, and y represents a recovery force. For the map (5.44), it was shown in [Cao & Wang (2010)] that the period-doubling bifurcation is dependent on the step size, while the saddle-node bifurcation is independent of the step size. It was also shown that the map (5.44) displays chaos in the sense of Marotto when the discrete step size is varied as a bifurcation parameter. We notice that map-based models are more complicated than continuous neuron models. In this section, we replace the function $\frac{1}{3}x^3 + x + \mu$ in (5.44) by the sine function in order to obtain the map:

$$f(x, y) = \begin{pmatrix} x + \delta \left(y - \sin x \right) \\ y + \delta\rho \left(a - x - by \right) \end{pmatrix} \tag{5.46}$$

The dynamics of map (5.46) are surprising and display several different chaotic attractors outside the range of parameters given in (5.45). Most of these attractors have complicated basins of attraction. The reason is that basin boundaries arise in dissipative dynamical systems when two or more attractors are present. In such situations, each attractor has a basin of initial conditions that lead asymptotically to that attractor. The sets that separate different basins are the basin boundaries. In these cases, the basin boundaries have very complicated fractal structures and, therefore, pose an additional impediment to predicting long-term behavior. For the map (5.46), we have calculated the attractors and their basins of attraction on a grid in ab-space where the system is chaotic. There is a wide variety of possible attractors, only some of which are shown in Figs. 5.11–5.20. Also, some of the basin boundaries are smooth, but others appear to be fractal, and this is not a result of numerical errors since the structure persists as the number of iterations for each initial condition is increased.

As with many other systems, maps such as the one in Eq. (5.46) typically have no direct application to particular physical systems, but they serve to exemplify the kinds of dynamical behavior, such as routes to chaos, that are common in physical chaotic systems. Thus, an analytical and numerical study is warranted.

The map under consideration does not characterize the dynamics of BVP oscillator, but it displays further phenomena and complicated behaviors when comparing it to the original map-based BVP model studied in [Cao & Wang (2010)]. The fixed points of map (5.46) are $(x, \sin x)$, where x is the solution of the equation $\sin x = \frac{a-x}{b}$. This equation is possible if and only if $-1 \le \frac{a-x}{b} \le 1$, that is $-b + a \le x \le b + a$ for $b > 0$. Such a solution can be found numerically for specific values of the bifurcation parameters. The Jacobian matrix of the map (5.46) is

$$J = \begin{pmatrix} 1 - \delta \cos x & \delta \\ -\delta\rho & 1 - \delta\rho b \end{pmatrix}$$

and its determinant is $\delta^2\rho - b\delta\rho + 1 + (b\delta^2\rho - \delta)\cos x$. Thus the map (5.46) is dissipative if and only if one of the following cases holds:

$$\begin{cases} b\delta^2\rho - \delta = 0, \delta^2\rho - b\delta\rho + 1 < 0. \\ b\delta^2\rho - \delta > 0, (\delta^2 - b\delta + b\delta^2)\rho + 1 - \delta \le 0. \\ b\delta^2\rho - \delta < 0, (\delta^2 - b\delta - b\delta^2)\rho + \delta + 1 \le 0. \end{cases}$$

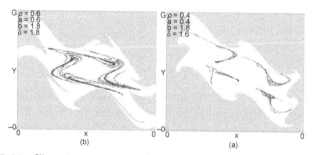

FIGURE 5.11: Chaotic attractors of map (5.46) with their basins of attraction (white) for (a) $\rho = 0.4, a = 0.4, b = 1.8, \delta = 1.6$. (b) $\rho = 0.6, a = 0.6, b = 1.8, \delta = 1.8$

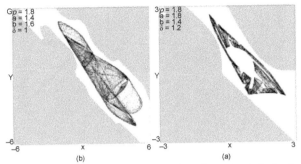

FIGURE 5.12: Chaotic attractors of map (5.46) with their basins of attraction (white) for (a) $\rho = 1.8, a = 1.8, b = 1.4, \delta = 1.2$. (b) $\rho = 1.8, a = 1.4, b = 1.6, \delta = 1$.

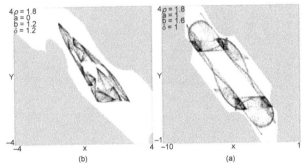

FIGURE 5.13: Chaotic attractors of map (5.46) with their basins of attraction (white) for (a) $\rho = 1.8, a = 1, b = 1.6, \delta = 1$. (b) $\rho = 1.8, a = 0, b = 1.2, \delta = 1.2$.

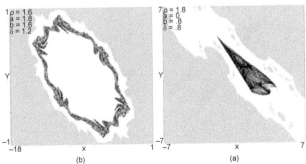

FIGURE 5.14: Chaotic attractors of map (5.46) with their basins of attraction (white) for (a) $\rho = 1.8, a = 0, b = 0.8, \delta = 0.8$. (b) $\rho = 1.6, a = 1.8, b = 1.6, \delta = 1.2$.

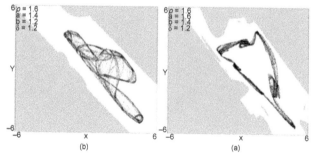

FIGURE 5.15: Chaotic attractors of map (5.46) with their basins of attraction (white) for (a) $\rho = 1.6, a = 1.6, b = 1.4, \delta = 1.2$. (b) $\rho = 1.6, a = 1.4, b = 1.2, \delta = 1.2$.

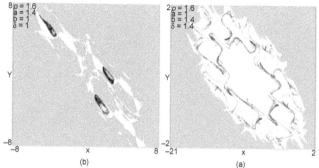

FIGURE 5.16: Chaotic attractors of map (5.46) with their basins of attraction (white) for (a) $\rho = 1.6, a = 1, b = 1.4, \delta = 1.4$. (b) $\rho = 1.6, a = 1.4, b = 1, \delta = 1$.

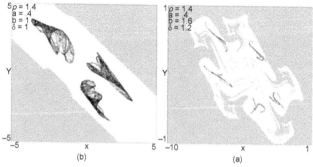

FIGURE 5.17: Chaotic attractors of map (5.46) with their basins of attraction (white) for (a) $\rho = 1.4, a = 0.4, b = 1.6, \delta = 1.2$. (b) $\rho = 1.4, a = 0.4, b = 1, \delta = 1$.

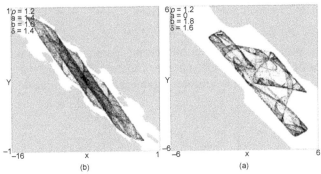

FIGURE 5.18: Chaotic attractors of map (5.46) with their basins of attraction (white) for (a) $\rho = 1.2, a = 0$, $b = 1.8$, $\delta = 1.6$. (b) $\rho = 1.2, a = 1.4$, $b = 1.6$, $\delta = 1.4$.

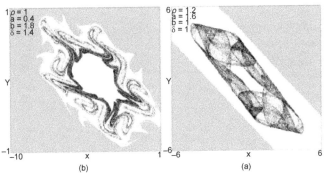

FIGURE 5.19: Chaotic attractors with their basins of attraction (white) for (a) $\rho = 1.2, a = 1.6, b = 1, \delta = 1$. (b) $\rho = 1, a = 0.4, b = 1.8, \delta = 1.4$.

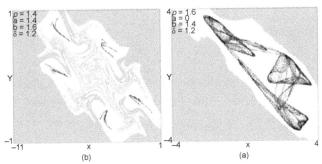

FIGURE 5.20: Chaotic attractors of map (5.46) with their basins of attraction (white) for (a) $\rho = 1.6, a = 0, b = 1.4$, $\delta = 1.2$. (b) $\rho = 1.4, a = 1.4$, $b = 1.6$, $\delta = 1.2$.

The proof is based on the fact that $\cos x < \frac{-\delta^2 \rho + b\delta\rho - 1}{\delta(b\delta\rho - 1)}$ holds if $\cos x < 1 \le \frac{-\delta^2 \rho + b\delta\rho - 1}{\delta(b\delta\rho - 1)}$ and $\cos x > \frac{-\rho\delta^2 + b\rho\delta - 1}{(b\delta^2\rho - \delta)}$ holds if $\cos x > -1 \ge \frac{-\rho\delta^2 + b\rho\delta - 1}{(b\delta^2\rho - \delta)}$.

To find some regions in the bifurcation parameters for which map (5.46) is not chaotic or hyperchaotic, we use Theorem 15 to find the upper and the lower bounds for all the Lyapunov exponents $l_i(x_0)$ of map (5.46) as follows: We have

$$
\begin{cases}
\lambda_{\max}\left(J^T J\right) = \xi_0 + \frac{1}{2}\delta\sqrt{\xi_1\xi_2} \\[2mm]
\lambda_{\min}\left(J^T J\right) = \xi_0 - \frac{1}{2}\delta\sqrt{\xi_1\xi_2}
\end{cases}
$$

where

$$
\begin{cases}
\xi_0 = \frac{1}{2}m^2\delta^2 - m\delta + \frac{1}{2}\delta^2 + \frac{1}{2}\delta^2\rho^2 - b\delta\rho + \frac{1}{2}b^2\delta^2\rho^2 + 1 \\[2mm]
\xi_1 = \left(b^2\rho^2 - 2bm\rho + m^2 + \rho^2 - 2\rho + 1\right) \\[2mm]
\xi_2 = b^2\delta^2\rho^2 + 2bm\delta^2\rho - 4b\delta\rho + m^2\delta^2 - 4m\delta + \delta^2\rho^2 + 2\delta^2\rho + \delta^2 + 4.
\end{cases}
$$

(5.47)

Here $m = \sin x$. We obtain

$$
\lambda_{\max}\left(J^T J\right) \le \xi_3 + \frac{1}{2}\delta\sqrt{\xi_4\xi_5} = N
$$

where

$$
\begin{cases}
\xi_3 = \frac{1}{2}\left(2\delta + 2\delta^2 + \delta^2\rho^2 + 2b\delta\rho + b^2\delta^2\rho^2 + 2\right) \\[2mm]
\xi_4 = 2\rho + b^2\rho^2 + 2b\rho + \rho^2 + 2 \\[2mm]
\xi_5 = 4\delta + 2\delta^2 + \delta^2\rho^2 + 2\delta^2\rho + 4b\delta\rho + b^2\delta^2\rho^2 + 2b\delta^2\rho + 4.
\end{cases}
$$

(5.48)

Thus, in the region

$$
N = \xi_3 + \frac{1}{2}\delta\sqrt{\xi_4\xi_5} \le 1
$$

the map (5.46) displays no chaotic attractors and only fixed points and periodic solutions can appear. Now, $\lambda_{\min}\left(J^T J\right) > 1$ is equivalent to the conditions:

$$
\begin{cases}
\frac{1}{2}\delta^2 m^2 - \delta m + \xi_6 > 0 \\[2mm]
\xi_7 m^2 + \xi_8 m + \xi_9 > 0.
\end{cases}
$$

(5.49)

Inequalities (5.49) can be regarded as polynomials of degree two in the term $\sin x$ where

$$
\begin{cases}
\xi_6 = \frac{1}{2}\delta^2 + \frac{1}{2}\delta^2\rho^2 - b\delta\rho + \frac{1}{2}b^2\delta^2\rho^2 \\[2mm]
\xi_7 = b\delta^3\rho\,(b\delta\rho - 2) \\[2mm]
\xi_8 = \left(2b\delta^4\rho^2 - 2b^2\delta^3\rho^2 + 4b\delta^2\rho - 2\delta^3\rho\right) \\[2mm]
\xi_9 = \left(\delta^4\rho^2 - 2b\delta^3\rho^2 - \delta^2\rho^2 + 2\delta^2\rho - \delta^2\right).
\end{cases}
\tag{5.50}
$$

If conditions (5.49) hold for all x, then map (5.46) is hyperchaotic. Attempts to find some values fail due to the complicated formulas of the coefficients $(\xi_i)_{0 \leq i \leq 9}$.

Notice that the above analysis still holds true if we replace the term $\sin x$ in map (5.46) by any bounded function $g(x)$ such that the resulting map has at least one fixed point in which bounded orbits are possible, that is, where the equation $g(x) - \frac{a-x}{b} = 0$ has at least one real solution.

Chapter 6

Control, Synchronization and Chaotification of Dynamical Systems

6.1 Introduction

Various methods of chaos synchronization of dynamical systems are well known in the current literature. Examples includes active control, backstepping design and sliding mode control [Pecora & Carroll (1990), Boccaletti, *et al.*, (2002), Nimbler & Mareels (1997), Ashwin (2003), Pham & Slotine (2005), Park (2005)].

This chapter is devoted to presenting some definitions and relevant results regarding different techniques of synchronization for both continuous time systems and discrete time systems. In Sec. 6.2, we present the notion of compound synchronization of different chaotic systems, i.e., compound synchronization of four different chaotic systems based on active backstepping control, where the chaotic Lorenz, Rossler and Chen systems are considered as master systems and the Lû system is considered as a slave system. In Sec. 6.3, the synchronization of 3-D continuous-time quadratic systems using a universal non-linear control law is presented. In this case, The proposed control law does not need any conditions on the considered systems, hence, it is a universal synchronization approach for general 3-D continuous-time quadratic systems. In Sec. 6.4, the notion of *co-existence* of certain types of synchronization and their inverse is defined with some relevant results. In Sec. 6.5, we present the synchronization of 4-D continuous-time quadratic systems using a universal non-linear control law which does not need any conditions on the considered systems. In Sec. 6.6, the *quasi-synchronization* of discrete time systems with different dimensions is presented where the master system is the 3-D generalized Hénon map and the slave system is the usual Hénon mapping. Section 6.7 is devoted to the chaotification of 3-D linear continuous-time systems using the *signum function feedback* giving a very simple chaotic system. As an example of chaos control, we present in Sec. 6.8 the example of controlling a 3-D cancer model with structured uncertainties. For discrete times systems, we give in Sec. 6.9, the technique used for controlling homoclinic chaotic attractor in the general two-dimensional piecewise smooth map. Sec. 6.10 is concerned with the creation of *robust chaos* in the general 2-D discrete mappings via the controller of a simple piecewise smooth function under some realizable conditions. Finally, a more general case of chaotifying stable n-D linear maps via the controller of any bounded function is presented and discussed in Sec. 6.11.

6.2 Compound synchronization of different chaotic systems

In this section, we present an example of the compound synchronization of four different chaotic systems, based on active backstepping control as shown in [Sivasamy (2017)], where the chaotic Lorenz, Rôssler and Chen systems are considered as master systems and the Lû system is considered as a slave system. The model can be described as follows:

$$x' = f_1(x) \tag{6.1}$$
$$y' = f_2(y) \tag{6.2}$$
$$z' = f_3(z) \tag{6.3}$$
$$w' = g(w) + u \tag{6.4}$$

where $x = (x_1, \cdots, x_n)^T$, $y = (y_1, \cdots, y_n)^T$, $z = (z_1, \cdots, z_n)^T$ and $w = (w_1, \cdots, w_n)^T$ are the state of the systems (6.1)-(6.4) respectively. The functions $f_1, f_2, f_3, g : \mathbb{R}^n \to \mathbb{R}^n$ are four continuous functions and $u = (u_1, \cdots, u_n) \in \mathbb{R}^n$ is the controller to be designed later. For simplicity, let $X = diag(x_1, x_2, x_3)$, $Y = diag(y_1, y_2, y_3)$, $Z = diag(z_1, z_2, z_3)$, $W = diag(w_1, w_2, w_3)$ and $e = diag(e_1, e_2, e_3)$.

Definition 47 *Compound synchronization of the master systems (6.1)-(6.3) and the slave system (6.4) will be achieved, if there exist non-zero constant matrices $A, B, C, D \in \mathbb{R}^{n \times n}$ with $D \neq 0$ such that*

$$\lim_{t \to \infty} \|e\| = \lim_{t \to \infty} \|DW - AX(BY + CZ)\|$$

where $\|.\|$ denotes the Euclidean norm.

Here, the constant matrices A, B, C and D are called *scaling matrices*. In the present synchronization method, the error system can be defined by multiplication of chaotic systems, contrary to the cases of the existing synchronization schemes where the error system can be expressed as a simple addition and subtraction between chaotic systems. If $A \neq 0$, $B = 0$ or $C = 0$, then the compound synchronization problem is transformed to functional projective synchronization where scaling functional becomes some chaotic systems. If $A = B = C = 0$, the compound synchronization problem is turned into a chaos control problem of system (6.4).

Let us consider the following three master systems: The chaotic Lorenz system:

$$\begin{aligned}
x_1' &= a_1(x_2 - x_1) \\
x_2' &= b_1 x_1 - x_2 - x_1 x_3 \\
x_3' &= x_1 x_2 - c_1 x_3
\end{aligned} \tag{6.5}$$

the chaotic Rossler system:

$$\begin{aligned}
y_1' &= -y_2 - y_3 \\
y_2' &= y_1 + a_2 y_2 \\
y_3' &= b_2 + y_3(y_1 - c_2)
\end{aligned} \tag{6.6}$$

and the chaotic Chen system:

$$\begin{aligned}
z_1' &= a_3(z_2 - z_1) \\
z_2' &= (c_3 - a_3)z_1 + c_3 z_2 - z_1 z_3 \\
z_3' &= z_1 z_2 - b_3 z_3
\end{aligned} \tag{6.7}$$

The controlled slave is the chaotic Lû system:

$$\begin{aligned}
w_1' &= a_4(w_2 - w_1) + u_1 \\
w_2' &= c_4 w_2 - w_1 w_3 + u_2 \\
w_3' &= w_1 w_2 - b_4 w_3 + u_3
\end{aligned} \tag{6.8}$$

The constant matrices are chosen as follows:

$$\begin{cases}
A = diag(\alpha_1, \alpha_2, \alpha_3) \\
B = diag(\beta_1, \beta_2, \beta_3) \\
C = diag(\gamma_1, \gamma_2, \gamma_3) \\
D = diag(\delta_1, \delta_2, \delta_3).
\end{cases}$$

Then the error system takes the form:

$$\begin{cases}
e_1 = \delta_1 w_1 - \alpha_1 x_1(\beta_1 y_1 + \gamma_1 z_1) \\
e_2 = \delta_2 w_2 - \alpha_2 x_2(\beta_2 y_2 + \gamma_2 z_2) \\
e_3 = \delta_3 w_3 - \alpha_3 x_3(\beta_3 y_3 + \gamma_3 z_3)
\end{cases} \tag{6.9}$$

Thus

$$\begin{cases}
e_1' = \delta_1 w_1' - \alpha_1 x_1'(\beta_1 y_1 + \gamma_1 z_1) - \alpha_1 x_1(\beta_1 y_1' + \gamma_1 z_1') \\
e_2' = \delta_2 w_2' - \alpha_2 x_2'(\beta_2 y_2 + \gamma_2 z_2) - \alpha_2 x_2(\beta_2 y_2' + \gamma_2 z_2') \\
e_3' = \delta_3 w_3' - \alpha_3 x_3'(\beta_3 y_3 + \gamma_3 z_3) - \alpha_3 x_3(\beta_3 y_3' + \gamma_3 z_3')
\end{cases} \tag{6.10}$$

Thus, we have

$$
\left\{
\begin{aligned}
e_1' &= \delta_1 a_4(w_2 - w_1) - \alpha_1 a_1(x_2 - x_1)(\beta_1 y_1 + \gamma_1 z_1) \\
&\quad - \alpha_1 x_1(\beta_1(-y_2 - y_3) + \gamma_1 a_3(z_2 - z_1)) \\[8pt]
e_2' &= \delta_2(c_4 w_2 - w_1 w_3) - \alpha_2(b_1 x_1 - x_2 - x_1 x_3)(\beta_2 y_2 + \gamma_2 z_2) \\
&\quad - \alpha_2 x_2(\beta_2(y_1 + a_2 y_2) + \gamma_2((c_3 - a_3)z_1 + c_3 z_2 - z_1 z_3)) \\[8pt]
e_3' &= \delta_3(w_1 w_2 - b_4 w_3) - \alpha_3(x_1 x_2 - c_1 x_3)(\beta_3 y_3 + \gamma_3 z_3) \\
&\quad - \alpha_3 x_3(\beta_3(b_2 + y_3(y_1 - c_2)) + \gamma_3(z_1 z_2 - b_3 z_3))
\end{aligned}
\right.
\tag{6.11}
$$

By simple transformation, the above equations become

$$
\left\{
\begin{aligned}
e_1' &= -a_4 e_1 + \tfrac{\delta_1 a_4}{\delta_2} e_2 + f_1 + \delta u_1 \\[8pt]
e_2' &= c_4 e_2 - \tfrac{\delta_2}{\delta_1 \delta_3} e_1 e_3 - \tfrac{\delta_2}{\delta_1 \delta_3}\alpha_3 x_3(\beta_3 y_3 + \gamma_3 z_3)e_1 \\
&\quad - \tfrac{\delta_2}{\delta_1 \delta_3}\alpha_1 x_1(\beta_1 y_1 + \gamma_1 z_1)e_3 + f_2 + \delta_2 u_2 \\[8pt]
e_3' &= -b_4 e_3 + \tfrac{\delta_3}{\delta_1 \delta_2} e_1 e_2 + \tfrac{\delta_3}{\delta_1 \delta_2}\alpha_1 x_1(\beta_1 y_1 + \gamma_1 z_1)e_2 \\
&\quad + \tfrac{\delta_3}{\delta_1 \delta_2}\alpha_2 x_2(\beta_2 y_2 + \gamma_2 z_2)e_1 + f_3 + \delta_3 u_3
\end{aligned}
\right.
\tag{6.12}
$$

where

$$
\left\{
\begin{aligned}
f_1 &= \tfrac{\delta_1 a_4}{\delta_2}\alpha_2 x_2(\beta_2 y_2 + \gamma_2 z_2) - a_4 \alpha_1 x_1(\beta_1 y_1 + \gamma_1 z_1) \\
&\quad - \alpha_1 a_1(\beta_1 y_1 + \gamma_1 z_1)(x_2 - x_1) + \alpha_1 \beta_1 x_1(y_2 + y_3) \\
&\quad + \alpha_1 \gamma_1 x_1 a_3(z_2 - z_1) \\[8pt]
f_2 &= -\tfrac{\delta_2}{\delta_1 \delta_3}\alpha_1 \alpha_3 x_1 x_3(\beta_1 y_1 + \gamma_1 z_1)(\beta_3 y_3 + \gamma_3 z_3) \\
&\quad + c_4 \alpha_2 x_2(\beta_2 y_2 + \gamma_2 z_2) - \alpha_2 x_2 \beta_2(y_1 + a_2 y_2) \\
&\quad - \alpha_2(\beta_2 y_2 + \gamma_2 z_2)(b_1 x_1 - x_2 - x_1 x_3) \\[8pt]
f_3 &= -\alpha_2 x_2 \gamma_2((c_3 - a_3)z_1 + c_3 z_2 - z_1 z_3) \\
&\quad + \tfrac{\delta_2}{\delta_1 \delta_3}\alpha_1 \alpha_2 x_1(\beta_1 y_1 + \gamma_1 z_1)x_2(\beta_2 y_2 + \gamma_2 z_2) \\
&\quad - b_4 \alpha_3 x_3(\beta_3 y_3 + \gamma_3 z_3) - \alpha_3 \beta_3 x_3(b_2 + y_3(y_1 - c_2)) \\
&\quad - \alpha_3(\beta_3 y_3 + \gamma_3 z_3)(x_1 x_2 - c_1 x_3) - \alpha_3 x_3 \gamma_3(z_1 z_2 - b_3 z_3)
\end{aligned}
\right.
$$

Thus, the following result was proved in [Sivasamy (2017)]:

Theorem 47 *The compound synchronization between the master systems (6.5)-(6.7) and the slave system (6.8) can be achieved, if the controllers are chosen as follows:*

$$\left\{ \begin{array}{l}
u_1 = \frac{1}{\delta_1}\{-f_1\} \\[2mm]
u_2 = \frac{1}{\delta_2}\Big\{ -f_2 - \frac{\delta_2}{\delta_1}\frac{(a_4-1)}{a_4}(c_4+1)v_1 - \frac{\delta_1}{\delta_2}a_4v_1 + \\[2mm]
\qquad \frac{\delta_2}{\delta_1\delta_3}\alpha_3 x_3(\beta_3 y_3 + \gamma_3 z_3)v_1 - (c_4 - a_4 + 2)v_2 \Big\} \\[4mm]
u_3 = \frac{1}{\delta_2}\Big\{ -f_3 + (b_4-1)v_3 - \frac{\delta_3}{\delta_1^2}\frac{(a_4-1)}{a_4}v_1^2 \\[2mm]
\quad -\frac{\delta_3}{\delta_1\delta_2}\alpha_2 x_2(\beta_2 y_2 + \gamma_2 z_2) - \frac{\delta_3}{\delta_1^2}\frac{(a_4-1)}{a_4}\alpha_1 \times x_1(\beta_1 y_1 + \gamma_1 z_1)v_1 \\[2mm]
\quad + \left(\frac{\delta_2}{\delta_1\delta_3} - \frac{\delta_3}{\delta_1\delta_2} \right)(v_1 + \alpha_1 x_1(\beta_1 y_1 + \gamma_1 z_1))v_2 \Big\}
\end{array} \right.$$

where

$$\left\{ \begin{array}{l}
v_1 = e_1 \\[3mm]
v_2 = e_2 - \frac{\delta_2}{\delta_1}\frac{(a_4-1)}{a_4}v_1 \\[3mm]
v_3 = e_3
\end{array} \right.$$

Proof 13 *The result is obtained by using the active backstepping control method. Let $v_1 = e_1$, its derivative is given by:*

$$v_1' = -a_4 v_1 + \frac{\delta_1 a_4}{\delta_2}e_2 + f_1 + \delta u_1$$

where $e_2 = h_2(v_1)$ is regarded as a virtual controller to be defined later. For the correct choice of $h_2(v_1)$ and u_1 to stabilize the v_1-system, we consider the following Lyapunov function:

$$W_1 = \frac{1}{2}v_1^2$$

The derivative of W_1 is given by

$$w_1' = v_1 v_1' = v_1\left(-a_4 v_1 + \frac{\delta_1 a_4}{\delta_2}e_2 + f_1 + \delta u_1 \right)$$

Then one can choose $h_2(v_1) = \frac{\delta_2}{\delta_1}\frac{(a_4-1)}{a_4}v_1$ and u_1 such that $W_1' = -v_1^2 < 0$ which implies that the v_1-system is asymptotically stable. Define the error between e_2 and the estimate virtual control function $h_2(v_1)$ as $v_2 = e_2 - h_2(v_1)$. Then we have the following (v_1, v_2)-subsystem

$$\left\{ \begin{array}{l}
v_1' = -v_1 + \frac{\delta_1 a_4}{\delta_2}v_2 \\[3mm]
v_2' = \frac{\delta_2}{\delta_1}\frac{(a_4-1)}{a_4}(c_4+1)v_1 - \frac{\delta_2}{\delta_1\delta_3}\alpha_3 x_3(\beta_3 y_3 + \gamma_3 z_3)v_1 \\[2mm]
\qquad -\frac{\delta_2}{\delta_1\delta_3}(v_1 + \alpha_1 x_1(\beta_1 y_1 + \gamma_1 z_1))e_3 \\[2mm]
\qquad +(c_4 - a_4 + 1)v_2 + f_2 + \delta_2 u_2
\end{array} \right. \qquad (6.13)$$

where $e_3 = h_3(v_2, v_1)$ is regarded as a virtual controller. For the choice of $h_3(v_2, v_1)$ and u_2 to stabilize (v_1, v_2)-system (6.13), one can select the following Lyapunov function:

$$W_2 = W_1 + \frac{1}{2}v_2^2$$

The derivative of W_2 is given by

$$W_2' = -v_1^2 + v_2\left(\frac{\delta_2}{\delta_1}\frac{(a_4-1)}{a_4}(c_4+1)v_1 - \frac{\delta_2}{\delta_1\delta_3}\alpha_3 x_3(\beta_3 y_3 + \gamma_3 z_3)v_1\right) +$$

$$v_2\left((c_4 - a_4 + 1)v_2 - \frac{\delta_2}{\delta_1\delta_3}(v_1 + \alpha_1 x_1(\beta_1 y_1 + \gamma_1 z_1))e_3 + f_2 + \delta_2 u_2\right)$$

We can choose $h_3(v_2, v_1) = 0$ and u_2 such that $W_2' = -v_1^2 - v_2^2 < 0$, then the (v_1, v_2)-system (6.13) is asymptotically stable. Define the error between e_3 and then estimate the virtual control function $h_3(v_2, v_1)$ as $v_3 = e_3 - h_3(v_2, v_1)$. Then one can have following (v_1, v_2, v_3)- subsystem

$$
\begin{cases}
v_1' = -v_1 + \frac{\delta_1 a_4}{\delta_2}v_2 \\[2mm]
v_2' = -\frac{\delta_1 a_4}{\delta_2}v_1 - v_2 - \frac{\delta_2}{\delta_1\delta_3}(v_1 + \alpha_1 x_1(\beta_1 y_1 + \gamma_1 z_1))v_3 \\[2mm]
v_3' = \frac{\delta_3}{\delta_1^2}\frac{(a_4-1)}{a_4}v_1^2 + \frac{\delta_3}{\delta_1^2}\frac{(a_4-1)}{a_4}\alpha_1 x_1(\beta_1 y_1 + \gamma_1 z_1)v_1 \\[1mm]
\quad + \frac{\delta_3}{\delta_1\delta_2}\alpha_2 x_2(\beta_2 y_2 + \gamma_2 z_2)v_1 - b_4 v_3 \\[1mm]
\quad + \frac{\delta_3}{\delta_1\delta_2}(v_1 + \alpha_1 x_1(\beta_1 y_1 + \gamma_1 z_1))v_2 + f_3 + \delta_3 u_3
\end{cases}
\tag{6.14}
$$

For the choice of u_3 to stabilize (v_1, v_2, v_3)-subsystem (6.14), we consider the following Lyapunov function:

$$W_3 = W_2 + \frac{1}{2}v_3^2$$

The derivative of W_3 is given by

$$W_3' = -v_1^2 - v_2^2 + v_3\left(\frac{\delta_3}{\delta_1^2}\frac{(a_4-1)}{a_4}v_1^2 + \frac{\delta_3}{\delta_1^2}\frac{(a_4-1)}{a_4}\alpha_1 x_1(\beta_1 y_1 + \gamma_1 z_1)v_1\right.$$

$$+ \frac{\delta_3}{\delta_1\delta_2}\alpha_2 x_2(\beta_2 y_2 + \gamma_2 z_2)v_1 + \frac{\delta_3}{\delta_1\delta_2}(v_1 + \alpha_1 x_1(\beta_1 y_1 + \gamma_1 z_1))v_2$$

$$\left. - b_4 v_3 + f_3 + \delta_3 u_3\right)$$

Then one can choose u_3 such that $W_3' = -v_1^2 - v_2^2 - v_3^2 < 0$, i.e., the (v_1, v_2, v_3)-subsystem (6.14) is asymptotically stable, that is, $\lim_{t\to\infty} v_i = 0$, $i = 1, 2, 3$. By using following transformations: $v_1 = e_1$, $v_2 = e_2 - \frac{\delta_2}{\delta_1}\frac{(a_4-1)}{a_4}v_1$ and $v_3 = e_3$, we have $\lim_{t\to\infty} e_i = 0$, $i = 1, 2, 3$, which implies that the master systems (6.5), (6.6) and (6.7) and the slave system (6.8) will reach compound synchronization.

The following corollaries are easily derived from the above theorem.

Now, suppose $A = diag(\alpha_1, \alpha_2, \alpha_3) \neq 0$, $B = diag(\beta_1, \beta_2, \beta_3) = 0$ and $C = diag(\gamma_1, \gamma_2, \gamma_3) \neq 0$. Then the error system can be written in the following form:

$$\begin{cases} e_1' = -a_4 e_1 + \frac{\delta_1 a_4}{\delta_2} e_2 + \bar{f}_1 + \delta u_1 \\ \\ e_2' = c_4 e_2 - \frac{\delta_2}{\delta_1 \delta_3} e_1 e_3 - \frac{\delta_2}{\delta_1 \delta_3} \alpha_1 \gamma_1 x_1 z_1 e_3 - \frac{\delta_2}{\delta_1 \delta_3} \alpha_3 \gamma_3 x_3 z_3 e_1 + \bar{f}_2 + \delta_2 u_2 \\ \\ e_3' = -b_4 e_3 + \frac{\delta_3}{\delta_1 \delta_2} e_1 e_2 + \frac{\delta_3}{\delta_1 \delta_2} \alpha_1 \gamma_1 x_1 z_1 e_2 + \frac{\delta_3}{\delta_1 \delta_2} \alpha_2 \gamma_2 x_2 z_2 e_1 + \bar{f}_3 + \delta_3 u_3 \end{cases}$$

where

$$\begin{cases} \bar{f}_1 = \frac{\delta_1 a_4}{\delta_2} \alpha_2 \gamma_2 x_2 z_2 - a_4 \alpha_1 \gamma_1 x_1 z_1 - \alpha_1 a_1 \gamma_1 z_1 (x_2 - x_1) \\ \qquad + \alpha_1 \gamma_1 x_1 a_3 (z_2 - z_1) \\ \\ \bar{f}_2 = c_4 \alpha_2 \gamma_2 x_2 z_2 - \frac{\delta_2}{\delta_1 \delta_3} \alpha_1 \alpha_3 \gamma_1 \gamma_3 x_1 z_1 x_3 z_3 - \alpha_2 x_2 \gamma_2 ((c_3 - a_3) z_1 \\ \qquad + c_3 z_2 - z_1 z_3) - \alpha_2 \gamma_2 z_2 (b_1 x_1 - x_2 - x_1 x_3) \\ \\ \bar{f}_3 = -b_4 \alpha_3 x_3 \gamma_3 z_3 + \frac{\delta_2}{\delta_1 \delta_3} \alpha_1 \alpha_2 \gamma_1 \gamma_2 x_1 z_1 x_2 z_2 - \alpha_3 \gamma_3 z_3 (x_1 x_2 - c_1 x_3) \\ \qquad - \alpha_3 x_3 \gamma_3 (z_1 z_2 - b_3 z_3) \end{cases}$$

Hence, the following corollaries are proved in [Sivasamy (2017)]:

The compound synchronization between the master systems (6.5), (6.6) and (6.7) and the slave system (6.8) can be achieved, if the controllers are chosen as follows:

$$\begin{cases} u_1 = \frac{1}{\delta_1}\{-\bar{f}_1\} \\ \\ u_2 = \frac{1}{\delta_2}\left\{-\bar{f}_2 - \frac{\delta_2}{\delta_1}\frac{(a_4-1)}{a_4}(c_4+1)v_1 + \frac{\delta_2}{\delta_1 \delta_3}\alpha_3 \gamma_3 x_3 z_3 v_1\right\} \\ \qquad + \frac{1}{\delta_2}\left\{-(c_4 - a_4 + 2)v_2 - \frac{\delta_1}{\delta_2} a_4 v_1\right\} \\ \\ u_3 = \frac{1}{\delta_3}\left(-\bar{f}_3 + (b_4 - 1)v_3 - \frac{\delta_3}{\delta_1^2}\frac{(a_4-1)}{a_4} v_1^2 - \frac{\delta_3}{\delta_1 \delta_2}\alpha_2 \gamma_2 x_2 z_2\right) \\ \qquad + \frac{1}{\delta_3}\left(-\frac{\delta_3}{\delta_1^2}\frac{(a_4-1)}{a_4}\alpha_1 \gamma_1 x_1 z_1 v_1 + \left(\frac{\delta_2}{\delta_1 \delta_3} - \frac{\delta_3}{\delta_1 \delta_2}\right)(v_1 + \alpha_1 \gamma_1 x_1 z_1)v_2\right) \end{cases}$$

where

$$\begin{cases} v_1 = e_1 \\ \\ v_2 = e_2 - \frac{\delta_2}{\delta_1}\frac{(a_4-1)}{a_4} v_1 \\ \\ v_3 = e_3. \end{cases}$$

Similarly, if $A = diag(\alpha_1, \alpha_2, \alpha_3) \neq 0$, $B = diag(\beta_1, \beta_2, \beta_3) \neq 0$ and $C = diag(\gamma_1, \gamma_2, \gamma_3) = 0$, then we have error system as follows:

$$\begin{cases} e_1' = -a_4 e_1 + \frac{\delta_1 a_4}{\delta_2} e_2 + \tilde{f}_1 + \delta_1 u_1 \\[2mm] e_2' = c_4 e_2 - \frac{\delta_2}{\delta_1 \delta_3} e_1 e_3 - \frac{\delta_2}{\delta_1 \delta_3} \alpha_3 \beta_3 x_3 y_3 e_1 - \frac{\delta_2}{\delta_1 \delta_3} \alpha_1 \beta_1 x_1 y_1 e_3 \\[1mm] \qquad + \tilde{f}_2 + \delta_2 u_2 \\[2mm] e_3' = -b_4 e_3 + \frac{\delta_3}{\delta_1 \delta_2} e_1 e_2 + \frac{\delta_3}{\delta_1 \delta_2} \alpha_1 \gamma_1 x_1 z_1 e_2 + \frac{\delta_3}{\delta_1 \delta_2} \alpha_2 \gamma_2 x_2 z_2 e_1 \\[1mm] \qquad + \tilde{f}_3 + \delta_3 u_3 \end{cases}$$

where

$$\begin{cases} \tilde{f}_1 = \frac{\delta_1 a_4}{\delta_2} \alpha_2 \beta_2 x_2 y_2 - a_4 \alpha_1 \beta_1 x_1 y_1 - \alpha_1 \beta_1 a_1 y_1 (x_2 - x_1) \\[1mm] \qquad + \alpha_1 \beta_1 x_1 (y_2 + y_3) \\[3mm] \tilde{f}_2 = c_4 \alpha_2 \beta_2 x_2 y_2 - \frac{\delta_2}{\delta_1 \delta_3} \alpha_1 \alpha_3 \beta_1 \beta_3 x_1 y_1 x_3 y_3 \\[1mm] \qquad - \alpha_2 \beta_2 y_2 (b_1 x_1 - x_2 - x_1 x_3) - \alpha_2 \beta_2 x_2 (y_1 + a_2 y_2) \\[3mm] \tilde{f}_3 = -b_4 \alpha_3 \beta_3 x_3 y_3 + \frac{\delta_2}{\delta_1 \delta_3} \alpha_1 \alpha_2 \beta_1 \beta_2 x_1 y_1 x_2 y_2 \\[1mm] \qquad - \alpha_3 \beta_3 y_3 (x_1 x_2 - c_1 x_3) - \alpha_3 \beta_3 x_3 (b_2 + y_3 (y_1 - c_2)) \end{cases}$$

The compound synchronization between the master systems (6.5), (6.6) and (6.7) and the slave system (6.8) can be achieved, if the controllers are chosen as follows:

$$\begin{cases} u_1 = \frac{1}{\delta_1} \{-\tilde{f}_1\} \\[3mm] u_2 = \frac{1}{\delta_2} \left\{ -\tilde{f}_2 - \frac{\delta_2}{\delta_1} \frac{(a_4-1)}{a_4} (c_4 + 1) v_1 + \frac{\delta_2}{\delta_1 \delta_3} \alpha_3 \beta_3 x_3 y_3 v_1 \right\} \\[1mm] \qquad + \frac{1}{\delta_2} \left\{ -(c_4 - a_4 + 2) v_2 - \frac{\delta_1}{\delta_2} a_4 v_1 \right\} \\[3mm] u_3 = \frac{1}{\delta_3} \left\{ -\tilde{f}_3 + (b_4 - 1) v_3 - \frac{\delta_3}{\delta_1^2} \frac{(a_4-1)}{a_4} v_1^2 - \frac{\delta_3}{\delta_1 \delta_2} \alpha_2 \beta_2 x_2 y_2 \right\} \\[1mm] \qquad + \frac{1}{\delta_3} \left\{ -\frac{\delta_3}{\delta_1^2} \frac{(a_4-1)}{a_4} \alpha_1 \beta_1 x_1 y_1 v_1 \right\} \\[1mm] \qquad + \frac{1}{\delta_3} \left\{ \left(\frac{\delta_2}{\delta_1 \delta_3} - \frac{\delta_3}{\delta_1 \delta_2} \right) (v_1 + \alpha_1 \beta_1 x_1 y_1) v_2 \right\} \end{cases}$$

where $v_1 = e_1$, $v_2 = e_2 - \frac{\delta_2}{\delta_1} \frac{(a_4-1)}{a_4} v_1$ and $v_3 = e_3$.

Also, if $A = diag(\alpha_1, \alpha_2, \alpha_3) = 0$, $B = diag(\beta_1, \beta_2, \beta_3) = 0$ and $C = diag(\gamma_1, \gamma_2, \gamma_3) = 0$, then the compound synchronization problem is reduced to a chaos control problem of the system (6.8).

The system (6.8) is asymptotically stable, if the controllers are chosen as follows:

$$\begin{cases} u_1 = 0 \\[3mm] u_2 = \frac{1}{\delta_2} \left\{ -\frac{\delta_2}{\delta_1} \frac{(a_4-1)}{a_4} (c_4 + 1) v_1 - (c_4 - a_4 + 2) v_2 - \frac{\delta_1}{\delta_2} a_4 v_1 \right\} \\[3mm] u_3 = \frac{1}{\delta_3} \left\{ (b_4 - 1) v_3 - \frac{\delta_3}{\delta_1^2} \frac{(a_4-1)}{a_4} v_1^2 + \left(\frac{\delta_2}{\delta_1 \delta_3} - \frac{\delta_3}{\delta_1 \delta_2} \right) v_1 v_2 \right\} \end{cases}$$

where $v_1 = e_1$, $v_2 = e_2 - \frac{\delta_2}{\delta_1}\frac{(a_4-1)}{a_4}v_1$ and $v_3 = e_3$.

Example 11 *If $a_1 = 10$, $b_1 = 28$, $c_1 = 2.667$, $a_2 = 0.2$, $b_2 = 0.2$, $c_2 = 5.7$, $a_3 = 35$, $b_3 = 3$, $c_3 = 28$, $a_4 = 36$, $b_4 = 3$, $c_4 = 20$ and $\alpha_i = \beta_i = \gamma_i = \delta_1 = 1$, $i = 1, 2, 3$ and the initial conditions of the systems (6.5), (6.6), (6.7) and (6.8) are $(0.5, 1.0, 1.5)$, $(1, 1, 2)$, $(-4, 5, -6)$ and $(-5, 25, -5)$ respectively. Then Fig. 6.1 displays the state trajectories of $x_i(y_i + z_i)$ and w_i, $i = 1, 2, 3$ for master systems (6.5), (6.6) and (6.7) and the slave system (6.8) which shows that the compound synchronization between different chaotic systems (6.5)-(6.7) remains chaotic. Also, the error trajectories e_i, $i = 1, 2, 3$, between master systems (6.5), (6.6) and (6.7) and the slave system (6.8) are given in Fig. 6.2 which implies that compound synchronization is achieved.*

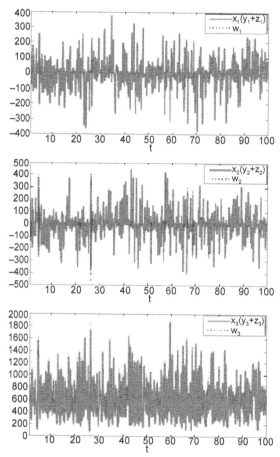

FIGURE 6.1: State trajectories of $x_i(y_i + z_i)$ and w_i, $i = 1, 2, 3$ for master systems (6.5), (6.6) and (6.7) and the slave system (6.8). [Sivasamy (2017)].

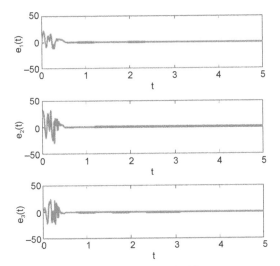

FIGURE 6.2: Error trajectories e_i, $i = 1, 2, 3$ between master systems (6.5), (6.6) and (6.7) and the slave system (6.8). [Sivasamy (2017)].

6.3 Synchronization of 3-D continuous-time quadratic systems using a universal non-linear control law

In this section, we apply non-linear control theory in order to synchronize two arbitrary 3-D continuous-time quadratic systems. The proposed control law does not need any conditions on the considered systems, hence, it is a universal synchronization approach for general 3-D continuous-time quadratic systems. Indeed, let us consider two arbitrary 3-D continuous-time quadratic systems. The one with variables x_1, y_1, and z_1 will be controlled to be the new system given by:

$$\begin{cases} x_1' = a_0 + a_1 x_1 + a_2 y_1 + a_3 z_1 + f_1(x_1, y_1, z_1) \\ y_1' = b_0 + b_1 x_1 + b_2 y_1 + b_3 z_1 + f_2(x_1, y_1, z_1) \\ z_1' = c_0 + c_1 x_1 + c_2 y_1 + c_3 z_1 + f_3(x_1, y_1, z_1) \end{cases} \quad (6.15)$$

where

$$\begin{cases} f_1(x_1, y_1, z_1) = a_4 x_1^2 + a_5 y_1^2 + a_6 z_1^2 + a_7 x_1 y_1 + a_8 x_1 z_1 + a_9 y_1 z_1 \\ f_2(x_1, y_1, z_1) = b_4 x_1^2 + b_5 y_1^2 + b_6 z_1^2 + b_7 x_1 y_1 + b_8 x_1 z_1 + b_9 y_1 z_1 \\ f_3(x_1, y_1, z_1) = c_4 x_1^2 + c_5 y_1^2 + c_6 z_1^2 + c_7 x_1 y_1 + c_8 x_1 z_1 + c_9 y_1 z_1 \end{cases} \quad (6.16)$$

and the one with variables $x_2, y_2,$ and z_2 as the response system

$$\begin{cases} x_2' = d_0 + d_1 x_2 + d_2 y_2 + d_3 z_2 + g_1 (x_2, y_2, z_2) + u_1 (t) \\[2mm] y_2' = r_0 + r_1 x_2 + r_2 y_2 + r_3 z_2 + g_2 (x_2, y_2, z_2) + u_2 (t) \\[2mm] z_2' = s_0 + s_1 x_2 + s_2 y_2 + s_3 z_2 + g_3 (x_2, y_2, z_2) + u_3 (t) \end{cases} \qquad (6.17)$$

where

$$\begin{cases} g_1 (x_2, y_2, z_2) = d_4 x_2^2 + d_5 y_2^2 + d_6 z_2^2 + d_7 x_2 y_2 + d_8 x_2 z_2 + d_9 y_2 z_2 \\[2mm] g_2 (x_2, y_2, z_2) = r_4 x_2^2 + r_5 y_2^2 + r_6 z_2^2 + r_7 x_2 y_2 + r_8 x_2 z_2 + r_9 y_2 z_2 \\[2mm] g_3 (x_2, y_2, z_2) = s_4 x_2^2 + s_5 y_2^2 + s_6 z_2^2 + s_7 x_2 y_2 + s_8 x_2 z_2 + s_9 y_2 z_2 \end{cases} \qquad (6.18)$$

Here $(a_i, b_i, c_i)_{0 \le i \le 9} \subset \mathbb{R}^{30}$ and $(d_i, r_i, s_i)_{0 \le i \le 9} \subset \mathbb{R}^{30}$ are bifurcation parameters, and $u_1(t), u_2(t), u_3(t)$ are the unknown (to be determined) non-linear controllers such that two systems (6.15) and (6.17) can be synchronized. Define the following quantities depending on the above two systems in which we can proceed with our proposed method:

$$\begin{cases} \xi_1 = a_1 + d_1 + a_4 (x_1 + x_2) + d_4 (x_1 + x_2) + a_7 y_1 + a_8 z_1 + d_7 y_2 + d_8 z_2 \\[2mm] \xi_2 = a_2 + d_2 + a_5 (y_1 + y_2) + d_5 (y_1 + y_2) + a_9 z_1 + d_9 z_2 \\[2mm] \xi_3 = a_3 + d_3 + a_6 (z_1 + z_2) + d_6 (z_1 + z_2) \\[2mm] \xi_4 = \eta_1 + \eta_2 + \eta_3 \\[2mm] \xi_5 = b_1 + r_1 + b_4 (x_1 + x_2) + r_4 (x_1 + x_2) + b_7 y_1 + b_8 z_1 + r_7 y_2 + r_8 z_2 \end{cases} \qquad (6.19)$$

and

$$\begin{cases} \xi_6 = b_2 + r_2 + b_5 (y_1 + y_2) + r_5 (y_1 + y_2) + b_9 z_1 + r_9 z_2 \\[2mm] \xi_7 = b_3 + r_3 + b_6 (z_1 + z_2) + r_6 (z_1 + z_2) \\[2mm] \xi_8 = \eta_4 + \eta_5 + \eta_6 \\[2mm] \xi_9 = c_1 + s_1 + c_4 (x_1 + x_2) + s_4 (x_1 + x_2) + c_7 y_1 + c_8 z_1 + s_7 y_2 + s_8 z_2 \\[2mm] \xi_{10} = c_2 + s_2 + c_5 (y_1 + y_2) + s_5 (y_1 + y_2) + c_9 z_1 + s_9 z_2 \\[2mm] \xi_{11} = c_3 + s_3 + c_6 (z_1 + z_2) + s_6 (z_1 + z_2) \\[2mm] \xi_{12} = \eta_7 + \eta_8 + \eta_9 \end{cases} \qquad (6.20)$$

where

$$
\begin{cases}
\eta_1 = d_4 x_1^2 + d_7 x_1 y_2 + d_8 x_1 z_2 + d_1 x_1 - a_4 x_2^2 - a_7 x_2 y_1 - a_8 x_2 z_1 \\[4pt]
\quad \eta_2 = -a_1 x_2 + d_5 y_1^2 + d_9 y_1 z_2 + d_2 y_1 - a_5 y_2^2 - a_9 y_2 z_1 - a_2 y_2 \\[4pt]
\quad\quad \eta_3 = d_6 z_1^2 + d_3 z_1 - a_6 z_2^2 - a_3 z_2 - a_0 + d_0 \\[4pt]
\eta_4 = r_4 x_1^2 + r_7 x_1 y_2 + r_8 x_1 z_2 + r_1 x_1 - b_4 x_2^2 - b_7 x_2 y_1 - b_8 x_2 z_1 \\[4pt]
\quad \eta_5 = -b_1 x_2 + r_5 y_1^2 + r_9 y_1 z_2 + r_2 y_1 - b_5 y_2^2 - b_9 y_2 z_1 - b_2 y_2 \qquad (6.21) \\[4pt]
\quad\quad \eta_6 = r_6 z_1^2 + r_3 z_1 - b_6 z_2^2 - b_3 z_2 - b_0 + r_0 \\[4pt]
\eta_7 = s_4 x_1^2 + s_7 x_1 y_2 + s_8 x_1 z_2 + s_1 x_1 - c_4 x_2^2 - c_7 x_2 y_1 - c_8 x_2 z_1 \\[4pt]
\quad \eta_8 = -c_1 x_2 + s_5 y_1^2 + s_9 y_1 z_2 + s_2 y_1 - c_5 y_2^2 - c_9 y_2 z_1 - c_2 y_2 \\[4pt]
\quad\quad \eta_9 = s_6 z_1^2 + s_3 z_1 - c_6 z_2^2 - c_3 z_2 - c_0 + s_0
\end{cases}
$$

The above quantities come from the formulation of the problem. The error states are $e_1 = x_2 - x_1, e_2 = y_2 - y_1$, and $e_3 = z_2 - z_1$. Then the error system is given by:

$$
\begin{cases}
e_1' = \xi_1 e_1 + \xi_2 e_2 + \xi_3 e_3 + \xi_4 + u_1(t) \\[4pt]
e_2' = \xi_5 e_1 + \xi_6 e_2 + \xi_7 e_3 + \xi_8 + u_2(t) \qquad\qquad (6.22) \\[4pt]
e_3' = \xi_9 e_1 + \xi_{10} e_2 + \xi_{11} e_3 + \xi_{12} + u_3(t)
\end{cases}
$$

We propose the following universal control law for the system (6.22):

$$
\begin{cases}
u_1(t) = -(\xi_1 + 1) e_1 - (\xi_2 + \xi_5) e_2 - \xi_4 \\[4pt]
u_2(t) = -(\xi_6 + 1) e_2 - (\xi_7 + \xi_{10}) e_3 - \xi_8 \qquad\qquad (6.23) \\[4pt]
u_3(t) = -(\xi_3 + \xi_9) e_1 - (\xi_{11} + 1) e_3 - \xi_{12}
\end{cases}
$$

Then the two 3-D continuous-time quadratic systems (6.15) and (6.17) approach synchronization for any initial condition. Indeed, the error system (6.22) becomes

$$
\begin{cases}
e_1' = -e_1 - \xi_5 e_2 + \xi_3 e_3 \\[4pt]
e_2' = \xi_5 e_1 - e_2 - \xi_{10} e_3 \qquad\qquad (6.24) \\[4pt]
e_3' = -\xi_3 e_1 + \xi_{10} e_2 - e_3
\end{cases}
$$

and if we consider the Lyapunov function $V = \frac{e_1^2 + e_2^2 + e_3^2}{2}$, then the asymptotic stability of the error system (6.24) is verified by Lyapunov stability theory since $\frac{dV}{dt} = -e_1^2 - e_2^2 - e_3^2 < 0$ for all $(a_i, b_i, c_i)_{0 \leq i \leq 9} \subset \mathbb{R}^{30}, (d_i, r_i, s_i)_{0 \leq i \leq 9} \subset \mathbb{R}^{30}$ and for all initial conditions. If the two systems (6.15) and (6.17) are chaotic, then the control law (6.23) also guarantees their synchronization for any initial condition. Also, any 3-D continuous-time quadratic chaotic system can be stabilized (controlled) to a stable 3-D continuous-time quadratic system that converges to an equilibrium point (to a 3-D continuous-time quadratic system that converges to a periodic solution). Furthermore, any 3-D continuous-time quadratic system can be chaotified to a chaotic 3-D continuous-time quadratic system.

Example 12 *As an example of this situation, consider the Lorenz system given by:*

$$\begin{cases} x_1' = a_1 x_1 - a_1 y_1 \\[2mm] y_1' = b_1 x_1 - y_1 - x_1 z_1 \\[2mm] z_1' = -c_3 z_1 + x_1 y_1 \end{cases} \qquad (6.25)$$

To apply the above method, we consider the one with variables $x_2, y_2,$ and z_2 as the response system:

$$\begin{cases} x_2' = d_1 x_2 - d_1 y_2 + u_1(t) \\[2mm] y_2' = r_1 x_2 - y_2 - x_2 z_2 + u_2(t) \\[2mm] z_2' = -s_3 z_2 + x_1 y_2 + u_3(t) \end{cases} \qquad (6.26)$$

Thus, the universal control law is given by:

$$\begin{cases} u_1(t) = -(\xi_1 + 1) e_1 - (\xi_2 + \xi_5) e_2 - \xi_4 \\[2mm] u_2(t) = e_2 - \xi_8 \\[2mm] u_3(t) = -\xi_9 e_1 - (\xi_{11} + 1) e_3 - \xi_{12} \end{cases} \qquad (6.27)$$

where

$$\begin{cases} \xi_1 = a_1 + d_1, \xi_2 = -a_1 - d_1, \xi_4 = d_1 x_1 - a_1 x_2 - d_1 y_1 + a_1 y_2 \\[2mm] \xi_5 = -z_1 - z_2, \xi_6 = -2, \xi_8 = -x_1 z_2 + x_2 z_1 - y_1 + y_2 \\[2mm] \xi_9 = y_1 + y_2, \xi_{11} = -c_3 - s_3, \xi_{12} = x_1 y_2 - x_2 y_1 - s_3 z_1 + c_3 z_2 \end{cases} \qquad (6.28)$$

If the two systems (6.25) and (6.26) are chaotic, then the control law (6.27) guaranties their synchronization for any initial condition.

6.4 Co-existence of certain types of synchronization and its inverse

In many cases, the co-existence of the full-state hybrid function projective synchronization (FSHFPS) and the inverse full-state hybrid function projective synchronization (IFSHFPS) is possible between some classes of three-dimensional master systems and four-dimensional slave systems. We consider the following master and slave systems

$$X'(t) = F(X(t)) \tag{6.29}$$
$$Y'(t) = G(Y(t)) + U \tag{6.30}$$

where $X(t) = (x_i(t))_{1 \leq i \leq n}$, $Y(t) = (y_i(t))_{1 \leq i \leq m}$ are the states of the master system and the slave system, respectively, $F : \mathbb{R}^n \to \mathbb{R}^n$, $G : \mathbb{R}^m \to \mathbb{R}^m$ and $U = (u_i)_{1 \leq i \leq m}$ is a vector controller.

Definition 48 *The master systems (6.29) and the slave system (6.30) are said to be full state hybrid function projective synchronized (FSHFPS), if there exists a controller $U = (u_i)_{1 \leq i \leq m}$ and differentiable functions $\alpha_{ij}(t) : \mathbb{R}^+ \to \mathbb{R}$, $i = 1, 2, ..., m$; $j = 1, 2, ..., n$, such that the synchronization errors:*

$$e_i(t) = y_i(t) - \sum_{j=1}^{n} \alpha_{ij}(t) x_j(t), \quad i = 1, 2, ..., m \tag{6.31}$$

satisfy $\lim_{t \to \infty} e_i(t) = 0$.

Definition 49 *The master systems (6.29) and the slave system (6.30) are said to be inverse full state hybrid function projective synchronized (IF-SHFPS), if there exists a controller $U = (u_i)_{1 \leq i \leq m}$ and differentiable functions $\beta_{ij}(t) : \mathbb{R}^+ \to \mathbb{R}$, $i = 1, 2, ..., n$; $j = 1, 2, ..., m$, such that the synchronization errors:*

$$e_i(t) = x_i(t) - \sum_{j=1}^{m} \beta_{ij}(t) y_j(t), \quad i = 1, 2, ..., n \tag{6.32}$$

satisfy $\lim_{t \to \infty} e_i(t) = 0$.

Assuming that the master system is given by:

$$x_i'(t) = f_i(X(t)), \quad i = 1, 2, 3 \tag{6.33}$$

where $X(t) = (x_i(t))_{1 \leq i \leq 3}$ is the state vector of the master system (6.33), $f_i : \mathbb{R}^3 \to \mathbb{R}$, $i = 1, 2, 3$. Also, the slave system is given by:

$$y_i'(t) = \sum_{j=1}^{4} b_{ij} y_j(t) + g_i(Y(t)) + u_i, \quad i = 1, 2, 3, 4 \tag{6.34}$$

where $Y(t) = (y_i)_{1 \leq i \leq 4}$ is the state vector of the slave system (6.34), $(b_{ij}) \in \mathbb{R}^{4 \times 4}$, $g_i : \mathbb{R}^4 \rightarrow \mathbb{R}$ are non-linear functions and u_i, $i = 1, 2, 3, 4$, are controllers to be designed.

Definition 50 *Let* $(\alpha_j(t))_{1 \leq j \leq 4}$, $(\beta_j(t))_{1 \leq j \leq 3}$, $(\gamma_j(t))_{1 \leq j \leq 4}$ *and* $(\theta_j(t))_{1 \leq j \leq 3}$ *be continuously differentiable and boundary functions, it is said that IFSHFPS and FSHFPS coexist in the synchronization of the master system (6.33) and the slave system (6.34), if there exist controllers* u_i, $= 1, 2, 3, 4$, *such that the synchronization errors:*

$$
\begin{cases}
e_1(t) = x_1(t) - \sum_{j=1}^{4} \alpha_j(t) y_j(t) \\[2mm]
e_2(t) = y_2(t) - \sum_{j=1}^{3} \beta_j(t) x_j(t) \\[2mm]
e_3(t) = x_3(t) - \sum_{j=1}^{4} \gamma_j(t) y_j(t) \\[2mm]
e_4(t) = y_4(t) - \sum_{j=1}^{3} \theta_j(t) x_j(t)
\end{cases}
\tag{6.35}
$$

satisfy $\lim_{t \rightarrow +\infty} e_i(t) = 0$, $i = 1, 2, 3, 4$.

The following result was proved in [Gasri (2018)]:

Theorem 48 *The coexistence of IFSHFPS and FSHFPS between the master system (6.33) and the slave system (6.34) will occur if*

$$\alpha_3(t) \gamma_1(t) - \alpha_1(t) \gamma_3(t) \neq 0$$

and the control law is designed as follows:

$$
\begin{cases}
u_1 = \sum_{i=1}^{4} P_i \left(\sum_{j=1}^{4} (b_{ij} - c_{ij}) e_j(t) - R_i \right) \\[2mm]
u_2 = \sum_{j=1}^{4} (b_{2j} - c_{2j}) e_j(t) - R_2 \\[2mm]
u_3 = \sum_{i=1}^{4} Q_i \left(\sum_{j=1}^{4} (b_{ij} - c_{ij}) e_j(t) - R_i \right) \\[2mm]
u_4 = \sum_{j=1}^{4} (b_{4j} - c_{4j}) e_j(t) - R_4
\end{cases}
\tag{6.36}
$$

where $(c_{ij})_{4 \times 4}$ *are control constants to be selected and*

$$
\begin{cases}
P_1 = \frac{\gamma_3(t)}{\alpha_3(t)\gamma_1(t)-\alpha_1(t)\gamma_3(t)}, \quad Q_1 = \frac{-\gamma_1(t)}{\alpha_3(t)\gamma_1(t)-\alpha_1(t)\gamma_3(t)} \\[3mm]
P_2 = \frac{\gamma_3(t)\alpha_2(t)-\alpha_3(t)\gamma_2(t)}{\alpha_3(t)\gamma_1(t)-\alpha_1(t)\gamma_3(t)}, \quad Q_2 = \frac{\alpha_1(t)\gamma_2(t)-\alpha_2(t)\gamma_1(t)}{\alpha_3(t)\gamma_1(t)-\alpha_1(t)\gamma_3(t)} \\[3mm]
P_3 = \frac{-\alpha_3(t)}{\alpha_3(t)\gamma_1(t)-\alpha_1(t)\gamma_3(t)}, \quad Q_3 = \frac{\alpha_1(t)}{\alpha_3(t)\gamma_1(t)-\alpha_1(t)\gamma_3(t)} \\[3mm]
P_4 = \frac{\gamma_3(t)\alpha_4(t)-\alpha_3(t)\gamma_4(t)}{\alpha_3(t)\gamma_1(t)-\alpha_1(t)\gamma_3(t)}, \quad Q_4 = \frac{\alpha_1(t)\gamma_4(t)-\alpha_4(t)\gamma_1(t)}{\alpha_3(t)\gamma_1(t)-\alpha_1(t)\gamma_3(t)}
\end{cases}
\tag{6.37}
$$

and

$$\begin{cases} R_1 = f_1(X(t)) - \sum_{j=1}^{4} \dot{\alpha}_j(t) y_j(t) \\ \quad - \sum_{i=1}^{4} \alpha_i(t) \left(\sum_{j=1}^{4} b_{ij} y_j(t) + g_i(Y(t)) \right) \\[2mm] R_2 = \sum_{j=1}^{4} b_{2j} y_j(t) + g_2(Y(t)) \\ \quad - \sum_{j=1}^{3} \dot{\beta}_j(t) x_j(t) - \sum_{j=1}^{3} \beta_j(t) \dot{x}_j(t) \\[2mm] R_3 = f_3(X(t)) - \sum_{j=1}^{4} \dot{\gamma}_j(t) y_j(t) \\ \quad - \sum_{i=1}^{4} \gamma_i(t) \left(\sum_{j=1}^{4} b_{ij} y_j(t) + g_i(Y(t)) \right) \\[2mm] R_4 = \sum_{j=1}^{4} b_{4j} y_j(t) + g_4(Y(t)) \\ \quad - \sum_{j=1}^{3} \dot{\theta}_j(t) x_j(t) - \sum_{j=1}^{3} \theta_j(t) \dot{x}_j(t) \end{cases} \tag{6.38}$$

The error system (6.35) can be differentiated as follows:

$$\begin{cases} e_1'(t) = x_1'(t) - \sum_{j=1}^{4} \dot{\alpha}_j(t) y_j(t) - \sum_{j=1}^{4} \alpha_j(t) \dot{y}_j(t) \\[2mm] e_2'(t) = y_2'(t) - \sum_{j=1}^{3} \dot{\beta}_j(t) x_j(t) - \sum_{j=1}^{3} \beta_j(t) \dot{x}_j(t) \\[2mm] e_3'(t) = x_3'(t) - \sum_{j=1}^{4} \dot{\gamma}_j(t) y_j(t) - \sum_{j=1}^{4} \gamma_j(t) \dot{y}_j(t) \\[2mm] e_4'(t) = y_4'(t) - \sum_{j=1}^{3} \dot{\theta}_j(t) x_j(t) - \sum_{j=1}^{3} \theta_j(t) \dot{x}_j(t) \end{cases} \tag{6.39}$$

Hence, the error system (6.39) can be written as follows:

$$\begin{cases} e_1'(t) = \sum_{j=1}^{4} \alpha_j(t) u_j + R_1 \\[2mm] e_2'(t) = u_2 + R_2 \\[2mm] e_3'(t) = \sum_{j=1}^{4} \gamma_j(t) u_j + R_3 \\[2mm] e_4'(t) = u_4 + R_4 \end{cases} \tag{6.40}$$

where R_i, $i = 1, 2, 3, 4$, are given by (6.38). By substituting the control law (6.36) into (6.40), the error system can be described as follows:

$$e_i'(t) = \sum_{j=1}^{4} (b_{ij} - c_{ij}) e_j(t), \quad i = 1, 2, 3, 4 \tag{6.41}$$

or in the compact form

$$e'(t) = (B - C) e(t) \tag{6.42}$$

where $B = (b_{ij})_{4 \times 4}$ and $C = (c_{ij})_{4 \times 4}$ is the control matrix. If the control matrix C is chosen such that all the eigenvalues of $B - C$ are strictly negative, then all solutions of the error system (6.42) go to zero as $t \to \infty$. Thus, the systems (6.33) and (6.34) are globally synchronized in 4D.

Example 13 *As an elementary example, one can choose the 3D master system as [Vaidyanathan (2016(a))]:*

$$\begin{cases} x_1' = a_1 \left(x_2 - x_1 \right) \\\\ x_2' = x_1 x_3 \\\\ x_3' = 50 - a_2 x_1^2 - a_3 x_3 \end{cases} \tag{6.43}$$

For $a_1 = 2.9$, $a_2 = 0.7$, $a_3 = 0.6$ the system (6.43) has a chaotic attractor. The salve system is described by

$$\begin{cases} y_1' = b_1 \left(y_2 - y_1 \right) + y_2 y_3 + y_4 + u_1 \\\\ y_2' = b_2 y_1 + y_4 - b_3 y_1 y_3 + u_2 \\\\ y_3' = -b_4 y_3 + b_5 y_1 y_2 + u_3 \\\\ y_4' = -y_1 - y_2 + u_4 \end{cases} \tag{6.44}$$

When the controllers $u_1 = u_2 = u_3 = u_4 = 0$. For $(b_1, b_2, b_3, b_4, b_5) = (18, 40, 5, 3, 4)$ the system (6.44) has a hyperchaotic attractor [Vaidyanathan, et al., (2016(d))]. The linear part B and the non-linear part g of the slave system (6.44) are given by:

$$B = \begin{pmatrix} -18 & 18 & 0 & 1 \\\\ 40 & 0 & 0 & 1 \\\\ 0 & 0 & -3 & 0 \\\\ -1 & -1 & 0 & 0 \end{pmatrix} \quad and \quad g = \begin{pmatrix} y_2 y_3 \\\\ -5 y_1 y_3 \\\\ 4 y_1 y_2 \\\\ 0 \end{pmatrix}$$

The synchronization errors between the master system (6.43) and the slave system (6.44) are defined as:

$$\begin{cases} e_1 = x_1 - \alpha_1 \left(t \right) y_1 - \alpha_2 \left(t \right) y_2 - \alpha_3 \left(t \right) y_3 - \alpha_4 \left(t \right) y_4 \\\\ e_2 = y_2 - \beta_1 \left(t \right) x_1 - \beta_2 \left(t \right) x_2 - \beta_3 \left(t \right) x_3 \\\\ e_3 = x_3 - \gamma_1 \left(t \right) y_1 - \gamma_2 \left(t \right) y_2 - \gamma_3 \left(t \right) y_3 - \gamma_4 \left(t \right) y_4 \\\\ e_4 = y_4 - \theta_1 \left(t \right) x_1 - \theta_2 \left(t \right) x_2 - \theta_3 \left(t \right) x_3 \end{cases} \tag{6.45}$$

where

$$\begin{cases} \alpha_1 \left(t \right) = \sin t, \alpha_2 \left(t \right) = 1, \alpha_3 \left(t \right) = \frac{1}{t+1}, \alpha_4 \left(t \right) = 2 \\\\ \beta_1 \left(t \right) = 3, \beta_2 \left(t \right) = \cos t, \beta_3 \left(t \right) = 4 \\\\ \gamma_1 \left(t \right) = e^{-t}, \gamma_2 \left(t \right) = 2, \gamma_3 \left(t \right) = 0, \gamma_4 \left(t \right) = \frac{1}{t^2+1} \\\\ \theta_1 \left(t \right) = \frac{t}{t+1}, \theta_2 \left(t \right) = 0, \theta_3 \left(t \right) = \sin 3t \end{cases}$$

Thus,

$$\alpha_3(t)\gamma_1(t) - \alpha_1(t)\gamma_3(t) = \frac{1}{e^t(t+1)} \neq 0$$

The coexistence of IFSHFPS and FSHFPS is achieved if the control matrix C is selected as follows:

$$C = \begin{pmatrix} 0 & 18 & 0 & 1 \\ 40 & 1 & 0 & 1 \\ 0 & 0 & 0 & 0 \\ -1 & -1 & 0 & 1 \end{pmatrix}$$

and the controllers u_i, $1 \leq i \leq 4$, are constructed according to (6.36) as follows:

$$\begin{cases} u_1 = -2e^t(-e_2 - R_2) + e^t(-3e_3 - R_3) - \frac{e^t}{t^2+1}(-e_4 - R_4) \\ \\ u_2 = -e_2 + 5y_1y_3 - 40y_1 - y_4 - R_2 \\ \\ u_3 = -(t+1)(-18e_1 - R_1) + e^t(t+1) \\ \left[-(2 + 2e_2 + 2R_2 + 3e_3 + R_3)\sin t + \left(\frac{\sin t}{t^2+1} - e^{-t}\right)(-e_4 - R_4)\right] \\ \\ u_4 = -e_4 + y_1 + y_2 - R_4 \end{cases}$$

where

$$\begin{cases} R_1 = 2.9(x_2 - x_1) - y_1\cos t + \frac{1}{(t+1)^2}y_3 \\ \quad - \sin t(18(y_2 - y_1) + y_2y_3) + \frac{1}{t+1}(4y_1y_2 - 3y_3) \\ \qquad\qquad -y_1 - y_2 \\ \\ R_2 = -5y_1y_3 + 40y_1 + y_4 + x_2\sin t - 8.7(x_2 - x_1) \\ \qquad\qquad -x_1x_3\cos t \\ \\ R_3 = 50 - 0.7x_1^2 - 0.6x_3 + e^{-t}y_1 + \frac{2t}{(t^2+1)^2}y_4 \\ \quad -e^{-t}(18(y_2 - y_1) + y_2y_3) + 10y_1y_3 + 80y_1 \\ \qquad -2y_4 + \frac{1}{t^2+1}(y_1 + y_2) \\ \\ R_4 = -y_1 - y_2 - \frac{t+1-t^2}{(t+1)^2}x_1 - 3x_3\cos 3t - \frac{2.9t}{t+1}(x_2 - x_1) \\ \quad - (50 - 0.7x_1^2 - 0.6x_3)\sin 3t \end{cases}$$

In this case, all the eigenvalues of $B - C$ have negative real parts. Thus, the error functions between systems (6.43) and (6.44) are given by:

$$e_1' = -18e_1 \tag{6.46}$$

$$e_2' = -e_2 \tag{6.47}$$

$$e_3' = -3e_3 \tag{6.48}$$

$$e_4' = -e_4 \tag{6.49}$$

and it converges to zero when $t \to \infty$.

Now, assuming that the master and the slave systems have the following forms:

$$x_i'(t) = \sum_{j=1}^{3} a_{ij} x_j(t) + f_i(X(t)), \quad i = 1, 2, 3 \tag{6.50}$$

$$y_i'(t) = g_i(Y(t)) + u_i, \quad i = 1, 2, 3, 4 \tag{6.51}$$

where $X(t) = (x_i)_{1 \leq i \leq 3}$, $Y(t) = (y_i)_{1 \leq i \leq 4}$ are the states of the master system and the slave system, respectively, $(a_{ij}) \in \mathbb{R}^{3 \times 3}$, $f_i : \mathbb{R}^3 \to \mathbb{R}$ are non-linear functions, $g_i : \mathbb{R}^4 \to \mathbb{R}$ and $u_i, = 1, 2, 3, 4$, are controllers to be constructed.

Thus, we have the following definition:

Definition 51 *Let $(\lambda_j(t))_{1 \leq j \leq 3}$, $(\mu_j(t))_{1 \leq j \leq 4}$ and $(\sigma_j(t))_{1 \leq j \leq 3}$ be continuously differentiable and boundary functions, it is said that FSHFPS and IF-SHFPS coexist in the synchronization of the master system (6.47) and the slave system (6.48), if there exist controllers $u_i, = 1, 2, 3$, such that the synchronization errors given by:*

$$\begin{cases} e_1(t) = y_1(t) - \sum_{j=1}^{3} \lambda_j(t) x_j(t) \\\\ e_2(t) = x_2(t) - \sum_{j=1}^{4} \mu_j(t) y_j(t) \\\\ e_3(t) = y_3(t) - \sum_{j=1}^{3} \sigma_j(t) x_j(t) \end{cases} \tag{6.52}$$

satisfy $\lim_{t \to +\infty} e_i(t) = 0, \ i = 1, 2, 3.$

The following result was proved in [Gasri (2018)]:

Theorem 49 *To achieve the coexistence of IFSHFPS and FSHFPS between the master system (6.47) and the slave system (6.48), we assume that $\mu_2(t) \neq 0$ and the control law is constructed as follows:*

$$\begin{cases} u_1 = \sum_{j=1}^3 (a_{1j} - l_{1j}) \, e_j(t) - R_1 \\[2mm] u_2 = -\frac{\mu_1(t)}{\mu_2(t)} \left(\sum_{j=1}^3 (a_{1j} - l_{1j}) \, e_j(t) - R_1 \right) \\[1mm] \qquad -\frac{1}{\mu_2(t)} \left(\sum_{j=1}^3 (a_{2j} - l_{2j}) \, e_j(t) - R_2 \right) \\[1mm] \qquad -\frac{\mu_3(t)}{\mu_2(t)} \left(\sum_{j=1}^3 (a_{3j} - l_{3j}) \, e_j(t) - R_3 \right) \\[2mm] u_3 = \sum_{j=1}^3 (a_{3j} - l_{3j}) \, e_j(t) - R_3 \\[2mm] u_4 = 0 \end{cases} \qquad (6.53)$$

where $(l_{ij})_{3\times 3}$ *are control constants to be determined, whereas* R_1, R_2 *and* R_3 *are chosen as follows*

$$\begin{cases} R_1 = g_1(Y(t)) - \sum_{j=1}^3 (a_{1j} - l_{1j}) \, e_j(t) - \sum_{j=1}^3 \dot{\lambda}_j(t) \, x_j(t) \\[1mm] \qquad - \sum_{i=1}^3 \lambda_i(t) \left(\sum_{j=1}^3 a_{ij} x_j(t) + f_i(X(t)) \right) \\[3mm] R_2 = \sum_{j=1}^3 a_{2j} x_j(t) + f_2(X(t)) - \sum_{j=1}^3 (a_{2j} - l_{2j}) \, e_j(t) \\[1mm] \qquad - \sum_{j=1}^4 \dot{\mu}_j(t) \, y_j(t) - \sum_{j=1}^4 \mu_j(t) \, g_j(Y(t)) \\[3mm] R_3 = g_3(Y(t)) - \sum_{j=1}^3 (a_{3j} - l_{3j}) \, e_j(t) - \sum_{j=1}^3 \dot{\sigma}_j(t) \, x_j(t) \\[1mm] \qquad - \sum_{i=1}^3 \sigma_i(t) \left(\sum_{j=1}^3 a_{ij} x_j(t) + f_i(X(t)) \right) \end{cases} \qquad (6.54)$$

Proof 14 *The error system (6.49), between master system (6.47) and the slave system (6.48) is given by:*

$$\begin{cases} e_1'(t) = y_1'(t) - \sum_{j=1}^3 \dot{\lambda}_j(t) \, x_j(t) - \sum_{j=1}^3 \lambda_j(t) \, \dot{x}_j(t) \\[1mm] e_2'(t) = x_2'(t) - \sum_{j=1}^4 \dot{\mu}_j(t) \, y_j(t) - \sum_{j=1}^4 \mu_j(t) \, \dot{y}_j(t) \\[1mm] e_3'(t) = y_3'(t) - \sum_{j=1}^3 \dot{\sigma}_j(t) \, x_j(t) - \sum_{j=1}^3 \sigma_j(t) \, \dot{x}_j(t) \end{cases} \qquad (6.55)$$

Hence, we get

$$\begin{cases} e_1'(t) = \sum_{j=1}^3 (a_{1j} - l_{1j}) \, e_j(t) + u_1 + R_1 \\[1mm] e_2'(t) = \sum_{j=1}^3 (a_{2j} - l_{2j}) \, e_j(t) - \sum_{j=1}^4 \mu_j(t) \, u_j + R_2 \\[1mm] e_3'(t) = \sum_{j=1}^3 (a_{3j} - l_{3j}) \, e_j(t) + u_3 + R_3 \end{cases}$$

where R_i, $i = 1, 2, 3$, *were given by (6.51). The control law (6.50) implies that the error dynamics between systems (6.47) and (6.48) are given by:*

$$e_i'(t) = \sum_{j=1}^3 (a_{ij} - l_{ij}) \, e_j(t), \quad i = 1, 2, 3 \qquad (6.56)$$

or in the compact form

$$e'(t) = (A - L) e(t) \tag{6.57}$$

where $e(t) = (e_i(t))_{1 \le i \le 3}$, $A = (a_{ij})_{3 \times 3}$, $L = (l_{ij})_{3 \times 3}$. *Let us consider the Lyapunov function* $V(e(t)) = e^T(t)e(t)$, *then we get*

$$
\begin{aligned}
V'(e(t)) &= e'^T(t)e(t) + e^T(t)\dot{e}(t) \\
&= e^T(t)(A - L)^T e(t) + e^T(t)(A - L)e(t) \\
&= e^T(t)\left[(A - L)^T + (A - L)\right]e(t)
\end{aligned}
$$

If the control matrix L *is chosen such that* $(A - L)^T + (A - L)$ *is a negative definite matrix, then* $V'(e(t)) < 0$. *Thus, from the Lyapunov stability theory described in Sec. 8.2, the zero solution of the error system (6.54) is globally asymptotically stable, i.e.,*

$$\lim_{t \to \infty} e_i(t) = 0, \quad i = 1, 2, 3 \tag{6.58}$$

Therefore, systems (6.47) and (6.48) are globally synchronized in 3D.

Example 14 *As an elementary example, one can choose the 3D master system as [Vaidyanathan (2016(c))]:*

$$
\begin{cases}
x_1' = x_2 \\
x_2' = x_3 \\
x_3' = -c_1 x_1 (1 - x_1) - x_2 + c_2 x_2^2
\end{cases}
\tag{6.59}
$$

If $(c_1, c_2) = (0.2, 0.01)$ *then system (6.56) has a chaotic attractor. The linear part* A *and the non-linear part* f *of the master system (6.56) are given by:*

$$
A = (a_{ij})_{3 \times 3} = \begin{pmatrix} 0 & 1 & 0 \\ 0 & 0 & 1 \\ -0.2 & -1 & 0 \end{pmatrix} \quad \text{and } f = \begin{pmatrix} 0 \\ 0 \\ 0.2x_1^2 + 0.01x_2^2 \end{pmatrix}
$$

The slave system is given by [Vaidyanathan (2016(b))]:

$$
\begin{cases}
y_1' = d_1 (y_2 - y_1) + y_2 y_3 - y_4 + u_1 \\
y_2' = d_2 y_2 - y_1 y_3 + y_4 + u_2 \\
y_3' = y_1 y_2 - d_3 y_3 + u_3 \\
y_4' = -d_4 (y_1 + y_2) + u_4
\end{cases}
\tag{6.60}
$$

When $u_1 = u_2 = u_3 = u_4 = 0$ and $(d_1, d_2, d_3, d_4) = (40, 20.5, 5, 2.5)$ the system (6.57) has an hyperchaotic attractor. The synchronization errors are given by:

$$
\begin{cases}
e_1 = y_1 - \lambda_1\left(t\right)x_1 - \lambda_1\left(t\right)x_1 - \lambda_1\left(t\right)x_1 \\[2mm]
e_2 = x_2 - \mu_1\left(t\right)y_1 - \mu_2\left(t\right)y_2 - \mu_3\left(t\right)y_3 - \mu_4\left(t\right)y_4 \\[2mm]
e_3 = y_3 - \sigma_1\left(t\right)x_1 - \sigma_2\left(t\right)x_2 - \sigma_3\left(t\right)x_3
\end{cases}
$$

where

$$
\begin{cases}
\lambda_1\left(t\right) = e^{-t}, \lambda_2\left(t\right) = \sin 2t, \lambda_3\left(t\right) = 0 \\[2mm]
\mu_1\left(t\right) = 0, \mu_2\left(t\right) = \frac{1}{\sqrt{t+1}}, \mu_3\left(t\right) = \frac{1}{1+\cos^2 t}, \mu_4\left(t\right) = 4 \\[2mm]
\sigma_1\left(t\right) = \frac{1}{\ln(t+1)}, \sigma_2\left(t\right) = \frac{1}{1+\sin^2 t}, \sigma_3\left(t\right) = 0
\end{cases}
$$

The control matrix L is given by:

$$
L = \begin{pmatrix}
0.1 & 1 & 0 \\
0 & 2 & 0 \\
-0.2 & -1 & 3
\end{pmatrix}
$$

The controllers u_1, u_2, u_3 and u_4 are designed as follows:

$$
\begin{cases}
u_1 = -e_1 - R_1 \\[2mm]
u_2 = -\left(\sqrt{t}+1\right)\left(-2e_2 - R_2\right) - \frac{\sqrt{t}+1}{1+\cos^2 t}\left(-3e_3 - R_3\right) \\[2mm]
u_3 = -3e_3 - R_3 \\[2mm]
u_4 = 0
\end{cases}
$$

where

$$
\begin{cases}
R_1 = 40\left(y_2 - y_1\right) + y_2 y_3 - y_4 + e_1 + e^{-t}x_1 - 2x_2 \cos 2t \\
\qquad\quad -x_2 e^{-t} - x_3 \sin 2t \\[2mm]
R_2 = x_3 + 2e_2 - \frac{y_2}{2\sqrt{t}\left(\sqrt{t}+1\right)^2} - y_3 \frac{2\sin t \cos t}{\left(1+\cos^2 t\right)} \\
\qquad\quad -\frac{1}{\sqrt{t}+1}\left(20.5y_2 - y_1 y_3 + y_4\right) - \frac{1}{1+\cos^2 t}\left(y_1 y_2 - 5.5y_3\right) \\
\qquad\qquad +10\left(y_1 + y_2\right) \\[2mm]
R_3 = y_1 y_2 - 5.5y_3 + 3e_3 + \frac{1}{(t+1)\ln^2(t+1)}x_1 + \frac{2\sin t \cos t}{(1+\sin^2 t)}x_2 \\
\qquad\quad -\frac{x_2}{\ln(t+1)} - \frac{x_3}{1+\sin^2 t}
\end{cases}
$$

The matrix $(A - L)^T + (A - L)$ is negative definite, then the error functions between systems (6.56) and (6.57) are given by:

$$e_1' = -0.1e_1 \tag{6.61}$$

$$e_2' = -2e_2 \tag{6.62}$$

$$e_3' = -3e_2 \tag{6.63}$$

and it converges to zero when $t \to \infty$.

6.5 Synchronization of 4-D continuous-time quadratic systems using a universal non-linear control law

In this section, we apply non-linear control theory in order to synchronize two arbitrary 4-D continuous-time quadratic systems. This control law is a universal synchronization approach since it does not need any conditions on the considered systems.

Let us consider two arbitrary 4-D continuous-time quadratic systems. The one with variables x_1, y_1, z_1, and u_1 will be controlled to be the new system given by

$$
\begin{cases}
x_1' = \mu_1 + \beta_{11}x_1 + \beta_{12}y_1 + \beta_{13}z_1 + \beta_{14}u_1 + f_1\,(x_1, y_1, z_1, u_1) \\[2mm]
y_1' = \mu_2 + \beta_{21}x_1 + \beta_{22}y_1 + \beta_{23}z_1 + \beta_{24}u_1 + f_2\,(x_1, y_1, z_1, u_1) \\[2mm]
z_1' = \mu_3 + \beta_{31}x_1 + \beta_{32}y_1 + \beta_{33}z_1 + \beta_{34}u_1 + f_3\,(x_1, y_1, z_1, u_1) \\[2mm]
u_1' = \mu_4 + \beta_{41}x_1 + \beta_{42}y_1 + \beta_{43}z_1 + \beta_{44}u_1 + f_4\,(x_1, y_1, z_1, u_1)
\end{cases}
\tag{6.64}
$$

where

$$
\begin{cases}
f_1 = a_4x_1^2 + a_5y_1^2 + a_6z_1^2 + a_7u_1^2 + a_8x_1y_1 + a_9x_1z_1 + p_1 \\[2mm]
f_2 = b_4x_1^2 + b_5y_1^2 + b_6z_1^2 + b_7u_1^2 + b_8x_1y_1 + b_9x_1z_1 + p_2 \\[2mm]
f_3 = c_4x_1^2 + c_5y_1^2 + c_6z_1^2 + c_7u_1^2 + c_8x_1y_1 + c_9x_1z_1 + p_3 \\[2mm]
f_4 = d_4x_1^2 + d_5y_1^2 + d_6z_1^2 + d_7u_1^2 + d_8x_1y_1 + d_9x_1z_1 + p_4 \\[2mm]
\quad p_1 = a_{10}y_1z_1 + a_{11}x_1u_1 + a_{12}z_1u_1 + a_{13}y_1u_1 \\[2mm]
\quad p_2 = b_{10}y_1z_1 + b_{11}x_1u_1 + b_{12}z_1u_1 + b_{13}y_1u_1 \\[2mm]
\quad p_3 = c_{10}y_1z_1 + c_{11}x_1u_1 + c_{12}z_1u_1 + c_{13}y_1u_1 \\[2mm]
\quad p_4 = d_{10}y_1z_1 + d_{11}x_1u_1 + d_{12}z_1u_1 + d_{13}y_1u_1
\end{cases}
\tag{6.65}
$$

and the one with variables x_2, y_2, z_2, and u_2 as the response system

$$
\begin{cases}
x_2' = \delta_1 + \rho_{11}x_2 + \rho_{12}y_2 + \rho_{13}z_2 + \rho_{14}u_2 + g_1\left(x_2, y_2, z_2, u_2\right) + v_1(t) \\[4pt]
y_1' = \delta_2 + \rho_{21}x_2 + \rho_{22}y_2 + \rho_{23}z_2 + \rho_{24}u_2 + g_2\left(x_2, y_2, z_2, u_2\right) + v_2(t) \\[4pt]
z_1' = \delta_3 + \rho_{31}x_2 + \rho_{32}y_2 + \rho_{33}z_2 + \rho_{34}u_2 + g_3\left(x_2, y_2, z_2, u_2\right) + v_3(t) \\[4pt]
u_1' = \delta_4 + \rho_{41}x_2 + \rho_{42}y_2 + \rho_{43}z_2 + \rho_{44}u_2 + g_4\left(x_2, y_2, z_2, u_2\right) + v_4(t)
\end{cases}
\tag{6.66}
$$

where

$$
\begin{cases}
g_1 = h_4x_2^2 + h_5y_2^2 + h_6z_2^2 + h_7u_2^2 + h_8x_2y_2 + h_9x_2z_2 + p_5 \\[4pt]
g_2 = m_4x_2^2 + m_5y_2^2 + m_6z_2^2 + m_7u_2^2 + m_8x_2y_2 + m_9x_2z_2 + p_6 \\[4pt]
g_3 = r_4x_2^2 + r_5y_2^2 + r_6z_2^2 + r_7u_2^2 + r_8x_2y_2 + r_9x_2z_2 + p_7 \\[4pt]
g_4 = s_4x_2^2 + s_5y_2^2 + s_6z_2^2 + s_7u_2^2 + s_8x_2y_2 + s_9x_2z_2 + p_8 \\[4pt]
p_5 = h_{10}y_2z_2 + h_{11}x_2u_2 + h_{12}z_2u_2 + h_{13}y_2u_2 \\[4pt]
p_6 = m_{10}y_2z_2 + m_{11}x_2u_2 + m_{12}z_2u_2 + m_{13}y_2u_2 \\[4pt]
p_7 = r_{10}y_2z_2 + r_{11}x_2u_2 + r_{12}z_2u_2 + r_{13}y_2u_2 \\[4pt]
p_8 = s_{10}y_2z_2 + s_{11}x_2u_2 + s_{12}z_2u_2 + s_{13}y_2u_2
\end{cases}
\tag{6.67}
$$

Here

$$
\begin{cases}
(\mu_i, \delta_i)_{1\leq i\leq 4} \in \mathbb{R}^8 \\[6pt]
(\beta_{ij}, \rho_{ij})_{1\leq i,j\leq 4} \in \mathbb{R}^{16} \\[6pt]
(a_i, b_i, c_i, d_i, h_i, m_i, r_i, s_i)_{4\leq i\leq 13} \in \mathbb{R}^{80}
\end{cases}
$$

are the bifurcation parameters, and $v_1(t), v_2(t), v_3(t)$, and $v_4(t)$ are the unknown non-linear controllers, such that two systems (6.59) and (6.61) can be synchronized. Firstly, we need the following quantities depending on the above two systems:

$$\left\{ \begin{aligned}
&\eta_1 = h_7 u_1^2 - a_9 u_1 x_2 + h_{13} u_1 y_2 - a_{12} u_1 z_2 + \rho_{14} u_1 - a_7 u_2^2 + h_9 u_2 x_1 \\[4pt]
&\eta_2 = -a_{13} u_2 y_1 + h_{12} u_2 z_1 - \beta_{14} u_2 + h_4 x_1^2 + h_8 x_1 y_2 + h_9 x_1 z_2 + \rho_{11} x_1 \\[4pt]
&\eta_3 = -a_4 x_2^2 - a_8 x_2 y_1 - a_9 x_2 z_1 - \beta_{11} x_2 + h_5 y_1^2 + h_{10} y_1 z_2 + \rho_{12} y_1 - a_5 y_2^2 \\[4pt]
&\eta_4 = -a_{10} y_2 z_1 - \beta_{12} y_2 + h_6 z_1^2 + \rho_{13} z_1 - a_6 z_2^2 - \beta_{13} z_2 - \mu_1 + \delta_1 \\[4pt]
&\eta_5 = m_7 u_1^2 - b_9 u_1 x_2 + m_{13} u_1 y_2 - b_{12} u_1 z_2 + \rho_{24} u_1 - b_7 u_2^2 + m_9 u_2 x_1 \\[4pt]
&\eta_6 = -b_{13} u_2 y_1 + m_{12} u_2 z_1 - \beta_{14} u_2 + m_4 x_1^2 + m_8 x_1 y_2 + m_9 x_1 z_2 + \rho_{21} x_1 \\[4pt]
&\eta_7 = -b_4 x_2^2 - b_8 x_2 y_1 - b_9 x_2 z_1 - \beta_{21} x_2 + m_5 y_1^2 + m_{10} y_1 z_2 + \rho_{22} y_1 - b_5 y_2^2 \\[4pt]
&\eta_8 = -b_{10} y_2 z_1 - \beta_{22} y_2 + m_6 z_1^2 + \rho_{23} z_1 - b_6 z_2^2 - \beta_{23} z_2 - \mu_2 + \delta_2
\end{aligned} \right.$$

$$(6.68)$$

$$\left\{ \begin{aligned}
&\eta_9 = r_7 u_1^2 - c_9 u_1 x_2 + r_{13} u_1 y_2 - c_{12} u_1 z_2 + \rho_{34} u_1 \\
&\qquad -c_7 u_2^2 + r_9 u_2 x_1 \\[4pt]
&\eta_{10} = -c_{13} u_2 y_1 + r_{12} u_2 z_1 - \beta_{34} u_2 + r_4 x_1^2 + r_8 x_1 y_2 \\
&\qquad +r_9 x_1 z_2 + \rho_{11} x_1 \\[4pt]
&\eta_{11} = -c_4 x_2^2 - c_8 x_2 y_1 - c_9 x_2 z_1 - \beta_{31} x_2 + r_5 y_1^2 + r_{10} y_1 z_2 \\
&\qquad +\rho_{32} y_1 - c_5 y_2^2 \\[4pt]
&\eta_{12} = -c_{10} y_2 z_1 - \beta_{32} y_2 + r_6 z_1^2 + \rho_{33} z_1 - c_6 z_2^2 - \beta_{33} z_2 \\
&\qquad -\mu_3 + \delta_3 \\[4pt]
&\eta_{13} = s_7 u_1^2 - d_9 u_1 x_2 + s_{13} u_1 y_2 - d_{12} u_1 z_2 + \rho_{44} u_1 - d_7 u_2^2 + s_9 u_2 x_1 \\[4pt]
&\eta_{14} = -d_{13} u_2 y_1 + s_{12} u_2 z_1 - \beta_{44} u_2 + s_4 x_1^2 + s_8 x_1 y_2 + s_9 x_1 z_2 + \rho_{41} x_1 \\[4pt]
&\eta_{15} = -d_4 x_2^2 - d_8 x_2 y_1 - d_9 x_2 z_1 - \beta_{41} x_2 + s_5 y_1^2 + s_{10} y_1 z_2 \\
&\qquad +\rho_{42} y_1 - d_5 y_2^2 \\[4pt]
&\eta_{16} = -d_{10} y_2 z_1 - \beta_{42} y_2 + s_6 z_1^2 + \rho_{43} z_1 - d_6 z_2^2 - \beta_{43} z_2 - \mu_4 + \delta_4
\end{aligned} \right.$$

$$(6.69)$$

$$\begin{cases} \xi_1 = \beta_{11} + \rho_{11} + (a_4 + h_4)(x_1 + x_2) + a_9 u_1 + a_8 y_1 \\ \qquad + a_9 z_1 + h_9 u_2 + h_8 y_2 + h_9 z_2 \\[6pt] \xi_2 = \beta_{12} + \rho_{12} + (a_5 + h_5)(y_1 + y_2) + a_{10} z_1 + z_2 h_{10} \\[6pt] \xi_3 = \beta_{13} + \rho_{13} + (a_6 + h_6)(z_1 + z_2) + u_1 a_{12} + u_2 h_{12} \\[6pt] \xi_4 = \beta_{14} + \rho_{14} + (a_7 + h_7)(u_1 + u_2) + y_1 a_{13} + y_2 h_{13} \\[6pt] \qquad\qquad \xi_5 = \eta_1 + \eta_2 + \eta_3 + \eta_4 \end{cases} \tag{6.70}$$

$$\begin{cases} \xi_6 = \beta_{21} + \rho_{21} + (b_4 + m_4)(x_1 + x_2) + b_9 u_1 + b_8 y_1 + b_9 z_1 \\ \qquad + m_9 u_2 + m_8 y_2 + m_9 z_2 \\[6pt] \xi_7 = \beta_{22} + \rho_{22} + (b_5 + m_5)(y_1 + y_2) + b_{10} z_1 + z_2 m_{10} \\[6pt] \xi_8 = \beta_{23} + \rho_{23} + (b_6 + m_6)(z_1 + z_2) + u_1 b_{12} + u_2 m_{12} \\[6pt] \xi_9 = \beta_{24} + \rho_{24} + (b_7 + m_7)(u_1 + u_2) + y_1 b_{13} + y_2 m_{13} \\[6pt] \qquad\qquad \xi_{10} = \eta_5 + \eta_6 + \eta_7 + \eta_8 \end{cases} \tag{6.71}$$

$$\begin{cases} \xi_{11} = \beta_{31} + \rho_{31} + (c_4 + r_4)(x_1 + x_2) + c_9 u_1 + c_8 y_1 \\ \qquad + c_9 z_1 + r_9 u_2 + r_8 y_2 + r_9 z_2 \\[6pt] \xi_{12} = \beta_{32} + \rho_{32} + (c_5 + r_5)(y_1 + y_2) + c_{10} z_1 + z_2 r_{10} \\[6pt] \xi_{13} = \beta_{33} + \rho_{33} + (c_6 + r_6)(z_1 + z_2) + u_1 c_{12} + u_2 r_{12} \\[6pt] \xi_{14} = \beta_{34} + \rho_{34} + (c_7 + r_7)(u_1 + u_2) + y_1 c_{13} + y_2 r_{13} \\[6pt] \qquad\qquad \xi_{15} = \eta_9 + \eta_{10} + \eta_{11} + \eta_{12} \end{cases} \tag{6.72}$$

$$\begin{cases} \xi_{16} = \beta_{41} + \rho_{41} + (d_4 + s_4)(x_1 + x_2) + d_9 u_1 + d_8 y_1 \\ \qquad + d_9 z_1 + s_9 u_2 + s_8 y_2 + s_9 z_2 \\[6pt] \xi_{17} = \beta_{42} + \rho_{42} + (d_5 + s_5)(y_1 + y_2) + d_{10} z_1 + z_2 s_{10} \\[6pt] \xi_{18} = \beta_{43} + \rho_{43} + (d_6 + s_6)(z_1 + z_2) + u_1 d_{12} + u_2 s_{12} \\[6pt] \xi_{19} = \beta_{44} + \rho_{44} + (d_7 + s_7)(u_1 + u_2) + y_1 d_{13} + y_2 s_{13} \\[6pt] \qquad\qquad \xi_{20} = \eta_{13} + \eta_{14} + \eta_{15} + \eta_{16} \end{cases} \tag{6.73}$$

The error states are given by:

$$\begin{cases} e_1 = x_2 - x_1 \\ e_2 = y_2 - y_1 \\ e_3 = z_2 - z_1 \\ e_4 = u_2 - u_1 \end{cases}$$

Then the error system is given by

$$\begin{cases} e_1' = \xi_1 e_1 + \xi_2 e_2 + \xi_3 e_3 + \xi_4 e_4 + \xi_5 + v_1(t) \\ e_2' = \xi_6 e_1 + \xi_7 e_2 + \xi_8 e_3 + \xi_9 e_4 + \xi_{10} + v_2(t) \\ e_3' = \xi_{11} e_1 + \xi_{12} e_2 + \xi_{13} e_3 + \xi_{14} e_4 + \xi_{15} + v_3(t) \\ e_4' = \xi_{16} e_1 + \xi_{17} e_2 + \xi_{18} e_3 + \xi_{19} e_4 + \xi_{20} + v_4(t) \end{cases} \quad (6.74)$$

The following universal control law for the system (6.69) is defined by:

$$\begin{cases} v_1(t) = -(1 + \xi_1) e_1 - \xi_5 \\ v_2(t) = -(\xi_2 + \xi_6) e_1 - (1 + \xi_7) e_2 - \xi_{10} \\ v_3(t) = -(\xi_3 + \xi_{11}) e_1 - (\xi_8 + \xi_{12}) e_2 - (1 + \xi_{13}) e_3 - \xi_{15} \\ v_4(t) = -(\xi_4 + \xi_{16}) e_1 - (\xi_9 + \xi_{17}) e_2 - (\xi_{14} + \xi_{18}) e_3 \\ \qquad\qquad - (1 + \xi_{19}) e_4 - \xi_{20} \end{cases} \quad (6.75)$$

Then the two 4-D, continuous-time, quadratic systems (6.59) and (6.61) approach synchronization for any initial condition, since the error system (6.69) becomes

$$\begin{cases} e_1' = -e_1 + \xi_2 e_2 + \xi_3 e_3 + \xi_4 e_4 \\ e_2' = -\xi_2 e_1 - e_2 + \xi_8 e_3 + \xi_9 e_4 \\ e_3' = -\xi_3 e_1 - \xi_8 e_2 - e_3 + \xi_{14} e_4 \\ e_4' = -\xi_4 e_1 - \xi_9 e_2 - \xi_{14} e_3 - e_4 \end{cases} \quad (6.76)$$

If we consider the Lyapunov function $V = \frac{e_1^2 + e_2^2 + e_3^2 + e_4^2}{2}$, then the Lyapunov stability theory implies the asymptotic stability of the error system (6.71) since:

$$\frac{dV}{dt} = -e_1^2 - e_2^2 - e_3^2 - e_4^2 < 0$$

for all

$$
\begin{cases}
(\mu_i, \delta_i)_{1 \le i \le 4} \in \mathbb{R}^8 \\[2mm]
(\beta_{ij}, \rho_{ij})_{1 \le i,j \le 4} \in \mathbb{R}^{16} \\[2mm]
(a_i, b_i, c_i, d_i, h_i, m_i, r_i, s_i)_{4 \le i \le 13} \in \mathbb{R}^{80}
\end{cases}
$$

and for all initial conditions. If the two systems (6.59) and (6.61) are chaotic, then the control law (6.70) also guarantees their synchronization for any initial condition. Also, any 4-D continuous-time quadratic chaotic system can be stabilized (resp. controlled) to a stable 4-D continuous-time quadratic system that converges to an equilibrium point (resp. to a 4-D continuous-time quadratic system that converges to a periodic solution). Also, any 4-D continuous-time quadratic system can be chaotified to a chaotic 4-D continuous-time quadratic system.

6.6 Quasi-synchronization of systems with different dimensions

In this example, the synchronization of two discrete chaotic systems with different dimensions is presented in [Boukhalfa & Laskri (2016)]. Because the synchronisation is incomplete, a *bi-control* or *quasi-synchronization* is used instead. In fact, various methods of chaos synchronization of dynamical systems are well known in the current literature. The majority of the works of synchronization between two systems of different dimensions are based on a function φ that increases or reduces the size of the synchronized system as appropriate. In this case, function φ requires some analytical conditions and the determination of a simple vector controller, followed by the use of the Lyapunov stability theory.

The master system is the 3-D generalized Hénon map described by:

$$
\begin{cases}
x_1(k+1) = -0.1x_3(k) - x_2^2(k) + 1.76 \\[2mm]
x_2(k+1) = x_1(k) \\[2mm]
x_3(k+1) = x_2(k)
\end{cases}
\tag{6.77}
$$

and the slave system is the Hénon map described by:

$$
\begin{cases}
y_1(k+1) = y_2(k) - ay_1^2(k) + 1 + u_1 \\[2mm]
y_2(k+1) = by_1(k) + u_2,
\end{cases}
\tag{6.78}
$$

where $(a, b) = (1.4, 0.3)$ and (u_1, u_2) is the vector controller.

Define the quasi-synchronization errors by

$$\begin{cases} e_1(k) = y_1(k) - x_1(k) \\ e_2(k) = y_2(k) - x_2(k) \end{cases} \tag{6.79}$$

The synchronization errors between master system (6.72) and slave system (6.73), can be derived as:

$$\begin{cases} e_1(k+1) = y_2(k) - ay_1^2(k) + 0.1x_3(k) + x_2^2(k) - 0.76 + u_1 \\ e_2(k+1) = by_1(k) - x_1(k) + u_2 \end{cases} \tag{6.80}$$

To attain synchronization between systems (6.72) and (6.73), we can choose the vector controller U as follows:

$$\begin{cases} u_1 = -\frac{1}{2}y_2(k) - \frac{1}{2}x_2(k) + ay_1^2(k) - 0.1x_3(k) - x_2^2(k) + 0.76 \\ u_2 = (1-b)x_1(k) \end{cases} \tag{6.81}$$

Then, the synchronization errors between systems (6.72) and (6.73) are simplified as:

$$\begin{cases} e_1(k+1) = \frac{1}{2}e_2(k) \\ e_2(k+1) = be_1(k) \end{cases} \tag{6.82}$$

At this end, we Consider the candidate Lyapunov function in order to study the stability of synchronization errors:

$$V(e(k)) = \sum_{i=1}^{3} e_i^2(k) \tag{6.83}$$

We obtain:

$$\begin{aligned} \Delta V(e(k)) &= \sum_{i=1}^{3} e_i^2(k+1) - \sum_{i=1}^{3} e_i^2(k) \\ &= \frac{1}{4}e_2^2(k) + b^2 e_1^2(k) - e_1^2(k) - e_2^2(k) \\ &= (b^2 - 1) e_1^2(k) - \frac{1}{4}e_2^2(k) < 0. \end{aligned}$$

then, by Lyapunov stability it is immediate that

$$\lim_{k\to\infty} e_i(k) = 0, \quad i = 1, 2 \tag{6.84}$$

We conclude that the systems (6.72) and (6.73) are quasi-synchronized, as shown in Fig. 6.3 and Fig. 6.4.

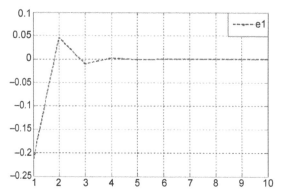

FIGURE 6.3: Synchronization error e_1 of (6.72) and (6.73). Reused with permission from [Boukhalfa and Laskri (2016)].

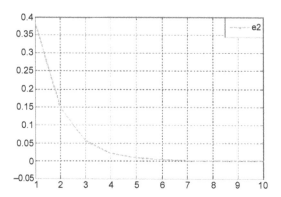

FIGURE 6.4: Synchronization error e_2 (6.72) and (6.73). Reused with permission from [Boukhalfa and Laskri (2016)].

In the second example, we consider the master system which is the Lozi map [Lozi (1978)] described by:

$$\begin{cases} x_1\,(k+1) = x_2\,(k) \\ \\ x_2\,(k+1) = 1 + x_1\,(k) - a\,|x_2\,(k)| \end{cases} \tag{6.85}$$

and the slave system is the 3-D generalized Hénon-like map given by:

$$\begin{cases} y_1(k+1) = 1 + y_3\,(k) - \alpha y_2^2\,(k) + u_1 \\ \\ y_2(k+1) = 1 + \beta y_2\,(k) - \alpha y_1^2\,(k) + u_2 \\ \\ y_3(k+1) = \beta y_1\,(k) + u_3 \end{cases} \tag{6.86}$$

where $(\alpha, \beta) = (1.4, 0.2)$ and (u_1, u_2, u_3) is the vector controller. The synchronization errors are defined by:

$$\begin{cases} e_1(k) = y_1(k) - x_1(k) \\ \\ e_2(k) = y_2(k) - x_2(k) \\ \\ e_3(k) = y_3(k) - x_1(k) \end{cases} \qquad (6.87)$$

The third error is arbitrary, then the synchronization errors between master system (6.80) and slave system (6.81), can be derived as:

$$\begin{cases} e_1(k+1) = 1 + y_3(k) - \alpha y_2^2(k) - x_2(k) + u_1 \\ \\ e_2(k+1) = \beta y_2(k) - \alpha y_1^2(k) - x_1(k) + a\,|x_2(k)| + u_2 \\ \\ e_3(k+1) = \beta y_1(k) - x_2(k) + u_3 \end{cases} \qquad (6.88)$$

To ensure synchronization between systems (6.80) and (6.81), we can choose the vector controller U as follows:

$$\begin{cases} u_1 = -\frac{1}{2}y_3(k) + x_2(k) - \frac{1}{2}x_3(k) + \alpha y_2^2(k) - 1 \\ \\ u_2 = -\beta x_2(k) + \alpha y_1^2(k) + x_1(k) - a\,|x_2(k)| \\ \\ u_3 = -\beta x_1(k) + x_2(k) \end{cases} \qquad (6.89)$$

The synchronization errors (6.83) can be written as:

$$\begin{cases} e_1(k+1) = \frac{1}{2}e_3(k) \\ \\ e_2(k+1) = \beta e_2(k) \\ \\ e_3(k+1) = \beta e_1(k) \end{cases} \qquad (6.90)$$

We choose the Lyapunov function:

$$V(e(k)) = \sum_{i=1}^{3} e_i^2(k), \qquad (6.91)$$

We obtain:

$$\begin{aligned} \Delta V(e(k)) &= \sum_{i=1}^{3} e_i^2(k+1) - \sum_{i=1}^{3} e_i^2(k) \\ \\ &= \frac{1}{4}e_3^2(k) + \beta^2 e_2^2(k) + \beta^2 e_1^2(k) - e_1^2(k) - e_2^2(k) - e_3^2(k) \\ \\ &= (\beta^2 - 1)\,e_2^2(k) + (\beta^2 - 1)\,e_1^2(k) - \frac{3}{4}e_3^2(k) < 0 \end{aligned}$$

By the Lyapunov stability, the errors tend to zero at infinity

$$\lim_{k \to \infty} e_i(k) = 0, \quad i = 1, 2, 3 \tag{6.92}$$

Hence, the systems (6.80) and (6.81) are quasi-synchronized as shown in Fig. 6.5.

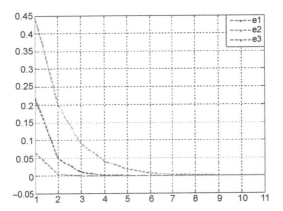

FIGURE 6.5: Synchronization errors: e_1, e_2, e_3 of (6.80) and (6.81). Reused with permission from [Boukhalfa and Laskri (2016)].

6.7   Chaotification of 3-D linear continuous-time systems using the signum function feedback

The anticontrol of chaos or chaotification is a reverse process of suppressing chaotic behaviors in a dynamical system. The aim of this process is to create or enhance the system complexity for some special applications, i.e., anticontrolling chaos is to generate some chaotic behaviors from a given system, which is not necessarily chaotic originally. There are many chaotification methods that are used to generate chaos in continuous-time systems, for example, the so-called differential-geometry control introduced in [Tang, *et al.*, (2001), Wang, *et al.*,(2000), Wang & Chen (2000)], time-delayed feedback given in [Wang, *et al.*, (2000)], and switching piecewise-linear control described in [Lû, *et al.*, (2002)], and an effective strategy of anticontrolling chaos in continuous-time systems has been discussed in [Ruoting, *et al.*, (2005)], from a homogeneity-based approach using the *p*-normal forms of non-linear systems.

In [Lû, *et al.*, (2002)] a switching piecewise-linear controller was presented as a chaos generator that can create chaos from a three-dimensional linear system within a wide range of parameter values. This switching piecewise-linear controller is defined by a relatively complicated law. In order to overcome this problem, we consider the simpler non-linear feedback in the form of

$$f(x) = \alpha sgn\,(x) \tag{6.93}$$

where $sgn(.)$ is the standard signum function that gives the sign of its argument and α is a real parameter. Feedback (6.88) is proposed for generating chaotic dynamics in a non-chaotic third-order linear autonomous continuous-time system, where $sgn\,(.)$ is the standard signum function that gives the sign of its argument. The signum function chaos generator is the simplest switching piecewise-linear controller, which can create chaos from a three-dimensional linear system within a wide range of parameter values. The importance of piecewise linear chaotic systems is that they are used to construct simple electronic circuits with several real applications, as in electronic engineers, in particular secure communications, computing and information processing, material engineering, and music, whereas much attention has been focused on effectively generating chaos via simple devices, such as circuitry design. In the end of this section, the resulting chaotic behaviors are verified by means of the largest Lyapunov exponent. Elementary examples show that the controlled system has a chaotic attractor with a different and special topological structure, characterized by two equilibria of saddle focus type and one piecewise-linear switching term which is the generator of chaos in this system. The electronic circuitry for the controlled system is also discussed, based on the extreme simplicity of the proposed non-linearity. Thus, the resulting system can be designed and implemented in a simple way.

The proposed chaotification method consists of applying the control law (6.88) to the general 3-D linear continuous-time autonomous system. Thus, we have the following switching piecewise-linear system given by:

$$\begin{cases} x' = a_0 + a_1 x + a_2 y + a_3 z + \alpha sgn(x) \\\\ y' = b_0 + b_1 x + b_2 y + b_3 z \\\\ z' = c_0 + c_1 x + c_2 y + c_3 z \end{cases} \tag{6.94}$$

where $(a_i, b_i, c_i)_{0 \leq i \leq 3} \in \mathbb{R}^{12}$ are the bifurcation parameters. It is clear that some previous works on piecewise linear circuits are special cases of system (6.89) which is of interest because it has the simplest form of non-linearities needed to generate chaos from a linear non-chaotic system. Due to the shape of the vector field of (6.89), the phase space can be divided into two linear regions denoted by $(E_i)_{1 \leq i \leq 2}$ as follow:

$$\begin{cases} E_1 = \left\{ (x, y, z) \in \mathbb{R}^3 : x < 0 \right\}, \\\\ E_2 = \left\{ (x, y, z) \in \mathbb{R}^3 : x \geq 0 \right\} \end{cases} \tag{6.95}$$

For generating chaos in system (6.89) we must assume that the resulting system has at least one isolated equilibrium point. We set

$$\begin{cases} d_0 = b_2c_3 - b_3c_2, d_1 = a_2b_3c_1 - a_2b_1c_3 + a_3b_1c_2 - a_3b_2c_1 \\[2mm] d_2 = a_0b_2c_3 - a_0b_3c_2 - a_2b_0c_3 + a_2c_0b_3 + b_0a_3c_2 - a_3b_2c_0 \\[2mm] d_3 = -b_2 - c_3, d_4 = b_2 + c_3, d_5 = b_2c_3 - a_3c_1 - a_2b_1 - b_3c_2 \\[2mm] d_6 = d_0 - d_5 + d_3d_4, d_7 = d_1 + d_3d_5 \end{cases} \tag{6.96}$$

Then, this case is possible if and only

$$d = d_0a_1 + d_1 \neq 0 \tag{6.97}$$

where in which the two equilibria are given by

$$\begin{cases} P_1 = \begin{pmatrix} \frac{\alpha b_2 c_3 - \alpha b_3 c_2 - a_0 b_2 c_3 + a_0 b_3 c_2 + a_2 b_0 c_3 - a_2 c_0 b_3 - b_0 a_3 c_2 + a_3 b_2 c_0}{d} \\[2mm] -\frac{\alpha b_1 c_3 - \alpha b_3 c_1 - a_0 b_1 c_3 + a_0 b_3 c_1 + a_1 b_0 c_3 - a_1 c_0 b_3 - b_0 a_3 c_1 + a_3 b_1 c_0}{d} \\[2mm] \frac{\alpha b_1 c_2 - \alpha b_2 c_1 - a_0 b_1 c_2 + a_0 b_2 c_1 + a_1 b_0 c_2 - a_1 b_2 c_0 - a_2 b_0 c_1 + a_2 b_1 c_0}{d} \end{pmatrix} \\[10mm] P_2 = \begin{pmatrix} -\frac{\alpha b_2 c_3 - \alpha b_3 c_2 + a_0 b_2 c_3 - a_0 b_3 c_2 - a_2 b_0 c_3 + a_2 c_0 b_3 + b_0 a_3 c_2 - a_3 b_2 c_0}{d} \\[2mm] \frac{\alpha b_1 c_3 - \alpha b_3 c_1 + a_0 b_1 c_3 - a_0 b_3 c_1 - a_1 b_0 c_3 + a_1 c_0 b_3 + b_0 a_3 c_1 - a_3 b_1 c_0}{d} \\[2mm] -\frac{\alpha b_1 c_2 - \alpha b_2 c_1 + a_0 b_1 c_2 - a_0 b_2 c_1 - a_1 b_0 c_2 + a_1 b_2 c_0 + a_2 b_0 c_1 - a_2 b_1 c_0}{d} \end{pmatrix} \end{cases} \tag{6.98}$$

The equilibrium point P_1 exists in E_1 if and only if $\frac{d_0}{d}\alpha - \frac{d_2}{d} < 0$, and the equilibrium point P_2 exists in E_2 if and only if $\frac{d_0}{d}\alpha + \frac{d_2}{d} \leq 0$. Thus, it is easy to prove the following results:

Lemme 6.1 *(a) The controlled system (6.89) has no equilibria if $\frac{d_0}{d}\alpha + \frac{d_2}{d} > 0$, i.e., $(d_1 + a_1d_0)(d_2 + \alpha d_0) > 0$.*

(b) The controlled system (6.89) has one equilibrium point if $\frac{d_0}{d}\alpha - \frac{d_2}{d} \geq 0$ or $\frac{d_0}{d}\alpha + \frac{d_2}{d} > 0$, i.e., $(d_1 + a_1d_0)(-d_2 + \alpha d_0) \geq 0$ or $(d_1 + a_1d_0)(d_2 + \alpha d_0) > 0$.

(c) The controlled system (6.89) has two equilibria P_1 and P_2 if $\frac{d_0}{d}\alpha - \frac{d_2}{d} < 0$, that is $(d_1 + a_1d_0)(-d_2 + \alpha d_0) < 0$.

The Jacobian matrix of the controlled system (6.89) at both equilibria P_1 and P_2 is the same and it is given by:

$$J = \begin{pmatrix} a_1 & a_2 & a_3 \\ b_1 & b_2 & b_3 \\ c_1 & c_2 & c_3 \end{pmatrix}$$

Thus, P_1 and P_2 have the same stability type. The exact value of the eigenvalues is obtained by using the Cardan method's for solving a cubic characteristic equation: $\lambda^3 + A\lambda^2 + B\lambda + C = 0$. By setting $\lambda = \frac{-A}{3} + w$, we have:

$w^3 + Pw + Q = 0$, where $P = \frac{-A^2}{3} + B$ and $Q = \frac{2A^3}{27} - \frac{AB}{3} + C$. We set $\Delta = 4P^3 + 27Q^2$, then if $\Delta > 0$, there is a unique negative real eigenvalue

$$\lambda_R = -\frac{A}{3} + \left(-\frac{Q}{2} + \sqrt{\frac{Q^2}{2} + \frac{P^3}{27}} \right)^{\frac{1}{3}} + \left(-\frac{Q}{2} - \sqrt{\frac{Q^2}{2} + \frac{P^3}{27}} \right)^{\frac{1}{3}}$$

along with two complex conjugate eigenvalues

$$(\lambda_C)^{\pm} = -\frac{A}{3} - \frac{w_R}{2} \pm \frac{i}{2}\sqrt{4P + 3(w_R)^2}$$

where

$$w_R = \left(-\frac{Q}{2} + \sqrt{\frac{Q^2}{2} + \frac{P^3}{27}} \right)^{\frac{1}{3}} + \left(-\frac{Q}{2} - \sqrt{\frac{Q^2}{2} + \frac{P^3}{27}} \right)^{\frac{1}{3}}$$

and if $\Delta < 0$, then there are three real eigenvalues

$$
\begin{cases}
\lambda_1 = -\frac{A}{3} + 2\sqrt{-\frac{P}{3}}\sin\left(\frac{\theta}{3}\right) \\[2mm]
\lambda_2 = -\frac{A}{3} + 2\sqrt{-\frac{P}{3}}\sin\left(\frac{\theta+2\pi}{3}\right) \\[2mm]
\lambda_3 = -\frac{A}{3} + 2\sqrt{-\frac{P}{3}}\sin\left(\frac{\theta+4\pi}{3}\right)
\end{cases}
$$

where

$$\theta = \arcsin\left(\sqrt{\frac{-27Q^2}{4P^3}} \right) \in [0, \pi]$$

The case $\Delta = 0$ corresponds to a measure-zero set of parameters. So, by a slight perturbation of parameters, without changing the behavior of the system, a system belonging to one of the two cases is obtained.

For $P_{1,2}$ resulting the following: $A = -a_1 + d_3$, $B = d_4 a_1 + d_5$, $C = -d_0 a_1 - d_1$. The Routh-Hurwitz conditions lead to the conclusion that the real parts of the roots are negative if and only if $A > 0, C > 0$ and $AB - C > 0$, i.e.,

$$
\begin{cases}
a_1 < d_3 \\[2mm]
d_0 a_1 < -d_1 \\[2mm]
-d_4 a_1^2 + d_6 a_1 + d_7 > 0
\end{cases}
\tag{6.99}
$$

which implies that the controlled system (6.89) does not have chaotic attractors in this case. Hence, the possible ranges for chaos in system (6.89) are included in the set

$$
\Omega = \left\{
\begin{array}{l}
(a_i, b_i, c_i)_{0 \le i \le 3} \in \mathbb{R}^{12} : \\[2mm]
a_1 \ge d_3, \text{or } d_0 a_1 \ge -d_1, \text{or } -d_4 a_1^2 + d_6 a_1 + d_7 \le 0
\end{array}
\right\}
\tag{6.100}
$$

with the conditions of existence of the equilibrium points given by (6.94). First, for the case of one equilibrium point, we have the following conditions:

$$
\begin{cases}
(d_1 + a_1 d_0)(-d_2 + \alpha d_0) \geq 0 \text{ and } a_1 \geq d_3, \text{ or} \\
(d_1 + a_1 d_0)(-d_2 + \alpha d_0) \geq 0 \text{ and } d_0 a_1 \geq -d_1, \text{ or} \\
(d_1 + a_1 d_0)(-d_2 + \alpha d_0) \geq 0 \text{ and } -d_4 a_1^2 + d_6 a_1 + d_7 \leq 0, \text{ or} \\
(d_1 + a_1 d_0)(d_2 + \alpha d_0) > 0 \text{ and } a_1 \geq d_3, \text{ or} \\
(d_1 + a_1 d_0)(d_2 + \alpha d_0) > 0 \text{ and } d_0 a_1 \geq -d_1, \text{ or} \\
(d_1 + a_1 d_0)(d_2 + \alpha d_0) > 0 \text{ and } -d_4 a_1^2 + d_6 a_1 + d_7 \leq 0
\end{cases}
\tag{6.101}
$$

Secondly, for the case of two equilibrium points, we have the following conditions:

$$
\begin{cases}
(d_1 + a_1 d_0)(-d_2 + \alpha d_0) < 0 \text{ and } a_1 \geq d_3, \text{ or} \\
(d_1 + a_1 d_0)(-d_2 + \alpha d_0) < 0 \text{ and } d_0 a_1 \geq -d_1, \text{ or} \\
(d_1 + a_1 d_0)(-d_2 + \alpha d_0) < 0 \text{ and } -d_4 a_1^2 + d_6 a_1 + d_7 \leq 0, \text{ or}
\end{cases}
\tag{6.102}
$$

In this case, to guarantee the chaotic nature of the resulting solutions of the controlled map (6.89), the range for the parameter α should be chosen in the set of parameters defined by (6.96) or (6.97).

In order to search for chaotic attractors in the controlled system (6.89), we assume that

$$
\begin{cases}
a_0 = a_2 = a_3 = 0, a_1 = a \\
b_0 = b_3 = 0, b_1 = -b, b_2 = b \\
c_0 = c_1 = 0, c_2 = 1, c_3 = -c.
\end{cases}
$$

In this case, the controlled system (6.98) below displays chaotic attractors in a very long range of bifurcation parameters, under some assumptions about them and with the application of the non-linear feedback controller (6.88) with $\alpha = -a$, i.e., the following controlled system:

$$
\begin{cases}
x' = ay - \alpha sgn(x) \\
y' = -bx + by \\
z' = y - cz
\end{cases}
\tag{6.103}
$$

The piecewise-linear system (6.98) has only with one switching non-linear term $sgn(x)$, and it can generate chaotic attractors with a very long domain of existence in the space of bifurcation parameters. The controlled system (6.98) has the natural symmetry $(x, y, z) \longrightarrow (-x, -y, -z)$ which persists for all values of parameters. Therefore for the controlled system (6.98), the divergence of the flow is given by:

$$\nabla V = \frac{\partial x'}{\partial x} + \frac{\partial y'}{\partial y} + \frac{\partial z'}{\partial z} = b - c$$

Then, If $b < c$ then system (6.98) has a bounded globally attracting ω-limit set. Then the system (6.98) is dissipative and all trajectories are ultimately confined to a specific subset having zero volume and the asymptotic motion settles onto an attractor; this result has been confirmed by some computer simulations. The only two equilibrium points of system (6.98) are given by $P_1 = \left(-1, -1, -\frac{1}{c}\right)$ and $P_2 = \left(1, 1, \frac{1}{c}\right)$ that exist for all values of a, b, and $c \neq 0$, one remarks that $P_1 \in E_1, P_2 \in E_2$. Thus, if there exists a chaotic attractor of the controlled system (6.98) then it has three possibilities: It surrounds the two equilibria P_1 and P_2 and its components are almost confined in the two corresponding regions E_1 and E_2 or it surrounds P_1 and its components are almost confined in the regions E_1, or it surrounds P_2 and its components are almost confined in the region E_2.

The Jacobian matrix of the controlled system (6.98) at both equilibria P_1 and P_2 is the same and it is given by:

$$J = \begin{pmatrix} 0 & a & 0 \\ -b & d & 0 \\ 0 & 1 & -c \end{pmatrix}$$

Thus, P_1 and P_2 have the same stability type. For $P_{1,2}$ resulting the following:

$$
\begin{cases}
A = c - b \\[2mm]
B = b(a - c) \\[2mm]
C = abc \\[2mm]
P = \frac{3ab - bc - b^2 - c^2}{3} \\[2mm]
Q = \frac{(b+2c)(9ab + bc - 2b^2 + c^2)}{27} \\[2mm]
\Delta = -b(b - 4a)(ab + bc + c^2)^2
\end{cases}
$$

Thus, one has $\Delta > 0$ if and only if $b(b - 4a) < 0$, i.e., The following cases holds:

(i) $b > 0$, and $b < 4a$: Firstly, if $a \geq 0$, one has $0 < b < 4a$. Secondly, if $a < 0$, the condition is impossible.

(ii) $b < 0$, and $b > 4a$: Firstly, if $a < 0$, one has $4a < b < 0$. Secondly, if $a \geq 0$, the condition is impossible.

The Routh-Hurwitz conditions lead to the conclusion that the real parts of the roots are negative if and only if $A > 0, C > 0$ and $AB - C > 0$. Thus, $P_{1,2}$ are stable if and only if $c > b$, and $abc > 0$ and $b(ab - bc + c^2) < 0$, which implies that the controlled system (6.98) does not have chaotic attractors in this case. Thus, the possible ranges for chaos in system (6.98) are included in the following set:

$$\Omega_{a,b,c} = \{(a, b, c) \in \mathbb{R}^3 : c \leq b, \text{or } abc \leq 0, \text{or } b(ab - bc + c^2) \geq 0\} \quad (6.104)$$

Indeed, for the parameter values $a = 20, b = 2, c = 3$, the controlled system (6.98) has the chaotic attractor shown in Fig. 6.6, and the eigenvalues corresponding to both equilibrium points are $\lambda_1 = -3, \lambda_2 = i\sqrt{39}+1, \lambda_3 = 1-i\sqrt{39}$, then there exist for all equilibrium points two conjugates complex eigenvalues, and in each case the equilibrium points are not stable, they are attracting in one direction but repelling in the other two directions. Hence, the two equilibria are of saddle-focus type, therefore the corresponding chaotic attractor governed by the controlled system (6.98) is characterized by two equilibria of saddle-focus type and one non-linearity.

The dynamical behaviors of the controlled system (6.98) are investigated numerically where the calculations of limit sets were performed using a fourth Runge-Kutta algorithm with a constant step size $\Delta t = 10^{-3}$. Then, to determine the long-time behavior and chaotic regions, we numerically computed the largest Lyapunov exponent. For $a = 20, b = 2$, and $c = 3$, the controlled system (6.98) has a chaotic attractor (with two wings) shown in Fig. 6.6 with $LE = 0,07874 \pm 0.009$. For $a = 25, b = 2$, and $c = 3$, the controlled system (6.98) has a chaotic attractor (with one wing) shown in Fig. 6.7 with $LE = 0,0841 \pm 0.011$.

Fix $b = 2, c = 3$, and vary $a \geq 0$, then, Fig. 6.8 shows the spectrum of the largest Lyapunov exponent of the controlled system (6.98) with respect to the parameter $a \geq 0$:

(i) when $0 \leq a < 11$, the system does not converge.

(ii) when $a \geq 11$, the maximum Lyapunov exponent is always positive, thus, the controlled system (6.98) has chaotic attractors in a large range of the parameter a as shown in Fig. 6.8. We remark, that for $a \geq 34.0436924$, the largest Lyapunov exponent is almost constant, and takes the values $LE \in [0.08595, 0.08619]$. This means that the controlled system (6.98) generates robust chaos defined by the absence of periodic windows and coexisting attractors in some neighborhood in the parameter space of a dynamical system [Zeraoulia and Sprott (2011(g))]. The fact that the attractor of system (6.98) is robust can be explained by the LCE spectrum. Indeed, this spectrum does not change under variation of initial conditions, which means that this attractor is the only one, and the basin of its attraction is the entire phase space as shown in Fig. 6.8. Also, this spectrum does not significantly change with variation of the parameters in the region of chaotic attractors for system (6.98). This result is confirmed also by plotting the chaotic attractor for several values of $a \geq 34.0436924$.

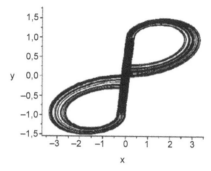

FIGURE 6.6: Projection onto the $x - y$ plane of the chaotic attractor obtained from system (6.98) for: $a = 20, b = 2, c = 3$.

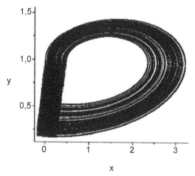

FIGURE 6.7: Projection onto the $x - y$ plane of the chaotic attractor obtained from system (6.98) for: $a = 25, b = 2, c = 3$.

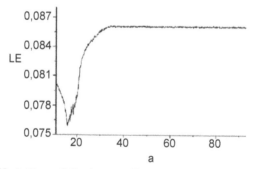

FIGURE 6.8: Variation of the largest Lyapunov exponent of system (6.98) versus the parameter $a \geq 11$, and $b = 2, c = 3$.

At this end, we note that the necessary non-linearity used in the controlled system (6.98) is robust, since the circuit elements should be confined to their linear regions of operation. Hence, in the controlled system (6.98) there is only one simpler piecewise-linear switching term and all the others are linear, this simplifies the circuitry realization of the controlled system (6.98), because the building of an electronic circuit (that describing the dynamics of this system) is possible for almost all cases of piecewise-linear systems [Kennedy & Kolumban (2000), Mahla & Palhares (1993)] with several applications in electronic engineers, in particular secure communications, music, etc...

6.8 Chaos control problem of a 3-D cancer model with structured uncertainties

In this section, we focus our attention on the chaos control problem of the 3-D cancer model [Itik & Salamci (2009)] with structured uncertainties given by:

$$
\begin{cases}
x' = x\,(1-x) - a_{12}xy - a_{13}xz \\[2mm]
y' = r_2 y\,(1-y) - a_{21}xy \\[2mm]
z' = \frac{r_3 xz}{x+k_3} - a_{31}xz - d_3 z
\end{cases}
\tag{6.105}
$$

The method of analysis is based on the Lyapunov stability theory. Taking into consideration the system's uncertainties, system (6.100) can be expressed as

$$
\begin{cases}
x' = x\,(1-x) - (b_{12} + \varepsilon_0)\,xy - (b_{13} + \varepsilon_1)\,xz \\[2mm]
y' = (s_2 + \varepsilon_2)\,y\,(1-y) - (b_{21} + \varepsilon_3)\,xy \\[2mm]
z' = \frac{(s_3 + \varepsilon_4)xz}{x+(v_3+\varepsilon_5)} - (b_{31} + \varepsilon_6)\,xz - (c_3 + \varepsilon_7)\,z
\end{cases}
\tag{6.106}
$$

where $(b_{12}, b_{13}, s_2, b_{21}, s_3, b_{31}, c_3, v_3) \in \mathbb{R}^8$ is the nominal value of $(a_{12}, a_{13}, r_2, a_{21}, r_3, a_{31}, d_3, k_3) \in \mathbb{R}^8$ and $(\varepsilon_0, \varepsilon_1, \varepsilon_2, \varepsilon_3, \varepsilon_4, \varepsilon_5, \varepsilon_6, \varepsilon_7) \in \mathbb{R}^8$ is a system structured uncertainty of parameter $(a_{12}, a_{13}, r_2, a_{21}, r_3, a_{31}, d_3, k_3) \in \mathbb{R}^8$. It is assumed that the uncertain unified chaotic system (6.101) still demonstrates chaotic behavior. To suppress the undesired chaotic dynamics of the uncontrolled system (6.100), the proposed method adds a control-input $u\,(t) = (u_1\,(t)\,, u_2\,(t)\,, u_3\,(t))$ to the system (6.101). The controlled system thus becomes

$$
\begin{cases}
x' = x\,(1-x) - (b_{12} + \varepsilon_0)\,xy - (b_{13} + \varepsilon_1)\,xz + u_1\,(t) \\[2mm]
y' = (s_2 + \varepsilon_2)\,y\,(1-y) - (b_{21} + \varepsilon_3)\,xy + u_2\,(t) \\[2mm]
z' = \frac{(s_3 + \varepsilon_4)xz}{x+(v_3+\varepsilon_5)} - (b_{31} + \varepsilon_6)\,xz - (c_3 + \varepsilon_7)\,z + u_3\,(t)
\end{cases}
\tag{6.107}
$$

The addition of $u(t)$ will be used in order to provide a robust controller to derive the system state to the zero point in the state space, even when the system is experiencing match and mismatch uncertainties. Consequentially, the chaos behavior of the system is suppressed. Let the control input be

$$
\begin{cases}
u_1(t) = -2x + x^2 \\[2mm]
u_2(t) = -(\varepsilon_2 + s_2 + 1)y + (\varepsilon_0 + b_{12})x^2 + (\varepsilon_2 + s_2)y^2 + (\varepsilon_3 + b_{21})xy \\[2mm]
u_3(t) = -\frac{(s_3 + \varepsilon_4)xz}{x + (v_3 + \varepsilon_5)} + (\varepsilon_7 + c_3 - 1)z + (\varepsilon_1 + b_{13})x^2 + (\varepsilon_6 + b_{31})xz
\end{cases}
$$

$$(6.108)$$

and then the controlled system (6.102) becomes

$$
\begin{cases}
x' = (-\varepsilon_0 - b_{12})xy + (-\varepsilon_1 - b_{13})xz - x \\[2mm]
y' = (\varepsilon_0 + b_{12})x^2 - y \\[2mm]
z' = (\varepsilon_1 + b_{13})x^2 - z
\end{cases}
$$

$$(6.109)$$

If we consider the Lyapunov function $V = \frac{x^2 + y^2 + z^2}{2}$, then it is easy to verify the asymptotic stability of the controlled system (6.104) by Lyapunov stability theory since we have $\frac{dV}{dt} = -(x^2 + y^2 + z^2) < 0$ for all $(b_{12}, b_{13}, s_2, b_{21}, s_3, b_{31}, c_3, v_3) \in \mathbb{R}^8$, $(\varepsilon_0, \varepsilon_1, \varepsilon_2, \varepsilon_3, \varepsilon_4, \varepsilon_5, \varepsilon_6, \varepsilon_7) \in \mathbb{R}^8$ and for all initial conditions.

6.9 Controlling homoclinic chaotic attractor

In this section, we give a simple method for controlling the homoclinic chaotic attractor generated by a single piecewise smooth map of the plane. The method of analysis is based on the construction of the simplest piecewise linear controller function. Indeed, let us consider the general two-dimensional piecewise smooth map of the form:

$$
f(x, y; \rho) =
\begin{cases}
f_1(x, y; \rho) = \begin{pmatrix} f_{11}(x, y; \rho) \\ f_{12}(x, y; \rho) \end{pmatrix}, & \text{if } x \le S(y, \rho), \\[5mm]
f_2(x, y; \rho) = \begin{pmatrix} f_{21}(x, y; \rho) \\ f_{22}(x, y; \rho) \end{pmatrix}, & \text{if } x > S(y, \rho).
\end{cases}
$$

$$(6.110)$$

which depends on a single parameter ρ. Let Γ_ρ, given by $x = S(y, \rho)$ denote a smooth curve that divides the phase plane into two regions R_1 and R_2, defined by:

$$R_1 = \{(x,y) \in \mathbb{R}^2, x \le S(y,\rho)\}, \tag{6.111}$$

$$R_2 = \{(x,y) \in \mathbb{R}^2, x > S(y,\rho)\}, \tag{6.112}$$

It is assumed that the functions f_1 and f_2 are both continuous and have continuous derivatives. The map f is continuous but its derivative is discontinuous at the line Γ_ρ, called the *border*. It is further assumed that the one-sided partial derivatives at the border are finite, and in each region R_1 and R_2 the map (6.105) has one fixed point $P_1 = (x_1, y_1)$ and $P_2 = (x_2, y_2)$, for certain value ρ_* of the parameter ρ.

Let us define the following quantities:

$$\begin{cases} \frac{\partial f_{11}}{\partial x}(P_1) = \beta_1, \ \frac{\partial f_{11}}{\partial y}(P_1) = \beta_2, \ \frac{\partial f_{12}}{\partial x}(P_1) = \beta_3, \ \frac{\partial f_{12}}{\partial y}(P_1) = \beta_4, \\[2mm] \frac{\partial f_{21}}{\partial x}(P_2) = \gamma_1, \ \frac{\partial f_{21}}{\partial y}(P_2) = \gamma_2, \ \frac{\partial f_{22}}{\partial x}(P_1) = \gamma_3, \ \frac{\partial f_{22}}{\partial y}(P_1) = \gamma_4. \end{cases} \tag{6.113}$$

In the case of two-dimensional piecewise smooth maps, it has been shown that it is possible to choose an appropriate coordinate transformation so that the choice of axis is independent of the parameter. In so doing, the normal form of the map (6.105) is given by:

$$f_N(x,y) = \begin{cases} \begin{pmatrix} \tau_1 & 1 \\ -\delta_1 & 0 \end{pmatrix} \begin{pmatrix} x \\ y \end{pmatrix} + \begin{pmatrix} 0 \\ 1 \end{pmatrix} \mu, \ \text{if } x < 0, \\[4mm] \begin{pmatrix} \tau_2 & 1 \\ -\delta_2 & 0 \end{pmatrix} \begin{pmatrix} x \\ y \end{pmatrix} + \begin{pmatrix} 0 \\ 1 \end{pmatrix} \mu, \text{if } x > 0. \end{cases} \tag{6.114}$$

where μ is a parameter and $\tau_i, \delta_i, i = 1, 2$ are the traces and determinants of the corresponding matrices of the linearized map of (6.105) in the two regions R_1 and R_2 evaluated at P_1 and P_2 respectively, that is:

$$\tau_1 = \beta_1 + \beta_4, \ \tau_2 = \gamma_1 + \gamma_4, \ \delta_1 = \beta_1\beta_4 - \beta_2\beta_3, \ \text{and} \ \delta_2 = \gamma_1\gamma_4 - \gamma_2\gamma_3, \tag{6.115}$$

Assume that:

$$\tau_1 > 1 + \delta_1, \ \text{and} \ \tau_2 < -(1 + \delta_2), \tag{6.116}$$

then there are no fixed points for the map (6.109) when $\mu < 0$, and there are two fixed points one each in R_1 and R_2, for $\mu > 0$, given by:

$$\begin{cases} P_1 = \left(\frac{\mu}{1-\tau_1+\delta_1}, \frac{-\mu\delta_1}{1-\tau_1+\delta_1} \right), \\[3mm] P_2 = \left(\frac{\mu}{1-\tau_2+\delta_2}, \frac{-\mu\delta_2}{1-\tau_2+\delta_2} \right) \end{cases}$$

Since as the parameter μ is varied through zero, the local bifurcation of the map (6.109) depends only on the values of $\tau_i, \delta_i, i = 1, 2$, then it suffices to study the bifurcations in the normal form (6.109). The map (6.109) has homoclinic chaotic attractors for some parameter space region, when the condition (6.111) holds, and:

$$1 > \delta_1 \ge 0, \ \text{and} \ 1 > \delta_2 \ge 0. \tag{6.117}$$

We give a method for controlling this robust homoclinic chaos, based on the simplest piecewise linear controller function defined by:

$$
U(x,y) =
\begin{cases}
\begin{pmatrix} u_1(x,y) = \alpha_1\,(x-x_1) + \alpha_2\,(y-y_1) \\ 0 \end{pmatrix}, & \text{if } x \le S(y,\rho), \\[2ex]
\begin{pmatrix} 0 \\ u_2(x,y) = \eta_1\,(x-x_2) + \eta_2\,(y-y_2) \end{pmatrix}, & \text{if } x > S(y,\rho).
\end{cases}
$$

(6.118)

where $(\alpha_i)_{1\le i\le 2}$ and $(\eta_i)_{1\le i\le 2}$ are parameters to be determined and $(x_i)_{1\le i\le 2}$ and $(y_i)_{1\le i\le 2}$ are defined above.

We prove the following result:

Theorem 50 *Consider the controlled map* $g(x,y;\rho) = f(x,y;\rho) + U(x,y)$ *and assume the following conditions:*

(1) The function f is continuously differentiable in both regions R_1 and R_2.

(2) The function f has only two fixed points P_1 in R_1, and P_2 in R_2, for some value ρ_* of the parameter ρ, such that:

$$
\begin{cases}
\beta_3 > 0, \ \gamma_1 > 0, \ \gamma_2 > 0 \\[2ex]
\alpha_2 > \dfrac{1+\alpha_1\beta_4+\delta_1-\alpha_1-\tau_1}{\beta_3} \\[2ex]
\dfrac{-(1+\eta_2\gamma_1+\delta_2+\eta_2+\tau_2)}{\gamma_2} < \eta_1 < \dfrac{1+\eta_2\gamma_1+\delta_2-\eta_2-\tau_2}{\gamma_2} \\[2ex]
\eta_2 > \dfrac{-(\delta_2+1)}{\gamma_1}
\end{cases}
$$

(6.119)

Then, the piecewise linear controller U defined in (6.113) transforms any homoclinic chaotic attractor generated by the map (6.105) into periodic orbit.

Proof 15 *The controlled map is given by:*

$$
g(x,y;\rho) =
\begin{cases}
g_1(x,y;\rho) = \begin{pmatrix} f_{11}(x,y;\rho) + u_1\,(x,y) \\ f_{12}(x,y;\rho) \end{pmatrix}, & \text{if } x \le S(y,\rho), \\[3ex]
g_2(x,y;\rho) = \begin{pmatrix} f_{21}(x,y;\rho) \\ f_{22}(x,y;\rho) + u_2\,(x,y) \end{pmatrix}, & \text{if } x > S(y,\rho).
\end{cases}
$$

(6.120)

From the first condition of Theorem 50, it is clear that the functions g_1 and g_2 are both continuous and have continuous derivatives, then the controlled map g is continuous but its derivative is discontinuous at the borderline $x = S(y,\rho)$. It is also clear that the one-sided partial derivatives at the border are finite according to the definition of the controller U, and in each region R_1 and R_2 the controlled map (6.115) has the same fixed points P_1 and P_2 as in the original map (6.105) for the same values ρ_* of the parameter ρ, since $u_1(P_1) = 0$, and $u_2(P_2) = 0$.

The normal form of the controlled map (6.115) is given by:

$$
g_N(x, y) = \begin{cases} \begin{pmatrix} \theta_1 & 1 \\ -\omega_1 & 0 \end{pmatrix} \begin{pmatrix} x \\ y \end{pmatrix} + \begin{pmatrix} 0 \\ 1 \end{pmatrix} \xi, \ \text{if } x < 0 \\[4mm] \begin{pmatrix} \theta_2 & 1 \\ -\omega_2 & 0 \end{pmatrix} \begin{pmatrix} x \\ y \end{pmatrix} + \begin{pmatrix} 0 \\ 1 \end{pmatrix} \xi, \text{if } x > 0 \end{cases} \tag{6.121}
$$

where ξ is a parameter and $\theta_i, \omega_i, i = 1, 2$ are the traces and determinants of the corresponding matrices of the linearized map of the controlled map (6.115) in the two regions R_1 and R_2 evaluated at P_1 and P_2 respectively.

In this case the values of $\theta_i, \omega_i, i = 1, 2$ are given by:

$$
\begin{cases} \theta_1 = \alpha_1 + \tau_1, \ \text{and } \theta_2 = \eta_2 + \tau_2 \\[3mm] \omega_1 = \alpha_1 \beta_4 - \alpha_2 \beta_3 + \delta_1, \ \text{and } \omega_2 = \eta_2 \gamma_1 - \eta_1 \gamma_2 + \delta_2 \end{cases} \tag{6.122}
$$

It is shown in [Banergee, et al., (1998)] that a periodic attractor occurs in the piecewise smooth map of the form (6.116) when:

$$
\theta_1 > 1 + \omega_1, \ \text{and } -(1 + \omega_2) < \theta_2 < (1 + \omega_2) \tag{6.123}
$$

In this case, and for $\xi > 0$, the fixed point in R_1 is a regular saddle and the one in R_2 is an attractor [Banergee, et al., (1998)]. This is like a saddle-node bifurcation occurring on the border. Since this region in the parameter space always has a periodic attractor for $\xi > 0$, we show that for certain values of $(\alpha_i)_{1 \le i \le 2}$ and $(\eta_i)_{1 \le i \le 2}$, the controlled map (6.116) has a periodic behavior using conditions (6.118), where the values $(\beta_i)_{1 \le i \le 4}$ and $(\gamma_i)_{1 \le i \le 4}$ are known.

Without loss of generality, one can suppose that:

$$
\beta_3 > 0, \ \gamma_2 > 0, \ \gamma_1 > 0 \tag{6.124}
$$

Thus, conditions (6.118) with assumptions (6.119) give the following inequalities:

$$
\begin{cases} \alpha_2 > \frac{1 + \alpha_1 \beta_4 + \delta_1 - \alpha_1 - \tau_1}{\beta_3} \\[3mm] \eta_1 > \frac{-(1 + \eta_2 \gamma_1 + \delta_2 + \eta_2 + \tau_2)}{\gamma_2} = \psi_1, \ \text{and } \eta_1 < \frac{1 + \eta_2 \gamma_1 + \delta_2 - \eta_2 - \tau_2}{\gamma_2} = \psi_2 \end{cases} \tag{6.125}
$$

In addition, if we suppose that:

$$
\eta_2 > \frac{-(\delta_2 + 1)}{\gamma_1} \tag{6.126}
$$

then, one can deduce the following conditions:

$$
\begin{cases}
\alpha_2 > \frac{1+\alpha_1\beta_4+\delta_1-\alpha_1-\tau_1}{\beta_3} \\[2mm]
\frac{-(1+\eta_2\gamma_1+\delta_2+\eta_2+\tau_2)}{\gamma_2} < \eta_1 < \frac{1+\eta_2\gamma_1+\delta_2-\eta_2-\tau_2}{\gamma_2}
\end{cases}
\tag{6.127}
$$

since $\psi_2 - \psi_1 = 2\gamma_2^{-1}\left(\delta_2 + \eta_2\gamma_1 + 1\right) > 0$, according to condition (6.121).

Example 15 *We present in this example the simplest piecewise smooth linear controller capable of transforming homoclinic chaos into periodic behavior in any piecewise smooth map of the plane: Since the parameter α_1 is free, one can choose $\alpha_1 = 0$ and, therefore, the use of inequality (6.120) gives $\alpha_2 > \frac{1+\delta_1-\tau_1}{\beta_3}$, with $\frac{1+\delta_1-\tau_1}{\beta_3} < 0$, since $\tau_1 > 1+\delta_1$. Thus, one can also choose $\alpha_2 = 0$. On the other hand, from (6.121) one has $\eta_2 > \frac{-(\delta_2+1)}{\gamma_1}$, and $\frac{-(\delta_2+1)}{\gamma_1} < 0$, this implies that one can choose also $\eta_2 = 0$, and in this case, the values of η_1 cannot be zero, since $\tau_2 < -\left(1+\delta_2\right)$ and $0 < \frac{-(1+\delta_2+\tau_2)}{\gamma_2} < \eta_1 < \frac{1+\delta_2-\tau_2}{\gamma_2}$.*

Finally, the simplest piecewise smooth linear controller is given by:

$$
U_{simplest}(x,y) =
\begin{cases}
\begin{pmatrix} 0 \\ 0 \end{pmatrix}, & \text{if } x \le S(y,\rho) \\[4mm]
\begin{pmatrix} 0 \\ \eta_1\,(x - x_2) \end{pmatrix}, & \text{if } x > S(y,\rho)
\end{cases}
\tag{6.128}
$$

6.10   Robustification of 2-D piecewise smooth mappings

The essential motivation of the present section is to create robust chaos in general 2-D discrete mappings via the controller of a simple piecewise smooth function under some realizable conditions (robustification). This method is presented in [Zeraoulia and Sprott (2011(e))].

   Let us consider the general two-dimensional map (not necessarily chaotic) of the form $f(x,y,\rho) = (f_1(x,y,\rho), f_2(x,y,\rho))$ where the function f is assumed to be locally continuously differentiable and depend on a single parameter ρ. The goal of the control is to create a border collision bifurcation leading to robust chaos in the simplest way, i.e., by considering the controlled map given by:

$$
g(x,y,\rho) =
\begin{pmatrix}
g_1(x,y,\rho) = f_1(x,y,\rho) + \alpha\,|x| \\[2mm]
g_2(x,y,\rho) = f_2(x,y,\rho) + \beta\,|x|
\end{pmatrix}
\tag{6.129}
$$

where α and β are real bifurcation parameters to be determined. The line $x = 0$ divides the phase plane into two regions $R_1 = \{(x, y) \in \mathbb{R}^2, x < 0\}$ and $R_2 = \{(x, y) \in \mathbb{R}^2, x > 0\}$. The controlled map (6.124) is a piecewise smooth map since the controller $u(x, y) = (\alpha |x|, \beta |x|)$ is also smooth. It is also clear that the functions f_1 and f_2 are both continuous and have continuous derivatives, thus, the controlled map g is continuous but its derivative is discontinuous at the borderline $x = 0$. Also, it is clear that the one-sided partial derivatives at the border are finite, and in each subregion R_1 and R_2, the controlled map (6.124) has only two fixed points P_1 and P_2 for values ρ_* of the parameter ρ by assuming that the equations:

$$\begin{cases} f_1(x, y, \rho) + \alpha |x| = x \\ \\ f_2(x, y, \rho) + \beta |x| = y \end{cases}$$

have only one root P_1 and P_2, respectively, for some ranges of the parameters ρ, α and β. The control design $u(x, y)$ leads to systems of linear inequalities for the unknown parameters α and β in terms of the sets $(\beta_i)_{1 \le i \le 4}$ and $(\gamma_i)_{1 \le i \le 4}$ assumed to be known and defined by

$$\begin{cases} \frac{\partial f_1(x,y,\rho)}{\partial x}(P_1) = \beta_1, \; \frac{\partial f_1(x,y,\rho)}{\partial y}(P_1) = \beta_2, \; \frac{\partial f_2(x,y,\rho)}{\partial x}(P_1) = \beta_3 \\ \\ \frac{\partial f_2(x,y,\rho)}{\partial y}(P_1) = \beta_4, \; \frac{\partial f_1(x,y,\rho)}{\partial x}(P_2) = \gamma_1, \; \frac{\partial f_1(x,y,\rho)}{\partial y}(P_2) = \gamma_2 \qquad (6.130) \\ \\ \frac{\partial f_2(x,y,\rho)}{\partial x}(P_1) = \gamma_3, \; \frac{\partial f_2(x,y,\rho)}{\partial y}(P_1) = \gamma_4 \end{cases}$$

The normal form of the controlled map (6.124) is given by:

$$N(x, y) = \begin{cases} \begin{pmatrix} \tau_1 & 1 \\ -\delta_1 & 0 \end{pmatrix} \begin{pmatrix} x \\ y \end{pmatrix} + \begin{pmatrix} 1 \\ 0 \end{pmatrix} \mu, \; \text{if } x < 0 \\ \\ \begin{pmatrix} \tau_2 & 1 \\ -\delta_2 & 0 \end{pmatrix} \begin{pmatrix} x \\ y \end{pmatrix} + \begin{pmatrix} 1 \\ 0 \end{pmatrix} \mu, \; \text{if } x > 0 \end{cases} \qquad (6.131)$$

where μ is the new bifurcation parameter of the map (6.124). The normal form (6.125) has two fixed points $P_L = \left(\frac{\mu}{1-\tau_1+\delta_1}, \frac{-\delta_1\mu}{1-\tau_1+\delta_1} \right) \in R_1$ and $P_R = \left(\frac{\mu}{1-\tau_2+\delta_2}, \frac{-\delta_2\mu}{1-\tau_2+\delta_2} \right) \in R_2$ (with eigenvalues $\lambda_{1,2}$ and $\omega_{1,2}$, respectively). The stability of these fixed points is determined by the eigenvalues of the corresponding Jacobian matrix, i.e., $\lambda = \frac{1}{2} \left(\tau \pm \sqrt{\tau^2 - 4\delta} \right)$. Here μ is a parameter and $\tau_i, \delta_i, i = 1, 2$ are the traces and determinants of the corresponding matrices of the linearized map in the two subregions R_1 and R_2 evaluated at P_1 and P_2, respectively, that is $A_1 = \begin{pmatrix} \beta_1 & \beta_2 \\ \beta_3 & \beta_4 \end{pmatrix} + \begin{pmatrix} -\alpha & 0 \\ -\beta & 0 \end{pmatrix}$ and $A_2 = \begin{pmatrix} \gamma_1 & \gamma_2 \\ \gamma_3 & \gamma_4 \end{pmatrix} + \begin{pmatrix} \alpha & 0 \\ \beta & 0 \end{pmatrix}$. In the case of the controlled map (6.125), the values of $\tau_i, \delta_i, i = 1, 2$ are given by $\tau_1 = -\alpha + \beta_1 + \beta_4, \tau_2 = \alpha + \gamma_1 + \gamma_4,$ $\delta_1 = \beta\beta_2 - \alpha\beta_4 + \beta_1\beta_4 - \beta_2\beta_3$, and $\delta_2 = \alpha\gamma_4 - \beta\gamma_2 + \gamma_1\gamma_4 - \gamma_2\gamma_3$. Robust homoclinic chaos occurs in the piecewise smooth map of the form (6.125) if one of the following conditions holds:

(a)

$$\begin{cases} \tau_1 > 1 + \delta_1, \text{ and } \tau_2 < -(1 + \delta_2) \\ \\ 0 < \delta_1 < 1, \text{ and } 0 < \delta_2 < 1 \end{cases} \qquad (6.132)$$

where the parameter range for a boundary crisis is given by:

$$C_1 : \delta_1\tau_2\lambda_1 - \delta_2\lambda_1\lambda_2 + \delta_2\lambda_2 - \delta_1\tau_2 + \delta_1\tau_1 - \delta_1 - \lambda_1\delta_1 > 0. \qquad (6.133)$$

Here, inequality (6.127) determines the condition for stability of the chaotic attractor of the map (6.125). The robust chaotic orbit persists as τ_1 is reduced below $1 + \delta_1$.

(b)

$$\begin{cases} \tau_1 > 1 + \delta_1, \text{ and } \tau_2 < -(1 + \delta_2) \\ \\ \delta_1 < 0, \text{ and } -1 < \delta_2 < 0 \\ \\ C_2 : \frac{\lambda_{1L} - 1}{\tau_1 - 1 - \delta_1} > \frac{\lambda_{2R} - 1}{\tau_2 - 1 - \delta_2} \end{cases} \qquad (6.134)$$

The condition for stability of the chaotic attractor is also determined by (6.127). If the third condition of (6.128) is not satisfied, then the condition for the existence of a chaotic attractor changes to the following inequality:

$$C_3 : \frac{\omega_2 - 1}{\tau_2 - 1 - \delta_1} < \frac{(\tau_1 - \delta_1 - \lambda_2)}{(\tau_1 - 1 - \delta_1)(\lambda_2 - \tau_2).} \qquad (6.135)$$

(c) The remaining ranges for the quantity $\tau_{1,2}, \delta_{1,2}$ can be determined in some cases using the same logic as in the above two cases, or there is no analytic condition for a boundary crisis, and it has to be determined numerically.

Thus, from the above three cases, we shall show that for certain values of α and β, the controlled map (6.125) has robust homoclinic chaos, where the values $(\beta_i)_{1 \le i \le 4}$ and $(\gamma_i)_{1 \le i \le 4}$ are assumed to be known. Indeed, for the first case (a), condition (6.126) gives the following inequalities:

$$\begin{cases} \phi_1 < \beta_2\beta < \min\{\phi_2, \phi_3\} \\ \\ \max\{\phi_4, \phi_5\} < \gamma_2\beta < \phi_6 \end{cases} \qquad (6.136)$$

where

$$\begin{cases} \phi_1 = \alpha\beta_4 - \beta_1\beta_4 + \beta_2\beta_3 \\[2mm] \phi_2 = \beta_1 - \alpha + \beta_4 + \alpha\beta_4 - \beta_1\beta_4 + \beta_2\beta_3 - 1 \\[2mm] \phi_3 = \alpha\beta_4 - \beta_1\beta_4 + \beta_2\beta_3 + 1 \\[2mm] \phi_4 = \alpha + \gamma_1 + \gamma_4 + \alpha\gamma_4 + \gamma_1\gamma_4 - \gamma_2\gamma_3 + 1 \\[2mm] \phi_5 = \alpha\gamma_4 + \gamma_1\gamma_4 - \gamma_2\gamma_3 - 1 \\[2mm] \phi_6 = \alpha\gamma_4 + \gamma_1\gamma_4 - \gamma_2\gamma_3. \end{cases}$$

The two inequalities in (6.130) hold if and only if $\phi_1 < \phi_2, \phi_1 < \phi_3, \phi_4 < \phi_6$, and $\phi_5 < \phi_6$, that is,

$$\alpha < \min\left\{\beta_1 + \beta_4 - 1, -\gamma_1 - \gamma_4 - 1\right\}. \tag{6.137}$$

The stability of the resulting homoclinic chaos can be tested using inequality (6.127). However, it is difficult to solve rigorously for this stability condition (or the conditions for the two remaining cases (b) and (c) above) due to the presence of complicated square formulas. In this case, one can use numerical estimates. The value of the parameter β in (6.130) depends mainly on the signs of β_2 and γ_2. If $\beta_2 > 0$ and $\gamma_2 > 0$, then (6.130) becomes

$$\begin{cases} \frac{\phi_1}{\beta_2} < \beta < \frac{\min\{\phi_2,\phi_3\}}{\beta_2} \\[3mm] \frac{\max\{\phi_4,\phi_5\}}{\gamma_2} < \beta < \frac{\phi_6}{\gamma_2}. \end{cases} \tag{6.138}$$

If $\beta_2 < 0$ and $\gamma_2 < 0$, then (6.130) becomes

$$\begin{cases} \frac{\phi_1}{\beta_2} > \beta > \frac{\min\{\phi_2,\phi_3\}}{\beta_2} \\[3mm] \frac{\max\{\phi_4,\phi_5\}}{\gamma_2} > \beta > \frac{\phi_6}{\gamma_2}. \end{cases} \tag{6.139}$$

If $\beta_2 > 0$ and $\gamma_2 < 0$, then (6.130) becomes

$$\begin{cases} \frac{\phi_1}{\beta_2} < \beta < \frac{\min\{\phi_2,\phi_3\}}{\beta_2} \\[3mm] \frac{\max\{\phi_4,\phi_5\}}{\gamma_2} > \beta > \frac{\phi_6}{\gamma_2}, \end{cases} \tag{6.140}$$

that is,

$$\max\left\{\frac{\phi_1}{\beta_2}, \frac{\phi_6}{\gamma_2}\right\} < \beta < \min\left\{\frac{\min\{\phi_2,\phi_3\}}{\beta_2}, \frac{\max\{\phi_4,\phi_5\}}{\gamma_2}\right\}. \tag{6.141}$$

If $\beta_2 < 0$ and $\gamma_2 > 0$, then (6.130) becomes

$$\begin{cases} \frac{\phi_1}{\beta_2} > \beta > \frac{\min\{\phi_2,\phi_3\}}{\beta_2} \\ \\ \frac{\max\{\phi_4,\phi_5\}}{\gamma_2} < \beta < \frac{\phi_6}{\gamma_2}, \end{cases} \tag{6.142}$$

that is,

$$\max\left\{\frac{\min\{\phi_2,\phi_3\}}{\beta_2}, \frac{\max\{\phi_4,\phi_5\}}{\gamma_2}\right\} < \beta < \min\left\{\frac{\phi_1}{\beta_2}, \frac{\phi_6}{\gamma_2}\right\}. \tag{6.143}$$

If $\beta_2 = 0$ and $\gamma_2 > 0$, then (6.130) becomes

$$\begin{cases} \phi_1 < 0 < \min\{\phi_2,\phi_3\} \\ \\ \frac{\max\{\phi_4,\phi_5\}}{\gamma_2} < \beta < \frac{\phi_6}{\gamma_2}. \end{cases} \tag{6.144}$$

The remaining cases (b) and (c) are similar.

Example 16 *As an example of the above analysis, let us consider the following 2-D generalized Hénon mapping:*

$$f(x,y) = \begin{pmatrix} 1 - 1.4\,(1-\rho)\,x^2 + y \\ \\ 0.3x \end{pmatrix} \tag{6.145}$$

Assuming that $\alpha = -1.4\rho$ and $\beta = 0$. Thus, the controlled version of the map (6.140) is given by

$$f(x,y) = \begin{pmatrix} 1 - 1.4\,(1-\rho)\,x^2 + y - 1.4\rho\,|x| \\ \\ 0.3x \end{pmatrix} \tag{6.146}$$

The controlled map (6.140) satisfies the conditions C_2 and C_3 given by (6.128) and (6.129), respectively. Indeed, consider the critical curves corresponding to the conditions (6.127),(6.128) and (6.129) as follows:

$$\begin{cases} C_1 : \frac{\lambda_1 - 1}{\tau_1 - 1 - \delta_1} - \frac{\omega_2 - 1}{\tau_2 - 1 - \delta_2} = 0 \\ \\ C_2 : (\tau_2 - \lambda_2)\,\lambda_1 + \tau_1 - \tau_2 - \delta_1 = 0 \\ \\ C_3 : \frac{\omega_2 - 1}{\tau_2 - 1 - \delta_2} - \frac{\delta_1(\tau_1 - \delta_1 - \lambda_2)}{(\tau_1 - 1 - \delta_1)(\delta_2\lambda_2 - \delta_1\tau_2)} = 0. \end{cases} \tag{6.147}$$

By applying Newton's method for finding roots of an algebraic equation with an error of 10^{-6} to system (6.140), Fig. 6.9 shows that the curve (C_2) has an intersection with the $\rho = 0$ axis at $\rho = 0.0866592234$, which means that conditions (6.128) hold for $\rho \in [0, 0.0866592234]$, while the curve (C_1) does *not intersect the $\rho = 0$ axis, then conditions (6.127) do not hold for all ρ, and the curve (C_3) also intersects the $\rho = 0$ axis once at $\rho = 0.493122734$, then condition (6.129) holds when $\rho \in [0.493122734, 1[$. We note that this property is absent for $\rho = 1$. Hence, the controlled map (6.140) displays robust homoclinic chaos when $\rho \in [0.493122734, 1]$. The Lyapunov exponents and the bifurcation diagram are shown in Fig. 6.10 and Fig. 6.12. An example of such a chaotic attractor is shown in Fig. 6.11.*

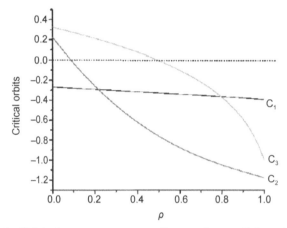

FIGURE 6.9: Critical curves corresponding to the conditions (6.127),(6.128) and (6.129) for the controlled map (6.140). [Zeraoulia & Sprott (2011(e))].

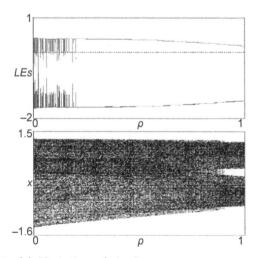

FIGURE 6.10: (a) Variation of the Lyapunov exponents of the controlled map (6.140) for $0 \leq \rho \leq 1$. (b) The bifurcation diagram of the controlled map (6.140) for $0 \leq \rho \leq 1$. [Zeraoulia & Sprott (2011(e))].

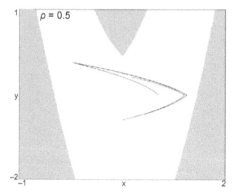

FIGURE 6.11: A robust chaotic attractor with its basin of attraction (in white) obtained from the controlled map (6.140) for $\rho = 0.5$. [Zeraoulia & Sprott (2011(e))].

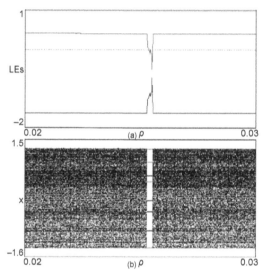

FIGURE 6.12: (a) Variation of the Lyapunov exponents of the controlled map (6.140) for $0.02 \leq \rho \leq 0.03$. (b) The bifurcation diagram of the controlled map (6.140) for $0.02 \leq \rho \leq 0.03$, showing a period-8 attractor obtained for $\rho = 0.025$. [Zeraoulia & Sprott (2011(e))].

6.11 Chaotifying stable n-D linear maps via the controller of any bounded function

A hyper-chaotification scheme, making all Lyapunov exponents of a controlled dynamical system positive via the controller of some simple triangular function, was given in [Li (2004)].

In this section, we derive a hyper-chaotification method by making all Lyapunov exponents of the controlled n-D discrete map positive via the controller of any non-linear function vector in $\mathbb{R}^n$ with only one non-zero component that possesses a bounded derivative with respect to the first component of the state variable.

We need the following preliminary results about the infinity norm of a matrix, and the eigenvalues of the sum of two matrices. Indeed, let $A = (a_{ij})_{1 \leq i,j \leq n}$ be a real matrix, hence, the infinity norm of A is given by:

$$\|A\|_\infty = \max_i \left\{ \sum_{j=1}^n |a_{ij}| \right\}$$

Proposition 6.1 *(a) If by Assuming that A_1 and its perturbation A_2 are real symmetric matrices. If the eigenvalues of A_1 and $A_1 + A_2$ are listed respectively in order $\lambda_1 \leq \lambda_2 \leq ... \leq \lambda_n$, and $\mu_1 \leq \mu_2 \leq ... \leq \mu_n$, then one has*

$$|\lambda_i - \mu_i| \leq \|A_2\|_\infty, i = 1, 2, .., n.$$

(b) For a real square matrix A, if δ is an arbitrary eigenvalue of A and $\lambda_{\min}, \lambda_{\max}$ are the smallest and largest eigenvalues, respectively, of $A^T A$, then one has

$$(0 \leq) \lambda_{\min} \leq \delta^2 \leq \lambda_{\max}$$

Now, consider the following n-dimensional dynamical system $X_{k+1} = f(X_k), X_k \in \mathbb{R}^n, k = 0, 1, 2, ...,$ where $f : \mathbb{R}^n \longrightarrow \mathbb{R}^n$, let $J(X)$ be the Jacobian of this system evaluated at X, and let the matrix $T_l(X_0) = J(X_{l-1}) J(X_{l-2}) ... J(X_1) J(X_0)$. Moreover, let $J_i(X_0, l)$ be the module of the ith eigenvalue of the lth matrix $T_l(X_0)$, where $i = 1, 2, .., n$ and $l = 0, 1, 2, ...$ Now, the Lyapunov exponents of a n-D discrete time systems are defined by $w_i(X_0) = \ln \left(\lim_{n \longrightarrow +\infty} J_i(X_0, l)^{\frac{1}{l}} \right), i = 1, 2, ..., n$. We use Theorem 15 about the bounds of Lyapunov exponents of a n-D discrete time. Let us consider the linear n-D dynamical system of the form:

$$\begin{cases} x_{k+1} = Ax_k + b, b, x_k \in \mathbb{R}^n \\ \\ x_0\text{-given} \end{cases} \tag{6.148}$$

where $x_k = \left(x_k^{(1)}, x_k^{(2)}, ..., x_k^{(n)} \right)$ is the state variable and $A = (a_{ij})_{1 \leq i,j \leq n}$ is a real matrix, such that $\sum_{j=1}^n |a_{1j}| > 2$. Let us consider the controlled map:

$$\begin{cases} x_{k+1} = Ax_k + b + \alpha h(x_k) \\ \\ x_0\text{-given}. \end{cases} \tag{6.149}$$

where α is a real parameter and the function $h : \mathbb{R}^n \longrightarrow \mathbb{R}^n$ is given by

$$h(x_k) = \begin{pmatrix} g\left(x_k^{(1)}\right) \\ 0 \\ \ldots \\ 0 \end{pmatrix} \tag{6.150}$$

where $g : \mathbb{R} \longrightarrow \mathbb{R}$, is a non-linear function with bounded derivative with respect to the first component $x_k^{(1)}$ of the vector x_k, i.e.,

$$\exists M > 0, \forall x_k^{(1)} \in \mathbb{R} : \left| g'\left(x_k^{(1)}\right) \right| = \left| \frac{\partial g\left(x_k^{(1)}\right)}{\partial x_k^{(1)}} \right| \le M$$

Assume also that the controlled map (6.143) has at least one fixed point, i.e.,

$$(A - I_n)x + b + \alpha h(x) = 0$$

has at least one real solution.

The advantages of the proposed controller (6.144) are that it is very simple and it depends on only one variable $x_k^{(1)}$ in the first direction with $(n - 1)$ zero components, and it is general in which the expression of the function g is not necessary to confirm the hyper-chaotic nature of the map (6.143).

The Jacobian matrix of the system (6.143) is given by:

$$J(x_k) = A + \alpha h'(x_k) = A + \alpha g'\left(x_k^{(1)}\right) B$$

where

$$B = \begin{pmatrix} 1 & 0 & \ldots & 0 \\ 0 & 0 & \ldots & 0 \\ \ldots & \ldots & \ldots & \ldots \\ 0 & 0 & \ldots & 0 \end{pmatrix}$$

with $\|B\|_\infty = 1, B^T = B$, and $B^2 = B$. We remark that

$$\|J(x_k)\| \le \|A\|_\infty + M |\alpha| = N(\alpha)$$

then the first condition of Theorem 15 is verified for some finite α. On the other hand, one remarks that

$$J^T(x_k) J(x_k) = A^T A + \alpha g'\left(x_k^{(1)}\right)\left(A^T B + BA\right) + \alpha^2 g'^2\left(x_k^{(1)}\right) B = A_1 + A_2$$

where

$$\begin{cases} A_1 = A^T A \\ \\ A_2 = \alpha g'\left(x_k^{(1)}\right)\left(A^T B + BA\right) + \alpha^2 g'^2\left(x_k^{(1)}\right) B \end{cases}$$

A_1 and A_2 are symmetrical since

$$
\begin{cases}
A_1^T = \left(A^T A\right)^T = A_1 \\[2mm]
A_2^T = \left(\alpha g'\left(x_k^{(1)}\right)\right)(A^T B + BA) + \alpha^2 g'^2 \left(x_k^{(1)}\right) B\Big)^T \\[2mm]
= \alpha g'\left(x_k^{(1)}\right)(BA + A^T B) + \alpha^2 g'^2 \left(x_k^{(1)}\right) B = A_2
\end{cases}
$$

Now, suppose that the eigenvalues of $A^T A$ and $J^T\left(x_k\right) J\left(x_k\right)$ are listed respectively in order:

$$
\begin{cases}
\lambda_{\min}\left(A^T A\right) = \lambda_1 \leq \lambda_2 \leq \ldots \leq \lambda_{\max}\left(A^T A\right) = \lambda_n \\[2mm]
\mu_{\min}\left(J^T\left(x_k\right) J\left(x_k\right)\right) = \mu_1 \leq \mu_2 \leq \ldots \leq \mu_{\max}\left(J^T\left(x_k\right) J\left(x_k\right)\right) = \mu_n
\end{cases}
$$

Then, by applying Proposition 6.1 for $i = 1$, we have:

$$
\mu_{\min}\left(J^T\left(x_k\right) J\left(x_k\right)\right) \geq \lambda_{\min}\left(A^T A\right) - \|A_2\|_\infty
$$

A simple calculation gives the expression of A_2 as follows:

$$
A_2 = \alpha g'\left(x_k^{(1)}\right)
\begin{pmatrix}
a_{11} + \alpha g'\left(x_k^{(1)}\right) & a_{12} & \ldots & a_{1n} \\
a_{12} & 0 & \ldots & 0 \\
\ldots & \ldots & \ldots & \ldots \\
a_{1n} & 0 & \ldots & 0
\end{pmatrix}
$$

Thus, the infinity norm of A_2 is given by:

$$
\|A_2\|_\infty = \max_i \left\{\Psi_1, \ldots, \Psi_n\right\}
$$

where

$$
\begin{cases}
\Psi_1 = \left|a_{11}\alpha g'\left(x_k^{(1)}\right) + \alpha^2\left(g'\right)^2\left(x_k^{(1)}\right)\right| + \displaystyle\sum_{j=2}^{n}\left|a_{1j}\alpha g'\left(x_k^{(1)}\right)\right| \\
\qquad\qquad\qquad \ldots \\
\Psi_n = \left|a_{1n}\alpha g'\left(x_k^{(1)}\right)\right|
\end{cases}
$$

Hence, one has:

$$
\|A_2\|_\infty = \left|a_{11}\alpha g'\left(x_k^{(1)}\right) + \alpha^2\left(g'\right)^2\left(x_k^{(1)}\right)\right| + \sum_{j=2}^{n}\left|a_{1j}\alpha g'\left(x_k^{(1)}\right)\right|
$$

Using the boundedness of the function g' we have:

$$
\begin{cases}
\|A_2\|_\infty \leq \alpha^2 M^2 + M \left(\sum_{j=1}^{n} |a_{1j}| \right) |\alpha| \\
\\
\mu_{\min} \left(J^T (x_k) J (x_k) \right) \geq \lambda_{\min} \left(A^T A \right) - \alpha^2 M^2 - M \left(\sum_{j=1}^{n} |a_{1j}| \right) |\alpha|
\end{cases}
$$

For the sake of simplicity, we can assume that $\alpha > 0$, then we have:

$$
\mu_{\min} \left(J^T (x_k) J (x_k) \right) \geq \lambda_{\min} \left(A^T A \right) - \alpha^2 M^2 - M \left(\sum_{j=1}^{n} |a_{1j}| \right) \alpha = \beta
$$

In what follows, we choose only values of $\alpha > 0$ such that $\beta > 1$, i.e., all the Lyapunov exponents of the controlled n-D map (6.143) are positive, that is, the following inequality:

$$
M^2 \alpha^2 + M \left(\sum_{j=1}^{n} |a_{1j}| \right) \alpha + 1 - \lambda_{\min} \left(A^T A \right) < 0
$$

If $\lambda_{\min} \left(A^T A \right) < 1$, then the proposed method cannot be applied here. Thus, we must assume that $\lambda_{\min} \left(A^T A \right) > 1$. Let

$$
\Delta = M^2 \left(\left(\sum_{j=1}^{n} |a_{1j}| \right)^2 + 4 \left(\lambda_{\min} \left(A^T A \right) - 1 \right) \right) > 0
$$

be the discriminant of the corresponding null equation in α. Thus, $\beta > 1$ if and only if $\alpha \in (\alpha_1, \alpha_2)$, where

$$
\begin{cases}
\alpha_1 = \dfrac{-M \sum_{j=1}^{n} |a_{1j}| - \sqrt{\Delta}}{2M^2} < 0 \\
\\
\alpha_2 = \dfrac{-M \sum_{j=1}^{n} |a_{1j}| + \sqrt{\Delta}}{2M^2} > 0.
\end{cases}
$$

Since we choose $\alpha > 0$, then we have $\mu_{\min} \left(J^T (x_k) J (x_k) \right) > 1$, for $\alpha \in (0, \alpha_2)$. Applying Theorem 15 we confirm that the controlled map (6.143) displays hyperchaos for $\alpha \in (0, \alpha_2)$.

Chapter 7

Boundedness of Some Forms of Quadratic Systems

7.1 Introduction

This chapter concerns some relevant methods that are used in order to prove the boundedness of certain forms of discrete time and continuous time systems. In Sec. 7.2, we present the main method (Lyapunov function) that is used for proving the boundedness of certain forms of 3-D quadratic continuous-time systems. In Sec. 7.3, we present a method that is used for proving the bounded jerky dynamics. This method is based on a result about the boundedness of solutions of a certain type of third-order non-linear differential equation with bounded delay. Another variant of this result is presented in Sec. 7.3.1 and it is used to prove the boundedness of general forms of jerky dynamics. In Sec. 7.3.2, some examples of bounded jerky and minimal chaotic attractors are given. Sec. 7.4, deals with some forms of bounded hyperjerky dynamics. In Sec. 7.5, we give some proofs that certain forms of 4-D systems are bounded, i.e., the generalized 4-D hyperchaotic model containing Lorenz-Stenflo and Lorenz-Haken systems. In Sec. 7.5.1, an estimation of the bounds for the Lorenz-Haken system is given with a detailed proof. The same result for the Lorenz–Stenflo system is presented in Sec. 7.5.2. In Sec. 7.6, we present some results about the boundedness of certain forms of 2-D Hénon-like mappings with unknown bounded function. In this case, the values of parameters and initial conditions domains for which the dynamics of this equation are bounded or unbounded are rigorously derived. Some numerical examples are also given and discussed. In Sec. 7.7, some examples of fully bounded chaotic attractors for all bifurcation parameters are presented and discussed.

7.2 Boundedness of certain forms of 3-D quadratic continuous-time systems

The boundedness of 3-D quadratic continuous-time autonomous systems has a crucial role in chaos control, chaos synchronization, and their applications

[Chen, 1999]. The estimation of the upper bound of a chaotic system is a difficult problem. There are many works that are focused on this topic. For example, it was shown in [Li, et al., 2005] that if $a > 0$, $b > 0$ and $c > 0$, then the Lorenz system

$$\begin{cases} x' = a\,(y - x) \\ y' = cx - y - xz \\ z' = -bz + xy \end{cases} \tag{7.1}$$

is contained in the sphere

$$\Omega = \left\{ (x, y, z) \in \mathbb{R}^3 : x^2 + y^2 + (z - a - c)^2 = R^2 \right\} \tag{7.2}$$

where

$$R^2 = \begin{cases} \frac{(a+c)^2 b^2}{4(b-1)}, & \text{if } a \geq 1, b \geq 2 \\ (a + c)^2, & \text{if } a > \frac{b}{2}, b < 2 \\ \frac{(a+c)^2 b^2}{4a(b-a)}, & \text{if } a < 1, b \geq 2a \end{cases} \tag{7.3}$$

It was shown in [Wen-Xin and Chen, (2007), Sun, (2009)] that for the system parameters in some specified regions, the solutions of the Chen system are globally bounded. This is achieved by constructing a suitable Lyapunov function. In [Li, et al., (2005)] a better upper bound in the current literature was derived for the Lorenz system for all positive values of its parameters. This estimation overcomes some problems related to the existence of singularities arising in the value of the upper bound given in [Leonov, et al., (1987)]. In [Sun, (2009)] the solution bounds of the generalized Lorenz chaotic system were investigated using the time-domain approach.

In [Zeraoulia and Sprott (2010(a))] a generalization of all the relevant results of the literature was given with a description of some of these bounds using multivariable function analysis. In particular, large regions in the bifurcation parameter space were found, for which this system is bounded.

The most general 3-D quadratic continuous-time autonomous system is given by:

$$\begin{cases} x' = a_0 + a_1 x + a_2 y + a_3 z + a_4 x^2 + a_5 y^2 + a_6 z^2 + a_7 xy + a_8 xz + a_9 yz \\ y' = b_0 + b_1 x + b_2 y + b_3 z + b_4 x^2 + b_5 y^2 + b_6 z^2 + b_7 xy + b_8 xz + b_9 yz \\ z' = c_0 + c_1 x + c_2 y + c_3 z + c_4 x^2 + c_5 y^2 + c_6 z^2 + c_7 xy + c_8 xz + c_9 yz. \end{cases} \tag{7.4}$$

where $(a_i, b_i, c_i)_{0 \le i \le 9} \in \mathbb{R}^{30}$ are bifurcation parameters. To estimate the bound for the general system (7.4), we consider the Lyapunov function $V(x, y, z)$ defined by:

$$V(x, y, z) = \frac{(x_0 - \alpha)^2 + (y_0 - \beta)^2 + (z_0 - \gamma)^2}{2} \tag{7.5}$$

where $(\alpha_1, \beta_1, \gamma_1) \in \mathbb{R}^3$ is any set of real constants for which the derivative of (7.5) along the solutions of (7.4) is given by:

$$\frac{dV}{dt} = -\omega(x_0 - \alpha_1)^2 - \varphi(y_0 - \beta_1)^2 - \phi(z_0 - \gamma_1)^2 + d \tag{7.6}$$

where

$$
\begin{cases}
d = \omega \alpha_1^2 + \varphi \beta_1^2 + \phi \gamma_1^2 - \beta b_0 - \gamma c_0 - \alpha a_0 \\[2mm]
\omega = \alpha a_4 - a_1 + \beta b_4 + \gamma c_4 \\[2mm]
\varphi = \alpha a_5 - b_2 + \beta b_5 + \gamma c_5
\end{cases}
$$

$$
\begin{cases}
\phi = \alpha a_6 - c_3 + \beta b_6 + \gamma c_6 \\[2mm]
\alpha_1 = \frac{a_0 - \alpha a_1 - \beta b_1 - \gamma c_1}{2\omega}, \text{ if } \omega \neq 0 \\[2mm]
\beta_1 = \frac{b_0 - \alpha a_2 - \beta b_2 - \gamma c_2}{2\varphi}, \text{ if } \varphi \neq 0 \\[2mm]
\gamma_1 = \frac{c_0 - \alpha a_3 - \beta b_3 - \gamma c_3}{2\phi}, \text{ if } \phi \neq 0
\end{cases}
\tag{7.7}
$$

If $\omega = 0$ or $\phi = 0$ or $\varphi = 0$, then there is no need to calculate α_1, β_1 and γ_1, respectively, and a condition relating $(a_i, b_i, c_i)_{0 \le i \le 9} \in \mathbb{R}^{30}$ to $(\alpha_1, \beta_1, \gamma_1) \in \mathbb{R}^3$ is obtained. If not, we have the formulas given by the last three equalities of (7.7). Equation (7.6) is possible only for systems of the form:

$$
\begin{cases}
x' = a_0 + a_1 x + a_2 y + a_3 z + a_5 y^2 + a_6 z^2 + a_7 xy + a_8 xz + a_9 yz \\[2mm]
y' = b_0 + b_1 x + b_2 y + b_3 z - a_7 x^2 + b_5 y^2 + b_6 z^2 - a_5 xy + b_8 xz + b_9 yz \\[2mm]
z' = c_0 + c_1 x + c_2 y + c_3 z - a_8 x^2 - b_9 y^2 - (a_9 + b_8) xy - a_6 xz - b_6 yz.
\end{cases}
\tag{7.8}
$$

with the formulas for b_1, c_1, and c_2 given by:

$$
\begin{cases}
b_1 = \alpha a_7 - \beta a_5 - a_2 + \gamma c_7 \\[2mm]
c_1 = \alpha a_8 + \beta b_8 - a_3 - \gamma a_6 \\[2mm]
c_2 = \alpha a_9 + \beta b_9 - b_3 - \gamma b_6
\end{cases}
\tag{7.9}
$$

To prove the boundedness of system (7.4), we assume that it is bounded and then we find its bound, i.e., assume that w, φ, ϕ, and d are strictly positive, i.e.,

$$
\begin{cases}
w\alpha_1^2 + \varphi\beta_1^2 + \phi\gamma_1^2 - \beta b_0 - \gamma c_0 - \alpha a_0 > 0 \\[2mm]
a_1 < \alpha a_4 + \beta b_4 + \gamma c_4 \\[2mm]
b_2 < \alpha a_5 + \beta b_5 + \gamma c_5 \\[2mm]
c_3 < \alpha a_6 + \beta b_6 + \gamma c_6
\end{cases}
\tag{7.10}
$$

Then if system (7.4) is bounded, the function (7.5) has a maximum value, and the maximum point (x_0, y_0, z_0) satisfies:

$$
\frac{(x_0 - \alpha_1)^2}{\frac{d}{w}} + \frac{(y_0 - \beta_1)^2}{\frac{d}{\varphi}} + \frac{(z_0 - \gamma_1)^2}{\frac{d}{\phi}} = 1.
\tag{7.11}
$$

Now consider the ellipsoid given by:

$$
\Gamma = \left\{ (x, y, z) \in \mathbb{R}^3 : \frac{(x - \alpha_1)^2}{\frac{d}{w}} + \frac{(y - \beta_1)^2}{\frac{d}{\varphi}} + \frac{(z - \gamma_1)^2}{\frac{d}{\phi}} = 1, w, \varphi, \phi, d > 0 \right\}
\tag{7.12}
$$

and define the functions:

$$
\begin{cases}
F(x, y, z) = G(x, y, z) + \lambda H(x, y, z) \\[2mm]
G(x, y, z) = x^2 + y^2 + z^2 \\[2mm]
H(x, y, z) = \frac{(x - \alpha_1)^2}{\frac{d}{w}} + \frac{(y - \beta_1)^2}{\frac{d}{\varphi}} + \frac{(z - \gamma_1)^2}{\frac{d}{\phi}} - 1
\end{cases}
\tag{7.13}
$$

where $\lambda \in \mathbb{R}$ is a finite parameter. Then we have $\max_{(x,y,z) \in \Gamma} G = \max_{(x,y,z) \in \Gamma} F$ and

$$
\begin{cases}
\frac{\partial F(x,y,z)}{\partial x} = -2d^{-1} \left(w\lambda\alpha_1 - (w\lambda + d) x \right) \\[2mm]
\frac{\partial F(x,y,z)}{\partial y} = -2d^{-1} \left(\varphi\lambda\beta_1 - (\varphi\lambda + d) y \right) \\[2mm]
\frac{\partial F(x,y,z)}{\partial z} = -2d^{-1} \left(\phi\lambda\gamma_1 - (\phi\lambda + d) z \right)
\end{cases}
\tag{7.14}
$$

We have the following cases according to the value of the parameter λ with respect to the values $-\frac{d}{w}, -\frac{d}{\varphi}$, and $\lambda \neq -\frac{d}{\phi}$ if $w, \varphi, \phi > 0$. Otherwise, a similar study can be done.

(i) If $\lambda \neq -\frac{d}{w}, \lambda \neq -\frac{d}{\varphi}$, and $\lambda \neq -\frac{d}{\phi}$, then

$$
(x_0, y_0, z_0) = \left(\frac{w\lambda\alpha_1}{d + w\lambda}, \frac{\varphi\lambda\beta_1}{d + \varphi\lambda}, \frac{\phi\lambda\gamma_1}{d + \phi\lambda} \right)
\tag{7.15}
$$

and

$$\max_{(x,y,z)\in\Gamma} G = \xi_1 \tag{7.16}$$

where

$$\xi_1 = \frac{\omega^2\lambda^2\alpha_1^2}{(d+\omega\lambda)^2} + \frac{\varphi^2\lambda^2\beta_1^2}{(d+\varphi\lambda)^2} + \frac{\phi^2\lambda^2\gamma_1^2}{(d+\phi\lambda)^2}. \tag{7.17}$$

In this case, there exists a parameterized family (in λ) of bounds given by (7.16) of system (7.4).

(ii) If $\lambda = -\frac{d}{\omega}, (\omega \neq 0, \varphi \neq 0, \phi \neq 0, \omega \neq \varphi, \omega \neq \phi), \lambda \neq -\frac{d}{\varphi}, \lambda \neq -\frac{d}{\phi}$, then

$$(x_0, y_0, z_0) = \left(\pm\sqrt{\frac{d}{\omega}\left(1 - \frac{\xi_2}{\xi_3}\right)} + \alpha_1, \frac{-\beta_1\varphi}{\omega-\varphi}, \frac{-\gamma_1\phi}{\omega-\phi}\right) \tag{7.18}$$

where

$$\begin{cases} \xi_2 = \frac{(\varphi\beta_1^2+\phi\gamma_1^2)(\omega-\phi)^2 d^3}{\omega^2} \\ \\ \xi_3 = \frac{(\phi-\omega)^2(\omega-\varphi)^2 d^4}{\omega^4} \end{cases} \tag{7.19}$$

with the condition

$$\xi_3 \geq \xi_2. \tag{7.20}$$

This confirms that the value x_0 in (7.18) is well defined and the conditions $\omega \neq 0, \varphi \neq 0, \phi \neq 0, \omega \neq \varphi$, and $\omega \neq \phi$ are formulated as follows:

$$a_1 \neq 0, b_2 \neq 0, c_3 \neq 0, b_2 \neq a_1, c_3 \neq a_1 \tag{7.21}$$

In this case, we have

$$\max_{(x,y,z)\in\Gamma} G = \left(\sqrt{\frac{d}{\omega}\left(1 - \frac{\xi_2}{\xi_3}\right)} + \alpha_1\right)^2 + \frac{\beta_1^2\varphi^2}{(\omega-\varphi)^2} + \frac{\gamma_1^2\phi^2}{(\omega-\phi)^2} \tag{7.22}$$

(iii) If $\lambda \neq -\frac{d}{\omega}, \lambda = -\frac{d}{\varphi} (\omega \neq 0, \varphi \neq 0, \phi \neq 0, \omega \neq \varphi, \varphi \neq \phi), \lambda \neq -\frac{d}{\phi}$, then

$$(x_0, y_0, z_0) = \left(-\frac{\alpha_1\omega}{\varphi-\omega}, \pm\sqrt{\frac{d}{\varphi}\left(1 - \frac{\xi_4}{\xi_5}\right)} + \beta_1, \frac{\gamma_1\phi}{\phi-\varphi}\right) \tag{7.23}$$

where

$$\begin{cases} \xi_4 = \left(2\omega\varphi\phi\alpha_1^2 - 2\omega\varphi\phi\gamma_1^2 - \omega\varphi^2\alpha_1^2 - \omega\phi^2\alpha_1^2 + \omega^2\phi\gamma_1^2 + \varphi^2\phi\gamma_1^2\right)\varphi^2 \\ \\ \xi_5 = (\phi-\varphi)^2 (\varphi-\omega)^2 d \end{cases} \tag{7.24}$$

with the condition

$$\xi_5 \geq \xi_4. \tag{7.25}$$

This confirms that the value y_0 in (7.23) is well defined and the conditions $\omega \neq 0, \varphi \neq 0, \phi \neq 0, \omega \neq \varphi$, and $\varphi \neq \phi$ are formulated by the first four equations of (7.10) and

$$c_3 \neq b_2 \tag{7.26}$$

In this case, we have

$$\max_{(x,y,z)\in\Gamma} G = \left(\sqrt{\frac{d}{\varphi}\left(1 - \frac{\xi_4}{\xi_5}\right)} + \beta_1 \right)^2 + \frac{\alpha_1^2\omega^2}{(\varphi - \omega)^2} + \frac{\gamma_1^2\phi^2}{(\varphi - \phi)^2} \tag{7.27}$$

(iv) If $\lambda \neq -\frac{d}{\omega}, \lambda \neq -\frac{d}{\varphi}, \lambda = -\frac{d}{\phi}$ $(\omega \neq 0, \varphi \neq 0, \phi \neq 0, \omega \neq \phi, \phi \neq \varphi)$, then

$$(x_0, y_0, z_0) = \left(\frac{-\alpha_1\omega}{\phi - \omega}, \frac{-\beta_1\varphi}{\phi - \varphi}, \pm\sqrt{\frac{d}{\phi}\left(1 - \frac{\xi_6}{\xi_7}\right)} + \gamma_1 \right) \tag{7.28}$$

where

$$\begin{cases} \xi_6 = \left(\omega\varphi^2\alpha_1^2 - 2\omega\varphi\phi\beta_1^2 - 2\omega\varphi\phi\alpha_1^2 + \omega\phi^2\alpha_1^2 + \omega^2\varphi\beta_1^2 + \varphi\phi^2\beta_1^2\right)\phi^2 \\ \\ \xi_7 = (\phi - \varphi)^2 (\phi - \omega)^2 d \end{cases} \tag{7.29}$$

with the condition

$$\xi_7 \geq \xi_6. \tag{7.30}$$

This confirms that the value z_0 in (7.28) is well defined and the conditions $\omega \neq 0, \varphi \neq 0, \phi \neq 0, \omega \neq \phi$, and $\phi \neq \varphi$ are formulated by the first four equations of (7.10).

Thus, the following result was proved in [Zeraoulia and Sprott (2010(a))]:

Theorem 51 *Assume that conditions (7.10), (7.20), and (7.21) hold. Then the general 3-D quadratic continuous-time system (7.8) is bounded, i.e., it is contained in the ellipsoid (7.12).*

Several results like this can be found in [Zehrour & Zeraoulia (2010(c))].

7.3   Bounded jerky dynamics

In this section, we rigorously prove the boundedness of the following 3-D quadratic continuous-time system

$$\begin{cases} x' = y \\ \\ y' = z \\ \\ z' = d_0 + d_1x + d_2y + d_3z + d_4y^2 + d_5z^2 + d_6yz \end{cases} \tag{7.31}$$

where $(d_i)_{0 \leq i \leq 6}$ are bifurcation parameters. The method of analysis is based on a recent result given in [Tunc (2010(b))] about the boundedness of solutions of a certain type of third-order non-linear differential equations with bounded delay of the form:

$$x''' (t) + f(t) x'' (t) + g(t) + h(t) = p(t) \tag{7.32}$$

where

$$\begin{cases} f(t) = f(x(t), x'(t)) \\[2mm] g(t) = g((x(t - r(t))), x'(t - r(t))) \\[2mm] h(t) = h(x(t - r(t))) \\[2mm] p(t) = p(t, x(t), x'(t), x(t - r(t)), x'(t - r(t)), x''(t)) \end{cases} \tag{7.33}$$

Here $0 \leq r(t) \leq \alpha$, and α is a positive constant. The functions r, f, g, h and p depend only on the arguments displayed explicitly and they are continuous for all values of their respective arguments which confirms the existence of a solution for system (7.32) [Elsgolts and Norkin (1973)]. Also, the derivatives $r'(t)$, h', $\frac{\partial g}{\partial x}$, $\frac{\partial g}{\partial y}$ exist and they are continuous and $r'(t) \leq \beta, 0 < \beta < 1$. In addition, if the functions $f(x(t - r(t)), y(t - r(t)))$, $g(x(t - r(t)), y(t - r(t)))$, $h(x(t - r(t)))$, and $p(t, x(t), y(t), x(t - r(t)), y(t - r(t)), z(t))$ satisfy a Lipschitz condition in $x(t), y(t), x(t - r(t)), y(t - r(t))$, and $z(t)$, then the solution of system (7.32) is unique [Elsgolts and Norkin (1973)]. The following result was proved in [Tunc (2010(b)]:

Theorem 52 *In addition to the basic assumptions imposed on the functions f, g, h and p appearing in Eq. (7.32), we assume that there are positive constants $a, b, c_0, c, \lambda, \rho, \mu, \gamma, L$ and M such that the following conditions hold:*
(i)

$$f(x, y) \geq a + 2\lambda$$

for all x and $y \neq 0$.
(ii)

$$g(x, 0) = 0, \frac{g(x, y)}{y} \geq b + 2\mu, y \neq 0, \left| \frac{\partial g}{\partial x} \right| \leq L, \left| \frac{\partial g}{\partial y} \right| \leq M$$

for all x and y.
(iii)

$$h(0) = 0, 0 < c_0 < \frac{\partial h}{\partial x} \leq c$$

for all x.

(iv)

$$ab - c > \frac{a^y}{y_0} \frac{\partial f}{\partial x}(x, \eta)\, d\eta + \frac{1^y}{y_0} \frac{\partial g}{\partial x}(x, \eta)\, d\eta \geq 0$$

for all x and $y \neq 0$.

(v)

$$p(t, x(t), y(t), x(t - r(t)), y(t - r(t)), z(t)) \leq q(t)$$

for all $t, x, y, x(t - r(t)), y(t - r(t))$ and z, where $\max q(t) < \infty$ and $q \in L^1(0, \infty), L^1(0, \infty)$ is the space of integrable Lebesgue functions.

Then, there exists a finite positive constant K such that the solution $x(t)$ of Eq. (7.32) defined by the initial functions:

$$x(t) = \phi(t), x'(t) = \phi'(t), x''(t) = \phi''(t)$$

satisfies the inequalities $|x(t)| \leq \sqrt{K}, |x'(t)| \leq \sqrt{K}, |x''(t)| \leq \sqrt{K}$, for all $t \geq t_0$, where $\phi \in C^2([t_0 - r, t_0], \mathbb{R})$, provided that

$$\alpha < \min\left\{\frac{4a\mu}{a(L + M + c) + 2\rho}, \frac{4\lambda}{L + M + c + 2\gamma}\right\}$$

As a remark, the application of Theorem 52 to some cases of 3-D quadratic continuous-time autonomous systems displays some discontinuities or singularities for the functions f, g, h and p in the differential equation (7.32). In particular, all the systems presented in [Zhou and Chen (2006)] have singularities in their corresponding functions f, g, h and p, so results like Theorem 55 cannot be applied here. In this section, we give an example of this type of system without any discontinuities or singularities, i.e., systems of the form (7.31) that can be rewritten in the form (7.32) with

$$\begin{cases} r(t) = 0 \\[2mm] f(x(t), x'(t)) = -d_3 \\[2mm] g(x(t), x'(t)) = -d_2 x' \\[2mm] h(x) = -d_1 x \\[2mm] p(t, x(t), x'(t), x''(t)) = d_5 (x'')^2 + d_4 (x')^2 + d_0 - d_6 x' x'' \end{cases} \quad (7.34)$$

i.e.,

$$x''' + x'' + \alpha_1 x' + 2\alpha_1 x = -\frac{1}{\alpha_1}(x')^2 \qquad (7.35)$$

where the transformation T is defined by $x' = -\alpha_1 y$ and $x'' = -\alpha_1 x - \alpha_1 z$.

Equation (7.35) is one of the minimal chaotic dynamical systems with

$$
\begin{cases}
r\left(t\right) = 0 \\[2mm]
f\left(x\left(t\right), x'\left(t\right)\right) = 1 \\[2mm]
g\left(x\left(t\right), x'\left(t\right)\right) = \alpha_1 x' \\[2mm]
h\left(x\right) = 2\alpha_1 x \\[2mm]
p\left(t, x\left(t\right), x'\left(t\right), x''\left(t\right)\right) = -\frac{1}{\alpha_1}\left(x'\right)^2
\end{cases}
\tag{7.36}
$$

In this case, we have $0 \leq r\left(t\right) = 0 \leq \alpha = 0$, and r, f, g, h and p are continuous for all values of their respective arguments. So the solution of system (7.35) exists. The derivatives $r'(t) = 0 \leq \beta$, for all $0 < \beta < 1$, $h' = -d_1, \frac{\partial g}{\partial x} = 0$, $\frac{\partial g}{\partial y} = -d_2$ exist and are continuous.

Now, we verify the conditions of Theorem 55, as follows:

(i) $f\left(x, y\right) = -d_3 \geq a + 2\lambda$ for all x and $y \neq 0$, i.e., $d_3 \leq 0$ and we can choose $a = -2\lambda - d_3 \geq 0$, so $\lambda \leq \frac{-d_3}{2}$, so we set $\lambda = 0$.

(ii) $g\left(x, 0\right) = 0, \frac{g(x,y)}{y} = -d_2 \geq b + 2\mu, y \neq 0$, i.e., $d_2 \leq 0$ and we can choose $b = -2\mu - d_2 \geq 0$, so $\mu \leq \frac{-d_2}{2}$, so we set $\mu = 0, \left|\frac{\partial g}{\partial x}\right| = 0 = L, \left|\frac{\partial g}{\partial y}\right| = |d_2| = M$ for all x and y.

(iii) $h\left(0\right) = 0, 0 < c_0 < \frac{\partial h}{\partial x} = -d_1 = c$ for all x, so we have $d_1 < 0$.

(iv) Since $\frac{\partial f}{\partial x} = \frac{\partial g}{\partial x} = 0$, we can choose $ab > c$ for all x and $y \neq 0$. In this case, we have $d_3 d_2 > -d_1$.

(v) $p\left(t, x, y, z\right) = d_5 z^2 + d_4 y^2 - d_6 yz + d_0 \leq q\left(t\right) = 0 = \max q\left(t\right) < \infty$, for all t, x, y, and z, if $d_6 = 0$ and $d_5 \leq 0, d_4 \leq 0, d_0 \leq 0$.

We have $q \equiv 0 \in L^1\left(0, \infty\right)$. However, if $4d_4 d_5 - d_6^2 > 0, d_0 = 0$ and $d_4 < 0$, then $p\left(t, x, y, z\right)$ has a relative maximum at $(0, 0)$, that is $p\left(t, x, y, z\right) \leq q\left(t\right) = 0$. Applying Theorem 52, we conclude that there exists a finite positive constant K such that the solution $x(t)$ of system (7.35), defined by any finite initial condition, satisfies the inequalities $|x\left(t\right)| \leq \sqrt{K}, |y\left(t\right)| \leq \sqrt{K}, |z\left(t\right)| \leq \sqrt{K}$, for all $t \geq t_0$. Here, the constant K was defined in [Tunc (2010(b))] by using a Lyapunov functional.

Example 17 *As an elementary example of such a situation, let $d_0 = 0.0, d_1 = -0.95, d_2 = -1.0, d_3 = -1.6, d_4 = -30.0, d_5 = -10.9$ and $d_6 = 0.0$. It is clear that this set of values satisfies the above conditions and the bounded chaotic attractor is shown in Fig. 7.1 with initial condition $x = y = z = -0.01$.*

7.3.1 Boundedness of general forms of jerky dynamics

In this section, we find regions in the bifurcation parameter space of certain forms of jerky dynamic systems where they are bounded. The method of

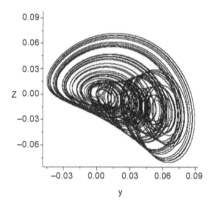

FIGURE 7.1: A bounded chaotic attractor obtained from system (7.35) with $d_0 = 0.0$, $d_1 = -0.95$, $d_2 = -1.0$, $d_3 = -1.6$, $d_4 = -30.0$, $d_5 = -10.9$ and $d_6 = 0.0$ and initial condition $x = y = z = -0.01$.

analysis is based on a result about the boundedness of solutions of a certain type of third-order non-linear differential equation with bounded delay. In particular, the boundedness of some chaotic attractors displayed by these systems is confirmed analytically.

It is known that any jerky dynamics $x''' = J(x, x', x'')$ can be recast in the functional form

$$x''' + g_1(x, x') x'' + q_1(x, x', x'') = 0 \tag{7.37}$$

and it was shown in [Eichhorn, *et al.*, (1998)] that:

Theorem 53 *Any dynamical system of the functional form*

$$\begin{cases} x' = c_1 + b_{11}x + b_{12}y + b_{13}z + n_1(x) \\[2mm] y' = c_2 + b_{21}x + b_{22}y + b_{23}z + n_2(x, y, z) \\[2mm] z' = c_3 + b_{31}x + b_{32}y + b_{33}z + n_3(x, y, z) \end{cases} \tag{7.38}$$

with n_i $(i = 1, 2, 3)$ being non-linear functions of the indicated arguments can be reduced to a jerky dynamics form $x''' = J(x, x', x'')$, if the conditions

$$b_{12}n_2(x, y, z) + b_{13}n_3(x, y, z) = f_1(x, b_{12}y + b_{13}z) \tag{7.39}$$

hold, with f_1 being an arbitrary function of the indicated arguments and

$$b_{12}^2 b_{23} - b_{13}^2 b_{32} + b_{12}b_{13}(b_{33} - b_{22}) \neq 0. \tag{7.40}$$

Assuming that

$$g_1(x, x') x'' + q_1(x, x', x'') = g(x, x') + h(x) - p(x, x', x''). \tag{7.41}$$

We will show that this assumption is still possible for certain types of functions n_i ($i = 1, 2, 3$) in system (7.38). Indeed, for the general case, it was shown in [Eichhorn, *et al.*, (1998)] that:

Proposition 7.1 *The jerky dynamic equation of the system (7.38) is given by*

$$x''' = J\left(T\left(x\right)\right) = g_1\left(x, x'\right)x'' + h_1\left(x, x'\right)x' + k\left(T\left(x\right)\right) \qquad (7.42)$$

where

$$k\left(T\left(x\right)\right) = \beta_1 + \beta_2 x + A_1 \begin{pmatrix} n_1\left(x\right) \\ n_2\left(x, T_2^{-1}\left(T\left(x\right)\right), T_3^{-1}\left(T\left(x\right)\right)\right) \\ n_3\left(x, T_2^{-1}\left(T\left(x\right)\right), T_3^{-1}\left(T\left(x\right)\right)\right) \end{pmatrix} \in \mathbb{R}$$

where $\beta_1, \beta_2 \in \mathbb{R}$ are real constants and A_1 is a constant column vector, and $T = (T_1, T_2, T_3)$ is an invertible transformation.

The T transformation was defined in [Eichhorn, *et al.*, (1998)] between system (7.38) and the jerky dynamics $x''' = J(x, x', x'')$. The inverse components $\left(T_i^{-1}\left(T\left(x\right)\right)\right)_{2 \leq i \leq 3}$ depend on x, x', x''. From [Eichhorn, *et al.*, (1998)] we have that:

Proposition 7.2 *The transformation $\left(T_i^{-1}\left(T\left(x\right)\right)\right)_{2 \leq i \leq 3}$ is linear in x' if and only if $b_{12} = b_{13} = 0$ (see equations (18b) and (18c) in [Eichhorn, et al., (1998)]), i.e., the functions $\left(n_i\left(x, T_2^{-1}\left(T\left(x\right)\right), T_3^{-1}\left(T\left(x\right)\right)\right)\right)_{1 \leq i \leq 3}$ depend only on x and x', that is $k\left(T\left(x\right)\right) = \beta_1 + \beta_2 x + m\left(x, x'\right).$*

Thus, we can choose $f\left(x, x'\right) = g_1\left(x, x'\right)$, $g\left(x, x'\right) = m\left(x, x'\right)$, $h\left(x\right) = \beta_1 + \beta_2 x$, and $p\left(x, x', x''\right) = -h_1\left(x, x'\right)x'$. Notice that this is not the only case possible because if $n_1 = n_2 = 0$ and if the function n_3 is quadratic in its corresponding arguments, then the above conditions hold for some parameter values.

The most general form of a quadratic jerky equation is given by:

$$x''' = \left(s_1 + s_2 x + s_3 x' + s_4 x''\right)x'' + \left(s_5 + s_6 x + s_7 x''\right)x' + \left(s_8 + s_9 x\right)x + s_{10} \qquad (7.43)$$

The same method shows the possibility of finding some ranges in the space $(s_i)_{1 \leq i \leq 10}$ such that all the solutions of system (7.43) are bounded. For the general case (7.38), the function $k\left(T\left(x\right)\right)$ contains the linear and non-linear terms in x' and x''. So, according to the shape of the functions f_1 defined in (7.39) and $\left(n_i\left(x, T_2^{-1}\left(T\left(x\right)\right), T_3^{-1}\left(T\left(x\right)\right)\right)\right)_{1 \leq i \leq 3}$, the form (7.43) may be possible or not, and the general conditions for a such a situation cannot be derived here. However, for the general 3-D quadratic continuous-time system, some corresponding jerky dynamics were derived in [Eichhorn, *et al.*, (1998)] for some minimal chaotic flows studied in [Sprott (1997(a))]. Thus, the following result was proved in [Zeraoulia & Sprott (2012(b))]:

Theorem 54 *System (7.38) is bounded if there exists a finite positive constant K such that the solution $x(t)$ of system (7.38), defined by any finite initial condition, satisfies the inequalities $|x(t)| \leq \sqrt{K}$, $|x'(t)| \leq \sqrt{K}$, $|x''(t)| \leq \sqrt{K}$ for all $t \geq t_0$.*

Proof 16 *Since the function n_1 in (7.38) depends only on x, the general 3-D quadratic continuous-time system can be transformed into a jerky dynamic if it has the following form:*

$$
\begin{cases}
x' = a_0 + a_1 x + a_2 y + a_3 z + a_4 x^2 \\[2mm]
y' = b_0 + b_1 x + b_2 y + b_3 z + b_4 x^2 + b_5 y^2 + b_6 z^2 + b_7 xy + b_8 xz + b_9 yz \\[2mm]
z' = c_0 + c_1 x + c_2 y + c_3 z + c_4 x^2 + c_5 y^2 + c_6 z^2 + c_7 xy + c_8 xz + c_9 yz.
\end{cases}
\tag{7.44}
$$

Assuming $a_2 \neq 0$, we have

$$
\begin{cases}
y = -\frac{1}{a_2}\left(-x' + a_0 + x a_1 + z a_3 + x^2 a_4\right) \\[2mm]
y' = -\frac{1}{a_2}\left(-x'' + x' a_1 + z' a_3 + 2x x' a_4\right) \\[2mm]
y' = \xi_1 z^2 + \xi_2 z + \xi_3 \\[2mm]
z' = \xi_4 z^2 + \xi_5 z + \xi_6
\end{cases}
\tag{7.45}
$$

where $(\xi_i(x, x'))_{1 \leq i \leq 6}$ are given by:

$$
\begin{cases}
\xi_1 = \frac{a_2^2 b_6 + a_3^2 b_5 - a_2 a_3 b_9}{a_2^2}, \; \xi_4 = \frac{a_2^2 c_6 + a_3^2 c_5 - a_2 a_3 c_9}{a_2^2} \\[3mm]
\xi_2 = \mu_1 x^2 + \mu_2 x + \mu_3 x' + \mu_4, \; \xi_5 = \mu_{14} x^2 + \mu_{15} x \\
\qquad\quad + \mu_{16} x' + \mu_{17} \\[3mm]
\xi_3 = \mu_5 x^4 + \mu_6 x^3 + \mu_7 x^2 x' + \mu_8 x^2 + \mu_9 x x' + \mu_{10} x \\
\qquad\quad + \mu_{11}(x')^2 + \mu_{12} x' + \mu_{13} \\[3mm]
\xi_6 = \mu_{18} x^4 + \mu_{19} x^3 + \mu_{20} x^2 x' + \mu_{21} x^2 + \mu_{22} x x' + \mu_{23} x \\
\qquad\quad + \mu_{24}(x')^2 + \mu_{25} x' + \mu_{26}
\end{cases}
\tag{7.46}
$$

and

$$
\begin{cases}
\mu_1 = \frac{2}{a_2^2} a_3 a_4 b_5 - \frac{1}{a_2} a_4 b_9, \mu_2 = b_8 - \frac{1}{a_2}\left(a_1 b_9 + a_3 b_7\right) + 2\frac{a_1}{a_2^2} a_3 b_5 \\[3mm]
\mu_3 = \frac{1}{a_2} b_9 - \frac{2}{a_2^2} a_3 b_5, \mu_4 = b_3 - \frac{1}{a_2}\left(a_3 b_2 + a_0 b_9\right) + 2\frac{a_0}{a_2^2} a_3 b_5
\end{cases}
$$

$$\left\{ \begin{array}{c} \mu_5 = \frac{1}{a_2^2} a_4^2 b_5, \mu_6 = 2\frac{a_1}{a_2^2} a_4 b_5 - \frac{1}{a_2} a_4 b_7, \mu_7 = -\frac{2}{a_2^2} a_4 b_5 \\[2mm] \mu_8 = b_4 - \frac{1}{a_2}\left(a_4 b_2 + a_1 b_7\right) + \frac{1}{a_2^2} b_5 \left(a_1^2 + 2a_0 a_4\right), \mu_9 = \frac{1}{a_2} b_7 - 2\frac{a_1}{a_2^2} b_5 \\[2mm] \mu_{10} = b_1 - \frac{1}{a_2}\left(a_1 b_2 + a_0 b_7\right) + 2a_0 \frac{a_1}{a_2^2} b_5, \mu_{11} = \frac{1}{a_2^2} b_5 \end{array} \right. \tag{7.47}$$

and

$$\left\{ \begin{array}{c} \mu_{12} = \frac{1}{a_2} b_2 - 2\frac{a_0}{a_2^2} b_5, \mu_{13} = b_0 - \frac{a_0}{a_2} b_2 + \frac{a_0^2}{a_2^2} b_5 \\[2mm] \mu_{14} = \frac{2}{a_2^2} a_3 a_4 c_5 - \frac{1}{a_2} a_4 c_9, \mu_{15} = c_8 - \frac{1}{a_2}\left(a_1 c_9 + a_3 c_7\right) + 2\frac{a_1}{a_2^2} a_3 c_5 \\[2mm] \mu_{16} = \frac{1}{a_2} c_9 - \frac{2}{a_2^2} a_3 c_5, \mu_{17} = c_3 - \frac{1}{a_2}\left(a_3 c_2 + a_0 c_9\right) + 2\frac{a_0}{a_2^2} a_3 c_5 \\[2mm] \mu_{18} = \frac{1}{a_2^2} a_4^2 c_5, \mu_{19} = 2\frac{a_1}{a_2^2} a_4 c_5 - \frac{1}{a_2} a_4 c_7, \mu_{20} = -\frac{2}{a_2^2} a_4 c_5 \\[2mm] \mu_{21} = c_4 - \frac{1}{a_2}\left(a_4 c_2 + a_1 c_7\right) + \frac{1}{a_2^2} c_5 \left(a_1^2 + 2a_0 a_4\right), \mu_{22} = \frac{1}{a_2} c_7 - 2\frac{a_1}{a_2^2} c_5 \\[2mm] \mu_{23} = c_1 - \frac{1}{a_2}\left(a_1 c_2 + a_0 c_7\right) + 2a_0 \frac{a_1}{a_2^2} c_5, \mu_{24} = \frac{1}{a_2^2} c_5 \\[2mm] \mu_{25} = \frac{1}{a_2} c_2 - 2\frac{a_0}{a_2^2} c_5, \mu_{26} = c_0 - \frac{a_0}{a_2} c_2 + \frac{a_0^2}{a_2^2} c_5. \end{array} \right. \tag{7.48}$$

We also have

$$x'' = \lambda_1 x^4 + \lambda_2 x^3 + \lambda_3 x^2 y + \lambda_4 x^2 z + \lambda_5 x^2 + \lambda_6 xy + \lambda_7 xz$$

$$+\lambda_8 x + \lambda_9 y^2 + \lambda_{10} yz + \lambda_{11} y + \lambda_{12} z^2 + \lambda_{13} z + \lambda_{14}$$

where

$$\left\{ \begin{array}{c} \lambda_1 = \left(\mu_{11} a_2 + \mu_{24} a_3\right) a_4^2 + \left(\mu_7 a_2 + \mu_{20} a_3\right) a_4 + \mu_5 a_2 + \mu_{18} a_3 \\[2mm] \lambda_2 = \varsigma_1 + \varsigma_2 \\[2mm] \lambda_3 = a_2 \left(\mu_7 a_2 + \mu_{20} a_3\right) + 2a_2 a_4 \left(\mu_{11} a_2 + \mu_{24} a_3\right) \\[2mm] \lambda_4 = \varsigma_3 + \varsigma_4 \\[2mm] \lambda_5 = \varsigma_5 + \varsigma_6 \\[2mm] \lambda_6 = a_2 \left(2a_4 + \mu_9 a_2 + \mu_{22} a_3\right) + 2a_1 a_2 \left(\mu_{11} a_2 + \mu_{24} a_3\right) \\[2mm] \lambda_7 = \varsigma_7 + \varsigma_8 \\[2mm] \lambda_8 = \varsigma_9 + \varsigma_{10} \end{array} \right. \tag{7.49}$$

and

$$\begin{cases}
\lambda_9 = a_2^2 \left(\mu_{11}a_2 + \mu_{24}a_3 \right) \\[2mm]
\lambda_{10} = a_2 \left(\mu_3 a_2 + \mu_{16} a_3 \right) + 2a_2 a_3 \left(\mu_{11} a_2 + \mu_{24} a_3 \right) \\[2mm]
\lambda_{11} = a_2 \left(a_1 + \mu_{12} a_2 + \mu_{25} a_3 \right) + 2a_0 a_2 \left(\mu_{11} a_2 + \mu_{24} a_3 \right) \\[2mm]
\lambda_{12} = a_3^2 \left(\mu_{11}a_2 + \mu_{24}a_3 \right) + \varsigma_1 a_2 + \varsigma_4 a_3 + a_3 \left(\mu_3 a_2 + \mu_{16} a_3 \right) \\[2mm]
\lambda_{13} = \varsigma_{11} + \varsigma_{12} \\[2mm]
\lambda_{14} = a_0 + a_0^2 \left(\mu_{11}a_2 + \mu_{24}a_3 \right) + \mu_{13}a_2 + \mu_{26}a_3 + a_0 \left(a_1 + \mu_{12} a_2 + \mu_{25} a_3 \right)
\end{cases}$$

$$(7.50)$$

where

$$\begin{cases}
\varsigma_1 = a_4 \left(2a_4 + \mu_9 a_2 + \mu_{22} a_3 \right) + \mu_6 a_2 + \mu_{19} a_3 + a_1 \left(\mu_7 a_2 + \mu_{20} a_3 \right) \\[2mm]
\varsigma_2 = 2a_1 a_4 \left(\mu_{11} a_2 + \mu_{24} a_3 \right) \\[2mm]
\varsigma_3 = \mu_1 a_2 + \mu_{14} a_3 + a_3 \left(\mu_7 a_2 + \mu_{20} a_3 \right) + a_4 \left(\mu_3 a_2 + \mu_{16} a_3 \right) \\[2mm]
\varsigma_4 = 2a_3 a_4 \left(\mu_{11} a_2 + \mu_{24} a_3 \right) \\[2mm]
\varsigma_5 = a_1 \left(2a_4 + \mu_9 a_2 + \mu_{22} a_3 \right) + \mu_8 a_2 + \mu_{21} a_3 + \left(a_1^2 + 2a_0 a_4 \right) \left(\mu_{11} a_2 + \mu_{24} a_3 \right) \\[2mm]
\varsigma_6 = a_0 \left(\mu_7 a_2 + \mu_{20} a_3 \right) + a_4 \left(a_1 + \mu_{12} a_2 + \mu_{25} a_3 \right) \\[2mm]
\varsigma_7 = a_3 \left(2a_4 + \mu_9 a_2 + \mu_{22} a_3 \right) + \mu_2 a_2 + \mu_{15} a_3 \\[2mm]
\varsigma_8 = a_1 \left(\mu_3 a_2 + \mu_{16} a_3 \right) + 2a_1 a_3 \left(\mu_{11} a_2 + \mu_{24} a_3 \right) \\[2mm]
\varsigma_9 = a_0 \left(2a_4 + \mu_9 a_2 + \mu_{22} a_3 \right) + \mu_{10} a_2 + \mu_{23} a_3 \\[2mm]
\varsigma_{10} = a_1 \left(a_1 + \mu_{12} a_2 + \mu_{25} a_3 \right) + 2a_0 a_1 \left(\mu_{11} a_2 + \mu_{24} a_3 \right) \\[2mm]
\varsigma_{11} = \mu_4 a_2 + \mu_{17} a_3 + a_0 \left(\mu_3 a_2 + \mu_{16} a_3 \right) \\[2mm]
\varsigma_{12} = a_3 \left(a_1 + \mu_{12} a_2 + \mu_{25} a_3 \right) + 2a_0 a_3 \left(\mu_{11} a_2 + \mu_{24} a_3 \right)
\end{cases}$$

$$(7.51)$$

The equation

$$y' = \xi_1 z^2 + \xi_2 z + \xi_3 = -\frac{1}{a_2} \left(-x'' + x' a_1 + z' a_3 + 2xx' a_4 \right)$$

gives

$$\omega_0 z^2 + r_1 z + r_2 = 0 \qquad (7.52)$$

where

$$
\begin{cases}
r_1 = \omega_1 x^2 + \omega_2 x + \omega_3 x' + \omega_4 \\[2mm]
r_2 = h_1(x, x') - \dfrac{x''}{a_2} \\[2mm]
h_1 = \omega_5 x^4 + \omega_6 x^3 + \omega_8 x^2 + \omega_{10} x + \omega_{13} + \omega_7 x^2 x' + \omega_9 x x' \\
\qquad\quad + \omega_{11}(x')^2 + \omega_{12} x'
\end{cases}
\tag{7.53}
$$

where

$$
\begin{cases}
\omega_0 = \xi_1 + \frac{\xi_4}{a_2} a_3,\ \omega_1 = \mu_1 + \frac{\mu_{14}}{a_2} a_3,\ \omega_2 = \mu_2 + \frac{\mu_{15}}{a_2} a_3 \\[2mm]
\omega_3 = \mu_3 + \frac{\mu_{16}}{a_2} a_3,\ \omega_4 = \mu_4 + \frac{\mu_{17}}{a_2} a_3,\ \omega_5 = \mu_5 + \frac{\mu_{18}}{a_2} a_3 \\[2mm]
\omega_6 = \mu_6 + \frac{\mu_{19}}{a_2} a_3,\ \omega_7 = \mu_7 + \frac{\mu_{20}}{a_2} a_3,\ \omega_8 = \mu_8 + \frac{\mu_{21}}{a_2} a_3 \\[2mm]
\omega_9 = \mu_9 + \frac{1}{a_2}\left(2a_4 + \mu_{22} a_3\right),\ \omega_{10} = \mu_{10} + \frac{\mu_{23}}{a_2} a_3 \\[2mm]
\omega_{11} = \mu_{11} + \frac{\mu_{24}}{a_2} a_3,\ \omega_{12} = \mu_{12} + \frac{1}{a_2}\left(a_1 + \mu_{25} a_3\right),\ \omega_{13} = \mu_{13} + \frac{\mu_{26}}{a_2} a_3
\end{cases}
\tag{7.54}
$$

Equation (7.52) has a real solution z if and only if its discriminant is positive, i.e.,

$$
r_1{}^2 - 4\omega_0 r_2 = \left(r_1{}^2 - 4\omega_0 r_2\right)(x, x', x'') \geq 0
$$

for all x, x', and x''. It is very hard to find sufficient conditions for a such an inequality. For the sake of simplicity, we can choose $\omega_0 = 0$, i.e.,

$$
b_6 = -a_3 \frac{a_2^2 c_6 + a_3^2 c_5 - a_2^2 b_9 + a_2 a_3 b_5 - a_2 a_3 c_9}{a_2^3}
$$

In this case, we have $z = -\frac{r_2}{r_1}$ with $r_1 \neq 0$, that is

$$
\begin{aligned}
r_1 &= \omega_1 x^2 + \omega_2 x + \omega_3 x' + \omega_4 \\[2mm]
&= w(x) + \left(\mu_3 + \frac{\mu_{16}}{a_2} a_3\right) x' + \mu_4 + \frac{\mu_{17}}{a_2} a_3 \neq 0 \\[3mm]
w(x) &= \left(\mu_1 + \frac{\mu_{14}}{a_2} a_3\right) x^2 + \left(\mu_2 + \frac{\mu_{15}}{a_2} a_3\right) x
\end{aligned}
$$

for all x and x'. We can assume that

$$
\begin{cases}
\omega_1 = \mu_1 + \frac{\mu_{14}}{a_2} a_3 = 0 \\[2mm]
\omega_2 = \mu_2 + \frac{\mu_{15}}{a_2} a_3 = 0 \\[2mm]
\omega_3 = \mu_3 + \frac{\mu_{16}}{a_2} a_3 = 0 \\[2mm]
\omega_4 = \mu_4 + \frac{\mu_{17}}{a_2} a_3 \neq 0
\end{cases}
$$

i.e.,

$$\begin{cases} a_4 = 0 \text{ or } b_9 = \frac{1}{a_2^2}\left(2a_3^2c_5 + 2a_2a_3b_5 - a_2a_3c_9\right) \text{ and} \\ \\ b_8 = \frac{a_1a_2^2b_9 - 2a_1a_3^2c_5 + a_2^2a_3b_7 + a_2a_3^2c_7 - a_2^2a_3c_8 - 2a_1a_2a_3b_5 + a_1a_2a_3c_9}{a_2^3} \\ \\ b_9 = \frac{1}{a_2^2}\left(2a_3^2c_5 + 2a_2a_3b_5 - a_2a_3c_9\right) \\ \\ a_2^2b_3 - a_3^2c_2 - a_2a_3b_2 + a_2a_3c_3 \neq 0 \end{cases}$$

Thus, we have the following two cases:
    (i) If

$$\begin{cases} a_4 = 0, b_8 = \frac{1}{a_2^2}a_3\left(a_2b_7 - a_2c_8 + a_3c_7\right) \\ \\ b_9 = \frac{1}{a_2^2}\left(2a_3^2c_5 + 2a_2a_3b_5 - a_2a_3c_9\right) \\ \\ a_2^2b_3 - a_3^2c_2 - a_2a_3b_2 + a_2a_3c_3 \neq 0 \end{cases}$$

(ii) If

$$\begin{cases} a_4 \neq 0 \\ \\ b_9 = \frac{1}{a_2^2}\left(2a_3^2c_5 + 2a_2a_3b_5 - a_2a_3c_9\right) \\ \\ b_8 = \frac{1}{a_2^2}a_3\left(a_2b_7 - a_2c_8 + a_3c_7\right) \\ \\ a_2^2b_3 - a_3^2c_2 - a_2a_3b_2 + a_2a_3c_3 \neq 0 \end{cases}$$

Thus, we have

$$x''' + \delta_1\left(\frac{r_2}{r_1}\right)^3 - \delta_2\left(\frac{r_2}{r_1}\right)^2 + \delta_3\left(\frac{r_2}{r_1}\right) - \delta_4 = 0, \qquad (7.55)$$

which can be rewritten in the form (7.41) with

$$\begin{cases} f\left(x, x'\right) = -\frac{\delta_3\omega_4^2 + 3\delta_1h_1^2\left(x, x'\right) - 2\delta_2\omega_4h_1\left(x, x'\right)}{\omega_4^3a_2} \\ \\ g + h - p = -\frac{\delta_1}{\omega_4^3a_2^3}\left(x''\right)^3 + \frac{-\delta_2\omega_4 + 3\delta_1h_1}{\omega_4^3a_2^2}\left(x''\right)^2 - \delta_4 + \frac{\delta_3}{\omega_4}h_1 \\ \qquad\qquad - \frac{\delta_2}{\omega_4^2}h_1^2 + \frac{\delta_1}{\omega_4^3}h_1^3 \end{cases} \qquad (7.56)$$

where

$$\begin{cases} \delta_1 = \xi_1\lambda_{10} + 2\xi_4\lambda_{12} - 2\xi_1\frac{\lambda_9}{a_2}a_3 - \xi_4\frac{\lambda_{10}}{a_2}a_3 \\ \\ \delta_2 = \varphi_1 + \varphi_2, \delta_3 = \varphi_3 + \varphi_4 + \varphi_5 + \varphi_6, \delta_4 = \varphi_7 + \varphi_8 \\ \qquad\qquad\qquad + \varphi_9 + \varphi_{10} \end{cases} \qquad (7.57)$$

where $\left(\varphi_i = \varphi_i\left((\psi_j)_{1\leq j\leq 77}\right)\right)_{1\leq i\leq 10}$ and $(\psi_j)_{1\leq j\leq 77}$ are given in the **Appendix A.**

We will give some conditions for the function f so that we can apply Theorem 52. The first assumption is $f(x,y) \geq a + 2\lambda$ for all x and $y \neq 0$. Since f is a polynomial function in x, x' of degree 8, then the simplest choice for this situation is when f is a constant function, i.e.,

$$\frac{\partial}{\partial x}\left(\frac{\delta_3\omega_4^2 + 3\delta_1 h_1^2(x,x') - 2\delta_2\omega_4 h_1(x,x')}{\omega_4^3 a_2}\right) \quad =$$

$$\frac{\partial}{\partial x'}\left(\frac{\delta_3\omega_4^2 + 3\delta_1 h_1^2(x,x') - 2\delta_2\omega_4 h_1(x,x')}{\omega_4^3 a_2}\right) \quad = \quad 0$$

We can choose

$$\begin{cases} f(x,x') = -d_3 \\ \\ g(x,x') = -d_2 x', h(x) = -d_1 x \\ \\ p(x,x',x'') = d_5(x'')^2 + d_4(x')^2 + d_0 - d_6 x' x'' \end{cases}$$

Hence

$$-d_2 x' - d_1 x - d_5(x'')^2 - d_4(x')^2 - d_0 + d_6 x' x''$$

$$= \quad -\frac{\delta_1}{\omega_4^3 a_2^3}(x'')^3 + \frac{-\delta_2\omega_4 + 3\delta_1 h_1}{\omega_4^3 a_2^2}(x'')^2 - \delta_4 + \frac{\delta_3}{\omega_4}h_1$$

$$\quad -\frac{\delta_2}{\omega_4^2}h_1^2 + \frac{\delta_1}{\omega_4^3}h_1^3$$

Since δ_2, δ_1, h_1, and δ_4 do not depend on x'', we must choose $\delta_1 = 0$, that is,

$$\begin{cases} -d_2 x' - d_1 x - d_5(x'')^2 - d_4(x')^2 - d_0 + d_6 x' x'' - s(x'') = 0 \\ \\ s(x'') = -\frac{\delta_2\omega_4}{\omega_4^3 a_2^2}(x'')^2 - \delta_4 + \frac{\delta_3}{\omega_4}h_1 - \frac{\delta_2}{\omega_4^2}h_1^2 \end{cases}$$

The last equation is a zero polynomial in x, x', x'', which can be solved (the unknowns here are the coefficients $(d_i)_{0\partial i\leq 6}$.) by setting all the coefficients to zero. This equation gives several relations between the parameters of system (7.38). We have $\lambda_{10} = 2a_3\frac{a_2 b_5 + a_3 c_5}{a_2}$. Thus, $\delta_1 = 0$ holds if $a_3 \neq 0, b_5 \neq -\frac{a_3 c_5}{a_2}$, then

$$b_6 \quad = \quad \frac{1}{2a_2^2}\frac{-2\xi_4\lambda_{12}a_2^3 - 2a_2^4 b_5 c_5 - 2a_2 a_3^3 b_5^2 + 2a_2^2 a_3^2 b_5 b_9}{a_3(a_2 b_5 + a_3 c_5)}$$

$$+\frac{1}{2a_2^2}\frac{2\xi_1\lambda_9 a_2^2 a_3 + \xi_4\lambda_{10}a_2^2 a_3 + 2a_2 a_3^3 c_5 b_9}{a_3(a_2 b_5 + a_3 c_5)}$$

since the right-hand side does not depend on b_6. If $a_3 = 0$, then the condition becomes $c_6 b_6 = 0$. Hence, system (7.38) is bounded. Thus, there exists a finite positive constant K such that the solution $x(t)$ of system (7.38) defined by any finite initial condition satisfies the inequalities $|x(t)| \leq \sqrt{K}$, $|x'(t)| \leq \sqrt{K}$, $|x''(t)| \leq \sqrt{K}$ for all $t \geq t_0$, that is $|x| \leq \sqrt{K}, |x'| = |a_4 x^2 + a_1 x + a_0 + y a_2 + z a_3| \leq \sqrt{K}$ and $|x''| \leq \sqrt{K}$, x'' is defined above. We remark that in some cases, it is possible to estimate the bounds for x, y, and z in system (7.38). For example if $a_3 = 0, a_4 = 0$, and $b_6 = 0$, i.e., systems of the form

$$\begin{cases} x' = a_0 + a_1 x + a_2 y \\\\ y' = b_0 + b_1 x + b_2 y + b_3 z + b_4 x^2 + b_5 y^2 + b_7 xy \\\\ z' = c_0 + c_1 x + c_2 y + c_3 z + c_4 x^2 + c_5 y^2 + c_6 z^2 + c_7 xy + c_8 xz + c_9 yz, \end{cases}$$
$$(7.58)$$

then

$$\begin{cases} -\frac{1}{a_2}\left(a_0 + a_1 x + \sqrt{K}\right) \leq y \leq -\frac{1}{a_2}\left(a_0 + a_1 x - \sqrt{K}\right) \\\\ \Lambda_1 \leq z \leq \Lambda_2, \ if \ \lambda_{13} > 0 \\\\ \qquad\qquad or \\\\ \Lambda_2 \leq z \leq \Lambda_1, \ if \ \lambda_{13} < 0 \\\\ \Lambda_1 = \dfrac{-\sqrt{K} - \left(\lambda_5 x^2 + \lambda_6 xy + \lambda_8 x + \lambda_9 y^2 + \lambda_{11} y + \lambda_{14}\right)}{\lambda_{13}} \\\\ \Lambda_2 = \dfrac{\sqrt{K} - \left(\lambda_5 x^2 + \lambda_6 xy + \lambda_8 x + \lambda_9 y^2 + \lambda_{11} y + \lambda_{14}\right)}{\lambda_{13}} \end{cases}$$

Here $\lambda_{13} = b_3 a_2 \neq 0$ from the above conditions. The quantity $\lambda_5 x^2 + \lambda_6 xy + \lambda_8 x + \lambda_9 y^2 + \lambda_{11} y + \lambda_{14}$ is bounded since x and y are also bounded.

7.3.2   Examples of bounded jerky chaos

The first example of the above case has the form (7.35) where the transformation T is defined by $x' = -\alpha_1 y$ and $x'' = -\alpha_1 x - \alpha_1 z$. Equation (7.35) is one of the minimal chaotic dynamical systems studied in [Sprott (1997(a))] (the case (I) in [Eichhorn, et al., (1998)]) with $r(t) = 0$, $f(x(t), x'(t)) = 1$, $g(x(t), x'(t)) = \alpha_1 x'$, $h(x) = 2\alpha_1 x$, and $p(t, x(t), x'(t), x''(t)) = -\frac{1}{\alpha_1}(x')^2$.
    Thus, the following result was proved in [Zeraoulia & Sprott (2012(b))]:

Proposition 7.3 *The minimal chaotic attractors displayed by system (7.35) are all bounded, provided that $\alpha_1 > 0, \alpha_1 > \frac{2\mu(2\lambda-1)}{2\lambda+1}$, i.e., there is a constant $K > 0$ such that:*

$$|x(t)| \leq \sqrt{K}, |y(t)| \leq \frac{\sqrt{K}}{\alpha_1}, |z(t)| \leq \left(\frac{\alpha_1+1}{\alpha_1}\right)\sqrt{K}$$

where μ and λ are constants with some conditions given below.

Proof 17 *In this case, we have $0 \leq r(t) = 0 \leq \alpha = 0$, and r, f, g, h and p are continuous for all values of their respective arguments. So the solution of system (7.35) exists. The derivatives $r'(t) = 0 \leq \beta$ for all $0 < \beta < 1$, $h' = 2\alpha_1, \frac{\partial g}{\partial x} = 0, \frac{\partial g}{\partial y} = \alpha_1$ exist and are continuous. Now we verify the conditions of Theorem 52:*

(i) $f(x,y) = 1 \geq a + 2\lambda$ for all x and $y \neq 0$, we can choose $a = 1 - 2\lambda \geq 0$, that is, $0 \leq \lambda \leq \frac{1}{2}$.

(ii) $g(x,0) = 0, \frac{g(x,y)}{y} = \alpha_1 \geq b + 2\mu, y \neq 0$, i.e., we can choose $b = \alpha_1 - 2\mu \geq 0$, that is, $0 \leq \mu \leq \frac{\alpha_1}{2}, \left|\frac{\partial g}{\partial x}\right| = 0 = L, \left|\frac{\partial g}{\partial y}\right| = |\alpha_1| = M$ for all x and y.

(iii) $h(0) = 0, 0 < c_0 < \frac{\partial h}{\partial x} = 2\alpha_1 = c$ for all x.

(iv) Since $\frac{\partial f}{\partial x} = \frac{\partial g}{\partial x} = 0$, we can choose $ab > c$ for all x and $y \neq 0$. In this case, we have $\alpha_1 > \frac{2\mu(2\lambda-1)}{2\lambda+1}$.

(v) $p(t,x,y,z) = -\frac{1}{\alpha_1}y^2 \leq q(t) = 0 = \max q(t) < \infty$, for all $t, x, y,$ and z if $\alpha_1 > 0$. We have $q \equiv 0 \in L^1(0,\infty)$. Finally, the minimal chaotic attractors displayed by system (7.35) are all bounded provided that $\alpha_1 > 0, \alpha_1 > \frac{2\mu(2\lambda-1)}{2\lambda+1}$, i.e., $|x(t)| \leq \sqrt{K}, |y(t)| \leq \frac{\sqrt{K}}{\alpha_1}, |z(t)| \leq \left(\frac{\alpha_1+1}{\alpha_1}\right)\sqrt{K}$. Thus, for $\alpha_1 = 0.2$, system (7.35) displays a bounded chaotic attractor.

The second example has the form (the case (Q) in [Eichhorn, *et al.*, (1998)]):

$$\begin{cases} x' = -z \\ y' = x - y \\ z' = \alpha_1 x + y^2 + \beta_1 z \end{cases} \tag{7.59}$$

which can be rewritten as

$$y''' - (\beta_1 - 1)y'' - (\beta_1 - \alpha_1)y' + \alpha_1 y = -y^2 \tag{7.60}$$

where the transformation T is defined by $y' = -y + x$ and $y'' = y - x - z$. Equation (7.60) is one of the minimal chaotic dynamical systems studied in [Sprott (1997(a))] (the case (Q) in [Eichhorn, *et al.*, (1998)]) with $r(t) = 0$, $f(y(t), y'(t)) = 1 - \beta_1$ $g(y(t), y'(t)) = -(\beta_1 - \alpha_1)y', h(y) = \alpha_1 y$, and $p(t, y(t), y'(t), y''(t)) = -y^2$. Thus, the following result was proved in [Zeraoulia & Sprott (2012(b))]:

Proposition 7.4 *The minimal chaotic attractors displayed by system (7.60) are all bounded, provided that*

$$
\begin{cases}
0 \leq \beta_1 \leq 1 \\[2mm]
\alpha_1 \geq \beta_1 \\[2mm]
\alpha_1 < \frac{(2\mu+\beta_1)(2\lambda+\beta_1-1)}{2\lambda+\beta_1}
\end{cases}
$$

i.e., there is a constant $K > 0$ such that $|y(t)| \leq \sqrt{K}$, $|x(t)| \leq \sqrt{K}$, and $|z(t)| \leq \sqrt{K}$, where μ and λ are constants with some conditions given below.

Proof 18 *In this case, we have $0 \leq r(t) = 0 \leq \alpha = 0$, and r, f, g, h and p are continuous for all values of their respective arguments. So the solution of system (7.60) exists. The derivatives $r'(t) = 0 \leq \beta$ for all $0 < \beta < 1$, $h' = \alpha_1, \frac{\partial g}{\partial x} = 0, \frac{\partial g}{\partial y} = -(\beta_1 - \alpha_1)$ exist and are continuous. Now we verify the conditions of the result in [Tunc (2010(a))] as follows:*

(i) $f(x,y) = 1 - \beta_1 \geq a + 2\lambda$ for all x and $y \neq 0$, we can choose $a = 1 - \beta_1 - 2\lambda \geq 0$, that is, $0 \leq \lambda \leq \frac{1-\beta_1}{2}$, with $0 \leq \beta_1 \leq 1$.

(ii) $g(x,0) = 0, \frac{g(x,y)}{y} = -(\beta_1 - \alpha_1) \geq b + 2\mu, y \neq 0$, i.e., we can choose $b = \alpha_1 - 2\mu - \beta_1 \geq 0$, that is, $0 \leq \mu \leq \frac{\alpha_1 - \beta_1}{2}, \alpha_1 \geq \beta_1, \left|\frac{\partial g}{\partial x}\right| = 0 = L, \left|\frac{\partial g}{\partial y}\right| = |\beta_1 - \alpha_1| = M$ for all x and y.

(iii) $h(0) = 0, 0 < c_0 < \frac{\partial h}{\partial x} = \alpha_1 = c$ for all x.

(iv) Since $\frac{\partial f}{\partial x} = \frac{\partial g}{\partial x} = 0$, we can choose $ab > c$ for all x and $y \neq 0$. In this case, we have $\alpha_1 < \frac{(2\mu+\beta_1)(2\lambda+\beta_1-1)}{2\lambda+\beta_1}$.

(v) $p(t,x,y,z) = -y^2 \leq q(t) = 0 = \max q(t) < \infty$, for all $t, x, y,$ and z. We have $q \equiv 0 \in L^1(0,\infty)$. Finally, the minimal chaotic attractors displayed by system (7.60) are all bounded, provided that $0 \leq \beta_1 \leq 1$, $\alpha_1 \geq \beta_1$, and $\alpha_1 < \frac{(2\mu+\beta_1)(2\lambda+\beta_1-1)}{2\lambda+\beta_1}$, i.e., there is a constant $K > 0$ such that $|y(t)| \leq \sqrt{K}$, $|x(t)| \leq \sqrt{K}$, and $|z(t)| \leq \sqrt{K}$.

7.3.3   Appendix A

In this appendix we give the expressions of the quantities $(\varphi_i = \varphi_i)_{1 \leq i \leq 10}$ and $(\psi_j)_{1 \leq j \leq 77}$ defined in Eq. (7.57) as follows:

$$
\begin{cases}
\varphi_1 = \psi_1 x^2 + \psi_2 x + \psi_3 x' + \psi_4 \\[2mm]
\varphi_2 = \psi_5 x^2 + \psi_6 x + \psi_7 x' + \psi_8 \\[2mm]
\varphi_3 = \psi_9 x^4 + \psi_{10} x^3 + \psi_{11} x^2 x' + \psi_{12} x^2 + \psi_{13} x x + \psi_{14} x + \psi_{15}(x')^2 \\
\qquad + \ 16 x' + \psi_{17}
\end{cases}
$$

$$\left\{ \begin{aligned} \varphi_4 &= \psi_{18}x^4 + \psi_{19}x^3 + \psi_{20}x^2x' + \psi_{21}x^2 + \psi_{22}xx' + \psi_{23}x + \psi_{24}(x')^2 \\ &\quad + \psi_{25}x' + \psi_{26} \\[2mm] \varphi_5 &= \psi_{27}x^3 + \psi_{28}x^2 + \psi_{29}xx' + \psi_{30}x + \psi_{31}x' \\[2mm] \varphi_6 &= \psi_{32}x^4 + \psi_{33}x^3 + \psi_{34}x^2x' + \psi_{35}x^2 + \psi_{36}xx' + \psi_{37}x \end{aligned} \right. \tag{7.61}$$

$$\left\{ \begin{aligned} \varphi_7 &= \psi_{38}x^5 + \psi_{39}x^4 + \psi_{40}x^3x' + \psi_{41}x^3 + \psi_{42}x^2x' + \psi_{43}x^2 + \chi_1 \\[2mm] \chi_1 &= \psi_{44}x(x')^2 + \psi_{45}xx' + \psi_{46}x + \psi_{47}(x')^2 + \psi_{48}x' + \psi_{49} \\[2mm] \varphi_8 &= \psi_{50}x^6 + \psi_{51}x^5 + \psi_{52}x^4x' + \psi_{53}x^4 + \psi_{54}x^3x' + \psi_{55}x^3 \\ &\quad + \psi_{56}x^2(x')^2 + \chi_2 \\[2mm] \chi_2 &= \psi_{57}x^2x' + \psi_{58}x^2 + \psi_{59}xx' + \psi_{60}(x')^2 + \psi_{61}x' \end{aligned} \right.$$

$$\left\{ \begin{aligned} \varphi_9 &= \psi_{62}x^6 + \psi_{63}x^5 + \psi_{64}x^4x' + \psi_{65}x^4 + \psi_{66}x^3x' + \psi_{67}x^3 \\ &\quad + \psi_{68}x^2(x')^2 + \chi_3 \\[2mm] \chi_3 &= \psi_{69}x^2x' + \psi_{70}x^2 + \psi_{71}x(x')^2 + \psi_{72}xx' + \psi_{73}x + \psi_{74}(x')^3 + \psi_{75}(x')^2 \\ &\quad + \psi_{76}x' + \psi_{77} \\[2mm] \varphi_{10} &= \left(-2\tfrac{\lambda_3}{a_2}a_4\right)x^3x' + \left(-2\lambda_3\tfrac{a_1}{a_2}\right)x^2x' + 2\tfrac{\lambda_3}{a_2}x(x')^2 + \left(-2\lambda_3\tfrac{a_0}{a_2}\right)xx' \end{aligned} \right. \tag{7.62}$$

$$\left\{ \begin{aligned} \psi_1 &= \mu_1\lambda_{10} + 2\lambda_{12}\mu_{14} - 2\tfrac{\lambda_9}{a_2}(\mu_1 a_3 + \xi_1 a_4) \\[2mm] \psi_2 &= \mu_2\lambda_{10} + 2\lambda_{12}\mu_{15} - 2\tfrac{\lambda_9}{a_2}(\xi_1 a_1 + \mu_2 a_3) \\[2mm] \psi_3 &= \mu_3\lambda_{10} + 2\lambda_{12}\mu_{16} + 2\tfrac{\lambda_9}{a_2}(\xi_1 - \mu_3 a_3) \\[2mm] \psi_4 &= \xi_1\lambda_{11} + \mu_4\lambda_{10} + \xi_4\lambda_{13} + 2\lambda_{12}\mu_{17} \\ &\quad - 2\tfrac{\lambda_9}{a_2}(\xi_1 a_0 + \mu_4 a_3) \end{aligned} \right.$$

$$\left\{ \begin{aligned} \psi_5 &= \lambda_3\xi_1 + \lambda_4\xi_4 - \tfrac{\lambda_{10}}{a_2}(\xi_4 a_4 + \mu_{14}a_3) \\[2mm] \psi_6 &= \xi_1\lambda_6 + \lambda_7\xi_4 - \tfrac{\lambda_{10}}{a_2}(\xi_4 a_1 + \mu_{15}a_3) \\[2mm] \psi_7 &= \tfrac{\lambda_{10}}{a_2}(\xi_4 - \mu_{16}a_3), \; \psi_8 = -\tfrac{\lambda_{10}}{a_2}(\xi_4 a_0 + \mu_{17}a_3) \\[2mm] \psi_9 &= 2\lambda_{12}\mu_{18} - 2\mu_1\tfrac{\lambda_9}{a_2}a_4, \; \psi_{10} = 2\lambda_{12}\mu_{19} - 2\tfrac{\lambda_9}{a_2}(\mu_1 a_1 + \mu_2 a_4) \end{aligned} \right. \tag{7.63}$$

$$\begin{cases} \psi_{11} = 2\lambda_{12}\mu_{20} + 2\frac{\lambda_9}{a_2}\left(\mu_1 - \mu_3 a_4\right) \\[2mm] \psi_{12} = \mu_1\lambda_{11} + 2\lambda_{12}\mu_{21} + \lambda_{13}\mu_{14} - 2\frac{\lambda_9}{a_2}\left(\mu_1 a_0 + \mu_2 a_1 + \mu_4 a_4\right) \\[2mm] \psi_{13} = 2\lambda_{12}\mu_{22} + 2\frac{\lambda_9}{a_2}\left(\mu_2 - \mu_3 a_1\right) \\[2mm] \psi_{14} = \mu_2\lambda_{11} + 2\lambda_{12}\mu_{23} + \lambda_{13}\mu_{15} - 2\frac{\lambda_9}{a_2}\left(\mu_2 a_0 + \mu_4 a_1\right) \\[2mm] \psi_{15} = 2\lambda_{12}\mu_{24} + 2\mu_3\frac{\lambda_9}{a_2} \end{cases}$$

$$\begin{cases} \psi_{16} = \lambda_7 + \mu_3\lambda_{11} + 2\lambda_{12}\mu_{25} + \lambda_{13}\mu_{16} + 2\frac{\lambda_9}{a_2}\left(\mu_4 - \mu_3 a_0\right) \\[2mm] \psi_{17} = \mu_4\lambda_{11} + \xi_3\lambda_{10} + 2\lambda_{12}\mu_{26} + \lambda_{13}\mu_{17} - 2\frac{\lambda_9}{a_2}\left(\mu_4 a_0 + \xi_3 a_3\right) \\[2mm] \psi_{18} = -\frac{\lambda_{10}}{a_2}\left(\mu_{14}a_4 + \mu_{18}a_3\right),\, \psi_{19} = -\frac{\lambda_{10}}{a_2}\left(\mu_{14}a_1 + \mu_{15}a_4 + \mu_{19}a_3\right) \\[2mm] \psi_{20} = -\frac{\lambda_{10}}{a_2}\left(\mu_{20}a_3 - \mu_{14} + \mu_{16}a_4\right) \end{cases}$$

$$(7.64)$$

$$\begin{cases} \psi_{21} = -\frac{\lambda_{10}}{a_2}\left(\mu_{14}a_0 + \mu_{21}a_3 + \mu_{15}a_1 + \mu_{17}a_4\right) \\[2mm] \psi_{22} = -\frac{\lambda_{10}}{a_2}\left(\mu_{22}a_3 - \mu_{15} + \mu_{16}a_1\right) \\[2mm] \psi_{23} = -\frac{\lambda_{10}}{a_2}\left(\mu_{15}a_0 + \mu_{23}a_3 + \mu_{17}a_1\right) \\[2mm] \psi_{24} = \frac{\lambda_{10}}{a_2}\left(\mu_{16} - \mu_{24}a_3\right) \\[2mm] \psi_{25} = -\frac{\lambda_{10}}{a_2}\left(\mu_{16}a_0 - \mu_{17} + \mu_{25}a_3\right) \end{cases}$$

$$\begin{cases} \psi_{26} = -\frac{\lambda_{10}}{a_2}\left(\mu_{17}a_0 + \mu_{26}a_3\right) \\[2mm] \psi_{27} = \mu_1\lambda_6,\, \psi_{28} = \mu_2\lambda_6 \\[2mm] \psi_{29} = 2\lambda_4 + \mu_3\lambda_6 - 2\frac{\lambda_3}{a_2}a_3 \\[2mm] \psi_{30} = \mu_4\lambda_6 \end{cases}$$

$$(7.65)$$

$$\begin{cases} \psi_{31} = -\frac{\lambda_6}{a_2}a_3,\, \psi_{32} = \lambda_4\mu_{14},\, \psi_{33} = \lambda_4\mu_{15} + \lambda_7\mu_{14} \\[2mm] \psi_{34} = \lambda_4\mu_{16},\, \psi_{35} = \lambda_3\xi_2 + \lambda_4\mu_{17} + \lambda_7\mu_{15} \\[2mm] \psi_{36} = \lambda_7\mu_{16},\, \psi_{37} = \lambda_7\mu_{17},\, \psi_{38} = \lambda_6\mu_5 + \lambda_7\mu_{18} \\[2mm] \psi_{39} = \lambda_6\mu_6 + \mu_5\lambda_{11} + \lambda_7\mu_{19} + \lambda_{13}\mu_{18},\, \psi_{40} = \lambda_6\mu_7 + \lambda_7\mu_{20} \end{cases}$$

$$(7.66)$$

$$\begin{cases} \psi_{41} = \lambda_6\mu_8 + \mu_6\lambda_{11} + \lambda_7\mu_{21} + \lambda_{13}\mu_{19} \\[2mm] \psi_{42} = 3\lambda_2 + \lambda_6\mu_9 + \mu_7\lambda_{11} + \lambda_7\mu_{22} + \mu_{20}\lambda_{13} \\[2mm] \psi_{43} = \lambda_6\mu_{10} + \mu_8\lambda_{11} + \lambda_7\mu_{23} + \lambda_{13}\mu_{21} \\[2mm] \psi_{44} = \lambda_6\mu_{11} + \lambda_7\mu_{24} \\[2mm] \psi_{45} = 2\lambda_5 + \lambda_6\mu_{12} + \mu_9\lambda_{11} + \lambda_7\mu_{25} + \lambda_{13}\mu_{22} \end{cases}$$

$$\begin{cases} \psi_{46} = \lambda_6\mu_{13} + \lambda_7\mu_{26} + \lambda_{11}\mu_{10} + \lambda_{13}\mu_{23} \\[2mm] \psi_{47} = \lambda_{11}\mu_{11} + \lambda_{13}\mu_{24} \\[2mm] \psi_{48} = \lambda_8 + \lambda_{11}\mu_{12} + \lambda_{13}\mu_{25} \\[2mm] \psi_{49} = \lambda_{11}\mu_{13} + \lambda_{13}\mu_{26} \\[2mm] \psi_{50} = \lambda_3\mu_5 + \lambda_4\mu_{18} \end{cases} \qquad (7.67)$$

$$\begin{cases} \psi_{51} = \lambda_3\mu_6 + \lambda_4\mu_{19},\ \psi_{52} = \lambda_3\mu_7 + \lambda_4\mu_{20} \\[2mm] \psi_{53} = \lambda_3\mu_8 + \lambda_4\mu_{21},\ \psi_{54} = 4\lambda_1 + \lambda_3\mu_9 + \lambda_4\mu_{22} \\[2mm] \psi_{55} = \lambda_3\mu_{10} + \lambda_4\mu_{23},\ \psi_{56} = \lambda_3\mu_{11} + \lambda_4\mu_{24} \\[2mm] \psi_{57} = \lambda_3\mu_{12} + \lambda_4\mu_{25} - \frac{\lambda_6}{a_2}a_4,\ \psi_{58} = \lambda_3\mu_{13} + \lambda_4\mu_{26} \\[2mm] \psi_{59} = -\lambda_6\frac{a_1}{a_2},\ \psi_{60} = \frac{\lambda_6}{a_2} \end{cases} \qquad (7.68)$$

$$\begin{cases} \psi_{61} = -\lambda_6\frac{a_0}{a_2},\ \psi_{62} = -2\mu_5\frac{\lambda_9}{a_2}a_4 - \lambda_{10}\frac{\mu_{18}}{a_2}a_4 \\[2mm] \psi_{63} = -2\frac{\lambda_9}{a_2}(\mu_5a_1 + \mu_6a_4) - \frac{\lambda_{10}}{a_2}(\mu_{18}a_1 + \mu_{19}a_4) \\[2mm] \psi_{64} = 2\frac{\lambda_9}{a_2}(\mu_5 - \mu_7a_4) + \frac{\lambda_{10}}{a_2}(\mu_{18} - \mu_{20}a_4) \\[2mm] \psi_{65} = -2\frac{\lambda_9}{a_2}(\mu_5a_0 + \mu_6a_1 + \mu_8a_4) \\ \qquad\quad - \frac{\lambda_{10}}{a_2}(\mu_{21}a_4 + \mu_{18}a_0 + \mu_{19}a_1) \\[2mm] \psi_{66} = -2\frac{\lambda_9}{a_2}(\mu_7a_1 - \mu_6 + \mu_9a_4) \\ \qquad\quad - \frac{\lambda_{10}}{a_2}(\mu_{20}a_1 - \mu_{19} + \mu_{22}a_4) \end{cases}$$

$$\left\{ \begin{array}{l} \psi_{67} = -2\frac{\lambda_9}{a_2}\left(\mu_6 a_0 + \mu_8 a_1 + \mu_{10} a_4\right) \\ \qquad -\frac{\lambda_{10}}{a_2}\left(\mu_{21} a_1 + \mu_{23} a_4 + \mu_{19} a_0\right) \\[2mm] \psi_{68} = 2\frac{\lambda_9}{a_2}\left(\mu_7 - \mu_{11} a_4\right) + \frac{\lambda_{10}}{a_2}\left(\mu_{20} - \mu_{24} a_4\right) \\[2mm] \psi_{69} = -2\frac{\lambda_9}{a_2}\left(\mu_7 a_0 - \mu_8 + \mu_9 a_1 + \mu_{12} a_4\right) \\ \qquad -\frac{\lambda_{10}}{a_2}\left(\mu_{20} a_0 - \mu_{21} + \mu_{22} a_1 + \mu_{25} a_4\right) \\[2mm] \psi_{70} = -2\frac{\lambda_9}{a_2}\left(\mu_8 a_0 + \mu_{10} a_1 + \mu_{13} a_4\right) \\ \qquad -\frac{\lambda_{10}}{a_2}\left(\mu_{21} a_0 + \mu_{23} a_1 + \mu_{26} a_4\right) \end{array} \right. \tag{7.69}$$

$$\left\{ \begin{array}{l} \psi_{71} = 2\frac{\lambda_9}{a_2}\left(\mu_9 - \mu_{11} a_1\right) + \frac{\lambda_{10}}{a_2}\left(\mu_{22} - \mu_{24} a_1\right) \\[2mm] \psi_{72} = -2\frac{\lambda_9}{a_2}\left(\mu_9 a_0 - \mu_{10} + \mu_{12} a_1\right) - \frac{\lambda_{10}}{a_2}\left(\mu_{22} a_0 - \mu_{23} + \mu_{25} a_1\right) \\[2mm] \psi_{73} = -2\frac{\lambda_9}{a_2}\left(\mu_{10} a_0 + \mu_{13} a_1\right) - \frac{\lambda_{10}}{a_2}\left(\mu_{23} a_0 + \mu_{26} a_1\right) \\[2mm] \psi_{74} = 2\lambda_9\frac{\mu_{11}}{a_2} + \lambda_{10}\frac{\mu_{24}}{a_2} \end{array} \right.$$

$$\left\{ \begin{array}{l} \psi_{75} = 2\frac{\lambda_9}{a_2}\left(\mu_{12} - \mu_{11} a_0\right) + \frac{\lambda_{10}}{a_2}\left(\mu_{25} - \mu_{24} a_0\right) \\[2mm] \psi_{76} = 2\frac{\lambda_9}{a_2}\left(\mu_{13} - \mu_{12} a_0\right) + \frac{\lambda_{10}}{a_2}\left(\mu_{26} - \mu_{25} a_0\right) \\[2mm] \psi_{77} = -\left(2\lambda_9\mu_{13}\frac{a_0}{a_2} + \lambda_{10}\mu_{26}\frac{a_0}{a_2}\right) \end{array} \right. \tag{7.70}$$

7.4   Bounded hyperjerky dynamics

The same logic applies for some hyperjerky dynamics. Indeed, the method of analysis is based on a result given in [Tunc (2010(a))] about the boundedness of solutions of a certain type of third-order non-linear differential equation, with bounded delay of the form:

$$x^{(4)}(t) + \varphi(t)\, x'''(t) + h(t)\, x''(t) + \phi(t) + f(t) = p(t) \tag{7.71}$$

where

$$
\begin{cases}
\varphi\left(t\right) = \varphi\left(x''\left(t\right)\right) \\[2mm]
h\left(t\right) = h\left(x''\left(t-r\right)\right) \\[2mm]
\phi\left(t\right) = \phi\left(x'\left(t-r\right)\right) \\[2mm]
f\left(t\right) = f\left(x\left(t-r\right)\right) \\[2mm]
p\left(t\right) = p\left(t, x\left(t\right), x'\left(t\right), x''\left(t\right), x'''\left(t\right), x\left(t-r\right), , x'\left(t-r\right), x''\left(t-r\right)\right)
\end{cases}
$$

Here $r > 0$. The φ, h, ϕ, f and p depend only on the arguments displayed explicitly. They are continuous for all values of their respective arguments and differentiable for all $t \in [0, +\infty)$, which confirms the existence of solutions for system (7.71). Also, the derivatives $\frac{\partial g}{\partial x'}$ and $\frac{\partial h}{\partial x}$ exist and are continuous. In addition, if the functions f_1, f_2, g, h and p satisfy a Lipschitz condition in $x\left(t\right), x'\left(t\right), x''\left(t\right), x'''\left(t\right), x\left(t-r\right), , x'\left(t-r\right)$, and $x''\left(t-r\right)$, $h\left(0\right) = \phi\left(0\right) = f\left(0\right) = 0$ and the derivatives $\frac{d\phi}{dx'} = \phi'\left(x'\right)$ and $\frac{df}{dx} = f'\left(x\right)$ exist and are also continuous. Let

$$
\begin{cases}
\varphi_1\left(z\right) = \begin{cases} \frac{1}{z}\int_0^z \varphi\left(\tau\right)d\tau, z \neq 0 \\[2mm] \varphi\left(0\right), z = 0 \end{cases} \\[6mm]
\phi_1\left(y\right) = \begin{cases} \frac{\phi(y)}{y}, y \neq 0 \\[2mm] \phi'\left(0\right), y = 0 \end{cases}
\end{cases}
$$

then the solution of system (7.71) is unique.

The following result was proved in [Tunc (2010(a)), The boundedness to non-linear differential equations of fourth order with delay, Non-linear Studies, 17(1)47-56, 2010.]:

Theorem 55 *In addition to the basic assumptions imposed on φ, h, ϕ, f and p, we assume the following conditions are satisfied: (i) There are positive constants $\alpha_1, \alpha_2, \alpha_3, \alpha_4, \Delta, L, d_1, d_2, d_3$ and ε such that $\alpha_1\alpha_2\alpha_3 - \alpha_3\phi'\left(y\right) - \alpha_1\alpha_4\varphi\left(z\right) \geq \Delta > 0$ for all y and z, in which $\varepsilon \leq \frac{\Delta}{2\alpha_1\alpha_2 D}, D = \alpha_1\alpha_2 + \frac{\alpha_2\alpha_3}{\alpha_4}$.*

(ii) $0 < \alpha_4 - \frac{\alpha_1\Delta}{4\alpha_3} < f'\left(x\right) \leq \alpha_4$ for all x;

(iii) $\phi'\left(y\right) \geq \alpha_3$ and $0 \leq \phi_1\left(y\right) - \alpha_3 < \frac{\Delta}{8\alpha_3}\sqrt{\frac{\alpha_4}{2\alpha_1\alpha_3}}$ for all y;

(iv) $0 \leq \frac{h(z)}{z} - \alpha_2 \leq \frac{\alpha_3}{8\alpha_4}\sqrt{\frac{\varepsilon\Delta}{\alpha_1}}$ for all z, $(z \neq 0)$ and $|h'\left(z\right)| \leq L$ for all z,

(v) $\varphi\left(z\right) \geq \alpha_1$ and $\varphi_1\left(z\right) - \varphi\left(z\right) < \frac{\Delta}{2\alpha_1^2\alpha_3}\sqrt{\frac{\alpha_4}{2\alpha_1\alpha_3}}$ for all z;

(vi) $|p\left(t, x, x\left(t-r\right), y, y\left(t-r\right), z, z\left(t-r\right)\right)| \leq q\left(t\right)$, where $\max q\left(t\right) < \infty$ and $q \in L^1\left(0, \infty\right), L^1\left(0, \infty\right)$ is the space of integrable Lebesgue functions.

Hence, there exists a finite positive constant K, such that the solution $x(t)$ of Eq. (7.71) defined by the initial functions

$$
\begin{cases}
x(t) = \phi(t) \\[2mm]
x'(t) = \phi'(t) \\[2mm]
x''(t) = \phi''(t) \\[2mm]
x'''(t) = \phi'''(t)
\end{cases}
$$

satisfies the inequalities

$$
|x(t)| \le K, |x'(t)| \le K, |x''(t)| \le K, |x'''(t)| \le K
$$

for all $t \ge t_0$, where $\phi \in C^3\left([t_0 - r, t_0], \mathbb{R}\right)$, provided that

$$
\begin{cases}
r < 2\min\{w_1, w_2, w_3\} \\[2mm]
w_1 = \dfrac{\varepsilon \alpha_3}{2(d_2\alpha_4 + d_2 L + d_2\alpha_1\alpha_2 + 2\lambda)} \\[3mm]
w_2 = \dfrac{\Delta}{8\alpha_1\alpha_3\left(\alpha_1\alpha_2 + L + \alpha_4 + 2\mu\right)} \\[3mm]
w_3 = \dfrac{\varepsilon \alpha_1}{d_1\alpha_4 + d_1 L + d_1\alpha_1\alpha_2 + 2\rho}
\end{cases}
\tag{7.72}
$$

with

$$
\begin{cases}
\lambda = \frac{\alpha_4}{2}(d_1 + d_2 + 1) > 0 \\[2mm]
\mu = \frac{\alpha_1\alpha_2}{2}(d_1 + d_2 + 1) > 0 \\[2mm]
\rho = \frac{L}{2}(d_1 + d_2 + 1) > 0
\end{cases}
$$

Now, assume that

$$
x^{(4)} + a_0 x''' + a_1 x'' + a_2 x' = g\left(t, x, x', x'', x'''\right)
\tag{7.73}
$$

and define

$$
g\left(t, x, x', x'', x'''\right) = p\left(t, x, x', x'', x'''\right) + a_1 x'' - h\left(x''\right)x'' - f(x)
$$

where the function g is a well known function, and search the expressions for h, f and p, verifying the above conditions. Assuming that

$$\begin{cases} p(x, x', x'') = 0 \\ g(x, x', x'') = x''(a_1 - x''b_1) - f(x) \\ h(x'') = b_1 x'' \\ f(x) = b_2 x \\ \varphi(t) = a_0 \\ \phi(t) = a_2 x'(t) \end{cases}$$

We have $\varphi_1(z) = a_0$ and $\phi_1(y) = a_2$.

(i) The above conditions are

$$\begin{cases} \alpha_1 \alpha_2 \alpha_3 - \alpha_3 a_2 - \alpha_1 \alpha_4 a_0 \geq \Delta > 0 \\ \varepsilon \leq \dfrac{\Delta}{2\alpha_1 \alpha_2 D} \\ D = \alpha_1 \alpha_2 + \dfrac{\alpha_2 \alpha_3}{\alpha_4} \\ a_2 \geq \alpha_3 \\ 0 \leq a_2 - \alpha_3 < \dfrac{\Delta}{8\alpha_3} \sqrt{\dfrac{\alpha_4}{2\alpha_1 \alpha_3}} \\ a_0 \geq \alpha_1 \\ 0 < \dfrac{\Delta}{2\alpha_1^2 \alpha_3} \sqrt{\dfrac{\alpha_4}{2\alpha_1 \alpha_3}} \end{cases}$$

(ii) $0 < \alpha_4 - \dfrac{\alpha_1 \Delta}{4\alpha_3} < f'(x) \leq \alpha_4$ for all x,

(iv) $0 \leq b_1 - \alpha_2 \leq \dfrac{\alpha_3}{8\alpha_4} \sqrt{\dfrac{\varepsilon \Delta}{\alpha_1}}$ for all z, $(z \neq 0)$ and $|b_1| \leq L$ for all z,

(vi) $0 \leq q(t) = 0$, where $\max q(t) < \infty$ and $q \in L^1(0, \infty)$.

Then there exists a finite positive constant K, such that the solution $x(t)$ of Eq. (7.71), defined by the initial functions $x(t) = \phi(t), x'(t) = \phi'(t), x''(t) = \phi''(t), x'''(t) = \phi'''(t)$, satisfies the inequalities $|x(t)| \leq K, |x'(t)| \leq K, |x''(t)| \leq K, |x'''(t)| \leq K$ for all $t \geq t_0$, where $\phi \in C^3([t_0 - r, t_0], \mathbb{R})$, provided that (7.72) holds true.

7.5　Boundedness of the generalized 4-D hyperchaotic model containing Lorenz-Stenflo and Lorenz-Haken systems

In this section, we give an example that demonstrates how one can prove that a 4-D hyperchaotic dynamical system is bounded. Namely, we study the boundedness of a solution for the generalized 4-D hyperchaotic model containing Lorenz-Stenflo and Lorenz-Haken systems:

$$
\begin{cases}
x' = (5\alpha + 1)y - (5\alpha + 1)x + \frac{15}{10}(1 - \alpha)w \\[2mm]
y' = (65\alpha + 26)x - y - \frac{25}{10}\alpha z + 6\alpha\frac{z^2}{x} - \alpha xw + (\alpha - 1)xz \\[2mm]
z' = \frac{-15}{10}\alpha x + \frac{25}{10}\alpha y - (\frac{3\alpha+7}{10})z - 6\alpha\frac{yz}{x} + (1 - \alpha)xy \\[2mm]
w' = -(\frac{2\alpha}{10} + 1)w + \alpha xy + (\alpha - 1)x
\end{cases}
\tag{7.74}
$$

This system is chaotic for all $0 \leq \alpha \leq 1$; with $\alpha = 0$ it reduces to the original Lorenz-Stenflo system [Stenflo (1996)] and with $\alpha = 1$ it is the original Lorenz-Haken system [Li, *et al.*, (2009)]. These results are obtained in [Zehrour and Zeraoulia (2012)] and they are considered as the generalizations of some results given in [Li, *et al.*, (2009), Wang, *et al.*, (2010)].

The main method for this analysis is based on the choice of an appropriate Lyapunov function.

Let us consider the following dynamical system

$$
X' = f(X) \tag{7.75}
$$

where $X \in \mathbb{R}^4$, $f : \mathbb{R}^4 \to \mathbb{R}^4$, $X = (x, y, z, w) \in \mathbb{R}^4$:, $t_0 \geq 0$ is the initial time, and $X(t, t_0, X_0)$ is a solution to system (7.75) satisfying $X(t, t_0, X_0) = X_0$ which, for simplicity, is denoted as $X(t)$. Assume $\Omega \in \mathbb{R}^4$ is a compact set. Define the distance between the solution $X(t, t_0, X_0)$ and the set Ω by

$$
\rho(X(t, t_0, X_0), \Omega) = \inf_{Y \in \Omega} \|X(t, t_0, X_0) - Y\|
$$

and denote

$$
\Omega_\varepsilon = \{X - \rho(X, \Omega) < 0\}
$$

Clearly, we have $\Omega \subset \Omega_\varepsilon$.

Definition 52 *Suppose that there is a compact set $\Omega \in \mathbb{R}^4$. If, for every $X_0 \in \mathbb{R}^4/X$, we have $\lim_{T\to\infty}\rho(X(t), \Omega) = 0$, that is, for any $\varepsilon > 0$, there is a $T > t_0$ such that for $t \geq T$, $X(t, t_0, X_0) \subset \Omega_\varepsilon$, then the set Ω is called an ultimate bound for system (1). If, for any $X_0 \in \Omega$ and all $t \geq t_0$, $X(t, t_0, X_0) \subset \Omega$, then the set Ω is called the positively invariant set for system (7.75).*

The bounds of system (7.75) are obtained based on multivariate function analysis concerned with locating maxima and minima. In particular, we find large regions in the bifurcation parameter space where system (7.75) is bounded. We note that when $\alpha = 0$, the Lorenz-Stenflo system describes finite-amplitude, low-frequency, short-wavelength, acoustic gravity waves in a rotational system [Stenflo (1996)]. Several results about the dynamics of system (7.75) have been reported in [Yu & Yang (1996), Yu, *et al.*, (1996), Zhou, *et al.*, (1997), Yu, (1999), Banerjee, *et al.*, (2001)]. In [Xavier & Rech (2010)], the precise locations for pitchfork and Hopf bifurcations of fixed points were determined, along with a numerical characterization of periodic and chaotic attractors.

The following result was proved in [Zehrour and Zeraoulia (2010(c))]:

Theorem 56 *When $0 < \alpha \leq 1$, the chaotic system (7.74) is contained in the sphere given by:*

$$\Omega = \left\{ (x, y, z, w) \in \mathbb{R}^4 : \frac{10}{15}x^2 + y^2 + z^2 + (w - (\frac{205\alpha + 80}{3\alpha}))^2 \leq l \right\} \quad (7.76)$$

Proof 19 *Let us consider the following Lyapunov function*

$$V = \frac{10}{15}x^2 + y^2 + z^2 + \left(w - \left(\frac{205\alpha + 80}{3\alpha} \right) \right)^2 \quad (7.77)$$

Then, the time derivative along the orbits of system (7.74) is given by

$$
\begin{aligned}
\frac{1}{2}\frac{dV}{dt} &= \frac{10}{15}xx' + yy' + zz' + (w - (\frac{205\alpha + 80}{3\alpha}))w' \\
&= \frac{10}{15}(5\alpha + 1)xy - \frac{10}{15}(5\alpha + 1)x^2 + (1 - \alpha)xw \\
&\quad + (65\alpha + 26)xy - y^2 - \frac{25}{10}\alpha yz + 6\alpha\frac{yz^2}{x} \\
&\quad - \alpha xyw + (\alpha - 1)xyz - \frac{15}{10}\alpha xz + \frac{25}{10}\alpha yz \\
&\quad - (\frac{3\alpha + 7}{10})z^2 - 6\alpha\frac{yz^2}{x} + (1 - \alpha)xyz - (\frac{2\alpha}{10} + 1)w^2 \\
&\quad + \alpha xyw + (\alpha - 1)xw + (\frac{205\alpha + 80}{3\alpha})(\frac{2\alpha}{10} + 1)w \\
&\quad - \alpha(\frac{205\alpha + 80}{3\alpha})xy - (\frac{205\alpha + 80}{3\alpha})(\alpha - 1)x
\end{aligned}
$$

$$
\begin{aligned}
\frac{1}{2}\frac{dV}{dt} &= -\frac{10}{15}(5\alpha + 1)x^2 - y^2 - (\frac{3\alpha + 7}{10})z^2 - \frac{15}{10}\alpha xz \\
&\quad - (\frac{3\alpha + 7}{10})z^2 - (\frac{2\alpha}{10} + 1)w^2 + (\frac{205\alpha + 80}{3\alpha})(\frac{2\alpha}{10} + 1)w \\
&\quad + (\frac{205\alpha + 80}{3\alpha})(1 - \alpha)x
\end{aligned}
$$

$$\frac{1}{2}\frac{dV}{dt} = -\left[\sqrt{\frac{10}{15}(5\alpha+1)}x - \frac{1}{2}\sqrt{\frac{15}{10(5\alpha+1)}}(\frac{205\alpha+80}{3\alpha})(1-\alpha)\right]^2$$

$$+\frac{1}{4}\frac{15}{10(5\alpha+1)}(\frac{205\alpha+80}{3\alpha})^2(1-\alpha)^2 - y^2 - (\frac{3\alpha+7}{10})z^2$$

$$-\frac{15}{10}\alpha xz - (\frac{2\alpha}{10}+1)\left[w - \frac{1}{2}(\frac{205\alpha+80}{3\alpha})\right]^2 +$$

$$\frac{1}{4}(\frac{2\alpha}{10}+1)(\frac{205\alpha+80}{3\alpha})^2$$

$$\frac{1}{2}\frac{dV}{dt} = -\left[\sqrt{\frac{10}{15}(5\alpha+1)}x - \frac{1}{2}\sqrt{\frac{15}{10(5\alpha+1)}}(\frac{205\alpha+80}{3\alpha})(1-\alpha)\right]^2$$

$$-y^2 - (\frac{3\alpha+7}{10})z^2 - \frac{15}{10}\alpha xz - (\frac{2\alpha}{10}+1)\left[w - \frac{1}{2}(\frac{205\alpha+80}{3\alpha})\right]^2$$

$$+\frac{1}{4}(\frac{205\alpha+80}{3\alpha})^2(\frac{25\alpha^2+22\alpha+25}{10(5\alpha+1)})$$

For $V' = 0$, we have the following four-dimensional sphere Γ:

$$\begin{cases} \check{s}^2 + y^2 + (\frac{3\alpha+7}{10})z^2 + \frac{15}{10}\alpha xz + (\frac{2\alpha}{10}+1)\hat{w}^2 = \frac{1}{4}(\frac{205\alpha+80}{3\alpha})^2(\frac{25\alpha^2+22\alpha+25}{10(5\alpha+1)}) \\[2mm] \check{s} = \sqrt{\frac{10}{15}(5\alpha+1)}x - \frac{1}{2}\sqrt{\frac{15}{10(5\alpha+1)}}(\frac{205\alpha+80}{3\alpha})(1-\alpha) \\[2mm] \hat{w} = w - \frac{1}{2}(\frac{205\alpha+80}{3\alpha}) \end{cases}$$

$$(7.78)$$

The corresponding matrix of the left-hand side quadratic from of Eq. (7.74) is given by

$$A = \begin{bmatrix} 1 & 0 & \frac{15}{20} & 0 \\ 0 & 1 & 0 & 0 \\ \frac{15}{20} & 0 & \frac{3\alpha+7}{10} & 0 \\ 0 & 0 & 0 & \frac{2\alpha}{10}+1 \end{bmatrix}$$

with positive eigenvalues given by:

$$\begin{cases} \lambda_1 = 1 \\[2mm] \lambda_2 = \frac{17}{20} + \frac{3}{20}\alpha + \frac{3}{20}\sqrt{(26 - 2\alpha + \alpha^2)} \\[2mm] \lambda_3 = \frac{17}{20} + \frac{3}{20}\alpha - \frac{3}{20}\sqrt{(26 - 2\alpha + \alpha^2)} \\[2mm] \lambda_4 = \frac{1}{5}\alpha + 1. \end{cases}$$

We have $\frac{17}{20} + \frac{3}{20}\alpha - \frac{3}{20}\sqrt{(26 - 2\alpha + \alpha^2)} = 0$, *only for* $\alpha = -0.4583333$. *Then it is positive for all* $0 < \alpha \leq 1$. *Hence, for all* $0 < \alpha \leq 1$, A *is a positive definite matrix. Thus, Eq. (7.78) is a four-dimensional ellipsoidal surface. Outside* Γ, *that is, when*

$$
\begin{cases}
\check{s}_1^2 + y^2 + (\frac{3\alpha+7}{10})z^2 + \frac{15}{10}\alpha xz + (\frac{2\alpha}{10} + 1)\hat{w}^2 > \frac{1}{4}(\frac{205\alpha+80}{3\alpha})^2(\frac{25\alpha^2+22\alpha+25}{10(5\alpha+1)}) \\[2mm]
\check{s}_1 = \frac{10}{15}\sqrt{(5\alpha + 1)}x + \frac{1}{2}\sqrt{\frac{15}{10(5\alpha+1)}}(\frac{205\alpha+80}{3\alpha})(1 - \alpha)
\end{cases}
$$

we have $V' < 0$, *then the surface* $V = const$ *is monotonously degressive. Since* Γ *is a closed set and* V *is continuous on* Γ, *the extreme values of* V *can be attained on* Γ. *Denote the maximum value of* V *as* l, *that is* $l = \max V_{(x,y,z,w)\in\Gamma}$. *Namely, if we denote* $\sqrt{\frac{15}{10}}x = \check{s}$, *Eq (7.77) becomes*

$$
V = \check{s}^2 + y^2 + z^2 + \left(w - \left(\frac{205\alpha + 80}{3\alpha}\right)\right)^2 \tag{7.79}
$$

and the ellipsoid surface Γ *becomes* Γ' *as follows:*

$$
\check{s}_1^2 + y^2 + \left(\frac{3\alpha + 7}{10}\right)z^2 + \frac{15}{10}\alpha xz + \left(\frac{2\alpha}{10} + 1\right)\hat{w}^2 \tag{7.80}
$$

$$
\tag{7.81}
$$

$$
= \frac{1}{4}\left(\frac{205\alpha + 80}{3\alpha}\right)^2\left(\frac{25\alpha^2 + 22\alpha + 25}{10(5\alpha + 1)}\right)
$$

Thus, l *is the maximum value of the square of the distance from the points on* Γ' *to point* $(0, 0, 0, \frac{205\alpha+80}{3\alpha})$, *which exists. So, it is possible to compute it when the parameters in system (7.74) are given. For the set* Ω *given by (7.76), we have* $\Gamma \subset \Omega$ *and*

$$
\lim_{t\to\infty} \rho(X(t), \Omega) = 0 \tag{7.82}
$$

using the proof by absurd, where $X(t) = (x(t), y(t), z(t), w(t))$. *If (7.82) does not hold, then the orbits of system (7.74) are outside the set* Ω, *therefore* $V' < 0$. *Thus, the function* $V(X(t))$ *monotonously decreases outside the set* Ω, *then we get*

$$
\lim_{t\to\infty} V(X(t)) = V^* > l \tag{7.83}
$$

Let

$$
s = \inf_{X\in D}(-V'(X(t))) \tag{7.84}
$$

where

$$
D = \{X(t) : V^* < V(X(t)) < V(X(t_0))\}
$$

where t_0 *is the initial time. Thus,* s *and* V^* *are positive constants, and*

$$
\frac{dV(X(t))}{dt} \leq -s \tag{7.85}
$$

As $t \to \infty$, we have

$$0 \le V(X(t)) \le V(X(t_0)) - s(t - t_0) \to -\infty \qquad (7.86)$$

which is a contradiction. Thus, (7.82) holds true, and Ω is the ultimate bound of the system (7.74). In fact Ω is also the positively invariant set. Indeed, suppose that V attains its maximum value on the surface Γ at point $P_0 = (\hat{x}_0, \hat{y}_0, \hat{z}_0, \hat{w}_0)$. Since $\Gamma \subset \Omega$, then for any point $X(t)$ on Ω and $X(t) \ne P_0$, we have $V'(X) < 0$, thus, any orbit $X(t)$ (such that $X(t) \ne P_0$) of system (7.74) will go into the set Ω. When $X(t) = P_0$, by the continuation theorem, the orbit $X(t)$ will also go into the set Ω. Thus, Ω is the positively invariant set of system (7.74).

7.5.1   Estimating the bounds for the Lorenz-Haken system

As a special case, for $\alpha = 1$, we find the chaotic Lorenz-Haken system given by:

$$\begin{cases} x' = -6x + 6y \\[2mm] y' = 91x - y - \frac{25}{10}z + 6\frac{z^2}{x} - xw \\[2mm] z' = -\frac{15}{10}x + \frac{25}{10}y - z - 6\frac{z}{x} \\[2mm] w' = -\frac{12}{10}w + xy \end{cases} \qquad (7.87)$$

Define the following Lyapunov function

$$V = \frac{10}{15}x^2 + y^2 + z^2 + (w - 95)^2 \qquad (7.88)$$

and the four-dimension ellipsoidal surface

$$\Gamma : (2x)^2 + y^2 + z^2 + \frac{15}{10}xz + \frac{12}{10}\left(w - \frac{95}{2}\right)^2 = \left(\frac{95}{2}\right)^2 \left(\frac{6}{5}\right) = \frac{5415}{2} \qquad (7.89)$$

To get

$$l = \max_{(x,y,z,w)\in\Gamma} \left[\frac{10}{15}x^2 + y^2 + z^2 + (w - 95)^2\right] \qquad (7.90)$$

let

$$\begin{cases} F = \frac{10x^2}{15} + y^2 + z^2 + (w - 95)^2 + \lambda g\,(x, y, z, w) \\[3mm] g\,(x, y, z, w) = (2x)^2 + y^2 + z^2 + \frac{15xz}{10} + \frac{12\left(w - \frac{95}{2}\right)^2}{10} - \frac{5415}{2} \end{cases} \qquad (7.91)$$

Thus, we have

$$
\begin{cases}
F_x = \frac{20}{15}x + 4\lambda x + \frac{15}{10}z = 0 \\[2mm]
F_y = 2y + 2\lambda y = 0 \\[2mm]
F_z = 2z + 2\lambda z + \frac{15}{10}x = 0 \\[2mm]
F_w = 2(w - 95) + \frac{24}{10}\lambda(w - \frac{95}{2}) = 0 \\[2mm]
(2x)^2 + y^2 + z^2 + \frac{15}{10}xz + \frac{12}{10}(w - \frac{95}{2})^2 = \frac{5415}{2}
\end{cases}
\tag{7.92}
$$

We have

$$
(x, y, z, w, \lambda) = (0, 0, 0, 95, 0) \ \text{ or } \ (x, y, z, w, \lambda) = \left(0, 0, 0, \frac{-190}{114}\right)
\tag{7.93}
$$

Then

$$
l = \max_{(x,y,z,w)\in\Gamma} \left[\frac{10}{15}x^2 + y^2 + z^2 + (w - 95)^2\right] = 95
\tag{7.94}
$$

Thus, the chaotic Lorenz-Haken system is contained in the ellipsoid surface

$$
\Omega = \left\{(x, y, z, w) \in \mathbb{R}^4 \colon \frac{10}{15}x^2 + y^2 + z^2 + (w - 95)^2 \le 95\right\}
\tag{7.95}
$$

7.5.2 Estimating the bounds for the Lorenz–Stenflo system

In this section, we find upper and lower bounds for the Lorenz-Stenflo system [Stenflo (1996)] given by:

$$
\begin{cases}
x' = -\sigma\,(x - y) + sw \\[2mm]
y' = -xz + rx - y \\[2mm]
z' = xy - bz \\[2mm]
w' = -x - \sigma w
\end{cases}
\tag{7.96}
$$

where σ, s, r, b are bifurcation parameters. The bounds for system (7.96) are obtained based on multivariable function analysis concerned with locating maxima and minima. In particular, we find large regions in the bifurcation parameter space $(\sigma, r, b, s) \in \mathbb{R}^4$ where system (7.96) is bounded. The Lorenz-Stenflo system (7.96) describes finite-amplitude, low-frequency, short-wavelength, acoustic gravity waves in a rotational system [Stenflo (1996)].

We have proved the following result:

Theorem 57 *The Lorenz-Stenflo system (7.96) is contained in part of the 4-D ellipsoid defined by*

$$\Omega = \left\{ (x, y, z, w) \in \mathbb{R}^4 : Q^2 < \frac{\frac{1}{s}x^2 + y^2 + \left(z - \left(r + \frac{\sigma}{s}\right)\right)^2 + w^2}{2} \le R^2 \right\}$$

for all $r \ge 0, b > 0, s \ge 1, 0 < \sigma \le s, \sigma < bs$, and all initial conditions, where $Q^2 = \frac{\left(2rs^2 + 2\sigma s\right)^2}{16s^4}$ and $R^2 = \frac{b^2(\sigma + rs)^2}{4(bs - \sigma)^2}$.

Proof 20 *To estimate the bound for the Lorenz-Stenflo system (7.96), we consider the Lyapunov function $V(x, y, z, w)$ defined by*

$$V(x, y, z, w) = \frac{\frac{1}{s}x^2 + y^2 + \left(z - \left(r + \frac{\sigma}{s}\right)\right)^2 + w^2}{2}$$

The derivative of V along the solutions of (7.96) is given by

$$\frac{dV}{dt} = -\frac{\sigma}{s}x^2 - y^2 - b\left(z - \frac{\sigma + rs}{2s}\right)^2 - \sigma w^2 + \frac{b\left(r + \frac{\sigma}{s}\right)^2}{4}$$

Let

$$H(x, y, z, w) = \frac{x^2}{\frac{b(\sigma + rs)^2}{4\sigma s}} + \frac{y^2}{\frac{b\left(r + \frac{\sigma}{s}\right)^2}{4}} + \frac{\left(z - \frac{\sigma + rs}{2s}\right)^2}{\frac{(\sigma + rs)^2}{4s^2}} + \frac{w^2}{\frac{b(\sigma + rs)^2}{4\sigma s^2}} - 1$$

Thus, in order to prove the boundedness of system (7.96), we assume that it is bounded, and then we find its bound, i.e., assume that σ, s, and b are strictly positive and $r \ge 0$. Then if system (7.96) is bounded, the function $\frac{dV}{dt}(x, y, z, w)$ has a maximum value, and the maximum point (x_0, y_0, z_0, w_0) satisfies $H(x_0, y_0, z_0, w_0) = 0$.

Now consider the 4-D ellipsoid defined by:

$$\Gamma = \left\{ (x, y, z, w) \in \mathbb{R}^4 : H(x, y, z, w) = 0, \sigma > 0, s > 0, b > 0, r \ge 0 \right\}$$

and define the function

$$F(x, y, z, w) = G(x, y, z, w) + \lambda H(x, y, z, w)$$

where

$$G(x, y, z, w) = x^2 + y^2 + z^2 + w^2$$

and $\lambda \in \mathbb{R}$ is a finite parameter. We have

$$
\begin{cases}
\max_{(x,y,z,w)\in\Gamma} G = \max_{(x,y,z,w)\in\Gamma} V \\[2mm]
\dfrac{\partial F}{\partial x} = \dfrac{2\left(br^2s^2+2brs\sigma+4\lambda s\sigma+b\sigma^2\right)}{b(\sigma+rs)^2}x \\[4mm]
\dfrac{\partial F}{\partial y} = \dfrac{2\left(br^2s^2+2brs\sigma+4\lambda s^2+b\sigma^2\right)}{b(\sigma+rs)^2}y \\[4mm]
\dfrac{\partial F}{\partial z} = \dfrac{2\left(r^2s^2+2rs\sigma+4\lambda s^2+\sigma^2\right)}{(\sigma+rs)^2}z - \dfrac{2\left(2r\lambda s^2+2\sigma\lambda s\right)}{(\sigma+rs)^2} \\[4mm]
\dfrac{\partial F}{\partial w} = \dfrac{2\left(br^2s^2+2brs\sigma+4\lambda s^2\sigma+b\sigma^2\right)}{b(\sigma+rs)^2}w.
\end{cases}
$$

In this case, the Hessian matrix of the function F is diagonal with the elements (eigenvalues)

$$
\begin{cases}
\dfrac{2\left(br^2s^2+2brs\sigma+4\lambda s\sigma+b\sigma^2\right)}{b(\sigma+rs)^2} \\[4mm]
\dfrac{2\left(br^2s^2+2brs\sigma+4\lambda s^2+b\sigma^2\right)}{b(\sigma+rs)^2} \\[4mm]
\dfrac{2\left(r^2s^2+2rs\sigma+4\lambda s^2+\sigma^2\right)}{(\sigma+rs)^2} \\[4mm]
\dfrac{2\left(br^2s^2+2brs\sigma+4\lambda s^2\sigma+b\sigma^2\right)}{b(\sigma+rs)^2}
\end{cases}
$$

Thus, the scalar function F has a maximum point if all eigenvalues of the corresponding Hessian matrix are strictly negative, that is,

$$
\lambda < \min\left(\dfrac{-b\left(\sigma+rs\right)^2}{4s\sigma}, \dfrac{-b\left(\sigma+rs\right)^2}{4s^2}, \dfrac{-\left(\sigma+rs\right)^2}{4s^2}, \dfrac{-b\left(\sigma+rs\right)^2}{4s^2\sigma}\right)
$$

If $s \geq 1$, $0 < \sigma \leq s$, and $0 < \sigma < bs$, we have

$$
\begin{cases}
\dfrac{-b(\sigma+rs)^2}{4s\sigma} - \left(\dfrac{-b(\sigma+rs)^2}{4s^2}\right) = \dfrac{-b(s-\sigma)(\sigma+rs)^2}{4s^2\sigma} \leq 0 \\[4mm]
\dfrac{-b(\sigma+rs)^2}{4s\sigma} - \left(\dfrac{-(\sigma+rs)^2}{4s^2}\right) = \dfrac{-(\sigma+rs)^2(-\sigma+bs)}{4s^2\sigma} \leq 0 \\[4mm]
\dfrac{-b(\sigma+rs)^2}{4s\sigma} - \left(\dfrac{-b(\sigma+rs)^2}{4s^2\sigma}\right) = \dfrac{-b(s-1)(\sigma+rs)^2}{4s^2\sigma} \leq 0.
\end{cases}
$$

Thus

$$
\lambda < \dfrac{-b\left(\sigma+rs\right)^2}{4s\sigma}
$$

Then the only critical point of F is $x_0 = 0$, $y_0 = 0$, $z_0 = \dfrac{2s(\sigma+rs)\lambda}{\sigma^2+4s^2\lambda+r^2s^2+2rs\sigma}$, *and* $w_0 = 0$, *and hence*

$$\max_{(x,y,z,w)\in\Gamma} G = \left(\frac{2s\sigma\lambda + 2rs^2\lambda}{\sigma^2 + 4s^2\lambda + r^2s^2 + 2rs\sigma}\right)^2 = f(\lambda)$$

In this case, there exists a parameterized family (in λ) of bounds of system (7.96). We remark that, for different values of λ, one can get different estimates for system (7.96). Some calculations lead to

$$f'(\lambda) = \frac{(\sigma+rs)^4 \, 8s^2\lambda}{(r^2s^2 + 2rs\sigma + 4\lambda s^2 + \sigma^2)^3}$$

We have $r^2s^2 + 2rs\sigma + 4\lambda s^2 + \sigma^2 < 0$ *for all* $-\infty < \lambda < \frac{-b(\sigma+rs)^2}{4s\sigma}$, *hence* $f'(\lambda) > 0$, *which means that $f(\lambda)$ is an increasing function, that is,*

$$\lim_{\lambda\to-\infty} \left(\frac{2s\sigma\lambda + 2rs^2\lambda}{\sigma^2 + 4s^2\lambda + r^2s^2 + 2rs\sigma}\right)^2 = Q^2 = \frac{(2rs^2 + 2\sigma s)^2}{16s^4}$$

$$< f(\lambda) < \frac{1}{4}b^2 \frac{(\sigma+rs)^2}{(-\sigma+bs)^2} = R^2$$

Finally, we have

$$\max_{(x,y,z,w)\in\Gamma}(x^2 + y^2 + z^2 + w^2) < \frac{b^2(\sigma+rs)^2}{4(-\sigma+bs)^2} = R^2$$

which is the upper bound for the Lorenz–Stenflo system (7.96). For the other values of $(\sigma, r, b, s) \in \mathbb{R}^4$, the same logic applies. We remark that if $\sigma \to bs$, then the upper bound converges to infinity. The volume of the resulting set in $\mathbb{R}^4$ is given by:

$$\frac{1}{4}b^2 \frac{(\sigma+rs)^2}{(-\sigma+bs)^2} - \left(\frac{(2rs^2 + 2\sigma s)^2}{16s^4}\right) = \frac{\sigma(\sigma+rs)^2(2bs-\sigma)}{4s^2(bs-\sigma)^2} > 0$$

since $bs > \sigma$.

Example 18 *To validate this result, we consider the numerical simulations reported in [Xavier & Rech (2010)] for $\sigma = 10, b = \frac{8}{3}, r = (1.2375s + 30)$, and $s \leq 2000$. Clearly, all the attractors (periodic and chaotic) are contained in the above 4-D ellipsoid. In that work of Xavier and Rech, the precise location for pitchfork and Hopf bifurcation of fixed points was determined along some numerical characterization of periodic and chaotic attractors. For example, if $s = 160$ and $r = 60$, then system (7.96) displays a chaotic attractor with*

$$\begin{cases} -17.693 \leq x \leq 16.985 \\\\ -37.834 \leq y \leq 36.253 \\\\ 6.215 \leq z \leq 65.145 \\\\ -1.24 \leq w \leq 1.264. \end{cases}$$

In this case, we have $R^2 = \frac{1}{4} \left(\frac{8}{3}\right)^2 \frac{(10+(60)(160))^2}{\left(-10+\left(\frac{8}{3}\right)(160)\right)^2} = 945.69.$ *Taking the maximum values of this attractor, i.e.,* $x = 16.985, y = 36.253, z = 65.145,$
and $w = 1.264,$ *it is easy to check that this point belongs to* Ω *since*
$Q^2 = \frac{\left(2(60)(160)^2+2(10)(160)\right)^2}{16(160)^4} = 901.88.$

7.6   Boundedness of 2-D Hénon-like mapping

This section investigates the dynamics of a 2-D Hénon-like mapping with
unknown bounded functions. The values of parameters and initial conditions
domains for which the dynamics of this equation are bounded or unbounded
are rigorously derived. The results given here are universal and cannot depend
on the expression of the non-linearity in the considered map.

The Hénon map [Hénon (1976)] is given by:

$$H(x,y) = \begin{pmatrix} 1 - ax^2 + by \\ \\ x \end{pmatrix} \tag{7.97}$$

Many works that model the original Hénon map are known in the current
literature [Benedicks and Carleson (1991), Maorotto (1979)]. Moreover, it is
possible to change the form of the Hénon map in order to obtain other, new
chaotic attractors with some interesting phenomena [Aziz Alaoui (2001), Aziz
Alaoui, *et al.*, (2001), Lozi (1978), Misiurewicz (1980), Cao & Liu (1978),
Zeraoulia (2005)]. This type of application is used in secure communications
using the notions of chaos.

A 2-D Hénon-like mapping with an unknown bounded function has the
following form:

$$g(x,y) = \begin{pmatrix} 1 - af(x) + by \\ \\ x \end{pmatrix} \tag{7.98}$$

where a, b are the control parameters, and $f : \mathbb{R} \longrightarrow \mathbb{R}$, is an unknown non-
linear bounded function that is not necessarily continuous. The map (7.98) can
be rewritten as a second order non-linear difference equation given by:

$$x_{n+1} = 1 - af(x_n) + bx_{n-1} \tag{7.99}$$

which, together with some specified values of initial conditions, defines a sequence $(x_n)_n$. Difference equations have a variety of applications, as in computer science and approximations in numerical analysis. The asymptotic behavior of solutions of a difference equation (7.99) generally depend on both
the parameter values and the initial conditions.

The main motivation of this section is to rigorously find regions in the $a - b$ plane and the initial conditions x_0 and x_1 (i.e., y_0 for the map (7.98)) for which the dynamics of the map (7.99) are bounded or unbounded.

In all proofs given here, we use the following standard results:

Theorem 58 *Let $(x_n)_n$, and $(z_n)_n$ be two real sequences, if $|x_n| \leq |z_n|$ and $\lim_{n \to +\infty} |z_n| = A < +\infty$, then $\lim_{n \to +\infty} |x_n| \leq A$, or if $|z_n| \leq |x_n|$, and $\lim_{n \to +\infty} |z_n| = +\infty$ then, $\lim_{n \to +\infty} |x_n| = +\infty$.*

We use this result in order to construct a sequence $(z_n)_n$ that satisfies the above conditions, for concluding that the difference equation (7.99) has bounded or unbounded orbits.

Theorem 59 *Suppose that f is a bounded function over its definition set; such that: $\sup_x |f(x)| = \delta$, then for every $n > 1$, and all values of a and b the sequence $(x_n)_n$ given in (7.99) satisfies the following inequality:*

$$|1 - x_n + bx_{n-2}| \leq |a| \, \delta \qquad (7.100)$$

Proof 21 *We have for every $n > 1$: $x_n = 1 - af(x_{n-1}) + bx_{n-2}$, then, one has:*

$$|-x_n + 1 + bx_{n-2}| = |af(x_{n-1})| \leq |a| \, \delta \qquad (7.101)$$

since $\sup_x |f(x)| = \delta$.

Theorem 60 *For every $n > 1$, and all values of a and b, and for all values of the initial conditions $(x_0, x_1) \in \mathbb{R}^2$, the sequence $(x_n)_n$ satisfies the following equalities:*

Firstly, if $b \neq 1$, then:

$$
\begin{cases}
x_n = \frac{b^{\frac{n-1}{2}} - 1}{b-1} + b^{\frac{n-1}{2}} x_1 - a \sum_{p=1}^{p=\frac{n-1}{2}} b^{p-1} f(x_{n-(2p-1)}), & \text{if } n \text{ is odd} \\[4mm]
x_n = \frac{b^{\frac{n}{2}} - 1}{b-1} + b^{\frac{n}{2}} x_0 - a \sum_{p=1}^{p=\frac{n}{2}} b^{p-1} f(x_{n-(2p-1)}), & \text{if } n \text{ is even}
\end{cases}
$$

$$(7.102)$$

Secondly, if $b = 1$, then:

$$
\begin{cases}
x_n = \frac{n-1}{2} + x_1 - a \sum_{p=1}^{p=\frac{n-1}{2}} f(x_{n-(2p-1)}), & \text{if } n \text{ is odd} \\[4mm]
x_n = \frac{n}{2} + x_0 - a \sum_{p=1}^{p=\frac{n}{2}} f(x_{n-(2p-1)}), & \text{if } n \text{ is even}
\end{cases}
$$

$$(7.103)$$

Proof 22 *We have for every $n > 1$, the following equalities:*

$$x_n = 1 - af(x_{n-1}) + bx_{n-2} \tag{7.104}$$

$$x_{n-2} = 1 - af(x_{n-3}) + bx_{n-4} \tag{7.105}$$

$$x_{n-4} = 1 - af(x_{n-5}) + bx_{n-6}... \tag{7.106}$$

Then, the results in (7.105) and (7.106) are obtained by successive substitutions of, for example, (7.104) in (7.105)..., for all $k = n - 2, n - 4, ..., 2$.

Theorem 61 *The fixed points (l, l) of the map (7.99) exist if one of the following conditions holds:*
(i) If $a \neq 0$, and $b \neq 1$, then l satisfies the following conditions:

$$\begin{cases} 1 - af(l) + (b-1)l = 0, \ and \ l \leq \frac{1+|a|\delta}{1-b}, \ if \ b > 1 \\ \\ \frac{1+|a|\delta}{1-b} \leq l, \ if \ b < 1 \end{cases} \tag{7.107}$$

(ii) If $b = 1$, and $a \neq 0$, then, l is given by: $f(l) = \frac{1}{a}$.
(iii) If $b \neq 1$, and $a = 0$, then, l is given by $l = \frac{1}{1-b}$.
(iv) If $a = 0$, and $b = 1$, then, there is no fixed points for the map (7.99).

Proof 23 *The proof is direct. We conclude that all fixed points of the map (7.99) are confined in the interval $\left]-\infty, \frac{1+|a|\delta}{1-b}\right]$, if $b > 1$, and in $\left[\frac{1+|a|\delta}{1-b}, +\infty\right[$, if $b < 1$. On the other hand, case (iii) gives a simple linear second-order difference equation $x_n = 1 + bx_{n-2}$.*

We have the following result about bounded orbits:

Theorem 62 *Consider the map (7.99) and assume that f is a bounded function. Then, for $|b| < 1$, and all $a \in \mathbb{R}$, and all initial conditions $(x_0, x_1) \in \mathbb{R}^2$, the orbits of the map (7.99) are bounded.*

Proof 24 *From equation (7.99) and the effect that f is a bounded function, one has the followings inequalities for all $n > 1$:*

$$|x_n| \leq 1 + |a|\delta + |bx_{n-2}| \tag{7.108}$$

$$|x_{n-2}| \leq 1 + |a|\delta + |bx_{n-4}| \tag{7.109}$$

$$... |x_2| \leq 1 + |a|\delta + |bx_0| \tag{7.110}$$

This implies from (7.108)-(7.110)... that:

$$|x_n| \leq 1 + |a|\delta + |bx_{n-2}|, \tag{7.111}$$

$$|x_n| \leq (1 + |a|)\,\delta + |b|\,(1 + |a|\,\delta + |bx_{n-4}|)\,, \qquad (7.112)$$

$$|x_n| \leq (1 + |a|\,\delta) + (1 + |a|\,\delta)\,|b| + |b|^2\,|x_{n-4}|\,,\,... \qquad (7.113)$$

Hence, from (7.113) one has:

$$|x_n| \leq (1 + |a|\,\delta) + (1 + |a|\,\delta)\,|b| + |b|^2\,(1 + |a|\,\delta) + |b|^3\,|x_{n-6}|\,,\,... \qquad (7.114)$$

Since $|b| < 1$, then the use of (7.102), the induction about some integer k and the use of the sum of a geometric growth formula permit us to get the following inequalities for every $n > 1$:

$$|x_n| \leq (1 + |a|\,\delta)\left(\frac{1 - |b|^k}{1 - |b|}\right) + |b|^k\,|x_{n-2k}| \qquad (7.115)$$

where k is the biggest integer j, such that $j \leq \frac{n}{2}$. Thus, one has the following two cases:

(1) if n is odd, i.e., $\exists m \in \mathbb{N}$, such that $n = 2m+1$, then the biggest integer $k \leq \frac{n}{2}$ is $k = \frac{n-1}{2}$, for which $(x_n)_n$ satisfies the following inequalities:

$$|x_{2m+1}| \leq (1 + |a|\,\delta)\left(\frac{1 - |b|^m}{1 - |b|}\right) + |b|^m\,|x_1| = z_m, \qquad (7.116)$$

(2) if n is even, i.e., $\exists m \in \mathbb{N}$, such that $n = 2m$, then, the biggest integer $k \leq \frac{n}{2}$ is $k = \frac{n}{2}$, for which x_n satisfies the following inequalities:

$$|x_{2m}| \leq (1 + |a|\,\delta)\left(\frac{1 - |b|^m}{1 - |b|}\right) + |b|^m\,|x_0| = u_m, \qquad (7.117)$$

Thus, since $|b| < 1$, then, the sequences $(z_m)_m$ and $(u_m)_m$ are bounded, i.e.,

$$\begin{cases} z_m \leq \frac{(1+|a|\delta)}{1-|b|} + \left||x_1| - \frac{(1+|a|\delta)}{1-|b|}\right|, \text{ for all } m \in \mathbb{N} \\[3mm] u_m \leq \frac{(1+|a|\delta)}{1-|b|} + \left||x_0| - \frac{(1+|a|\delta)}{1-|b|}\right|, \text{ for all } m \in \mathbb{N} \end{cases} \qquad (7.118)$$

Thus, Formula (7.102) and inequalities (7.118) give the following bounds for the sequence $(x_n)_n$:

$$|x_n| \leq \max \left\{ \begin{array}{c} \frac{(1+|a|\delta)}{1-|b|} + \left||x_0| - \frac{(1+|a|\delta)}{1-|b|}\right|, \\[3mm] \frac{(1+|a|\delta)}{1-|b|} + \left||x_1| - \frac{(1+|a|\delta)}{1-|b|}\right| \end{array} \right\}$$

Finally, for all values of a and all values of b, satisfying $|b| < 1$, and all values of the initial conditions $(x_0, x_1) \in \mathbb{R}^2$, one has that all orbits of the map (7.99) are bounded, i.e., in the subregion of $\mathbb{R}^4$:

$$\Omega_1 = \left\{(a, b, x_0, x_1) \in \mathbb{R}^4 /\ |b| < 1\right\} \qquad (7.119)$$

We have the following result about unbounded orbits:

Theorem 63 *Consider the map (7.99) and assume that f is a bounded function. Then, the map (7.99) possesses unbounded orbits in the following subregions of $\mathbb{R}^4$:*

$$\Omega_2 = \left\{ (a, b, x_0, x_1) \in \mathbb{R}^4 / \ |b| > 1, \ and \ both \ |x_0|, |x_1| > \frac{|a|\,\delta + 1}{|b| - 1} \right\} \quad (7.120)$$

and

$$\Omega_3 = \left\{ (a, b, x_0, x_1) \in \mathbb{R}^4 / \ |b| = 1, \ and \ |a| < \frac{1}{\delta} \right\} \quad (7.121)$$

Proof 25 *(a) For every $n > 1$, we have: $x_n = 1 - af(x_{n-1}) + bx_{n-2}$ then, $|bx_{n-2} - af(x_{n-1})| = |x_n - 1|$ then, $||bx_{n-2}| - |af(x_{n-1})|| \leq |x_n - 1|$, (We use the inequalities: $|x| - |y| \leq ||x| - |y|| \leq |x - y|$), this implies that*

$$|bx_{n-2}| - |af(x_{n-1})| \leq |x_n| + 1, \quad (7.122)$$

Since $|f(x_{n-1})| \leq |\delta|$, this implies that $-|af(x_{n-1})| \geq -|a|\,\delta$, and $|bx_{n-2}| - |af(x_{n-1})| \geq |bx_{n-2}| - |a|\,\delta$, finally, one has from (7.123) that:

$$|bx_{n-2}| - (|a|\,\delta + 1) \leq |x_n|, \quad (7.123)$$

Then, by induction as in the previous section, one has:

$$|x_n| \geq \begin{cases} \left(\frac{-(|a|\delta+1)}{|b|-1} + |x_1| \right) |b|^{\frac{n-1}{2}} + \frac{|a|\delta+1}{|b|-1}, \ if \ n \ is \ odd \\[3mm] \left(\frac{-(|a|\delta+1)}{|b|-1} + |x_0| \right) |b|^{\frac{n}{2}} + \frac{|a|\delta+1}{|b|-1}, \ if \ n \ is \ even \end{cases} \quad (7.124)$$

Thus, if $(|b| > 1$, and both $|x_0|, |x_1| > \frac{|a|\delta+1}{|b|-1})$, then, $\lim_{n \to +\infty} |x_n| = +\infty$.
(b) For $b = 1$, one has:

$$|x_n| \geq \begin{cases} (1 - |a|\,\delta) \left(\frac{n-1}{2} \right) + x_1, \ if \ n \ is \ odd \\[3mm] (1 - |a|\,\delta) \left(\frac{n}{2} \right) + x_0, \ if \ n \ is \ even \end{cases} \quad (7.125)$$

Hence, if $|a| < \frac{1}{\delta}$, then $\lim_{n \to +\infty} x_n = +\infty$.
For $b = -1$, one has from Theorem 60 the inequalities:

$$x_n \leq \begin{cases} - \left(\frac{n-1}{2} \right) + x_1 + \left| \sum_{p=1}^{p=\frac{n-1}{2}} a \, (-1)^{p-1} f(x_{n-(2p-1)}) \right|, \ if \ n \ is \ odd \\[5mm] - \left(\frac{n}{2} \right) + x_0 + \left| \sum_{p=1}^{p=\frac{n}{2}} a \, (-1)^{p-1} f(x_{n-(2p-1)}) \right|, if \ n \ is \ even \end{cases}$$

$$\quad (7.126)$$

Since $\left| a \ (-1)^{p-1} \ f(x_{n-(2p-1)}) \right| \leq |a| \, \delta$, *then one has:*

$$
x_n \leq
\begin{cases}
(|a| \, \delta - 1) \left(\frac{n-1}{2} \right) + x_1, & \text{if } n \text{ is odd} \\[2mm]
(|a| \, \delta - 1) \left(\frac{n}{2} \right) + x_0, & \text{if } n \text{ is even}
\end{cases}
\tag{7.127}
$$

Thus, if $|a| < \frac{1}{\delta}$, *then* $\lim_{n \longrightarrow +\infty} x_n = -\infty$. *There is no similar proof for the following subregions of* $\mathbb{R}^4$:

$$
\Omega_4 = \left\{ (a, b, x_0, x_1) \in \mathbb{R}^4 \ / \ |b| > 1, \text{ and both } |x_0|, |x_1| \leq \frac{|a| \, \delta + 1}{|b| - 1} \right\} \tag{7.128}
$$

and

$$
\Omega_5 = \left\{ (a, b, x_0, x_1) \in \mathbb{R}^4 \ / \ |b| = 1, \text{ and } |a| \geq \frac{1}{\delta} \right\}. \tag{7.129}
$$

Example 19 *If we choose the function* $f(x) = \sin(x)$, *then the map (7.99) with sine function is capable of generating multi-fold strange attractors, as shown in Fig. 7.3 [Zeraoulia and Sprott (2008(b))]; obtained via period doubling bifurcations routes to chaos as shown in Fig. 7.2. In this case, we have* $\delta = 1$, *and if we set* $a = 4$ *and* $|b| < 1$, *then the orbits of the map (7.99) with sine function are all bounded, i.e., there is bounded stable fixed point, periodic and chaotic orbits.*

The same phenomena are observed for the function $f(x) = \cos(x)$.

Example 20 *If* $f(x) = \operatorname{sgn}(x)$; *(sgn(.) is the standard signum function that gives the sign of its argument), then we show numerically that the map (7.99) with the signum function converges to a period-2 orbit for all* $|b| < 1$ *and* $a = 4$ *as shown in Fig. 7.4.*

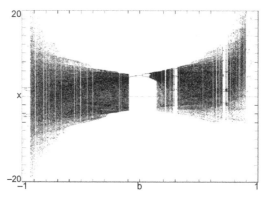

FIGURE 7.2: Bifurcation diagram of the map (7.99) with sine function obtained for $-1 < b < 1$ and $a = 4$.

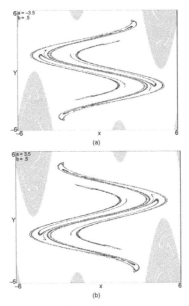

(a)

(b)

FIGURE 7.3: Different multi-fold chaotic attractors (with their bassins of attraction in white) obtained from map (7.99) with a sine function, observed for the initial condition $(x_0, y_0) = (0.01, 0.01)$ and (a) $a = -3.5, b = 0.5$ (b) $a = 3.5, b = 0.5$

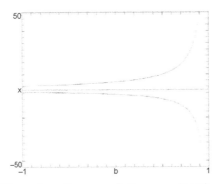

FIGURE 7.4: Bifurcation diagram of the map (7.99) with signum function obtained for $-1 < b < 1$ and $a = 4$.

7.7 Examples of fully bounded chaotic attractors

In this section, we consider a class of 2-D mappings that display fully bounded chaotic attractors for all bifurcation parameters. In the current literature, there are some cases where the boundedness of a map was proved rigorously in some regions of the bifurcation parameters space. In [Zeraoulia and Sprott 2008(b)], for example, it was proved that the two-dimensional discrete mapping given by (7.99) is bounded for all $|b| < 1$ and unbounded for all and $|b| > 1$. This partial boundedness of the above map is due to the presence of the terms by and x.

In this section, we consider the following simple family of 2-D discrete mappings:

$$\begin{pmatrix} x \\ y \end{pmatrix} \rightarrow \begin{pmatrix} a_0 + a_1 \sin x + a_2 \cos y \\ a_3 + a_4 \cos x + a_5 \sin y \end{pmatrix} \tag{7.130}$$

$$\begin{pmatrix} x \\ y \end{pmatrix} \rightarrow \begin{pmatrix} a_0 + a_1 \sin y + a_2 \cos x \\ a_3 + a_4 \cos y + a_5 \sin x \end{pmatrix} \tag{7.131}$$

$$\begin{pmatrix} x \\ y \end{pmatrix} \rightarrow \begin{pmatrix} a_0 + a_1 \cos x + a_2 \sin y \\ a_3 + a_4 \sin x + a_5 \cos y \end{pmatrix} \tag{7.132}$$

$$\begin{pmatrix} x \\ y \end{pmatrix} \rightarrow \begin{pmatrix} a_0 + a_1 \cos y + a_2 \sin x \\ a_3 + a_4 \sin y + a_5 \cos x \end{pmatrix} \tag{7.133}$$

These maps are studied in [Zeraoulia (2011(f))]. It is clear that the systems (7.131)-(7.134) are not symmetric in general, but they are bounded for all bifurcation parameters values. Generally, this main property avoids some problems related to the existence of unbounded solutions. For each map, there are 10 terms (six parameters and four functions) and the map is simple if it displays chaotic attractor with the minimum number of terms. In Fig. 7.5, we show a fully bounded chaotic attractor of the map:

$$\begin{pmatrix} x \\ y \end{pmatrix} \rightarrow \begin{pmatrix} -2 - 4 \sin x + 2 \cos y \\ -3 \cos x - \sin y \end{pmatrix} \tag{7.134}$$

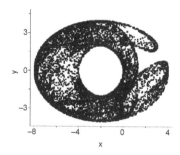

FIGURE 7.5: A chaotic attractor obtained from system (7.135).

From our numerical experiences, we observe that these bounded chaotic attractors persist for small values of their bifurcation parameters and it seems that their orientation mainly depends on the sign of these parameters. The aperiodicity of these attractors can be seen from the calculation of the power spectrum of the time series (one can choose the x-component which confirms that the spectrum is broadband).

Chapter 8

Some Forms of Globally Asymptotically Stable Attractors

8.1   Introduction

This chapter is devoted to presenting some definitions and relevant results of some forms of globally asymptotically stable discrete time mappings and continuous time systems. In Sec. 8.2, we present the important definitions and results of the direct Lyapunov stability for ordinary differential equations. The necessary conditions for the exponential stability of non-linear time-varying are presented in Sec. 8.3. In Sec. 8.4, we present and prove the Lasalle's invariance principle, since the two first sections of this chapter are mainly dependant on it. In Sec. 8.5, we present an overview of the extension of the direct Lyapunov method, presented in Sec. 8.2, to systems of equations of perturbed motion with fractional-like derivatives.

In Sec. 8.6, we eliminate any bounded behavior (chaotic or not) in some forms of n-D discrete mapping and drive them globally asymptotically to their fixed points. The method of analysis is based on the explicit construction of a non-negative semi-definite Lyapunov function that guaranties the stabilization of such a system. In Sec. 8.7, we construct a family of superstable n-D mappings, i.e., the superstability of a dynamical motion is defined with existence of a minus infinity Lyapunov exponent, meaning that this motion is attractive. In the same direction, we present in Sec. 8.8 a construction of globally superstable 1-D quadratic mappings. In Sec. 8.9, we present a construction of globally superstable 3-D quadratic mappings. In Sec. 8.10, we present and discuss some relevant results regarding the hyperbolic theory of dynamical systems. In particular, the consequences of the uniform hyperbolicity are reported in Sec. 8.11. A classification of singular-hyperbolic attracting sets is given in Sec. 8.11.1. The basic idea is that these systems accumulate their singularities in a pathological way, explained using matrices with entries in $\{0, 1\}$, or have no Lorenz-like singularities. Generally, since proving hyperbolicity is a hard problem that has many factors involved in its definition, it is useful to use the notion of *structural stability* introduced in Sec. 8.12 since hyperbolicity implies structural stability. In Sec. 8.12.2, we give examples of some forms of structurally stable of 3-D quadratic mappings.

Now, in Sec. 8.13, a construction of globally asymptotically stable partial differential systems is presented in some detail. Here, the considered system is the *spatially SVIR model of infectious disease epidemics* with immigration of individuals. In the same direction, a similar construction of a globally stable system of delayed differential equations is given in Sec. 8.14. The model under consideration is the *delay differential equation of hepatitis B virus infection* with immune response. At the end of this chapter, we describe the stabilization by the Jurdjevic-Quinn method in Sec. 8.15. This method of stabilization is used in several domains for improving stability performances.

8.2 Direct Lyapunov stability for ordinary differential equations

Let us consider the general ordinary differential equations of the form:

$$x'(t) = f(x(t)) \tag{8.1}$$

where $x : \mathbb{R} \to \mathbb{R}^n$ and $f : \mathbb{R} \times \mathbb{R}^n \to \mathbb{R}^n$ are functions and t represents time. Here $x'(t)$ are the derivatives of each of the component functions, so if $x(t) = (x_1(t), ..., x_n(t))$ where $x_i(t)$ are real-valued functions on $\mathbb{R}$ then

$$x'(t) = (x_1'(t), ..., x_n'(t)).$$

Assuming that system (8.1) has a continuous unique solution with the initial condition $x(t_0) = x_0$ where t_0 is a constant and x_0 is a constant n-vector. On one hand, a solution will tend to a constant as t goes to infinity or to a continuous curve through n-space (called phase space). If a constant function c satisfies the differential equation (8.1), then the derivative $x'(t) = f = 0$ for $x_0 = c$ and c is called an *equilibrium point*. Periodic solutions satisfy $x(t) = x(t+n\tau)$ for some τ and any integer n. On the other hand, complicated solutions, which may or may not be stable, can also exist. The Lyapunov stability of these solutions is a qualitative notion, meaning that if one perturbs the initial condition slightly as $x(t_0) = x_0 + \varepsilon$, then the new solution will be close to the original system.

Definition 53 *A solution $x(t)$ is stable if there exists $\delta > 0$ such that if $\|x_0 - x_0\| < \delta$ then $\|x(t) - x(t)\| < \delta$, for all $t \geq t_0$, where $x(t)$ is the solution to*

$$x'(t) = f(t, x(t))$$

with $x(t_0) = x_0$ and $x(t_0) = x_0$.

In some cases, if we change the initial condition slightly, then the solution still converges to the exact solution. This notion is called *asymptotic stability*, which is more desirable in practice:

Definition 54 *A solution $x(t)$ is asymptotically stable if it is stable and there is a $\rho > 0$ such that if $\|x_0 - \bar{x}_0\| < \rho$ then*

$$\lim_{t \to \infty} \|x(t) - \bar{x}(t)\| = 0.$$

The Lyapunov's First Method works by characterizing solutions of system (8.1). Indeed, let us consider the linear differential equations

$$x'(t) = A(t)x(t) \tag{8.2}$$

where $A : \mathbb{R}^n \to \mathbb{R}^n$ is linear and $x : \mathbb{R} \to \mathbb{R}^n$. Then we have the following result:

Theorem 64 *If $\Phi'(t) = A(t)\Phi(t)$ with Φ an invertible $n \times n$ matrix, then all solutions of (8.2) are linear combinations of the columns of Φ and, in particular, we can choose some $\Phi(t)$ such that $\Phi(t_0) = I$, where I is the identity matrix, so that $x(t) = \Phi(t)x_0$ is a solution to (8.2).*

Here Φ is called the *fundamental matrix* of the differential equation (8.2) that depends only on the matrix function $A(t)$. If A is an $n \times n$ real matrix then the operator norm of A is defined by

$$\|A\| = \sum_{i,j=1}^{n} |a_{ij}|$$

We have $\|Ax\| \le \|A\|\|x\|$ where $\|A\|$ is the operator norm and the other two norms are the usual Euclidean vector norm in $\mathbb{R}^n$. Thus, we have the following result:

Theorem 65 *All solutions of (8.2) are stable if and only if they are bounded.*

Proof 26 *We know that all solutions of system (8.2) are of the form $\Phi(t)x_0$ where Φ is the fundamental matrix, such that $\Phi(t_0) = I$. Let $x(t)$ and $\bar{x}(t)$ be solutions satisfying $x(t_0) = x_0$ and $\bar{x}(t_0) = \bar{x}_0$ respectively. Assuming that all solutions to (8.2) are bounded, we can find an $M > 0$ such that $\|\Phi(t)\| < M$ for all $t > 0$. Given some $\varepsilon > 0$, if $\|x_0 - \bar{x}_0\| < \frac{\varepsilon}{M}$ then*

$$
\begin{aligned}
\|x(t) - \bar{x}(t)\| &= \|\Phi(t)x_0 - \Phi(t)\bar{x}_0\| \\
&\le \|\Phi(t)\| \, \|x_0 - \bar{x}_0\| \\
&< \varepsilon.
\end{aligned}
$$

Conversely, assuming that all solutions are stable, the particular solution $x(t) \equiv 0$ is stable, so for every $\varepsilon > 0$ there exists $\delta > 0$ such that if $\|\bar{x}_0\| < \delta$ then $\|\bar{x}(t)\| = \|\Phi(t)\bar{x}_0\| < \varepsilon$. In particular, let $\bar{x}_0^{(i)} = \frac{\delta}{2}e_i$ where e_i is the ith standard basis vector of $\mathbb{R}^n$. Then $\|\Phi(t)\bar{x}_0^{(i)}\| = \|\phi_i(t)\|\frac{\delta}{2} < \varepsilon$, where ϕ_i is the ith column of Φ. We can sum these over all i and get $\|\Phi(t)\| \le \sum_{i=1}^{n} \|\phi_i(t)\| \le \frac{2n\varepsilon}{\delta} = k$. Thus, $\|x(t)\| = \|\Phi(t)x_0\| \le k\|x_0\|$ and $x(t)$ is bounded.

At this stage, the Lyapunov's first method consists of showing that solutions of linear systems (8.2) are bounded.

Consider the constant linear equation:

$$x'(t) = A\,x(t) \qquad (8.3)$$

where A is a constant matrix. If all eigenvalues of A have negative real parts, then every solution to (8.3) is asymptotically stable. If we consider the following non-linear equation

$$x'(t) = A\,x(t) + f(t, x(t)) \qquad (8.4)$$

Then, if f satisfies: (1) $f(t,x)$ is continuous for $\|x\| < a$, $0 \le t < \infty$ and (2) $\lim_{\|x\| \to 0} \frac{\|f(x,t)\|}{\|x\|} = 0$, then the solution $x(t) \equiv 0$ is asymptotically stable.

Now, the Lyapunov's second or direct method does not require a characterization of the solutions in order to determine stability. This method uses a supplementary function called a *Lyapunov function* in order to determine properties of the asymptotic behavior of solutions.

We will need the following definitions:

Definition 55 *A function $V(t,x)$ is positive (negative) definite if there exists a real-valued function $V(r)$ such that:*
(1) $V(r)$ is strictly increasing on $0 \le r \le a$ and $V(0) = 0$ and
(2) $V(t,x) \ge V(\|x\|)$ $[V(t,x) \le -V(\|x\|)]$ for all $(t,x) \in \{(t,x) : t_0 \le t < \infty, \|x\| \le b < a\}$.

Definition 56 *The time derivative of V is given by*

$$V'(t, x(t)) = \sum_{i=1}^{n} \frac{\partial V}{\partial x_i} f_i(t, x) + \frac{\partial V}{\partial t}$$

Definition 57 *A real function $V(t,x)$ is said to admit an infinitesimal upper bound if there exists $h > 0$ and a continuous, real-valued, strictly increasing function with $\psi(0) = 0$ such that*

$$|V(t,x)| \le \psi(\|x\|) \text{ for } \|x\| < h \text{ and } t \ge t_0.$$

If one can find a function V verifying the above conditions, then one can find stable solutions to the differential equation (8.1). However, there is no general way to construct V from the differential equation (8.1).

Theorem 66 *If a continuous real-valued function $V(t,x)$ exists such that:*
(1) $V(t,x)$ is positive definite and (2) $V'(t,x)$ is non-positive
then $x(t) \equiv 0$ is a stable solution of (8.1).

Proof 27 *We know that $V(t,x)$ is positive definite, hence there exists a strictly increasing function $\varphi(r)$ such that $0 < \varphi(\|x\|) \le V(t,x)$ for $0 < \|x\| <$*

b and *t* > t_0. Given $\varepsilon > 0$ let $m_\varepsilon = \min_{\|x\|=\varepsilon} \varphi(\|x\|) = \varphi(\varepsilon)$. We have $m_\varepsilon > 0$. Since V is continuous and $V(t,0) = 0$ for all *t* (by positive definiteness), we can choose a $\delta > 0$ such that $V(t_0, x_0) < m_\varepsilon$ if $\|x_0\| < \delta$. Now $V'(t, x(t)) \leq 0$ for $t_1 \geq t_0$ and $\|x_0\| < \delta$ implies $V(t_1, x(t_1)) \leq V(t_0, x(t_0)) = V(t_0, x_0) < m_\varepsilon$. Now, suppose for some $t_1 \geq t_0$ that $\|x(t_1)\| \geq \varepsilon$ whenever $\|x_0\| < \delta$. Then $V(t_1, x(t_1)) \geq \varphi(\|x(t_1)\|) \geq \varphi(\varepsilon) = m_\varepsilon$ which is a contradiction, so for every $\varepsilon > 0$ there exists $\delta > 0$ such that $\|x(t)\| < \varepsilon$ if $\|x_0\| < \delta$ for $t \geq t_0$, i.e., $x(t) \equiv 0$ is stable.

The function V is called a *Lyapunov function*, which is the energy in physical systems.

Theorem 67 *If a continuous function $V(t, x)$ exists satisfying:*
(1) $V(t, x)$ is positive definite,
(2) $V(t, x)$ admits an infinitesimal upper bound
(3) $V'(t, x)$ is negative definite
Then the solution $x(t) \equiv 0$ of (8.1) is asymptotically stable.

Proof 28 *By Theorem 66, the solution is stable. Suppose that $x(t)$ is not asymptotically stable, so for every $\varepsilon > 0$ there exists $\delta > 0$ and $\lambda > 0$ such that any non-zero solution $x(t)$ where $x(t_0) = x_0$ satisfies $\lambda \leq \|x(t)\| < \varepsilon$ if $t \geq t_0$ and $\|x_0\| < \delta$. Since $V'(t, x)$ is negative definite, there is a strictly increasing function $\varphi(r)$ vanishing at the origin, such that $V'(t, x) \leq -\varphi(\|x\|)$. We have $\|x(t)\| \geq \lambda > 0$ for $t \geq t_0$ so there is a $d > 0$ such that $V'(t, x(t)) \leq -d$. This implies that*

$$V(t, x(t)) = V(t_0, x_0) + \int_{t_0}^{t} V'(s, x(s))ds$$

$$\leq V(t_0, x_0) - (t - t_0)d < 0$$

for large t, contradicting the assumption that $V(t, x)$ is positive definite. Hence, no such λ exists and since $V(t, x(t))$ is positive definite and decreasing with respect to t, we have $\lim_{t \to \infty} V(t, x(t)) = 0$ which implies that $\lim_{t \to \infty} \|x(t)\| = 0$ since $V(t, x(t)) \leq \psi(\|x(t)\|)$ where $\psi(r)$ is strictly increasing and vanishes at the origin (by the second hypothesis). Thus, as $V(t, x(t)) \to 0$, $\psi(\|x(t)\|) \to 0$, and $\|x(t)\| \to 0$ (because ψ is strictly increasing and vanishes at 0). Thus, $x(t) \equiv 0$ is asymptotically stable.

Also, the instability in some cases can be obtained by using the direct method:

Theorem 68 *If a continuous real-valued function $V(t, x)$ exists on some set $S = \{(t, x) : t \geq t_1 \text{ and } \|x\| < a\}$ satisfying:*
(1) $V(t, x)$ admits an infinitesimal upper bound.
(2) $V'(t, x)$ is positive definite on S.

(3) There exists a $T > t_1$ such that if $t_0 \geq T$ and $h > 0$ then there exists $c \in \mathbb{R}^n$ such that $\|c\| < h$ and $V(t_0, c) > 0$.

Then the solution $x(t) \equiv 0$ is not stable.

Proof 29 *Suppose $x(t)$ is a solution of (8.1) not identically zero. Then, by uniqueness, $\|x(t)\| \neq 0$ for all $t \geq t_1$. If $x(t)$ is defined at t_0 then for every $t \geq t_0$ for which $x(t)$ is defined we have*

$$V(t, x(t)) - V(t_0, x(t_0)) = \int_{t_0}^{t} V'(s, x(s)) ds > 0$$

since $V'(t, x(t))$ is positive definite. Let $t_0 \geq T$ and $\varepsilon > 0$. By the third assumption, there exists a c such that $\|c\| < \min(a, \varepsilon)$ and $V(t_0, c) > 0$. Since V admits an infinitesimal upper bound and is continuous, there is a $\lambda \in (0, a)$ such that $\|x\| < \lambda$ for $t \geq t_1$. Therefore, $|V(t, x)| < V(t_0, c)$. Let $x(t)$ satisfy the initial condition $x(t_0) = c$. Then $V(t, x(t) > V(t_0, x(t_0)) > 0$ (from the positive definiteness argument), so $\|x(t)\| \geq \lambda$. The second hypothesis implies the existence of a strictly increasing function $\varphi(r)$ such that $V'(t, x) \geq \varphi(\|x\|)$. Let

$$\mu = \min_{\|x\| \in [\lambda, a]} \varphi(\|x\|) = \varphi(\lambda).$$

then

$$V'(t, x(t)) \geq \varphi(\|x(t)\|) \geq \mu.$$

Using the integral inequality, we get

$$V(t, x(t)) \geq V(t_0, x(t_0)) + (t - t_0)\mu.$$

Thus, we can make $V(t, x(t))$ arbitrarily large. Since $V(t, x)$ admits an infinitesimal upper bound, there exists a strictly increasing function $\psi(r)$ such that $|V(t, x(t))| < \psi(\|x\|)$ and $\psi(0) = 0$. So, since $V(t, x(t))$ can be arbitrarily large, $\psi(\|x(t)\|)$ can also be arbitrarily large, thus, $\|x(t)\|$ becomes arbitrarily large and the solution is not stable.

Notice that asymptotic stability in the Lyapunov sense does not always work well for periodic systems. Also, for certain classes of Hamiltonian systems the Lyapunov analysis is not valid. This is the result of the fact that the points will not get closer together, i.e., if $x(t)$ is a solution to the autonomous system (8.1) then $x(t + h)$ is also a solution for any real number h. Indeed, Let $\bar{x}(t) = x(t + h)$ and $s = t + h$. Then $\bar{x}'(t) = \frac{dx}{ds}\frac{ds}{dt} = x'(s) = f(x(s)) = f(x(t + h)) = f(\bar{x}(t))$ so $\bar{x}(t) = x(t + h)$ is a solution to (8.1), that is, the translations of solutions are also solutions. In fact, non-trivial periodic solutions $x(t) \neq x(t + \varepsilon)$ for some small $\varepsilon > 0$ cannot be asymptotically stable.

Theorem 69 *If $x(t)$ is a non-trivial periodic solution of an autonomous system (8.1) then $x(t)$ is not asymptotically stable.*

Proof 30 *Suppose $x(t)$ is a non-trivial periodic solution. Thus, there exists t_0 such that $f(x(t_0)) \neq 0$ because if not, $x'(t) = 0$ for all t and the solution is constant (trivially periodic). Assuming that $x(t)$ is asymptotically stable. Hence, for a given $\varepsilon > 0$ there exists $\hat{\delta} > 0$ such that $\|x(t) - \bar{x}(t)\| < \varepsilon$ if $\|x_0 - \bar{x}_0\| < \hat{\delta}$ where $x(t)$ satisfies $x(t_0) = x_0$ and $\bar{x}(t)$ is a solution of (8.1) satisfying the initial condition $\bar{x}(t_0) = \bar{x}_0$. By the asymptotic stability assumption we can find a $\rho > 0$ such that $\lim_{t \to \infty} \|x(t) - \bar{x}(t)\| = 0$ if $\|x_0 - \bar{x}_0\| < \rho$. Let $\delta = \min(\hat{\delta}, \rho)$. By the continuity of $x(t)$ we can fix some small $\bar{t}$ so that $\|x(t_0) - x(t_0 + \bar{t})\| < \delta$. Since $x(t + \bar{t})$ is a solution to the differential equation, then $\lim_{t \to \infty} \|x(t) - x(t + \bar{t})\| = 0$. We know that $x'(t_0) \neq 0$ so we can find $r > 0$ so that $\|x(t_0) - x(t_0 + \bar{t})\| > r$. Also $x(t)$ is periodic, so assume it has period τ. Then $\|x(t_0 + n\tau) - x(t_0 + \bar{t} + n\tau)\| > r > 0$ for all n. Hence, the limit cannot go to zero, which contradicts the assumption of asymptotic stability. Thus, $x(t)$ cannot be asymptotically stable.*

The asymptotic stability of time-independent Hamiltonian systems has the same problem as before. Let us consider the $2n$ dimensional system given by:

$$\begin{cases} x'(t) = H_y(x(t), y(t)) \\ \\ y'(t) = -H_x(x(t), y(t)) \end{cases} \qquad (8.5)$$

where $x(t)$ and $y(t)$ are both functions from $\mathbb{R}$ to $\mathbb{R}^n$ and $H(x,y)$ is a real valued function with continuous partial derivatives.

Theorem 70 *If $(x(t), y(t))$ is a solution to a the time-independent Hamiltonian system (8.5) then it is not asymptotically stable.*

Proof 31 *Suppose $(x(t), y(t))$ is an asymptotically stable solution of (8.5). Then, we can find t_0 and $\delta > 0$ such that if $(\bar{x}(t), \bar{y}(t))$ is a solution and if $\|(x(t_0), y(t_0)) - (\bar{x}(t_0), \bar{y}(t_0))\| < \delta$ then $\lim_{t \to \infty} \|(x(t), y(t)) - (\bar{x}(t), \bar{y}(t))\| = 0$. But we have*

$$\frac{dH(x,y)}{dt} = \frac{\partial H}{\partial x}\frac{dx}{dt} + \frac{\partial H}{\partial y}\frac{dy}{dt} = H_x(H_y) + H_y(-H_x) = 0$$

so $H(x,y)$ is constant on any solution $(x(t), y(t))$. By the continuity of H, it must be constant on some δ-neighborhood N of $(x(t_0), y(t_0))$, so every point in N is an equilibrium point. Hence, the solution $(x(t), y(t))$ cannot be asymptotically stable because for initial condition $(\bar{x}(t_0), \bar{y}(t_0)) \in N$, the solution $(\bar{x}(t), \bar{y}(t))$ will not converge to $(x(t), y(t))$.

Generally, there is no method for finding a candidate Lyapunov function to establish asymptotic stability, but there are some facts on the subject:

- For asymptotically stable linear systems, one always can find a quadratic Lyapunov function.

- In physical models, the total energy of the system is a candidate Lyapunov function.

- For systems with polynomial right hand sides, polynomial functions are candidate Lyapunov functions.

- In many cases, one can start with a desired negative definite V', and then find the corresponding positive definite function V.

The last fact is demonstrated in the following result:

Theorem 71 *Suppose $A \in \mathbb{R}^{n \times n}$. The origin is an asymptotically stable equilibrium point of the system $x' = Ax$ if and only if for every positive definite $Q = Q^T \in \mathbb{R}^{n \times n}$ there exists a positive definite $P = P^T \in \mathbb{R}^{n \times n}$ such that $PA + A^T P = -Q$. Moreover, such P is unique.*

Proof 32 *Sufficiency is immediate. Conversely suppose that all eigenvalues of A have strictly negative real parts and $Q = Q^T$ is positive definite. Define*

$$P = \int_0^\infty e^{tA^T} Q e^{tA} \, dt \tag{8.6}$$

This integral (8.6) converges absolutely because the norm of the integrand is bounded by a function of the form $C t^N e^{-\alpha t}$ for some positive constants C, N, α. $P = P^T$ is symmetric since the integrand is symmetric at every $t \geq 0$. In order to prove that P is positive definite, suppose $x \in \mathbb{R}^n$ is such that $x^T P x = 0$. Then

$$0 = \int_0^\infty \left(e^{tA} x \right)^T Q \left(e^{tA} x \right) \, dt$$

Since Q is positive definite this implies that for all $t \geq 0$: $e^{tA} x = 0$. Since for every t, e^{tA} is invertible, this implies $x = 0$ proving that P is positive definite. In order to prove that the function $V(x) = x^T P x$ has the desired derivative along solutions, we have first

$$V'(x) = x'^T P x + x^T P x' = x^T (PA + A^T P) x$$

Now calculate

$$
\begin{aligned}
PA + A^T P &= \int_0^\infty \left(e^{tA^T} Q e^{tA} A + A^T e^{tA^T} Q e^{tA} \right) dt \\
&= \int_0^\infty \frac{d}{dt} \left(e^{tA^T} Q e^{tA} \right) dt \\
&= e^{tA^T} Q e^{tA} \Big|_0^\infty = -Q
\end{aligned}
$$

To show uniqueness, suppose $P, \tilde{P} \in \mathbb{R}^{n \times n}$ are both positive definite, symmetric and satisfy

$$PA + A^T P = \tilde{P} A + A^T \tilde{P} = -Q$$

Then $(P - \tilde{P})A + A^T(P - \tilde{P}) = 0$, *and hence for every* $t \geq 0$ *we have:*

$$e^{tA^T}\left((P - \tilde{P})A + A^T(P - \tilde{P})\right)e^{tA} = \frac{d}{dt}\left(e^{tA^T}(P - \tilde{P})e^{tA}\right) = 0$$

Since the derivative vanishes identically, the product $e^{tA^T}(P - \tilde{P})e^{tA}$ *must be a constant function of* t, *and in particular, must at every time be equal to its value at* $t = 0$. *Therefore for all* $t \geq 0$, *we have:*

$$e^{tA^T}(P - \tilde{P})e^{tA} = e^{0 \cdot A^T}(P - \tilde{P})e^{0 \cdot A} = P - \tilde{P}$$

However, the origin is asymptotically stable, as t *goes to infinity, then* e^{tA} *goes to zero, and therefore*

$$(P - \tilde{P}) = \lim_{t \to \infty} e^{tA^T}(P - \tilde{P})e^{tA} = 0 \cdot (P - \tilde{P}) \cdot 0 = 0.$$

The Lyapunov methods can be adapted for instability. Indeed, we have the following result:

Theorem 72 (adapted from Chetaev 1934) *Suppose* $E \subseteq \mathbb{R}^n$ *is open,* $0 \in E$, $f : E \mapsto \mathbb{R}^n$ *and* $V : E \mapsto \mathbb{R}$ *are continuously differentiable,* $f(0) = 0$, $V(0) = 0$. *Suppose that* U *is an open neighborhood of* 0 *with compact closure* $F = \overline{U} \subseteq E$ *such that the restriction of* $V' = (DV)f$ *to* $F \cap V^{-1}(0, \infty)$ *is strictly positive. If for every open neighborhood* $W \subseteq \mathbb{R}^n$ *of* 0 *the set* $W \cap V^{-1}(0, \infty)$ *is non-empty, then the origin is an unstable equilibrium of* $x' = f(x)$.

Proof 33 *Assume the above hypotheses. Let* W *be an open neighborhood of* $0 \in \mathbb{R}^n$. *There exists* $z \in W \cap V^{-1}(0, \infty)$. *Define* $a = V(z) > 0$. *The set* $K = F \cap V^{-1}[a, \infty)$ *is compact, and it is non-empty since* $z \in K$. *By using the continuity of* V *and* V' *there exists* $m = \min_{x \in K} V'(x)$ *and* $M = \max_{x \in K} V(x)$. *Since* $K = F \cap V^{-1}[a, \infty) \subseteq F \cap V^{-1}(0, \infty)$ *it follows that* $m > 0$. *Let* $I \subseteq [0, \infty)$ *be the maximal interval of existence for* $\phi : I \mapsto E$ *such that* $\phi(0) = z$, *and* $\phi' = f \circ \phi$. *Define the set*

$$T = \{t \in I : \text{for all } s \in [0, t], \ \phi(s) \in K\}$$

For all $t \in T$, *we have:*

$$\frac{d}{dt}(V \circ \phi)(t) = (V' \circ \phi)(t) \geq m$$

Hence, for all $t \in T$, *we have:*

$$V(\phi(t)) = V(\phi(0)) + \int_0^t V'(\phi(s)) \, ds \geq a + mt$$

Since V *is bounded above by* M, *then* T *is also bounded above. Let* t_0 *be the least upper bound of* T. *Since* $K \subseteq E$, $I \supseteq [0, t_0]$ *and* $\phi(t_0)$ *is well-defined,* *then t_0 lies in the interior of I. Since for all $t \in [0, t_0)$, we have $\phi(t) \in K$, and for every $\delta > 0$ there exists $t \in I$, $t_0 < t < t_0 + \delta$ such that $\phi(t) \notin K$, it follows that $\phi(t_0)$ lies on the boundary of the set K. The boundary of K is contained in the union $\partial K \subseteq V^{-1}(a) \cup (F \setminus U)$. Since*

$$V(\phi(t_0)) \geq a + mt_0 > a$$

it follows that $\phi(t_0) \notin V^{-1}(a)$. Thus, we get $\phi(t_0) \in (F \setminus U)$, and, in particular, $\phi(t_0) \notin U$. Since W is arbitrary, this shows that the origin is an unstable equilibrium of f.

8.3 Exponential stability of non-linear time-varying

In [Linh & Phat (2001)] sufficient conditions for the exponential stability of a class of non-linear time-varying systems are given by introducing a class of Lyapunov-like functions $V(t, x)$ that can be extended to the systems with non-smooth Lyapunov functions. This result is applicable to a wide range of stabilization problems of non-linear time-varying control systems. In order to present this method, we need the following definitions: Let $\mathbb{R}^n$ be the n-dimensional Euclidean vector space; $\mathbb{R}^+$ is the set of all non-negative real numbers; $\|x\|$ is the Euclidean norm of a vector $x \in \mathbb{R}^n$. let us consider the non-linear system described by the time-varying differential equations:

$$\begin{cases} x'(t) = f(t, x(t)), & t \geq 0, \\ x(t_0) = x_0, & t_0 \geq 0 \end{cases} \tag{8.7}$$

where $x(t) \in \mathbb{R}^n$, $f(t, x) : \mathbb{R}^+ \times \mathbb{R}^n \to \mathbb{R}^n$ is a given non-linear function satisfying $f(t, 0) = 0$ for all $t \in \mathbb{R}^+$. The conditions of the existence of solutions to (8.7) are assumed.

Definition 58 *The zero solution of system (8.7) is exponentially stable if any solution $x(t, x_0)$ of (8.7) satisfies*

$$\|x(t, x_0)\| \leq \beta(\|x_0\|, t_0)e^{-\delta(t - t_0)}, \quad \forall t \geq t_0,$$

where $\beta(h, t) : \mathbb{R}^+ \times \mathbb{R}^+ \to \mathbb{R}^+$ is a non-negative function increasing in $h \in \mathbb{R}^+$, and δ is a positive constant.

If $\beta(.)$ does not depend on t_0, the zero solution is called uniformly exponentially stable. Let us consider a non-linear time-varying control system

$$x'(t) = f(t, x(t), u(t)), \quad t \geq 0, \tag{8.8}$$

where $x \in \mathbb{R}^n$, $u \in \mathbb{R}^m$, $f(t, x, u) : \mathbb{R}^+ \times \mathbb{R}^n \times \mathbb{R}^m \to \mathbb{R}^n$.

Definition 59 *Control system (8.8) is exponentially stabilizable by the feed-back control* $u(t) = h(x(t))$, *where* $h(x) : \mathbb{R}^n \to \mathbb{R}^m$, *if the closed-loop system:*

$$x'(t) = f(t, x(t), h(x(t)))$$

is exponentially stable.

Let $D \subset \mathbb{R}^n$ be an open set containing the origin, and let $V(t,x) : \mathbb{R}^+ \times D \to R$ be a given function. Then, we define $W = \mathbb{R}^+ \times D$ and

$$D_f^+ V(t, x) = \limsup_{h \to 0^+} \frac{V(t+h, x+hf) - V(t,x)}{h}$$

where $f(.)$ is the right-hand side function of (8.7). $D_f^+ V$ is called the upper Dini derivative of $V(.)$ along the trajectory of (8.7). Let $x(t)$ be a solution of (8.7) and denote by $d^+ V(t,x)$ the upper right-hand derivative of $V(t, x(t))$, i.e.,

$$d^+ V(t, x(t)) = \limsup_{h \to 0^+} \frac{V(t+h, x(t+h)) - V(t, x(t))}{h}$$

Definition 60 *A function* $V(t,x) : \mathbb{R}^+ \times \mathbb{R}^n \to R$ *is Lipschitzian in* x *(uni-formly in* $t \in \mathbb{R}^+$*) if there is a number* $L > 0$ *such that for all* $t \in \mathbb{R}^+$,

$$|V(t, x_1) - V(t, x_2)| \leq L \|x_1 - x_2\|, \quad \forall (x_1, x_2) \in \mathbb{R}^n \times \mathbb{R}^n.$$

Assuming that $V(t,x)$ is continuous in t and Lipschitzian in x (uniformly in t) with the Lipschitz constant $L > 0$. In which case, $d^+ V$ and $D_f^+ V$ are related as follows:

$$V(t+h, x(t+h)) - V(t, x(t))$$

$$= V(t+h, x(t+h)) - V(t+h, x + hf(t,x))$$
$$+ V(t+h, x + hf(t,x)) - V(t, x(t)).$$

Also

$$\limsup_{h \to 0^+} \frac{V(t+h, x + hf(t,x))) - V(t, x(t))}{h}$$

$$\leq \limsup_{h \to 0^+} \frac{V(t+h, x(t+h)) - V(t, x(t))}{h}$$

$$+ L \{ \lim_{h \to 0^+} \frac{\|x(t+h) - x(t)\|}{h} - f(t, x(t)) \},$$

which gives

$$d^+ V(t, x) \leq \limsup_{h \to 0^+} \frac{V(t+h, x + hf)) - V(t, x)}{h} = D_f^+ V(t, x). \qquad (8.9)$$

It was shown in [McShane (1947)] that if $D_f^+ V(t,x) \leq 0$ and consequently, $d^+ V(t,x) \leq 0$, the function $V(t,x(t))$ is a non-increasing function of t, then $V(t,x)$ is non-increasing along a solution of (8.7).

Also, we need the following comparison theorem [Lakshmikantham, *et al.*, (1989), Yoshizawa (1966)]:

Theorem 73 *Consider a scalar differential equation*

$$u'(t) = g(t,u), \quad t \geq 0, \tag{8.10}$$

where $g(t,u)$ is continuous in (t,u). Let $u(t)$ be the maximal solution of (8.10) with $u(t_0) = u_0$. If a continuous function $v(t)$ with $v(t_0) = u_0$ satisfies

$$d^+ v(t) \leq g(t,u(t)), \quad \forall t \geq t_0,$$

then

$$v(t) - v(t_0) \leq \int_{t_0}^{t} g(s,u(s))ds, \quad \forall t \geq t_0.$$

Let us set

$$D_f V(t,x) = \frac{dV(t,x)}{dt} + \frac{dV(t,x)}{dx} f(t,x).$$

Definition 61 *(a) A function $V(t,x) : W \to R$ is called a Lyapunov-like function for (8.7) if $V(t,x)$ is continuously differentiable in $t \in \mathbb{R}^+$ and in $x \in D$, and there exist positive numbers $\lambda_1, \lambda_2, \lambda_3, K, p, q, r, \delta$ such that*

$$\lambda_1 \|x\|^p \leq V(t,x) \leq \lambda_2 \|x\|^q, \quad \forall (t,x) \in W, \tag{8.11}$$

$$D_f V(t,x) \leq -\lambda_3 \|x\|^r + K e^{-\delta t}, \quad \forall t \geq 0, x \in D \setminus \{0\} \tag{8.12}$$

(b) A function $V(t,x) : W \to R$ is called a generalized Lyapunov-like function for (8.7) if $V(t,x)$ is continuous in $t \in \mathbb{R}^+$ and Lipschitzian in $x \in D$ (uniformly in t) and there exist positive functions $\lambda_1(t), \lambda_2(t), \lambda_3(t)$, where $\lambda_1(t)$ is non-decreasing, and there exist positive numbers K, p, q, r, δ such that

$$D_f^+ V(t,x) \leq -\lambda_3(t)\|x\|^r + K e^{-\delta t}, \quad \forall t \geq 0, x) \in D \setminus \{0\} \tag{8.13}$$

$$D_f^+ V(t,x) \leq -\lambda_3(t)\|x\|^r + K e^{-\delta t}, \quad \forall t \geq 0, x) \in D \setminus \{0\} \tag{8.14}$$

The following result about the exponential stability of (8.7), with the existence of a uniform Lyapunov function was shown in [Sun, *et al.*, (1996)]:

Theorem 74 *Assume that (8.7) admits a Lyapunov-like function, where $p = q = r$. Then the system (8.7) is uniformly exponentially stable if*

$$\delta > \frac{\lambda_3}{\lambda_2}$$

The following two theorems about sufficient conditions for the exponential stability of (8.7) with a more general Lyapunov-like function are proved in [Linh & Phat (2001)]:

Theorem 75 *The system (8.7) is uniformly exponentially stable if it admits a Lyapunov-like function and the following two conditions hold for all $(t, x) \in W$:*

$$\delta > \frac{\lambda_3}{[\lambda_2]^{r/q}} \tag{8.15}$$

and there exist $\gamma > 0$ such that

$$V(t, x) - V(t, x)^{r/q} \leq \gamma e^{-\delta t} \tag{8.16}$$

Proof 34 *Consider any initial time $t_0 \geq 0$, and let $x(t)$ be any solution of (8.7) with $x(t_0) = x_0$. We set*

$$\begin{cases} Q(t, x) = V(t, x)e^{M(t-t_0)} \\ \\ M = \frac{\lambda_3}{[\lambda_2]^{r/q}} \end{cases}$$

Then we get

$$Q'(t, x(t)) = D_f V(t, x)e^{M(t-t_0)} + MV(t, x)e^{M(t-t_0)}$$

By (8.12), then for all $t \geq t_0, x \in D$, we have

$$Q'(t, x) \leq (-\lambda_3)\|x\|^r + Ke^{-\delta t})e^{M(t-t_0)} + MV(t, x)e^{M(t-t_0)}$$

By the condition (8.11) we have $\|x\|^q \geq \frac{V(t,x)}{\lambda_2}$, i.e.,

$$-\|x\|^r \leq -[\frac{V(t, x)}{\lambda_2}]^{r/q}$$

Thus, we have

$$Q'(t, x) \leq \{-V(t, x)^{r/q}\frac{\lambda_3}{[\lambda_2]^{r/q}} + Ke^{-\delta t}\}e^{M(t-t_0)} + MV(t, x)e^{M(t-t_0)}.$$

Since

$$\frac{\lambda_3}{[\lambda_2]^{r/q}} = M, \quad \forall t \geq 0,$$

we have

$$Q'(t, x) \leq M\{V(t, x) - V(t, x)^{r/q}\}e^{M(t-t_0)} + Ke^{(M-\delta)(t-t_0)}.$$

Using (8.16), we obtain

$$Q'(t, x) \leq (K + M\gamma)e^{(M-\delta)(t-t_0)}$$

Integrating both sides of the above inequality from t_0 to t, we obtain

$$Q(t, x(t)) - Q(t_0, x_0) \leq \int_{t_0}^{t} (K + M\gamma)e^{()M-\delta)(s-t_0)}ds$$

$$= (K + M\gamma)\frac{1}{M - \delta}\{e^{(M-\delta)(t-t_0)} - 1\}$$

Setting $\delta_1 = -(M - \delta)$, by (8.15) we have $\delta_1 > 0$ and

$$Q(t, x(t)) \leq Q(t_0, x_0) + \frac{K + M\gamma}{\delta_1} - \frac{K + M\gamma}{\delta_1}e^{(M-\delta)(t-t_0)}$$

$$\leq Q(t_0, x_0) + \frac{K + M\gamma}{\delta_1}$$

Since $Q(t_0, x_0) = V(t_0, x_0) \leq \lambda_2\|x_0\|^q$, we have

$$Q(t, x(t)) \leq \lambda_2\|x_0\|^q + \frac{K + M\gamma}{\delta_1}.$$

Define

$$\lambda_2\|x_0\|^q + \frac{K + M\gamma}{\delta_1} = \beta(\|x_0\|) > 0$$

we have

$$Q(t, x(t)) \leq \beta(\|x_0\|), \quad \forall t \geq t_0. \tag{8.17}$$

We get

$$\begin{cases} \lambda_1\|x(t)\|^p \leq V(t, x(t)) \\ \\ \|x(t)\| \leq \left(\frac{V(t,x(t))}{\lambda_1}\right)^{\frac{1}{p}} \end{cases}$$

Substituting $V(t, x) = Q(t, x)/e^{M(t-t_0)}$ into the last inequality, we obtain

$$\|x(t)\| \leq \left(\frac{Q(t, x(t))}{e^{M(t-t_0)}\lambda_1}\right)^{\frac{1}{p}} \tag{8.18}$$

Combination of (8.17) and (8.18) gives

$$\|x(t)\| \leq \{\frac{\beta(\|x_0\|)}{e^{M(t-t_0)}\lambda_1}\}^{\frac{1}{p}} = \{\frac{\beta(\|x_0\|)}{\lambda_1}\}^{\frac{1}{p}}e^{-\frac{M}{p}(t-t_0)}, \quad \forall t \geq t_0, \tag{8.19}$$

Inequality (8.19) shows that (8.7) is uniformly exponentially stable.

Example 21 *As an elementary example, let us consider the non-linear differential equation given by $x' = -\frac{1}{4}x^{\frac{3}{5}} + xe^{-2t}$, $t \geq 0$. The Lyapunov function is defined by: $V(t, x) : \mathbb{R}^+ \times D \to \mathbb{R}^+$, $V(t, x) = x^6$, where $D = \{x : |x| \leq 1\}$. We have $|x|^7 \leq V(t, x) \leq |x|^6$, $\forall x \in D$. Condition (8.11) holds with $\lambda_1 = \lambda_2 = 1$, $p = 7$, $q = 6$. Also, we have $V'(t, x) =$*

$6x^5x' = 6x^5(-\frac{1}{4}x^{\frac{3}{5}} + xe^{-2t}) = -\frac{3}{2}x^{\frac{28}{5}} + 6x^6e^{-2t}$. *Therefore,* $V'(t,x) \leq -\frac{3}{2}x^{\frac{28}{5}} + 6e^{-2t}$, $\forall x \in D$. *Also, Conditions (8.15), (8.16) of Theorem 66 are also satisfied with* $\lambda_3 = 3/2$, $K = 6$, $\delta = 2$, $r = 28/5$. *Moreover, we also have* $V(t,x) - V(t,x)^{r/q} = x^6 - x^{\frac{28}{5}} = x^{\frac{28}{5}}(x^{\frac{2}{5}} - 1) \leq 0 \leq e^{-2t}$, $\forall x \in D$. *Finally, the system is exponentially stable.*

Theorem 76 *The System (8.7) is exponentially stable if it admits a generalized Lyapunov-like function and the following two conditions hold for all* $(t,x) \in W$:

$$\delta > \inf_{t \in \mathbb{R}^+} \frac{\lambda_3(t)}{[\lambda_2(t)]^{r/q}} > 0 \tag{8.20}$$

and there exists $\gamma > 0$ *such that*

$$V(t,x) - [V(t,x)]^{r/q} \leq \gamma e^{-\delta t} \tag{8.21}$$

Proof 35 *Let us the function* $Q(t,x(t)) = V(t,x(t))e^{M(t-t_0)}$, *where*

$$M = \inf_{t \in \mathbb{R}^+} \frac{\lambda_3(t)}{[\lambda_2(t)]^{r/q}}$$

We have $M < \delta$ *and*

$$D_f^+ Q(t,x) = D_f^+ V(t,x)e^{M(t-t_0)} + MV(t,x(t))e^{M(t-t_0)}$$

The same logic used in the proof of Theorem 66 gives us

$$D_f^+ Q(t,x) \leq (-\lambda_3(t)\|x\|^r + Ke^{-\delta t})e^{M(t-t_0)} + MV(t,x)e^{M(t-t_0)}$$

By the assumption, $\lambda_2(t) > 0$ *for all* $t \in \mathbb{R}^+$ *and (8.13) we have*

$$\|x\|^q \geq \frac{V(t,x)}{\lambda_2(t)}$$

equivalently

$$-\|x\|^r \leq -[\frac{V(t,x)}{\lambda_2(t)}]^{r/q}$$

Thus, we have

$$D_f^+ Q(t,x) \leq \{-V(t,x)^{r/q}\frac{\lambda_3(t)}{[\lambda_2(t)]^{r/q}} + Ke^{-\delta t}\}e^{M(t-t_0)} + MV(t,x)e^{M(t-t_0)}$$

Since

$$\frac{\lambda_3(t)}{[\lambda_2(t)]^{r/q}} \geq M, \quad \forall t \geq 0,$$

and by the condition (8.21) we obtain

$$D_f^+ Q(t,x) \leq M\{V(t,x) - V(t,x)^{r/q}\}e^{M(t-t_0)} + Ke^{(M-\delta)(t-t_0)}$$

$$\leq M\gamma e^{-\delta t}e^{M(t-t_0)} + Ke^{-\delta t}e^{M(t-t_0)}$$

$$= (K + M\gamma)e^{-\delta t}e^{M(t-t_0)}$$

$$\leq (K + M\gamma)e^{-\delta(t-t_0)}e^{M(t-t_0)}$$

Thus, $D_f^+ Q(t,x) \leq (K+M\gamma)e^{(M-\delta)(t-t_0)}$. Hence, in applying Theorem 75 to the case

$$\begin{cases} v(t) = Q(t,x(t)) \\ \\ g(t,u(t)) = (K + M\gamma)e^{(M-\delta)(t-t_0)} \end{cases}$$

we obtain

$$Q(t,x(t)) - Q(t_0,x_0) \leq \int_{t_0}^t (K+M\gamma)e^{(M-\delta)(s-t_0)}ds$$

$$= (K+M\gamma)\frac{1}{M-\delta}\{(e^{(M-\delta)(t-t_0)} - 1\}$$

Setting $\delta_1 = -(M-\delta)$, by condition (8.20) we have $\delta_1 > 0$ and

$$Q(t,x(t)) \leq Q(t_0,x_0) + \frac{K+M\gamma}{\delta_1} - \frac{K+M\gamma}{\delta_1}e^{(M-\delta)(t-t_0)}$$

$$\leq Q(t_0,x_0) + \frac{K+M\gamma}{\delta_1}.$$

Since $Q(t_0,x_0) = V(t_0,x_0) \leq \lambda_2(t_0)\|x_0\|^q$, we get

$$Q(t,x(t)) \leq \lambda_2(t_0)\|x_0\|^q + \frac{K+M\gamma}{\delta_1}$$

Letting

$$\lambda_2(t_0)\|x_0\|^q + \frac{K+M\gamma}{\delta_1} = \beta(\|x_0\|,t_0) > 0$$

we have

$$Q(t,x(t)) \leq \beta(\|x_0\|,t_0), \quad \forall t \geq t_0 \tag{8.22}$$

Furthermore, from condition (8.21), it follows that

$$\lambda_1(t)\|x(t)\|^p \leq V(t,x(t)), \|x(t)\| \leq \left(\frac{V(t,x(t))}{\lambda_1(t)}\right)^{\frac{1}{p}}$$

Since $\lambda_1(t)$ is non-decreasing, $\lambda_1(t) \geq \lambda_1(t_0)$, we have

$$\|x(t)\| \leq \left(\frac{V(t, x(t))}{\lambda_1(t_0)} \right)^{\frac{1}{p}}$$

Substituting

$$V(t, x) = \frac{Q(t, x)}{e^{M(t-t_0)}}$$

into the last inequality, we obtain

$$\|x(t)\| \leq \left(\frac{Q(t, x(t))}{e^{M(t-t_0)}\lambda_1(t_0)} \right)^{\frac{1}{p}} \tag{8.23}$$

Combining (8.22) and (8.23),

$$\|x(t)\| \leq \left(\frac{\beta(\|x_0\|, t_0)}{e^{M(t-t_0)}\lambda_1(t_0)} \right)^{\frac{1}{p}} = \left(\frac{\beta(\|x_0\|, t_0)}{\lambda_1(t_0)} \right)^{\frac{1}{p}} e^{-\frac{M}{p}(t-t_0)}, \quad \forall t \geq t_0 \tag{8.24}$$

Finally, relation (8.24) shows that system (8.7) is exponentially stable.

Example 22 *As an elementary example, let us consider the system $x' = -\frac{1}{5}x^{1/3} + xe^{-2t}$. The Lyapunov function is $V(t, x) = |x|^5$, where $D = \{x : |x| \leq 1\}$. We have:*

$$V(t, x) = \begin{cases} x^5, & \text{if } x \geq 0 \\ -x^5, & \text{if } x < 0 \end{cases}$$

Then we calculate

$$D_f^+ V(t, x) = \begin{cases} 5x^4(-\frac{1}{5}x^{1/3} + xe^{-2t}) = -x^{-13/3} + 5x^5 e^{-2t}, & \text{if } x \geq 0 \\ -5x^4(-\frac{1}{5}x^{1/3} + xe^{-2t}) = x^{14/3} - 5x^5 e^{-2t}, & \text{if } x < 0 \end{cases}$$

Thus,

$$D_f^+ V(t, x) = -|x|^{13/3} + 5|x|^5 e^{-2t} \leq -|x|^{13/3} - 5e^{-2t}, \quad \forall x \in D$$

Hence, the conditions hold for $\lambda_1 = \lambda_2 = \lambda_3 = 1$, $p = q = K = 5$, $\delta = 2$, $r = 13/3$, $\delta > \lambda_3/\lambda_2^{r/q} = 1$. Condition (8.21) of Theorem 76 is also true because:

$$V(t, x) - [V(t, x)]^{r/q} = |x|^5 - |x|^{13/3} = |x|^{13/3}(|x|^{2/3} - 1) \leq 0, \quad \forall x \in D.$$

Then the system is uniformly exponentially stable.

At this end, let us consider the non-linear control system (8.8) and assume that $f(t,0,0) = 0$, for all $t \geq 0$. The system (8.8) is asymptotically stabilizable by a feedback control $u(t) = h(x,t)$, where $h(x) : \mathbb{R}^n \to \mathbb{R}^m$, $h(0) = 0$, if the zero solution of the system without control

$$
\begin{cases}
x'(t) = f(t, x(t), h(x)), & t \geq 0 \\
\\
x(t_0) = x_0, & t_0 \geq 0
\end{cases}
\tag{8.25}
$$

is asymptotically stable in the Lyapunov sense. If the zero solution of (8.25) is exponentially stable, we say that (8.8) is exponentially stabilizable.

The following result was proved in [Lee & Markus (1967)]:

Theorem 77 *Consider time-invariant system (8.25). If there exists a function $h(x) : \mathbb{R}^n \to \mathbb{R}^m$, $h(0) = 0$, where $h(x)$ is continuously differentiable in x and a positive definite function $V(x) : \mathbb{R}^n \to \mathbb{R}^+$, which is continuously differentiable in x such that:*

(i) $V(x) \to \infty$ as $\|x\| \to \infty$.

(ii) $\frac{\partial V}{\partial x_i} f^i(x, h(x)) < 0$, $i = 1, 2, \ldots, n$, for all $x \neq 0$.

Then the system is asymptotically stabilizable by the feedback control $u(t) = h(x(t))$.

The following result was proved in [Linh & Phat (2001)]:

Theorem 78 *(a) Assume that there is a function $h(x) : \mathbb{R}^n \to \mathbb{R}^m$, $h(0) = 0$ with $h(x)$ continuous in x, such that system (8.25) admits a Lyapunov-like function satisfying (8.15) and (8.16). Then the non-linear control system (8.8) is exponentially stabilizable by feedback $u(t) = h(x(t))$.*

(b) Assume that there is a function $h(x) : \mathbb{R}^n \to \mathbb{R}^m$ with $h(0) = 0$ and $h(x)$ continuous in x, such that system (8.25) admits a generalized Lyapunov-like function satisfying (8.20) and (8.21). Then the non-linear control system (8.8) is exponentially stabilizable by feedback control $u(t) = h(x(t))$.

8.4   Lasalle's Invariance Principle

To prove the Lasalle's invariance principle theorem, we need the following result:

Lemme 8.1 *Suppose $\Omega \subseteq \mathbb{R}^n$ is open, $f : \Omega \mapsto \mathbb{R}^n$ is locally Lipschitz continuous, and $V : \Omega \mapsto \mathbb{R}$ is continuously differentiable and bounded from below, and $V' \leq 0$. Suppose that $p \in \Omega$ and for all $t \geq 0$, $\Phi(t,p)$ is defined and is contained in Ω. Then for all $q \in \omega(p)$ (the ω-limit set of p), $V'(q) = 0$.*

Proof 36 *Assuming that Ω, f, V and p verifying the conditions and $\omega(p) \neq \emptyset$. Suppose that $q_1 = q_3 \in \omega(p)$, $q_2 \in \omega(p)$ and fix any $\varepsilon > 0$. Since V is continuous at q_1 and q_2, there exists $\delta > 0$ such that for all $x \in \Omega$, for $i = 1, 2$, if $\|x - q_i\| < \delta$ then*

$$|V(x) - V(q_i)| < \frac{\varepsilon}{4}$$

Since $q_1, q_2, q_3 \in \omega(p)$, then there exists $0 \leq t_1 \leq t_2 \leq t_3$ such that for $i = 1, 2, 3$,

$$|\Phi(t_i, p) - q_i\| < \delta$$

and hence for $i = 1, 2, 3$,

$$|V(\Phi(t_i, p)) - V(q_i)| < \frac{\varepsilon}{4}$$

This implies that

$$|V(\Phi(t_3, p)) - V(\Phi(t_1, p))| \leq |V(\Phi(t_3, p)) - V(q_3)| + |V(q_1) - V(\Phi(t_1, p))| < \frac{\varepsilon}{2}$$

The fact that V' does not change sign implies that

$$
\begin{aligned}
|V(\Phi(t_2, p)) - V(\Phi(t_1, p))| &= |\int_{t_1}^{t_2} V'(\Phi(s, p))\, ds| \\
&\leq |\int_{t_1}^{t_3} V'(\Phi(s, p))\, ds| \\
&\leq |V(\Phi(t_3, p)) - V(\Phi(t_1, p))| < \frac{\varepsilon}{2}
\end{aligned}
$$

Therefore

$$
\begin{aligned}
|V(q_2) - V(q_1)| &\leq |V(q_2) - V(\Phi(t_2, p))| + |V(\Phi(t_2, p)) - V(\Phi(t_1, p))| \\
&\quad + |V(\Phi(t_1, p)) - V(q_1)| \\
&< \frac{\varepsilon}{4} + \frac{\varepsilon}{2} + \frac{\varepsilon}{4} = \varepsilon
\end{aligned}
$$

Since $\varepsilon > 0$ is arbitrary, then V is constant on the set $\omega(p)$. Since $\omega(p)$ is invariant, i.e., for every $q \in \omega(p)$ and all $t \in \mathbb{R}$, we have $\Phi(t, q) \in \omega(p)$, then for every $q \in \omega(p)$ and all $t \in \mathbb{R}$, $V(\Phi(t, q)) = V(q)$ and therefore

$$V'(q) = \left.\frac{dV}{dt}\right|_{t=0} (V(\Phi(t, q)) - V(q)) = 0$$

proving that V' vanishes identically on the set $\omega(p)$.

Now, we can prove Lasalle's invariance principle theorem:

Theorem 79 (Lasalle's Invariance Principle) *Suppose $\Omega \subseteq \mathbb{R}^n$ is open, $f : \Omega \mapsto \mathbb{R}^n$ is locally Lipschitz continuous, and $V : \Omega \mapsto \mathbb{R}$ is continuously differentiable. Suppose $K \subseteq \Omega$ is compact, invariant, and for all $x \in K$, $V'(x) \leq 0$. Let $E = (V')^{-1}(0)$ and let $M \subseteq E$ be the largest invariant set contained in E. Then every solution starting in K approaches M as $t \to \infty$.*

Proof 37 *Suppose K, f, V, E and M verify the mentioned conditions. Suppose $p \in K$ and consider the curve $t \mapsto \Phi(t, p)$. Since V is continuous, it is bounded from below on the compact set K. Since K is invariant by hypothesis, for all $t \geq 0$, $\Phi(t, p) \in K$. Thus, using that K is compact, and therefore bounded, the ω-limit set $\omega(p)$ is non-empty. Since K is closed, $\omega(p) \subseteq K$. Using Lemma 8.1, for all $q \in \omega(p)$, $V'(q) = 0$ and therefore $\omega(p) \subseteq \{y : V'(y) = 0\} = (V')^{-1}(0) = E$.*

Since ω-limit sets are invariant, $\omega(p) \subseteq M$. Since ω-limit sets are attracting, then for every $\epsilon > 0$ there exists $T \in \mathbb{R}$ such that for every $t > T$ there exists $q \in \omega(p)$ such that $\|\Phi(t, p) - q\| < \epsilon$. Since $\omega(p) \subseteq M$, $q \in M$. Hence, the solution curve starting at an arbitrary $p \in K$ approaches M as $t \to \infty$.

In the next section, we will present an overview of the extension of the direct Lyapunov method presented in Sec. 8.3 to systems of equations of perturbed motion with fractional-like derivatives.

8.5 Direct Lyapunov-type stability for fractional-like systems

Let $q \in (0, 1]$, $\mathbb{R}_+ = [0, \infty)$, $t_0 \in \mathbb{R}_+$ and given a continuous function $x(t) : [t_0, \infty) \to \mathbb{R}$. In [Anatoliy & MartynyukIvanka (2018)] a definition of a fractional derivative named *fractional-like derivative* is given by:

Definition 62 *For any $q \in (0, 1]$ the fractional-like derivative $\mathcal{D}_{t_0}^q(x(t))$ of the function $x(t)$ of order $0 < q \leq 1$ is defined by*

$$\mathcal{D}_{t_0}^q(x(t)) = \lim_{\theta \to 0} \frac{x(t + \theta(t - t_0)^{1-q}) - x(t)}{\theta}$$

If $t_0 = 0$, then $\mathcal{D}_{t_0}^q(x(t))$ has the form:

$$\mathcal{D}_0^q(x(t)) = \lim_{\theta \to 0} \frac{x(t + \theta t^{1-q}) - x(t)}{\theta}$$

In the case $t_0 = 0$, we will denote $\mathcal{D}_0^q(x(t)) = \mathcal{D}^q(x(t))$.
If $\mathcal{D}^q(x(t))$ exists on an open interval of the type $(0, b)$, then

$$\mathcal{D}^q(x(0)) = \lim_{t \to 0^+} \mathcal{D}^q(x(t)).$$

If the fractional-like derivative of $x(t)$ of order q exists on (t_0, ∞), then the function $x(t)$ is said to be q-differentiable on the interval (t_0, ∞).

The following result is proved in [Khalil, et al., (2014)]:

Proposition 8.1 *Let $q \in (0, 1]$ and $x(t), y(t)$ be q-differentiable at a point $t > 0$. Then:*

(a) $\mathcal{D}^q_{t_0}(ax(t) + by(t)) = a\mathcal{D}^q_{t_0}(x(t)) + b\mathcal{D}^q_{t_0}(y(t))$ for all $a, b \in \mathbb{R}$;

(b) $\mathcal{D}^q_{t_0}(t^p) = pt^{p-q}$ for any $p \in \mathbb{R}$;

(c) $\mathcal{D}^q_{t_0}(x(t)y(t)) = x(t)\mathcal{D}^q_{t_0}(y(t)) + y(t)\mathcal{D}^q_{t_0}(x(t))$;

(d)

$$\mathcal{D}^q_{t_0}\left(\frac{x(t)}{y(t)}\right) = \frac{y(t)\mathcal{D}^q_{t_0}(x(t)) - x(t)\mathcal{D}^q_{t_0}(y(t))}{y^2(t)};$$

(e) $\mathcal{D}^q_{t_0}(x(t)) = 0$ for any $x(t) = \lambda$, where λ is an arbitrary constant.

The following result is proved in [Abdeljawad (2015), Pospíšil & Pospıılova Škripkova(2016)]:

Proposition 8.2 *Let $h(y(t)) : (t_0, \infty) \to \mathbb{R}$. If $h(\cdot)$ is differentiable with respect to $y(t)$ and $y(t)$ is q-differentiable, where $0 < q \le 1$, then for any $t \in \mathbb{R}_+$, $t \ne t_0$ and $y(t) \ne 0$*

$$\mathcal{D}^q_{t_0}h(y(t)) = h'(y(t))\mathcal{D}^q_{t_0}(y(t)),$$

where $h'(t)$ is a partial derivative of h.

Definition 63 *The fractional-like integral of order $0 < q \le 1$ with a lower limit t_0 is defined by [Khalil, et al., (2014)]:*

$$I^q_{t_0}x(t) = \int_{t_0}^t (s - t_0)^{q-1}x(s)ds.$$

Proposition 8.3 *[Khalil, et al., (2014)]: Let the function $x(t) : (t_0, \infty) \to \mathbb{R}$ be q-differentiable for $0 < q \le 1$. Then for all $t > t_0$,*

$$I^q_{t_0}(\mathcal{D}^q_{t_0}x(t)) = x(t) - x(t_0).$$

To describe the fractional-like derivatives of Lyapunov-type functions, let us consider a system of differential equations with fractional-like derivative of the state vector

$$\begin{cases} \mathcal{D}^q_{t_0}x(t) = f(t, x(t)) \\ \\ x(t_0) = x_0 \end{cases} \tag{8.26}$$

where $x \in \mathbb{R}^n$, $f \in C(\mathbb{R}_+ \times \mathbb{R}^n, \mathbb{R}^n)$, $t_0 \ge 0$. Assuming that for $(t_0, x_0) \in int(\mathbb{R}_+ \times \mathbb{R}^n)$ the initial value problem (IVP) (8.26) has a solution $x(t, t_0, x_0) \in C^q([t_0, \infty), \mathbb{R}^n)$ for all $t \ge t_0$. In addition, it is assumed that $f(t, 0) = 0$ for all $t \ge t_0$.

Let for equation (8.26) a Lyapunov-type function $V(t, x) \in C^q(\mathbb{R}_+ \times \mathbb{R}^n, \mathbb{R}_+)$ be constructed in some way such that $V(t, 0) = 0$ for all $t \in \mathbb{R}^n$. Introduce the notation $B_r = \{x \in \mathbb{R}^n : \|x\| < r\}$, $r > 0$.

Definition 64 *Let V be a continuous and q-differentiable function (scalar or vector), $V : R_+ \times B_r \to R^s$ ($s = 1$ or $s = m$, respectively), and $x(t, t_0, x_0)$ be the solution of the IVP (8.26), which exists and is defined on $R_+ \times B_r$. Then for $(t, x) \in R_+ \times B_r$ the expression:*

(1)

$$
\begin{cases}
{}^+\mathcal{D}^q_{t_0} V(t, x) = \lim\sup_{\theta \to 0^+} \dfrac{\mathbf{V} - V(t,x)}{\theta} \\
\mathbf{V} = V(t + \theta(t - t_0)^{1-q}, x(t + \theta(t - t_0)^{1-q}, t, x))
\end{cases}
\tag{8.27}
$$

is the upper right fractional-like derivative of the Lyapunov function,

(2)

$$
{}_+\mathcal{D}^q_{t_0} V(t, x) = \liminf_{\theta \to 0^+} \frac{\mathbf{V} - V(t, x)}{\theta}
$$

is the lower right fractional-like derivative of the Lyapunov function,

(3)

$$
{}^-\mathcal{D}^q_{t_0} V(t, x) = \limsup_{\theta \to 0^-} \frac{\mathbf{V} - V(t, x)}{\theta}
$$

is the upper left fractional-like derivative of the Lyapunov function,

(4)

$$
{}_-\mathcal{D}^q_{t_0} V(t, x) = \liminf_{\theta \to 0^-} \frac{\mathbf{V} - V(t, x)}{\theta}
$$

is the lower left fractional-like derivative of the Lyapunov function.

Definition 64 is based on the following result proved in [Yoshizawa (1966)]:

Lemme 8.2 *Let $V(t, x)$ be continuous, q-differentiable and locally Lipschitz with respect to its second variable x on $\mathbb{R}_+ \times B_r$. Then the fractional-like derivative of the function $V(t, x)$ with respect to the solution $x(t, t_0, x_0)$ is defined by*

$$
{}^+\mathcal{D}^q_{t_0} V(t, x) = \limsup_{\theta \to 0^+} \frac{V(t + \theta(t - t_0)^{1-q}, x + \theta(t - t_0)^{1-q} f(t, x)) - V(t, x)}{\theta}
$$

$$\tag{8.28}$$

where $(t, x) \in \mathbb{R}_+ \times B_r$.

If $V(t, x(t)) = V(x(t))$, $0 < q \leq 1$, the function V is differentiable on x, and the function $x(t)$ is q-differentiable on t for $t > t_0$, then

$$
{}^+\mathcal{D}^q_{t_0} V(t, x) = V'(x(t)) \mathcal{D}^q_{t_0} x(t),
$$

where V' is a partial derivative of the function V.

By relations (8.27) and (8.28), we obtain the result by Yoshizawa [Yoshizawa (1966)] for a fractional-like derivative of the function $V(t, x)$ in the form

$$
{}^+\mathcal{D}^q_{t_0} V(t, x(t, t_0, x_0)) = {}^+\mathcal{D}^q_{t_0} V(t, x) \big|\big|_{(8.26)}
$$

Definition 65 *If the function $V(t,x)$ together with one of its fractional-like derivatives resolves the problem of stability (instability) of the solutions of (8.26), we will call $V(t,x)$ a Lyapunov function for the fractional-like system (8.26) .*

The following Lemma as proved in [Anatoliy & MartynyukIvanka (2018)]:

Lemme 8.3 *Let $x \in \mathbb{R}$, $y \in \mathbb{R}^n$ and P is an $n \times n$ constant matrix. Then for the functions $V_1 = x^2(t)$, $V_2 = y^T(t)y(t)$, and $V_3 = y^T(t)Py(t)$ the following estimates hold:*

    (a) ${}_{t_0}^c D_t^q(x^2(t)) \leq^+ \mathcal{D}_{t_0}^q(x^2(t))$ for $x \in \mathbb{R}$.
    (b) ${}_{t_0}^c D_t^q(y^T(t)y(t)) \leq^+ \mathcal{D}_{t_0}^q(y^T(t)y(t))$ for $y \in \mathbb{R}^n$.
    (c) ${}_{t_0}^c D_t^q(y^T(t)Py(t)) \leq^+ \mathcal{D}_{t_0}^q(y^T(t)Py(t))$, for $y \in \mathbb{R}^n$.

Lemma 8.3 implies that the fractional-like derivative of a Lyapunov-type function is an upper bound of the Caputo fractional derivatives of this Lyapunov function.

It was shown in [Lakshmikantham, *et al.*, (2009), Stamova & Stamov (2017)] that the Lyapunov-type stability definitions for a fractional-like system (8.26) are the same as for ordinary differential equations (8.1) and differential equations with Caputo's fractional derivatives.

The following result regarding the Direct Lyapunov's method for a fractional-like system (8.26) was shown in [Anatoliy & MartynyukIvanka (2018)]. The method of analysis is based on the *Hahn class of functions:*

$$K = \{a \in C[\mathbb{R}_+, \mathbb{R}_+] : a(u) \text{ is strictly increasing and } a(0) = 0\}$$

Theorem 80 *Assume that for the fractional-like system (8.26) there exists a q-differentiable function $V(t,x)$, $V(t,0) = 0$ for $t \geq t_0$ and functions $a, b \in K$ such that:*

    (1) $V(t,x) \geq a(\|x\|)$, $(t,x) \in \mathbb{R}_+ \times B_r$,
    (2) $V(t,x) \leq b(\|x\|)$, $(t,x) \in \mathbb{R}_+ \times B_r$,
    (3)

$$^+\mathcal{D}_{t_0}^q(V(t,x(t))) \leq 0 \quad for \ (t,x) \in \mathbb{R}_+ \times B_r. \tag{8.29}$$

Then the state $x = 0$ of (8.26) is uniformly stable.

Proof 38 *Let $x(t) = x(t, t_0, x_0)$ be the solution of (8.26) for $(t_0, x_0) \in (\mathbb{R}_+ \times B_r)$ defined for all $t \geq t_0$. Let $t_0 \in \mathbb{R}_+$ and $0 < \varepsilon < r$ be given. By conditions (1), (2) of Theorem 80 we can choose $\delta = \delta(\varepsilon) > 0$ so that*

$$b(\delta) < a(\varepsilon) \tag{8.30}$$

We will prove the following fact: If $\|x_0\| < \delta$, then $\|x(t)\| < \varepsilon$ for all $t \geq t_0$. Assuming by contradiction that this is not true, then there exists a solution $x(t, t_0, x_0) = x(t)$ of (8.26) such that for $\|x_0\| < \delta$ there is $t_1 > t_0$ for which

$$\|x(t_1)\| = \varepsilon, \quad \|x(t)\| < \varepsilon \quad for \ all \ t \in [t_0, t_1).$$

By Proposition 8.3 and condition (8.29), the Lyapunov relation

$$V(t, x(t)) - V(t_0, x_0) = I_{t_0}^q ({}^+\mathcal{D}_{t_0}^q (V(t, x(t))))$$

becomes

$$V(t, x(t)) - V(t_0, x_0) \leq 0. \tag{8.31}$$

For $t = t_1$ we have from (8.31),

$$a(\varepsilon) \leq V(t_1, x(t_1)) \leq V(t_0, x_0) \leq b(\|x_0\|) < a(\varepsilon). \tag{8.32}$$

and this in contradiction with condition (8.30).

Also, the following result was proved in [Anatoliy & MartynyukIvanka (2018)]:

Theorem 81 *Let the condition of Theorem 80 be satisfied and instead of (8.29) the following estimate holds*

$$^+\mathcal{D}_{t_0}^q (V(t, x(t))) \leq -d(\|x\|) \tag{8.33}$$

for $(t, x) \in \mathbb{R}_+ \times B_r$, where $d \in K$. Hence, the state $x = 0$ of system (8.26) is uniformly asymptotically stable.

Proof 39 *All conditions of Theorem 81 are satisfied, so the state $x = 0$ is uniformly stable. Let $0 < \varepsilon < r$ and $\delta = \delta(\varepsilon) > 0$ be the same as in Theorem 81. For $\varepsilon_0 \leq r$ let $\delta_0 = \delta_0(\varepsilon_0) > 0$ and consider the solution $x(t, t_0, x_0)$ with initial data $t_0 \in \mathbb{R}_+$ and $\|x_0\| < \delta_0$. Let for $t_0 < t \leq t_0 + T(\varepsilon)$, where $T(\varepsilon) \geq (qb(\delta_0)/d(\delta(\varepsilon)))^{1/q}$ for $x(t)$ we have $\|x(t)\| \geq \delta(\varepsilon)$. In fact, this is impossible under the conditions of Theorem 82. From the Lyapunov relation we obtain:*

$$V(t, x(t)) - V(t_0, x_0) = I_{t_0}^q ({}^+\mathcal{D}_{t_0}^q (V(t, x(t))) \tag{8.34}$$

$$\leq -I_{t_0}^q (d(\|x(t)\|)) = -\int_{t_0}^t (s - t_0)^{q-1} d(\|x(s)\|) ds$$

From (8.34) we obtain

$$V(t, x(t)) \leq V(t_0, x_0) - \int_{t_0}^t (s - t_0)^{q-1} d(\|x(s)\|) ds \leq b(\delta_0) - d(\delta(\varepsilon)) \frac{(t - t_0)^q}{q} \tag{8.35}$$

For $t = t_0 + T(\varepsilon)$ by (8.35) we have

$$0 < a(\delta(\varepsilon)) \leq V(t_0 + T(\varepsilon), x(t_0 + T(\varepsilon)))$$

$$\leq b(\delta_0) - d(\delta(\varepsilon)) \frac{T(\varepsilon)^q}{q} \leq 0$$

which is a contradiction. Thus, there exists $t_1 \in [t_0, t_0 + T(\varepsilon)]$ such that $\|x(t_1)\| < \delta(\varepsilon)$. Hence, $\|x(t)\| < \varepsilon$ for all $t \geq t_0 + T(\varepsilon)$ as far as $\|x_0\| < \delta_0$ and $\lim \|x(t)\| = 0$ as $t \to \infty$ uniformly on $t_0 \in \mathbb{R}_+$.

The following theorem proved in [Anatoliy & MartynyukIvanka (2018)] gives conditions for the instability of the state $x = 0$ of system (8.26):

Theorem 82 *Let there be, for the system (8.26), a q-differentiable function $V(t, x) : \mathbb{R}_+ \times B_\varepsilon \to \mathbb{R}$, such that on $[t_0, \infty) \times G(h)$, where $G(h) \subset B_\varepsilon$, $t_0 \geq 0$, the following conditions are satisfied:*
(1) $0 < V(t, x) \leq c < \infty$ for some constant c.
(2) $^+\mathcal{D}^q_{t_0} V(t, x)|_{(8.26)} \geq a(V(t, x))$, where $a \in K$, $0 < q \leq 1$.
(3) the state $x = 0$ belongs to $\partial G(h)$.
(4) $V(t, x) = 0$ for $[t_0, \infty) \times (\partial G(h) \cap B_\varepsilon)$.
Then the state $x = 0$ of system (8.26) is unstable.

Proof 40 *Condition (3) implies that for any $\delta > 0$ there exists a $x_0 \in G(h) \cap B_\delta$ such that $V(t_0, x_0) > 0$. For the solution $x(t) = x(t, t_0, x_0)$ while $x(t) \in G(h)$ from conditions (8.26). We have*

$$c \geq V(t, x(t)) - V(t_0, x_0) \geq I^q_{t_0} a(V(t, x(t))) \geq V(t_0, x_0) + a(V(t_0, x_0))\frac{(t - t_0)^q}{q}$$
$$(8.36)$$

This inequality implies that the solution $x(t)$ must leave the domain $G(h)$ at some moment $t_1 > t_0$. Clearly, condition (4) is satisfied, then $x(t)$ can not leave the domain $G(h)$ across the boundary $\partial G(h)$, because $G(h) \subset B_\varepsilon$. Therefore $x(t)$ will leave B_ε, i.e., $\|x(t_1)\| \geq \varepsilon$.

From Theorem 82, we have the following results:
(A) Suppose that all conditions of Theorem 82 hold and conditions (1) and (2) are replaced by the following conditions, respectively:
(1*) $0 < V(t, x) \leq b(\|x\|)$.
(2*) $^+\mathcal{D}^q_{t_0} V(t, x) \geq a(\|x\|)$, where $a, b \in K$.
Then the state $x = 0$ of system (8.26) is unstable.
(B) Suppose that all conditions of Theorem 82 hold and condition (2) is replaced by

$$^+\mathcal{D}^q_{t_0} V(t, x) = \lambda V(t, x) + W(x(t)), t \in [t_0, \infty), \quad x \in G(h), \ \lambda > 0, \quad (8.37)$$

where the function W is continuous and $W(x) \geq 0$. Then the state $x = 0$ of system (8.26) is unstable.

Proof 41 *Relation (8.37) can be represented in the integral form given by:*

$$V(t, x(t)) = V(t_0, x(t_0)) \exp\left(\lambda \frac{(s - t_0)^q}{q}\right) + \int_{t_0}^t \exp\left(\lambda \frac{(s - t_0)^q}{q}\right)$$

$$\times \exp\left(-\lambda \frac{(s - t_0)^q}{q}\right) (s - t_0)^{q-1} W(x(s)) ds$$

Since the second term is positive by the conditions of the above item (B), for any $0 < q \leq 1$ we have:

$$V(t, x(t)) \geq V(t_0, x(t_0)) \exp\left(\lambda \frac{(t-t_0)^q}{q}\right), \quad t \geq t_0, \qquad (8.38)$$

At this end, assuming that the initial state of the solution $x(t) = x(t, t_0, x_0)$ is $x_0 \in U$ a neighborhood of the origin $x = 0$. Since for any $t \geq t_0$ the estimate (8.38) is satisfied with respect to the solution $x(t)$, then for $t \to \infty$ the function $V(t, x(t))$ increases while, by the conditions of Theorem 83, it is bounded. Hence, for $x(t)$ there exists t^* such that $x(t^*)$ will leave B_r. Thus, the state $x = 0$ of system (8.26) is instable.

8.6 Construction of globally asymptotically stable n-D discrete mappings

Stabilizing dynamical systems (chaotic or not) has a crucial importance in several sciences [Khalil (1996), Krichman, et al., (2001), Parmananda (1999), Oh-Jong, et al., (2001)]. Several methods are well known in the current literature. For example, see [Krichman, et al., (2001), Krstik, et al., (1995), Wei-hong (2001-2002), Parmananda (1999), Oh-Jong, et al., (2001), Chang, et al., (1998), Iggidr & Bensoubaya (1996)]. One of the most common methods used for stabilizing dynamical systems are the Lyapunov functions [Khalil (1996), Iggidr & Bensoubaya (1996), LaSalle (1986)].

In this section, we eliminate any bounded behavior (chaotic or not) in some forms of n-D discrete mapping and drive them globally asymptotically to their fixed points. The method of analysis is based on the explicit construction of a non-negative semi-definite Lyapunov function that guaranties the stabilization of such a system. Indeed, let us consider the general n-D discrete mappings of the form:

$$g(x_k, \mu) = \begin{cases} x_{k+1}^1 = x_k^i f_1(x_k, \mu), \\[2mm] x_{k+1}^2 = x_k^i f_2(x_k, \mu), \\[2mm] \cdots\cdots, \\[2mm] x_{k+1}^n = x_k^i f_n(x_k, \mu), \end{cases} \qquad (8.39)$$

where $\mu = (\mu_1, \mu_2, ..., \mu_j) \in \mathbb{R}^j$ are bifurcation parameters and $x_k = (x_k^1, x_k^2, ..., x_k^n) \in \mathbb{R}^n$ is the state variable and $(f_k)_{1 \leq k \leq n}$ are real continuous functions, defined as follows: $f_k : \mathbb{R}^n \times \mathbb{R}^j \longrightarrow \mathbb{R}$. The origin $(0, 0, ..., 0) \in \mathbb{R}^n$ is a fixed point for the map (8.39), i.e., $g(0, 0, ..., 0, \mu) = (0, 0, ..., 0)$. On the other hand, assume that for all μ in a subset $\Omega_\mu \subset \mathbb{R}^j$, the solutions of system (8.39) are bounded.

First, we recall some usual notations and standard definitions:

Definition 66 *(a) The ω-limit set of vector $x \in \mathbb{R}^n$ is defined by $f^+(x) = \{f^k(x), k \in \mathbb{N}\}$.*

(b) A ball in $\mathbb{R}^n$ of radius δ is given by $B_\delta = \{x \in \mathbb{R}^n, \|x\| \leq \delta\}$.

(c) A set G^* is positively invariant under the action of a map g if and only if: $g(G^*) \subset G^*$.

(d) Let G^* be a closed positively invariant set such that $(0, 0, ..., 0) \in G^*$, then the origin is said to be:

(1) G^*-stable if $\forall \varepsilon > 0$, $\exists \delta > 0$ $g^+ (B_\delta \cap G^*) \subset B_\varepsilon$.

(2) G^*-asymptotically stable if it is G^*-stable and there exists $\delta > 0$ such that: $\lim_{n \longrightarrow +\infty} g^n(x) = 0, \forall x \in B_\delta \cap G^*$.

The controlled system associated with system (8.39) is defined by:

$$g_c(x_k, \mu) = \begin{cases} x_{k+1}^1 = x_k^i f_1(x_k, \mu) \\ x_{k+1}^2 = x_k^i f_2(x_k, \mu) \\ \quad\quad \cdots\cdots \\ x_{k+1}^n = x_k^i f_n(x_k, \mu) \end{cases} + U(x_k, \mu) \qquad (8.40)$$

where the controller U is given by:

$$U(x_k, \mu) = \begin{pmatrix} 0 \\ 0 \\ \cdots\cdots \\ -x_k^i f_i(x_k, \mu) + h(x_k) x_k^i \\ \cdots\cdots \\ 0 \end{pmatrix} \qquad (8.41)$$

where the function $f_i(x_k, \mu)$ is given in the ith component of the uncontrolled map (8.39), and all the components of the controller U are zero, expect the ith component, and $h : \mathbb{R}^n \longrightarrow \mathbb{R}$ is any real continuous function satisfying $|h(x)| < 1$, for all $x \in \mathbb{R}^n$. Hence, the controlled system (8.40) is given explicitly by:

$$x_{k+1} = g_c(x_k, \mu) = \begin{cases} x_{k+1}^1 = x_k^i f_1(x_k, \mu) \\[2mm] x_{k+1}^2 = x_k^i f_2(x_k, \mu) \\[2mm] \qquad \cdots \cdots \\[2mm] x_{k+1}^i = h(x_k) x_k^i \\[2mm] \qquad \cdots \cdots \cdots \\[2mm] x_{k+1}^n = x_k^i f_n(x_k, \mu) \end{cases} \qquad (8.42)$$

In this case, the origin $(0, 0, ..., 0) \in \mathbb{R}^n$ is also a fixed point for the controlled map (8.42), and it has the same form given in (8.39).

The following result is proved:

Theorem 83 *Let us consider the map (8.39) and assume that for all $\mu \in \Omega_\mu \subset \mathbb{R}^j$, all its solutions are bounded. Then, for any continuous function $h : \mathbb{R}^n \longrightarrow \mathbb{R}$ verify:*

$$|h(x)| < 1, \text{ for all } x \in \mathbb{R}^n, \qquad (8.43)$$

the controller function (8.41) makes the map (8.39) globally asymptotically stable.

Proof 42 *The idea of our proof is the construction of a non-negative semi-definite Lyapunov function $V(x)$ for which the origin is globally asymptotically stable, as shown in [Iggidr & Bensoubaya (1996)], that is:*

(1) $V(x) \in C^0(\mathbb{R}^n, \mathbb{R}^+)$, and $V(x) \geq 0$ for all $x \in \mathbb{R}^n - \{(0, ..., 0)\}$, and $V(0, ..., 0) = 0$.

(2) $\Delta V(x) = V(g(x)) - V(x) \leq 0$, for all $x \in \mathbb{R}^n - \{(0, ..., 0)\}$, and $\Delta V(0, ..., 0) = 0$.

(3) The origin $(0, ..., 0)$ is G^*-globally asymptotically stable where G^* is the largest positively invariant set contained in $G = \{x \in \mathbb{R}^n : V(g(x)) - V(x) = 0\}$.

(4) All the solutions of the controlled system (8.42) are bounded for all $\mu \in \Omega_\mu$.

This result show that the global asymptotic stability of the controlled system (8.42) reduced to the invariant set G^* imply its global asymptotic stability in $\mathbb{R}^n$, which is a consequence of LaSalle Invariance Principle [LaSalle (1986)].

Let us consider the following scalar function:

$$V(x) = \left(x_k^i\right)^2 \qquad (8.44)$$

Firstly, we remark that: $V(x) \in C^0(\mathbb{R}^n, \mathbb{R}^+)$, and $V(x) \geq 0$ for all $x \in \mathbb{R}^n - \{(0, 0, .., 0)\}$, and $V(0, 0, .., 0) = 0$, then V is a non-negative semi-definite Lyapunov function on $\mathbb{R}^n$. Secondly, by using the condition (8.43), the first difference is negative for all $x \in \mathbb{R}^n - \{(0, ..., 0)\}$ and it is given by:

$$\Delta V(x) = V(g(x)) - V(x) = \left(h^2(x) - 1\right) x_i^2 \leq 0, \qquad (8.45)$$

and $\Delta V (0, ..., 0) = 0$. Thus, ΔV is semi negative-definite on $\mathbb{R}^n$. On the other hand, we have $G = \{(x_1, x_2, ..., x_i = 0, ..., x_n) \in \mathbb{R}^n\}$, let G^* be the largest positively invariant set contained in G, i.e., $g_c(G^*) \subset G^*$. We remark that for all $(x_1, x_2, ..., x_i = 0, ..., x_n) \in G$, then $g_c(x_1, .., x_i = 0, ..., x_n) = (0, 0, .., 0)$, thus, $G^* = \{(0, 0, .., 0)\}$. Hence, the origin $(0, ..., 0)$ is G^*-globally asymptotically stable since:

(a) The origin is G^*-stable because: $\forall \varepsilon > 0, \exists \delta > 0 : g_c^+ (B_\delta \cap G^*) = g_c^+ (0, 0, .., 0) = (0, 0, .., 0) \in B_\varepsilon$,

(b) There exists $\delta > 0$ such that:

$$\lim_{n \longrightarrow +\infty} g_c^n (x) = 0, \forall x \in B_\delta \cap G^*. \tag{8.46}$$

since $B_\delta \cap G^* = \{(0, 0, .., 0)\}$, for all $\delta > 0$. Thus, the zero solution of the controlled system (8.42) is globally asymptotically stable.

The proposed procedure has some advantages:

(1) There is an explicit and very simple universal control law.

(2) There are no numerical or analytic calculations.

(3) The proposed procedure does not depend on the behavior of the uncontrolled map, i.e., The knowledge of the behavior of the uncontrolled map (8.39) is not necessary for applying the proposed new simple control law.

The procedure also has some disadvantages:

(1) The proposed procedure is applicable only to some forms of the maps given in (8.39).

(2) There are no analytic tools for confirming the boundedness of the solutions of the uncontrolled system (8.39), i.e., The finding of Ω_μ is not easy. Hence, it is necessary to find the bounded states of the map g numerically.

Example 23 *To validate the above analysis, let us consider the rational chaotic map given by:*

$$g(x, y) = \begin{pmatrix} x \left(\frac{-a}{1+y^2} + 1 \right) \\ x \end{pmatrix} \tag{8.47}$$

where $\mu = a \in \mathbb{R}$ is the bifurcation parameter. We have $f_1 (x, y, a) = \frac{-a}{1+y^2} + 1$, $f_2 (x, y, a) = 1$, $i = 1$. For $a = 4$, the map (8.47) has the chaotic attractor shown in Fig. 8.1(a), and the time series x is shown in Fig. 8.1(b). Applying the proposed control law (8.41) with $h (x, y) = \frac{1}{2}$ in order to drive map (8.47) globally asymptotically to its fixed point $(0, 0)$ and obtain the controlled map:

$$g_c(x, y) = \begin{pmatrix} \frac{x}{2} \\ x \end{pmatrix}, \tag{8.48}$$

then, the map (8.48) converges globally asymptotically to its fixed point $(0, 0)$ as shown in Fig. 8.2.

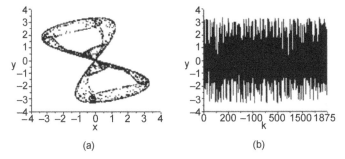

(a)   (b)

FIGURE 8.1: (a) The chaotic attractor obtained from the uncontrolled map (8.47) for: $a = 4$, and the initial condition $x = y = 0.1$. (b) The corresponding time series x.

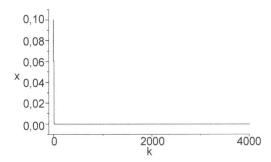

FIGURE 8.2: The time series x for the controlled map (8.48): The system converges to its fixed point $(0, 0)$.

8.7 Construction of superstable n-D mappings

The superstability of a dynamical motion is defined with the existence of a minus infinity Lyapunov exponent, this means that this motion is attractive. There are several methods for constructing 1-D polynomial mappings with attracting cycles or superstable cycles [Zhang & Agarwal (2003), Liu, *et al.*, (2003)] based on Lagrange and Newton interpolations. Superstable phenomena in some 1-D maps embedded in circuits and systems are studied in [Matsuoka & Saito (2005-2006-2007)], these maps are obtained from the study of non-autonomous piecewise constant circuit and biological models [Matsuoka & Saito (2006), Matsuoka, *et al.*, (2005-2006), Keener & Glass (1984)]. Rich dynamical behaviors can be seen in the presence of superstability [Matsuoka & Saito (2006), Matsuoka, *et al.*, (2005-2006)], especially, the attractivity of the motion that guarantees its stability.

In this section, we construct a family of superstable n-D mappings, in the sense that all the states of a member in this family are superstable, i.e., they have a minus infinity Lyapunov exponent for all the parameters of bifurcation. Indeed, let us consider the n-D map given by:

$$f(X, \rho) = \begin{pmatrix} f_1(x_1, x_2, ..., x_{n-1}, \rho) \\ f_2(x_1, x_2, ..., x_{n-1}, \rho) \\ ..... \\ f_n(x_1, x_2, ..., x_{n-1}, x_n, \rho) \end{pmatrix} \tag{8.49}$$

where $X = (x_1, x_2, ..., x_n) \in \mathbb{R}^n$ and $\rho \in \mathbb{R}^k$ is the bifurcation parameter. Assuming that the vector field of the map (8.49) is continuously differentiable in a region of interest, and it has at least one fixed point, bounded states are possible, i.e., The system of algebraic equations given by:

$$\begin{cases} f_1(x_1, x_2, ..., x_{n-1}, \rho) = x_1 \\ f_2(x_1, x_2, ..., x_{n-1}, \rho) = x_2 \\ ..... \\ f_n(x_1, x_2, ..., x_{n-1}, x_n, \rho) = x_n \end{cases} \tag{8.50}$$

has at least one real solution.

In what follows, we will give a rigorous proof that the n-D maps given by equation (8.49) have a minus infinity Lyapunov exponent for all $\rho \in \mathbb{R}^k$. By construction, the first $n-1$ components of the vector field of the map (8.49) do not depend on the last variable x_n, and the last component f_n must be depend on the variable x_n in which the map (8.49) is not trivial.

The Jacobian matrix of the map (8.49) is given by

$$Df(x, \rho) = \begin{pmatrix} f_{11} & f_{12} & \cdots & f_{1n-1} & 0 \\ f_{21} & f_{22} & \cdots & f_{2n-1} & 0 \\ \cdots & \cdots & \cdots & \cdots & 0 \\ f_{(n-1)1} & f_{(n-1)2} & \cdots & f_{(n-1)(n-1)} & 0 \\ f_{n1} & f_{n2} & \cdots & f_{n(n-1)} & f_{nn} \end{pmatrix} \tag{8.51}$$

where $f_{ij} = \frac{\partial f_i}{\partial x_j}(X_*, \rho) = f_{ij}(X_*, \rho), i, j = 1, 2, ..., n$, and $f_{nn} = f_{nn}(X, \rho)$, where $X_* = (x_1, x_2, ..., x_{n-1}) \in \mathbb{R}^{n-1}$. On the other hand, let us consider the following n-D dynamical system:

$$X_{l+1} = g(X_l), X_l \in \mathbb{R}^n, l = 0, 1, 2, ... \tag{8.52}$$

where the function $g: \mathbb{R}^n \longrightarrow \mathbb{R}^n$ is the vector field associated with system (8.52) and $X_l = \left(x_1^l, x_2^l, ..., x_n^l \right) \in \mathbb{R}^n$. Let $J\left(X_l\right)$ be its Jacobian, evaluated at $X_l \in \mathbb{R}^n$, $l = 0, 1, 2, ...$, and define the matrix:

$$T_N\left(X_0\right) = J\left(X_{N-1}\right) J\left(X_{N-2}\right) ... J\left(X_1\right) J\left(X_0\right). \tag{8.53}$$

Moreover, let $J_i(X_0, m)$ be the modulus of the ith eigenvalue of the Nth matrix $T_N\left(X_0\right)$, where $i = 1, 2, ..., n$ and $N = 0, 1, 2, ...$

Now, the Lyapunov exponents for a n-D discrete-time system are defined by:

$$l_i\left(X_0\right) = \ln\left(\lim_{N \longrightarrow +\infty} J_i(X_0, N)^{\frac{1}{N}} \right), i = 1, 2, ...n. \tag{8.54}$$

Based on this definition, we will give a rigorous proof that the family of n-D maps given by equation (8.49) is superstable for all its bifurcation parameters and all initial conditions. We have

$$T_N\left(X_0\right) \quad = \tag{8.55}$$

$$\prod_{i=N-1}^{i=0} Df\left(X_i, \rho\right) = \begin{pmatrix} h_{11} & ... & h_{1n-1} & 0 \\ h_{21} & ... & h_{2n-1} & 0 \\ ... & ... & ... & 0 \\ h_{(n-1)1} & ... & h_{(n-1)(n-1)} & 0 \\ h_{n1} & ... & h_{n(n-1)} & \prod_{i=0}^{i=N-1} f_{nn}\left(X_i, \rho\right) \end{pmatrix}$$

where the functions (h_{ij}), $1 \leq i \leq n, 1 \leq j \leq n - 1$, depend on the variables $x_1, x_2, ..., x_n$, and ρ. Thus, the characteristic polynomial of $T_N\left(X_0\right)$ has the form:

$$P\left(\lambda\right) = \prod_{i=0}^{i=N-1} f_{nn}\left(X_i, \rho\right) \det H \tag{8.56}$$

where H is the $n - 1 \times n - 1$ matrix given by:

$$H = \begin{pmatrix} h_{11} & h_{12} & ... & h_{1n-1} \\ h_{21} & h_{22} & ... & h_{2n-1} \\ ... & ... & ... & ... \\ h_{(n-1)1} & h_{(n-1)2} & ... & h_{(n-1)(n-1)} \end{pmatrix} \tag{8.57}$$

One of the eigenvalues of the Nth matrix $T_N\left(X_0\right)$ given by equation (8.53) is $\lambda = 0$, if

$$\prod_{i=0}^{i=N-1} f_{nn}\left(X_i, \rho\right) = 0 \tag{8.58}$$

for some $\rho \in \mathbb{R}^k$. Obviously, there is infinitely many choices of the function $f_n(x_1, x_2, ..., x_{n-1}, x_n, \rho)$ that satisfy equation (8.58). In this section, assuming that the function f_n has the form:

$$f_n(X, \rho) = g(X_*, \rho) + m(x_n, \rho) \tag{8.59}$$

Thus, the equation (8.58) becomes

$$\prod_{i=0}^{i=N-1} \frac{\partial m(x_n^i, \rho)}{\partial x_n} = 0 \tag{8.60}$$

Now, let us consider the following 1-D map in the variable x_n given by

$$x_n^{j+1} = \frac{\partial m(x_n^j, \rho)}{\partial x_n} = N(x_n^j, \rho) \tag{8.61}$$

We choose the function $m(x_n^j, \rho) = -\frac{2}{3}a(x_n^j)^2(2x_n^j - 3)$, where we can assume that the parameter a is the first component of the vector $\rho \in \mathbb{R}^k$, i.e., $\rho = (a, \rho_1, \rho_2, ..., \rho_k)$, hence, the map (8.61) is the standard logistic map, which has a superstable fixed point at $a = 2$. Thus, at this point the Lyapunov exponent is given by:

$$\mu = \lim_{N \to +\infty} \frac{1}{N} \ln \prod_{j=0}^{j=N-1} |4x_n^j - 2| = -\infty$$

This implies that the condition (8.58) holds. Finally, the following family of n-D mappings is globally superstable:

$$f(X, \rho) = \begin{pmatrix} f_1(x_1, x_2, ..., x_{n-1}, (2, \rho_1, \rho_2, ..., \rho_k)) \\ f_2(x_1, x_2, ..., x_{n-1}, (2, \rho_1, \rho_2, ..., \rho_k)) \\ .... \\ g(x_1, x_2, ..., x_{n-1}, (2, \rho_1, \rho_2, ..., \rho_k)) - \frac{2}{3}ax_n^2(2x_n - 3) \end{pmatrix} \tag{8.62}$$

for all its bifurcation parameters.

Example 24 *To see this numerically, let us consider the 3-D quadratic map given by:*

$$f(x, y, z, a, b) = \begin{pmatrix} axy \\ 1 - x^2 \\ bx - \frac{2}{3}az^2(2z - 3) \end{pmatrix} \tag{8.63}$$

Here, we have $n = 3, \rho = (2, a, b) \in \mathbb{R}^3$, the vector field of the map (8.63) is continuously differentiable for all $(x, y, z) \in \mathbb{R}^3$. The map (8.63) has the fixed points $\left(\frac{z - 2az^2 + \frac{4}{3}az^3}{b}, 1 - \left(\frac{z - 2az^2 + \frac{4}{3}az^3}{b} \right)^2, z \right)$, where z is the real solutions of the equations $z - 2az^2 + \frac{4}{3}az^3 = 0$, or $\left(a \left(\frac{z - 2az^2 + \frac{4}{3}az^3}{b} \right)^2 - a + 1 \right) = 0$. We have at least one fixed point $(0, 1, 0)$, thus, all conditions of the above analysis holds. Finally, the map (8.63) is globally superstable. The corresponding superstable attractors are shown in Fig. 8.3. In particular, for the chaotic attractor obtained for $a = 1.98$ and $b = 0.3$ the Lyapunov spectrum is given by: $l_1 = 0.4716, l_2 = 0$, and $l_3 = -\infty$.

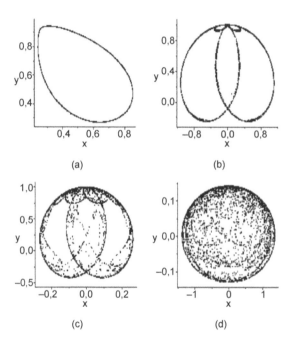

FIGURE 8.3: Superstable attractors obtained from the map (8.63) with $b = 0.3$, and (a) Periodic orbit for $a = 1.6$, (b) Quasi-periodic orbit for $a = 1.8$, (c) Chaotic orbit for $a = 1.85$, (d) Chaotic orbit for $a = 1.98$.

8.8 Examples of globally superstable 1-D quadratic mappings

In the sequel, we give an example of a globally superstable 1-D quadratic mappings. Indeed, the most general 2-D quadratic map is given by:

$$f(x,y) = \begin{pmatrix} a_0 + a_1 x + a_2 y + a_3 x^2 + a_4 y^2 + a_5 xy \\ \\ b_0 + b_1 x + b_2 y + b_3 x^2 + b_4 y^2 + b_5 xy \end{pmatrix} \qquad (8.64)$$

where $(a_i, b_i)_{0 \leq i \leq 5} \in \mathbb{R}^{12}$ are the bifurcation parameters. Some special cases of the map (8.64) can be used in potential applications. Assuming that the map (8.64) has at least one fixed point, so bounded states are possible. The Lyapunov exponents for a 2-D discrete-time system are defined by:

$$l_i(X_0) = \ln\left(\lim_{N \longrightarrow +\infty} J_i(X_0, N)^{\frac{1}{N}}\right), i = 1, 2. \qquad (8.65)$$

The Jacobian matrix of the map (8.64) is given by

$$J(x,y) = \begin{pmatrix} a_1 + 2a_3 x + a_5 y & a_2 + 2a_4 y + a_5 x \\ b_1 + 2b_3 x + b_5 y & b_2 + 2b_4 y + b_5 x \end{pmatrix} \qquad (8.66)$$

The determinant of Jacobian matrix (8.66) of the map (8.64) is given by

$$\det J(x,y) = \xi_1 x^2 + \xi_2 y^2 + \xi_3 xy + \xi_4 x + \xi_5 y + \xi_6 \qquad (8.67)$$

where

$$\begin{cases} \xi_1 = 2a_3 b_5 - 2b_3 a_5 \\ \\ \xi_2 = 2a_5 b_4 - 2a_4 b_5 \\ \\ \xi_3 = 4a_3 b_4 - 4a_4 b_3 \\ \\ \xi_4 = 2a_3 b_2 - 2a_2 b_3 + a_1 b_5 - b_1 a_5 \\ \\ \xi_5 = 2a_1 b_4 - 2b_1 a_4 - a_2 b_5 + b_2 a_5 \\ \\ \xi_6 = a_1 b_2 - a_2 b_1. \end{cases} \qquad (8.68)$$

If one finds ranges of the parameters $(a_i, b_i)_{0 \leq i \leq 5} \in \mathbb{R}^{12}$ such that $\det J(x,y) = 0$, for all $(x,y) \in \mathbb{R}^2$, then the matrix $J(x,y)$ is singular for all $(x,y) \in \mathbb{R}^2$, and one can deduce that it has at least a zero eigenvalue, which means also that the matrix $T_N(X_0)$ has a zero eigenvalue, this implies that the map (8.64) has a minus infinity Lyapunov exponent. Since $\det J(x,y)$ is a polynomial function, the only possible case for a vanishing determinant is when $\xi_i = 0, i = 1, 2, ..., 6$, i.e.,

$$\begin{cases} a_3 b_5 = b_3 a_5 \\[2mm] a_5 b_4 = a_4 b_5 \\[2mm] a_3 b_4 = a_4 b_3 \\[2mm] a_1 b_2 = a_2 b_1 \\[2mm] 2a_3 b_2 - 2a_2 b_3 + a_1 b_5 - b_1 a_5 = 0 \\[2mm] 2a_1 b_4 - 2b_1 a_4 - a_2 b_5 + b_2 a_5 = 0 \end{cases} \tag{8.69}$$

According to the classification given in [Zeraoulia and Sprott (2010(d))] one can determine the possible forms of globally superstable quadratic maps of the plane using the conditions (8.69) as follows:

For the class of one non-linearity we have two forms of superstable 1-D quadratic mappings given by:

$$x_{k+1} = a_0 + a_1 x_k + a_2 b_0 + a_3 x_k^2 \tag{8.70}$$

$$\begin{cases} x_{k+1} = a_0 + a_1 x_k + a_2 \alpha_k + a_4 \alpha_k^2 \\[2mm] \alpha_k = \left(b_0 \left(\frac{1-b_2^k}{1-b_2} \right) + b_2^k y_0 \right) \\[2mm] a_1 b_2 = 0 \end{cases} \tag{8.71}$$

and one superstable linear map of the form

$$x_{k+1} = a_0 + a_2 b_0 + (a_1 + a_5 b_0) x_k \tag{8.72}$$

For the class of two non-linearities we have four forms of superstable 1-D quadratic mappings given by:

$$x_{k+1} = a_0 + a_2 b_0 + a_4 b_0^2 + a_1 x_k + a_3 x_k^2 \tag{8.73}$$

$$x_{k+1} = a_0 + a_2 b_0 + (a_1 + a_5 b_0) x_k + a_3 x_k^2 \tag{8.74}$$

$$\begin{cases} x_{k+1} = a_0 + a_1 x_k + \frac{a_3 b_2}{b_3} y_k + a_3 x_k^2 \\[2mm] y_{k+1} = b_0 + b_1 x_k + b_2 y_k + b_3 x_k^2 \\[2mm] (b_1 a_3 - a_1 b_3) b_2 = 0 \end{cases} \tag{8.75}$$

If we set $y_k = a x_k + b$, then one has $a = \frac{b_3}{a_3}$ and $b = a_3^{-1} (b_0 a_3 - a_0 b_3)$. Thus, the 2-D quadratic map given by (8.75) can be rewritten as

$$\begin{cases} x_{k+1} = a_0 - a_0 b_2 + b_0 a_3 \frac{b_2}{b_3} + (a_1 + b_2)\, x_k + a_3 x_k^2 \\[2mm] (b_1 a_3 - a_1 b_3)\, b_2 = 0 \end{cases} \tag{8.76}$$

The same logic applies, then the fourth case is given by:

$$x_{k+1} = a_0 + b_0 a_2 - a_0 \frac{a_2}{a_5} b_5 + \left(a_1 - a_0 b_5 + \frac{a_2}{a_5} b_5 + b_0 a_5 \right) x_k + b_5 x_k^2 \tag{8.77}$$

For the case of three non-linearities, we have one possible case given by:

$$\begin{cases} x_{k+1} = a_0 + a_1 x_k + \frac{a_1 a_5}{2a_3} y_k + a_3 x_k^2 + \frac{a_5^2}{4a_3} y_k^2 + a_5 x_k y_k \\[2mm] y_{k+1} = b_0 + \frac{2b_2 a_3}{a_5} x_k + b_2 y_k \end{cases} \tag{8.78}$$

If $b_2 = 0$, then map (8.78) is clearly a 1-D quadratic map. For $b_2 \neq 0$, the same logic as above applies and one has, under some conditions, a 1-D quadratic map of the form:

$$x_{k+1} = \frac{4a_0 a_3 + 2 b a_1 a_5 + b^2 a_5^2}{4a_3} + \frac{(a_1 + b a_5)(2a_3 + a a_5)\, x_k}{2a_3} + \frac{(2a_3 + a a_5)^2\, x_k^2}{4a_3} \tag{8.79}$$

where a and b are solutions of a second degree polynomial equation.

For the case of five non-linearities, there are no 2-D maps that satisfy conditions (8.69). Finally, for the case of six non-linearities, we have 2 cases given by:

$$\begin{cases} x_{k+1} = a_0 + a_1 x_k + a_2 y_k + a_3 x_k^2 + a_4 y_k^2 + a_5 x_k y_k \\[2mm] y_{k+1} = b_0 + a_5^{-1} b_5 \left(a_1 x_k + a_2 y_k + a_3 x_k y_k + a_3 x_k^2 + a_4 y_k^2 \right) \end{cases} \tag{8.80}$$

with the condition

$$a_5^2 - 4a_3 a_4 \neq 0 \tag{8.81}$$

or equivalently

$$x_{k+1} = a_0 + b a_2 + b^2 a_4 + (a_1 + a a_2 + b a_5)\, x_k + 2 a b a_4 x_k + \left(a_3 + a a_5 + a^2 a_4 \right) x_k^2 \tag{8.82}$$

where a is a solution a five order polynomial equation with real coefficients in term of $(a_i, b_i)_{0 \leq i \leq 5} \in \mathbb{R}^{12}$, and

$$b = \frac{a_1 b_5 - a a_1 a_5 + a a_2 b_5 - a^2 a_2 a_5}{-a_3 b_5 - 2 a a_4 b_5 + a a_5^2 + 2 a^2 a_4 a_5}$$

Note that a real value of a exists because the polynomial function has an odd degree. The second case is given by:

$$
\begin{cases}
x_{k+1} = a_0 + a_1 x_k + a_2 y_k + a_3 x_k^2 + \frac{a_5^2}{4a_3} y_k^2 + a_5 x_k y_k \\[2mm]
y_{k+1} = b_0 + \frac{1}{4} a_5^{-1} a_3^{-1} b_5 \left(4 a_1 a_3 x_k + 4 a_2 a_3 y_k + 4 a_3^2 x_k^2 + a_5^2 y_k^2 + 4 a_3^2 x_k y_k \right)
\end{cases}
\tag{8.83}
$$

The same method as above gives the form:

$$
\begin{aligned}
x_{k+1} \;=\; & \frac{4 a_0 a_3 + 4 b a_2 a_3 + b^2 a_5^2}{4 a_3} + \frac{\left(2 a_1 a_3 + 2 a a_2 a_3 + 2 b a_3 a_5 + a b a_5^2 \right) x_k}{2 a_3} \\[2mm]
& + \frac{\left(2 a_3 + a a_5 \right)^2 x_k^2}{4 a_3}
\end{aligned}
\tag{8.84}
$$

Example 25 *As a test of the previous analysis, let us consider the following example:*

$$
f(x, y) = \begin{pmatrix} a_0 - 0.3x^2 - 0.6y^2 \\[2mm] b_0 + x^2 + 2y^2 \end{pmatrix}
\tag{8.85}
$$

or equivalently

$$
x_{k+1} = a_0 - 4 a_0 b_0 - \frac{20}{3} a_0^2 - \frac{3}{5} b_0^2 + \left(\frac{40}{3} a_0 + 4 b_0 \right) x - 6.9667 x^2
\tag{8.86}
$$

Different superstable dynamical behaviors of the map (8.85) are shown in Fig. 8.4 where regions of unbounded, periodic, and chaotic solutions in the $a_0 b_0$-plane for the map (8.85) are obtained using 10^6 iterations for each point.

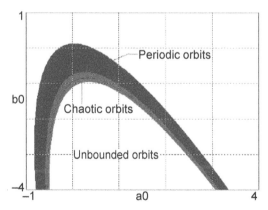

FIGURE 8.4: Regions of unbounded (in white) and bounded superstable periodic attractors and bounded superstable chaotic attractors in the a_0-b_0 plane for the map (8.85).

8.9 Construction of globally superstable 3-D quadratic mappings

The most general 3-D quadratic map is given by

$$f(x,y,z) = \begin{pmatrix} a_0 + a_1 x + a_2 y + a_3 z + a_4 x^2 + a_5 y^2 + a_6 z^2 + a_7 xy \\ + a_8 xz + a_9 yz \\ \\ b_0 + b_1 x + b_2 y + b_3 z + b_4 x^2 + b_5 y^2 + b_6 z^2 + b_7 xy \\ + b_8 xz + b_9 yz \\ \\ c_0 + c_1 x + c_2 y + c_3 z + c_4 x^2 + c_5 y^2 + c_6 z^2 + c_7 xy \\ + c_8 xz + c_9 yz \end{pmatrix}$$

(8.87)

where $(a_i, b_i, c_i)_{0 \leq i \leq 9} \in \mathbb{R}^{30}$ are the bifurcation parameters.

The global superstability of a system is allowed when all bounded orbits of this system are superstable. This property is probably rare in n-D dynamical systems with $n \geq 2$. First of all, assume that the map (8.87) has at least one fixed point. The Lyapunov exponents for a 3-D discrete-time system are defined by:

$$l_i(X_0) = \ln\left(\lim_{N \to +\infty} J_i(X_0, N)^{\frac{1}{N}} \right), i = 1, 2, 3.$$

(8.88)

In this section, we will find sufficient conditions for (8.87). Indeed, the Jacobian matrix of the map (8.87) is given by

$$J = \begin{pmatrix} j_{11} & j_{12} & j_{13} \\ j_{21} & j_{22} & j_{23} \\ j_{31} & j_{32} & j_{33} \end{pmatrix}.$$

(8.89)

where

$$\begin{cases} j_{11} = a_1 + 2xa_4 + ya_7 + za_8, j_{12} = a_2 + 2ya_5 + xa_7 + za_9 \\ \\ j_{13} = a_3 + xa_8 + 2za_6 + ya_9, j_{21} = b_1 + 2xb_4 + yb_7 + zb_8 \\ \\ j_{22} = b_2 + 2yb_5 + xb_7 + zb_9, j_{23} = b_3 + xb_8 + 2zb_6 + yb_9 \\ \\ j_{31} = c_1 + 2xc_4 + yc_7 + zc_8, j_{32} = c_2 + 2yc_5 + xc_7 + zc_9 \\ \\ j_{33} = c_3 + xc_8 + 2zc_6 + yc_9 \end{cases}$$

The determinant of Jacobian matrix (8.89) is given by:

$$\det J(x,y,z) = \psi_1 + \Phi_1(x,y,z) + \Phi_2(x,y,z) + \Phi_3(x,y,z)$$

(8.90)

where

$$
\begin{cases}
\Phi_1\left(x,y,z\right) = \xi_1 x + \xi_2 y + \xi_3 z + \xi_4 xy + \xi_5 xz + \xi_6 yz + \xi_7 xyz + \xi_8 x^2 \\[2mm]
\Phi_2\left(x,y,z\right) = \xi_9 y^2 + \xi_{10} z^2 + \xi_{11} x^3 + \xi_{12} y^3 + \xi_{13} z^3 + \left(\xi_{14} + \xi_{18}\right) yx^2 \\[2mm]
\Phi_3\left(x,y,z\right) = \xi_{15} y^2 x + \left(\xi_{16} + \xi_{19}\right) zx^2 + \xi_{17} z^2 x + \xi_{20} zy^2 + \xi_{21} z^2 y.
\end{cases}
\tag{8.91}
$$

and

$$
\begin{cases}
\xi_1 = \sum_{i=2}^{i=4} \psi_i, \xi_2 = \sum_{i=5}^{i=7} \psi_i, \xi_3 = -\sum_{i=8}^{i=10} \psi_i, \xi_4 = \sum_{i=11}^{i=15} \psi_i \\[4mm]
\xi_5 = -\sum_{i=16}^{i=20} \psi_i, \xi_6 = -\sum_{i=21}^{i=26} \psi_i, \xi_7 = -2\sum_{i=27}^{i=28} \psi_i, \xi_8 = \sum_{i=29}^{i=31} \psi_i \\[4mm]
\xi_9 = -\sum_{i=32}^{i=34} \psi_i, \xi_{10} = \sum_{i=35}^{i=37} \psi_i, \xi_{11} = -2\psi_{38}, \xi_{12} = 2\psi_{39} \\[4mm]
\xi_{13} = -2\psi_{40}, \xi_{14} = -4\psi_{41}, \xi_{15} = 2\sum_{i=42}^{i=43} \psi_i, \xi_{16} = 4\psi_{44} \\[4mm]
\xi_{17} = -2\sum_{i=45}^{i=46} \psi_i, \xi_{18} = -2\psi_{47}, \xi_{19} = 2\psi_{48} \\[4mm]
\xi_{20} = -2\sum_{i=49}^{i=50} \psi_i, \xi_{21} = 2\sum_{i=51}^{i=52} \psi_i
\end{cases}
\tag{8.92}
$$

where

$$
\begin{cases}
\psi_1 = a_1 b_2 c_3 - a_1 b_3 c_2 - a_2 b_1 c_3 + a_2 c_1 b_3 + b_1 a_3 c_2 - a_3 b_2 c_1 \\[2mm]
\psi_2 = 2a_2 b_3 c_4 - 2a_2 b_4 c_3 - 2a_3 b_2 c_4 + 2a_3 c_2 b_4 + 2b_2 a_4 c_3 - 2a_4 b_3 c_2 \\[2mm]
\psi_3 = a_1 b_2 c_8 - a_1 b_3 c_7 - a_1 c_2 b_8 + a_1 c_3 b_7 - a_2 b_1 c_8 + a_2 c_1 b_8 + b_1 a_3 c_7 \\[2mm]
\psi_4 = b_1 c_2 a_8 - b_1 c_3 a_7 - a_3 c_1 b_7 - b_2 c_1 a_8 + c_1 b_3 a_7 \\[2mm]
\psi_5 = 2a_1 c_3 b_5 - 2a_1 b_3 c_5 + 2b_1 a_3 c_5 - 2b_1 a_5 c_3 - 2a_3 c_1 b_5 + 2c_1 b_3 a_5
\end{cases}
$$

$$
\begin{cases}
\psi_6 = a_1b_2c_9 - a_1c_2b_9 - a_2b_1c_9 + a_2c_1b_9 + a_2b_3c_7 - a_2c_3b_7 + b_1c_2a_9 \\[4pt]
\psi_7 = -a_3b_2c_7 + a_3c_2b_7 - b_2c_1a_9 + b_2c_3a_7 - b_3c_2a_7 \\[4pt]
\psi_8 = 2a_1c_2b_6 - 2a_1b_2c_6 + 2a_2b_1c_6 - 2a_2c_1b_6 - 2b_1c_2a_6 + 2b_2c_1a_6 \\[4pt]
\psi_9 = a_1b_3c_9 - a_1c_3b_9 - a_2b_3c_8 + a_2c_3b_8 - b_1a_3c_9 + b_1c_3a_9 \\[4pt]
\psi_{10} = +a_3c_1b_9 - a_3c_2b_8 - b_2c_3a_8 - c_1b_3a_9 + b_3c_2a_8 + a_3b_2c_8
\end{cases}
\tag{8.93}
$$

and

$$
\begin{cases}
\psi_{11} = 4a_3b_4c_5 - 4a_3b_5c_4 - 4a_4b_3c_5 + 4a_4c_3b_5 + 4b_3a_5c_4 - 4a_5b_4c_3 \\[4pt]
\psi_{12} = 2a_1b_5c_8 - 2a_1c_5b_8 - 2b_1a_5c_8 + 2b_1c_5a_8 + 2c_1a_5b_8 - 2c_1b_5a_8 \\[4pt]
\psi_{13} = -2a_2b_4c_9 + 2a_2c_4b_9 + 2b_2a_4c_9 - 2b_2c_4a_9 - 2a_4c_2b_9 + 2c_2b_4a_9 \\[4pt]
\psi_{14} = a_1b_7c_9 - a_1c_7b_9 - a_2b_7c_8 + a_2b_8c_7 - b_1a_7c_9 + b_1a_9c_7 + b_2a_7c_8 \\[4pt]
\psi_{15} = -b_2a_8c_7 + c_1a_7b_9 - c_1b_7a_9 - c_2a_7b_8 + c_2a_8b_7
\end{cases}
$$

$$
\begin{cases}
\psi_{16} = 4a_2b_4c_6 - 4a_2c_4b_6 - 4b_2a_4c_6 + 4b_2a_6c_4 + 4a_4c_2b_6 - 4c_2b_4a_6 \\[4pt]
\psi_{17} = 2a_1b_6c_7 - 2a_1b_7c_6 - 2b_1a_6c_7 + 2b_1a_7c_6 + 2c_1a_6b_7 - 2c_1a_7b_6 \\[4pt]
\psi_{18} = -2a_3b_4c_9 + 2a_3c_4b_9 + 2a_4b_3c_9 - 2a_4c_3b_9 - 2b_3c_4a_9 + 2b_4c_3a_9 \\[4pt]
\psi_{19} = a_1b_8c_9 - a_1b_9c_8 - b_1a_8c_9 + b_1a_9c_8 + a_3b_7c_8 - a_3b_8c_7 \\[4pt]
\psi_{20} = c_1a_8b_9 - c_1a_9b_8 - b_3a_7c_8 + b_3a_8c_7 + c_3a_7b_8 - c_3a_8b_7
\end{cases}
\tag{8.94}
$$

and

$$
\begin{cases}
\psi_{21} = 4a_1b_6c_5 - 4a_1b_5c_6 + 4b_1a_5c_6 - 4b_1a_6c_5 - 4c_1a_5b_6 \\[4pt]
\psi_{22} = 4c_1a_6b_5 - 2a_2b_6c_7 + 2a_2b_7c_6 + 2b_2a_6c_7 - 2b_2a_7c_6 \\[4pt]
\psi_{23} = -2c_2a_6b_7 + 2a_5c_3b_8 - b_2a_8c_9 + c_3b_7a_9 \\[4pt]
\psi_{24} = 2c_2a_7b_6 + 2a_3b_5c_8 - 2a_3c_5b_8 - 2b_3a_5c_8 + 2b_3c_5a_8 \\[4pt]
\psi_{25} = -2c_3b_5a_8 + a_2b_8c_9 - a_2b_9c_8 - a_3b_7c_9 + a_3c_7b_9
\end{cases}
$$

$$\begin{cases} \psi_{26} = b_2 a_9 c_8 + b_3 a_7 c_9 - b_3 a_9 c_7 + c_2 a_8 b_9 - c_2 a_9 b_8 - c_3 a_7 b_9 \\[2mm] \psi_{27} = 4a_4 b_6 c_5 - 4a_4 b_5 c_6 + 4a_5 b_4 c_6 - 4a_5 c_4 b_6 - 4b_4 a_6 c_5 + 4a_6 b_5 c_4 \\[2mm] \psi_{28} = a_7 b_8 c_9 - a_7 b_9 c_8 - a_8 b_7 c_9 + a_8 c_7 b_9 + b_7 a_9 c_8 - a_9 b_8 c_7 \\[2mm] \psi_{29} = 2a_2 c_4 b_8 - 2a_2 b_4 c_8 + 2a_3 b_4 c_7 - 2a_3 c_4 b_7 + 2b_2 a_4 c_8 - 2b_2 c_4 a_8 \\[2mm] \psi_{30} = -2a_4 b_3 c_7 - 2a_4 c_2 b_8 + 2a_4 c_3 b_7 + 2b_3 c_4 a_7 + 2c_2 b_4 a_8 - 2b_4 c_3 a_7 \\[2mm] \psi_{31} = a_1 b_7 c_8 - a_1 b_8 c_7 - b_1 a_7 c_8 + b_1 a_8 c_7 + c_1 a_7 b_8 - c_1 a_8 b_7 \end{cases} \tag{8.95}$$

and

$$\begin{cases} \psi_{32} = 2a_1 c_5 b_9 - 2a_1 b_5 c_9 + 2b_1 a_5 c_9 - 2b_1 c_5 a_9 + 2a_3 b_5 c_7 \\[2mm] \psi_{33} = -2a_3 c_5 b_7 - 2c_1 a_5 b_9 + 2c_1 b_5 a_9 - 2b_3 a_5 c_7 + 2b_3 a_7 c_5 + 2a_5 c_3 b_7 \\[2mm] \psi_{34} = -2c_3 b_5 a_7 + a_2 b_7 c_9 - a_2 c_7 b_9 - b_2 a_7 c_9 + b_2 a_9 c_7 + c_2 a_7 b_9 - c_2 b_7 a_9 \\[2mm] \psi_{35} = 2a_1 c_6 b_9 - 2a_1 b_6 c_9 + 2a_2 b_6 c_8 - 2a_2 c_6 b_8 + 2b_1 a_6 c_9 - 2b_1 c_6 a_9 \\[2mm] \psi_{36} = -2b_2 a_6 c_8 + 2b_2 a_8 c_6 - 2c_1 a_6 b_9 + 2c_1 b_6 a_9 + 2c_2 a_6 b_8 - 2c_2 b_6 a_8 \\[2mm] \psi_{37} = a_3 b_8 c_9 - a_3 b_9 c_8 - b_3 a_8 c_9 + b_3 a_9 c_8 + c_3 a_8 b_9 - c_3 a_9 b_8 \end{cases}$$

$$\begin{cases} \psi_{38} = a_4 b_8 c_7 - a_4 b_7 c_8 + b_4 a_7 c_8 - b_4 a_8 c_7 - c_4 a_7 b_8 + c_4 a_8 b_7 \\[2mm] \psi_{39} = a_5 c_7 b_9 - a_5 b_7 c_9 + b_5 a_7 c_9 - b_5 a_9 c_7 - a_7 c_5 b_9 + c_5 b_7 a_9 \\[2mm] \psi_{40} = a_6 b_9 c_8 - a_6 b_8 c_9 + b_6 a_8 c_9 - b_6 a_9 c_8 - a_8 c_6 b_9 + c_6 a_9 b_8 \\[2mm] \psi_{41} = a_4 c_5 b_8 - a_4 b_5 c_8 + a_5 b_4 c_8 - a_5 c_4 b_8 - b_4 c_5 a_8 + b_5 c_4 a_8 \\[2mm] \psi_{42} = 2a_4 b_5 c_9 - 2a_4 c_5 b_9 - 2a_5 b_4 c_9 + 2a_5 c_4 b_9 + 2b_4 c_5 a_9 - 2b_5 c_4 a_9 \\[2mm] \psi_{43} = -a_5 b_7 c_8 + a_5 b_8 c_7 + b_5 a_7 c_8 - b_5 a_8 c_7 - a_7 c_5 b_8 + c_5 a_8 b_7 \\[2mm] \psi_{44} = a_4 b_7 c_6 - a_4 b_6 c_7 + b_4 a_6 c_7 - b_4 a_7 c_6 - a_6 c_4 b_7 + c_4 a_7 b_6 \end{cases} \tag{8.96}$$

and

$$
\left\{
\begin{aligned}
\psi_{45} &= 2a_4b_6c_9 - 2a_4c_6b_9 - 2b_4a_6c_9 + 2b_4c_6a_9 + 2a_6c_4b_9 - 2c_4b_6a_9 \\
\psi_{46} &= a_6b_7c_8 - a_6b_8c_7 - a_7b_6c_8 + a_7c_6b_8 + b_6a_8c_7 - a_8b_7c_6 \\
\psi_{47} &= a_4c_7b_9 - a_4b_7c_9 + b_4a_7c_9 - b_4a_9c_7 - c_4a_7b_9 + c_4b_7a_9 \\
\psi_{48} &= a_4b_9c_8 - a_4b_8c_9 + b_4a_8c_9 - b_4a_9c_8 - c_4a_8b_9 + c_4a_9b_8
\end{aligned}
\right.
$$

$$
\left\{
\begin{aligned}
\psi_{49} &= 2a_5b_7c_6 - 2a_5b_6c_7 + 2a_6b_5c_7 - 2a_6c_5b_7 - 2b_5a_7c_6 + 2a_7b_6c_5 \\
\psi_{50} &= a_5b_8c_9 - a_5b_9c_8 - b_5a_8c_9 + b_5a_9c_8 + c_5a_8b_9 - c_5a_9b_8 \\
\psi_{51} &= 2a_5b_6c_8 - 2a_5c_6b_8 - 2a_6b_5c_8 + 2a_6c_5b_8 + 2b_5a_8c_6 - 2b_6c_5a_8 \\
\psi_{52} &= a_6b_7c_9 - a_6c_7b_9 - a_7b_6c_9 + a_7c_6b_9 + b_6a_9c_7 - b_7c_6a_9
\end{aligned}
\right.
$$

$$(8.97)$$

The first test for the non-invertibility of the 3-D quadratic map (8.87) is the location of the term ψ_1, i.e., if $\psi_1 \neq 0$. The map (8.87) is invertible when the following condition holds true:

$$(a_2b_3 - a_3b_2)\, c_1 + (b_1a_3 - a_1b_3)\, c_2 + (a_1b_2 - a_2b_1)\, c_3 \neq 0 \qquad (8.98)$$

In order to simplify our search for non-invertible mappings of the form (8.87), assume that $c_1 = c_2 = c_3 = 0$, then one has the following system:

$$
f\left(x, y, z\right) = \left(
\begin{array}{c}
a_0 + a_1x + a_2y + a_3z + a_4x^2 + a_5y^2 + a_6z^2 + a_7xy \\
+a_8xz + a_9yz \\[4pt]
b_0 + b_1x + b_2y + b_3z + b_4x^2 + b_5y^2 + b_6z^2 + b_7xy \\
+b_8xz + b_9yz \\[4pt]
c_0 + c_4x^2 + c_5y^2 + c_6z^2 + c_7xy + c_8xz + c_9yz
\end{array}
\right)
$$

$$(8.99)$$

If one finds ranges of the parameters $(a_i, b_i, c_i)_{1 \leq i \leq 9} \in \mathbb{R}^{27}$ such that $\det J\left(x, y, z\right) = 0$, for all $(x, y, z) \in \mathbb{R}^3$, then the matrix $J\left(x, y, z\right)$ is singular for all $(x, y, z) \in \mathbb{R}^3$, and one can deduce that it has at least a zero eigenvalue, which means also that the matrix $T_N\left(X_0\right)$ has a zero eigenvalue, this implies that the map (8.87) has a minus infinity Lyapunov exponent. Since $\det J\left(x, y, z\right)$ is a polynomial function, the only possible case for a vanishing determinant is when $\psi_j\left(a_i, b_i, c_i\right)_{1 \leq i \leq 9} = 0, j = 1, 2, ..., 52$, i.e., We define the following subsets of $\mathbb{R}^{27}$:

$$\Omega_j = \left\{ (a_i, b_i, c_i)_{1 \leq i \leq 9} \in \mathbb{R}^{27} : \psi_j\left(a_i, b_i, c_i\right)_{1 \leq i \leq 9} = 0 \right\}, , j = 1, 2, ..., 52$$

$$(8.100)$$

Thus, one can obtain that if there are vectors $(a_i, b_i, c_i)_{1 \le i \le 9} \in \mathbb{R}^{27}$, such that $(a_i, b_i, c_i)_{1 \le i \le 9} \neq (0, 0, ..., 0)$, i.e., $\cap_{j=1}^{j=52} \Omega_j \neq \emptyset$, then $\det J(x, y, z) = 0$, for all $(x, y, z) \in \mathbb{R}^3$. However, the systems of equations $\psi_j (a_i, b_i, c_i)_{1 \le i \le 9} = 0, j = 1, 2, ..., 52$, can be rewritten as

$$AC = O \tag{8.101}$$

where $A = A\left((a_i, b_i)_{1 \le i \le 9}\right)$ is a 52×9 matrix, and

$$C = (c_1, c_2, c_3, c_4, c_5, c_6, c_7, c_8, c_9)$$

is a 9×1 matrix of unknowns, and O is null vector of $\mathbb{R}^{52}$. The classical method of Fontené-Rouché can be used in order to solve this system of equations by introducing the so-called principal determinant. However, theoretically, this is very hard because it is very difficult to determine the set of principal unknowns.

As a final result, we have the following Theorem:

Theorem 84 *There is at least one non-trivial globally superstable 3-D quadratic map.*

Proof 43 *It suffices to prove that there is at least one non-null vector $(a_i, b_i, c_i)_{1 \le i \le 9} \in \mathbb{R}^{27}$, such that the set $\cap_{j=1}^{j=52} \Omega_j$ is not empty. Indeed, the vector $a_0 = 1, a_i = 0, i \neq 0, 5, a_5 = a, b_i = 0, i \neq 1, 3, b_1 = 1, b_i = 0, i \neq 2, c_2 = 1,$ i.e., the map is given by:*

$$f(x, y, z) = \begin{cases} 1 - ay^2 \\[1em] x + bz \\[1em] y \end{cases} \tag{8.102}$$

8.10   Hyperbolicity of dynamical systems

In this section, we present and discuss some relevant results of the hyperbolic theory of dynamical systems. To do this, we need some important concepts and definitions. Let M be a closed 3-manifold and f be a C^1 vector field with generating flow f_t, $t \in R$. Given $p \in M$, one has the following definitions:

Definition 67 *(a) The ω-limit set of p, denoted by $\omega(p)$ is the set of points $x \in M$ such that $x = \lim_{n \to \infty} f_{t_n}(p)$ for some sequence of real numbers $t_n \to \infty$.*

(b) A compact invariant set Λ of f is isolated if there exists an open set $\Lambda \subset U$ such that $\Lambda = \cap_{t \in \mathbb{R}} f_t(U)$.

(c) If $f_t(U) \subset U$ for $t > 0$, then Λ is an attracting set.

(d) The topological basin of an attracting set Λ is the set given by

$$W^s(\Lambda) = \left\{ x \in M : \lim_{t \to +\infty} dist\,(f_t(x), \Lambda) = 0 \right\}. \qquad (8.103)$$

(e) A compact invariant set Λ of f is transitive if $\Lambda = \omega(p)$ for some $p \in \Lambda$.

(f) A compact invariant set Λ of f is attracting if it realizes as $\cap_{t>0} f_t(U)$ for some compact neighborhood U called the basin of attraction.

(g) An attractor is a transitive attracting set.

(h) An attractor, or repeller, is proper if it is not the whole manifold M.

(l) An invariant set of f is non-trivial if it is neither a periodic orbit nor a singularity.

In the remaining study, we assume that all basins of attraction are smooth manifolds with a boundary transverse to f.

Definition 68 *(a) A compact invariant set Λ of f is hyperbolic if there exist positive constants K and λ and a continuous invariant tangent bundle decomposition*

$$T_\Lambda M = E_\Lambda^s \oplus E_\Lambda^f \oplus E_\Lambda^u \qquad (8.104)$$

such that

(i) E_Λ^s is contracting, i.e.,

$$\left\| Df_t(x) \big|_{E_x^s} \right\| \le Ke^{-\lambda t}, \forall t > 0, \forall x \in \Lambda \qquad (8.105)$$

(ii) E_Λ^f is expanding, i.e.,

$$\left\| Df_{-t}(x) \big|_{E_x^u} \right\| \le Ke^{-\lambda t}, \forall t > 0, \forall x \in \Lambda \qquad (8.106)$$

(iii) E_Λ^f is tangent to f.

By the stable manifold theory [Hirsch, *et al.*, (1977)] the stable and unstable manifolds $W_f^s(p)$ and $W_f^u(p)$ associated to a point $p \in \Lambda$ exist. Here, the sets $W_f^s(p)$ and $W_f^u(p)$ are respectively tangent to the subspaces $E_p^s \oplus E_p^f$ and $E_p^f \oplus E_p^u$ of $T_p M$. In particular, $W_f^s(p)$ and $W_f^u(p)$ are well-defined if p belongs to a hyperbolic periodic orbit of f. If O is an orbit of f, then one can write $W_f^{s(u)}(O) = W_f^{s(u)}(p)$ for some $p \in O$ and $W_f^{s(u)}(O)$ does not depend on the point p. If $\dim E_\Lambda^s = \dim E_\Lambda^s = 1$, then the set Λ is of saddle type and the manifolds $W_f^s(p)$ and $W_f^u(p)$ are two-dimensional submanifolds of M, and the maps $p \in \Lambda \to W_f^s(p)$ and $p \in \Lambda \to W_f^u(p)$ are continuous on compact parts of Λ.

We have the following definitions of hyperbolic attractors and repellers:

Definition 69 *(b) A hyperbolic attractor is an attractor which is simultaneously a hyperbolic set.*

(c) A hyperbolic repeller is a hyperbolic attractor for the time-reversed vector field.

(d) A closed orbit of f is hyperbolic if it is hyperbolic as a compact invariant set of f.

Non-trivial hyperbolic attractor Λ is often called *hyperbolic strange attractor* and this is equivalent to the fact that $E_\Lambda^s \neq 0$ for the corresponding hyperbolic splitting of Λ given by (8.104).

Definition 70 *We say that f is uniformly hyperbolic on Λ, or Λ is a uniformly hyperbolic invariant set of f if T_Λ splits into a direct sum $T_\Lambda = E_\Lambda^s \oplus E_\Lambda^u$ of two Df-invariant subbundles and there are constants $c > 0$ and $0 < \lambda < 1$ such that*

$$\begin{cases} \left\| Tf^n|_{E_\Lambda^s} \right\| < c\lambda^n \\ \left\| Tf^{-n}|_{E_\Lambda^u} \right\| < c\lambda^n \end{cases} \qquad (8.107)$$

hold for all $n \geq 0$. Here $\|.\|$ denotes a metric on M.

Again, the stable manifold theory confirms the existence of the strong stable manifold $W_f^{ss}(p)$ associated to $p \in \Lambda$. This manifold is tangent to the subspace E_p^s of $T_p M$, and for all $p \in \Lambda$, one has

$$W_f^s(p) =_{t \in \mathbb{R}} W_f^{ss}(f_t(p)) \qquad (8.108)$$

If p is regular ($f(p) \neq 0$) then $W_f^s(p)$ is a well-defined two-dimensional submanifold of M, and the map $p \in \Lambda \to W_f^s(p)$ is continuous on compact parts of Λ at the regular points p of Λ.

The notion of *quasi-hyperbolicity* (a purely topological condition) was introduced in order to avoid some problems related to the notion of uniform hyperbolicity. The definition was given by using the fact that the differential of f induces a dynamical system $Df : TM \to TM$. By restricting it to the invariant set T_Λ, one obtains $Df : T_\Lambda \to T_\Lambda$:

Definition 71 *(a) An orbit of Df is called a trivial orbit if it is contained in the zero section of the bundle T_Λ.*

(b) The map f is quasi-hyperbolic on Λ if $Df : T_\Lambda \to T_\Lambda$ has no non-trivial bounded orbit.

In fact, hyperbolicity implies quasi-hyperbolicity and the converse is not true in general. If $f|_\Lambda$ is chain recurrent, these two notions coincide [Sacker & Sell (1974)]:

Theorem 85 *Assume that $f|_\Lambda$ is chain recurrent, that is, $\mathcal{R}(f|_\Lambda) = \Lambda$. Then f is uniformly hyperbolic on Λ if and only if f is quasi-hyperbolic on it.*

The *spectral decomposition theorem* of Smale is one of the most useful tools for the study of global dynamics of hyperbolic diffeomorphism. This important result uses the concept of *dominated splitting* as a weaker form of hyperbolicity. Let

$$m(A) = \inf_{v \neq 0} \frac{\|Av\|}{\|v\|} \qquad (8.109)$$

denote the minimum norm of a linear operator A. The notation $E_\Lambda \prec F_\Lambda$ indicate that the bundle F_Λ dominate the bundle E_Λ. Hence, we have the following definition:

Definition 72 *(a) Let Λ be a compact invariant set of f. A continuous invariant splitting $T_\Lambda M = E_\Lambda \oplus F_\Lambda$ over Λ is dominated if*
  (i) The splitting is non-trivial, i.e., $E_x \neq 0$ and $F_x \neq 0$ for every $x \in \Lambda$.
  (ii) There exist positive constants K and λ such that

$$\frac{\|Df_t(x)\,|_{E_x}\|}{m\left(Df_t(x)\,|_{F_x}\right)} \leq Ke^{-\lambda t}, \forall t > 0, \forall x \in \Lambda. \qquad (8.110)$$

  (b) An invariant splitting $T_\Lambda M = E_1 \oplus ... \oplus E_k$ is dominated if $\oplus_1^j E_i \prec \oplus_{j+1}^k E_i$ for every j.
  (c) The splitting $T_\Lambda M = E_1 \oplus ... \oplus E_k$ is called codimension one dominated splitting if all subbundles are one dimensional.

The dominated splittings are not unique, as shown by the case of the toral automorphism with a center and expanding subspace, or the case of any partially hyperbolic set.

The importance dominated splitting can be seen from the following result proved in [Pujals & Sambarino (2000(a))]:

Theorem 86 *Surface diffeomorphisms that can not be C^1-approximated by another exhibiting homoclinic tangencies, has the property that its Limit set has dominated splitting.*

Another result can be found in [Pujals (2002)] using the following definition:

Definition 73 *The point x is Lyapunov stable (in the future) if given $\epsilon > 0$ there exists $\delta > 0$ such that $f^n(B_\delta(x)) \subset B_\epsilon(f^n(x))$ for any positive integer n.*

Theorem 87 *Let $f : M \to M$ be a C^2-diffeomorphism of a finite dimensional compact riemannian manifold M and let there be a set having a codimension one dominated splitting. Then, there exists a neighborhood V such that if $f^n(x) \in V$ for any positive integer n, and x is Lyapunov stable, one of the following holds:*
  (1) $\omega(x)$ is a periodic orbit,
  (2) $\omega(x)$ is a periodic curve normally attractive supporting and irrational rotation.

The following result was proved in [Smale (1969)]:

Theorem 88 *If $f : M \to M$ is uniformly hyperbolic, the non-wandering set splits into a finite disjoint union $\Omega(f) = \Lambda_1 \cup .... \cup \Lambda_N$ of compact invariant sets Λ_i isolated and transitive. The α-limit set of every orbit is contained in some Λ_i, and analogously for the ω-limit set.*

The generalized term for the geometric Lorenz attractor is *partially hyperbolic attractors* [Morales (1998)]. These attractors have rich and complicated dynamics, resembling both the hyperbolic expanding and geometric Lorenz attractors [Bonatti, *et al.* (2003)]:

Definition 74 *(c) A compact invariant set Λ is partially hyperbolic if it exhibits a dominated splitting $T_\Lambda M = E_\Lambda \oplus F_\Lambda$ such that E_Λ is contracting, i.e.,*

$$\left\| Df_t(x) \big|_{E_x^s} \right\| \leq K e^{-\lambda t}, \forall t > 0, \forall x \in \Lambda. \qquad (8.111)$$

(d) The central direction E^{cu} in a partially hyperbolic splitting $T_\Lambda M = E_\Lambda \oplus F_\Lambda$ is volume expanding if there are constants $T > 0, k > 1$ so that

$$\left| \det \left(Df_T \big|_{F_\Lambda} \right) \right| > k \qquad (8.112)$$

(e) [Bonatti, et al. (2003), p. 105, or [Morales (1998)]: A singular-hyperbolic set Λ of f is a partially hyperbolic set with hyperbolic singularities and volume expanding central subbundle E_Λ^c , i.e.,

$$\left| \det \left(Df_t(x) \big|_{E_x^c} \right) \right| \geq K^{-1} e^{\lambda t}, \forall t > 0, \forall x \in \Lambda. \qquad (8.113)$$

The generalized term for Lorenz-type dynamics is *singular hyperbolicity*. Indeed, if $P_f(\Lambda)$ and $Sing_f(\Lambda)$ are the sets of periodic orbits and of singularities of f in Λ, respectively. Hence, one has the following definition:

Definition 75 *(a) A compact invariant singular set Λ of a C^1 vector fields f is singular hyperbolic if it is partially hyperbolic with volume expanding central direction and any $\sigma \in Sing_f(\Lambda)$ is hyperbolic.*

(b) A singular-hyperbolic attracting set is a singular-hyperbolic set which is simultaneously an attracting set.

(c) A singular-hyperbolic repeller is a singular-hyperbolic attractor for the time-reversed vector field.

An example of a partially hyperbolic set is the suspended horseshoe [de Melo & Palis (1982)] which is not singular hyperbolic. The contracting Lorenz attractor [Palis & Takens (1993)] is also an example of a partially hyperbolic transitive set which is not singular hyperbolic.

The basins of attraction of singular-hyperbolic sets have some important properties.

Definition 76 *(a) An attractor of a vector field f with generating flow f_t is a transitive set equal to $\cap_{t>0} f_t(U)$ for some neighborhood U called basin of attraction.*

(b) An attractor is singular-hyperbolic if it has singularities (all hyperbolic) and is partially hyperbolic with volume expanding direction.

(c) A singular basic set is a singular-hyperbolic set which is both isolated and transitive.

Secondly, we define the so called *handlebody* [Hempel (1976)], used to characterize the basin of attraction of singular-hyperbolic:

Definition 77 *(a) A handlebody of genus $n \in \mathbb{N}$ is a compact 3-manifold V containing a disjoint collection of n properly embedded 2-cells, such that the result of cutting V along these disks is a 3-cell, or a handlebody of genus $n \in \mathbb{N}$ (or a cube with n handles) is a compact 3-manifold with a boundary V containing a disjoint collection of n properly embedded 2-cells, such that the result of cutting V along these disks is a 3-cell .*

(b) If n is the genus of the handlebody, then the number of singularities of the attractors is $n - 1$.

We have the following examples:

- Any orientable handlebody of genus ≤ 1 cannot be the basin of attraction of any singular-hyperbolic attractor [Morales (2006)].

- The geometric Lorenz attractor [Afraimovich, *et al.* (1982), Guckenheimer and Williams (1979)] is an example of a vector field with an attractor having a genus-two handlebody as a basin of attraction.

- All partially hyperbolic attractors with hyperbolic singularities and volume expanding central subbundle are a generalization of the geometric Lorenz attractors including all C^1 robustly transitive sets with singularities on 3-manifolds up to flow reversing attractors [Morales, *et al.* (1998)].

- The geometric Lorenz attractor and singular horseshoe singular transitive sets.

- The basin of attraction of one-dimensional hyperbolic attractors for diffeomorphisms on m-dimensional manifolds with $m \geq 4$ is either $G \times \mathbb{R}^{m-3}$ for some open proper subset G of $\mathbb{S}^3$ or else an m-dimensional handlebody.

- The 3-ball and the solid torus $\mathbb{D}^2 \times \mathbb{S}^1$ are the only handlebodies which can be realized as the basin of attraction of a hyperbolic attractor of a vector field on a closed 3-manifold (the attractor is a singularity or a closed orbit) [Morales (2006)].

We need the following definitions regarding Lorenz and transitive mappings:

Definition 78 *(a) (Lorenz mappings) A mapping $f : [0,1] \to [0,1]$ is called n-Lorenz mapping if there are $n \in \mathbb{N}$ and $C = \{c_0, ..., c_n\}$ with $0 < c_n < c_{n-1}, ..., c_1 < c_0 = \frac{1}{2}$ such that the following properties hold:*
    (1) C is the set of discontinuity points of f,
    (2) $\lim_{x \to \bar{c}_i} f(x) = 1$ for all $i = \{1, ..., n\}$,
    (3) $\lim f(x) = r_i^+$ exists for all $i = \{1, ..., n\}$,
    (4) f is C^1 on $[0,1] \backslash C$ and there exists $\lambda > 2$ such that $f'(x) \geq \lambda$ if $x \in [0,1] \backslash C$.
    (b) The mapping f is transitive if it has a dense forward orbit.

The n-Lorenz mapping is a direct generalization of the classical Lorenz mappings (0-Lorenz mappings) in the interval $[0,1]$ and it has $(n+1)$ discontinuity points [Morales, 2007].

We have proof of the following result in [Guckenheimer & Williams (1979)] as an extension of the classical Guckenheimer-Williams conditions for transitivity of Lorenz mappings:

Lemme 8.4 *All n-Lorenz mappings with a sufficiently small norm are transitive.*

Definition 79 *(a) (Heegaard splittings) A Heegaard splitting of a closed 3-manifold M is a decomposition $M = V_1 \cup V_2$, where $V_1, V_2 \subset M$ are handle-bodies such that $V_1 \cap V_2 = \partial V_1 = \partial V_2$.*
    (b) The genus of the Heegaard splitting $M = V_1 \cup V_2$ is the genus of V_1 (or V_2).

In fact, every closed 3-manifold M has a Heegaard splitting [Hempel (1976)]:

Lemme 8.5 *Every closed 3-manifold with a genus-one Heegaard splitting has a genus-two Heegaard splitting.*

Theorem 89 *Every orientable handlebody of genus $n \geq 2$ can be realized as the basin of attraction of a singular-hyperbolic attractor [Morales (2007)] .*

Every closed orientable 3-manifold supports a vector field whose non-wandering set consists of a singular-hyperbolic attractor and a singular-hyperbolic repeller.

The following result was proved in [Morales (2007)]:

Theorem 90 *For every orientable handlebody V of genus $n \geq 2$, there exists a C^∞ vector field f in $\mathbb{R}^3$ such that V is the basin of attraction of a singular-hyperbolic attractor of f.*

Theorem 90 implies the following consequences [Morales (2007)]:

(a) For every $(n \geq 1)$ there exists a C^∞ vector field in $\mathbb{R}^3$ exhibiting a singular-hyperbolic attractor with just n singularities.

(b) Every closed orientable 3-manifold supports a vector field whose non-wandering set is the union of a singular-hyperbolic attractor and a singular-hyperbolic repeller.

(c) There are open sets of C^r vector fields without hyperbolic attractors or hyperbolic repellers on every closed 3-manifold, $r \geq 1$.

Theorem 90 extends the classification for hyperbolic systems given in [Christy (1993), Zhirov (1995-1996)] and suggests a classification of singular-hyperbolic attractors on closed 3-manifolds extending [Christy (1993)]:

Theorem 91 *(a) A three dimensional isolating neighborhood M of a two dimensional hyperbolic attractor A is determined, up to homeomorphism, by the tidy swaddled graph B.*

(b) The flow f on a hyperbolic attractor A is determined, up to topological equivalence on the neighborhood M, by B.

Examples of singular hyperbolic sets include the following case:

- The Lorenz attractor [Tucker (1999)] and its geometric models [Afraimovich, *et al.* (1977), Guckenheimer & Williams (1979), Williams (1979)].

- The singular horseshoe [Labarca & Pacifico (1986)].

- Attractors arising from certain resonant double homoclinic loops [Morales, *et al.* (2005)].

- Attractors obtained from certain singular cycles [Morales & Pujals (1997)].

- Attractors obtained from certain models across the boundary of uniform hyperbolicity [Morales, *et al.* (2000)].

- A class of vector fields on 3-manifolds containing the hyperbolic ones and the geometric Lorenz attractor were constructed in [Morales (1999)] by imposing to a partially hyperbolic set to have volume expanding the central direction: They have Lorenz-like singularities accumulated by periodic orbits and they cannot be approximated by flows with non-hyperbolic critical elements.

The following result was proved in [Morales (1999)]:

Theorem 92 *Let Λ be a singular hyperbolic set of f and assume that Λ is non-hyperbolic. Then, Λ has at least one singularity accumulated by regular f-orbits in Λ. In addition, the following holds either for f or $-f$: each attached singularity σ of Λ is Lorenz-like and satisfies $\Lambda \cap W^{ss}(\sigma) = \{\sigma\}$.*

The notion of *singular Axiom A* is necessary to state an important corollary of Theorem 92:

Definition 80 *(a) A C^1 vector fields f is a singular Axiom A if:*
(1) The non-wandering set $\Omega(f)$ of f is a disjoint union of basic or singular basic sets,
(2) The non-wandering set $\Omega(f)$ is the closure of the critical elements of f .
(b) A singular Axiom A vector fields f has no cycles if there is no set of regular orbits linking its basic or singular basic sets in a cyclic way.

We need the following definition:

Definition 81 *A vector field f is said to be topologically Ω-stable if nearby vector fields (in the C^1 topology on the space of vector fields) have non-wandering sets homeomorphic to the non-wandering set of f.*

As a result from Definition 81 and Theorem 92, Axiom A vector field (without cycles) is a singular Axiom A (without cycles).

Let f be a singular Axiom A vector field in M . Then,
(a) f is Axiom A if and only if there is no attached Lorenz-like singularities either for f or $-f$,
(b) f is Ω-stable if and only if f is Axiom A without cycles [Palis & Takens (1993)].

Note that the classification given in [Hayashi (1992)] claiming that the C^1 interior of the set of diffeomorphisms in M whose periodic points are hyperbolic are Axiom A without cycles is not for flows because the Lorenz attractor is not a hyperbolic set. Hence, the following result was proved in [Morales, *et al.* (1999)]:

Theorem 93 *A C^r singular Axiom A flow without cycles is in $C^r(M)$ for any $r \geq 1$.*

The inverse of Theorem 93 is not true, i.e., A flow in $C^r(M)$ is not necessarily a singular Axiom A, because the singular cycle given in [Bamon, *et al.*, (1993)] is in $C^1(M)$ but it is not singular Axiom A.

The following conjecture was proposed in [Morales, *et al.*, (1999)]:

Conjecture 94 *Singular Axiom A flows without cycles to form an open and dense subset in $C^1(M)$.*

As an example of singular Axiom A, vector field in $C^r(M)$ with a singular basic set L is given by

$$L : \begin{cases} x' = \lambda_1 x \\ y' = -\lambda_2 y \\ z' = -\lambda_3 z \end{cases} \tag{8.114}$$

which is equivalent to the geometric Lorenz attractor. The system (8.114) induce a flow given by:

$$L_t\,(x,y,z) = (xe^{\lambda_1 t}, ye^{-\lambda_2 t}, ze^{-\lambda_3 t}) \tag{8.115}$$

For C^r-flows on 3-manifolds $(r \geq 1)$, the notion of singular hyperbolicity is a generalization of hyperbolicity in the sense of Smale for C^1-robustly transitive sets with singularities. The expanding geometric Lorenz attractor is an example of this situation. However, the contracting geometric Lorenz attractor and the singular horseshoe (either expanding or contracting) are not C^1 robust singular transitive sets.

In order to state Theorem 95 below [Arroyo (2007)], we need to introduce the notion of linear Poincaré flow.

Definition 82 *Let N be the normal bundle of f, defined on the set of regular points $\tilde{U} = U - S(\Lambda)$, by the orthogonal complement of $f(p) \in T_pU$. For each $t \in \mathbb{R}$, the tangent map of f_t restricted to N induces an automorphism $L_t : N \to N$ which covers f_t. The family $\{L_t\}$, of automorphisms of N, is called the Linear Poincaré Flow associated to f.*

We set $Sing\,(f) = Sing_f(M)$, then one has the following definition:

Definition 83 *(a) Let $\sigma \in Sing(f)$ be a singularity of a vector field f contained in an invariant compact set Λ of f . We say that σ is attached to Λ if σ is accumulated by periodic orbits of f in Λ.*

(b) We say that $\sigma \in Sing(f)$ is Lorenz-like if the eigenvalues of $Df\,(\sigma)$ are real and satisfy $\lambda_2 < \lambda_3 < 0 < -\lambda_3 < \lambda_1$.

(c) The set $Sing\,(f)$ is Lorenz-like if any $\sigma \in Sing\,(f)$ is a hyperbolic, non-resonant satisfy $\Lambda \cap W^{ss}(\sigma) = \{\sigma\}$ and $\lambda_2 < \lambda_3 < 0 < -\lambda_3 < \lambda_1$, and both unstable separatrices accumulates on $Sing\,(f)$.

(d) The linear Poincaré flow has dominated splitting on U if there is a splitting $N = E \oplus F$ such that $L_t\,(E\,(x)) = E(f_t(x))$ and $L_{-t}\,(F\,(x)) = F(f_t(x))$, for any $x \in U$ and any $t \in \mathbb{R}$, and there are $C > 0$ and $\rho < 0$ such that: $\|L_t|E(x)\|\,\|L_{-t}|_{F(f_t(x))}\| \leq C\exp(\rho t), \forall t \geq 0$.

(e) The linear Poincaré flow has a contracting direction if there is $C > 0$ and $\lambda < 0$ such that for any $x \in \Lambda$ and any $t > 0$: $\|L_t|_{E(x)}\| < C\exp(\lambda t)$.

Note that hyperbolic sets cannot have attached singularities and an attached singularity of Λ is accumulated by regular (non-singular) f-orbits in Λ and the converse is false. Also, it was proved in [Doering (1987)] that attached singularities cannot coexist with robustness of transitivity of M. Then, one has the following result that has been proved in [Morales, *et al.* (1998)]:

Theorem 95 *Let Λ be a C^1 robust singular transitive set of a C^1 vector fields f . Then, either for f or $-f$, Λ is a singular hyperbolic set which is an attractor and any of its singularities are Lorenz-like.*

It was proved in [Morales, *et al.* (2004)] that any robust invariant attractor of a 3-dimensional flow that contains some equilibrium must be singular hyperbolic. In particular, any robust invariant set of a 3-dimensional flow containing some equilibrium is a singular hyperbolic attractor or repeller, and in the absence of equilibria, robustness implies uniform hyperbolicity. Let σ be a hyperbolic saddle-type singularity of f with real distinct eigenvalues, we denote by $W^{ss}(\sigma)$ (resp. $W^{uu}(\sigma)$) the strong stable (resp. unstable) manifold of σ. Hence, the following results have been proved in [Morales, *et al.* (2004)]:

Theorem 96 *(a) A robust transitive set containing singularities of a flow on a closed 3-manifold is either a proper attractor or a proper repeller.*

(b) Let Λ be a robust singular transitive set of a C^1 vector fields f. Then either for $g = f$ or $g = -f$, every $\sigma \in Sing_g(\Lambda)$ is Lorenz-like for g and satisfies $W_g^{ss}(\sigma) \cap \Lambda = \{\sigma\}$.

(c) Robust attractors of a C^1 vector field f containing singularities are singular hyperbolic sets for f.

As a direct result of Theorem 96, the following results were proved in [Morales, *et al.* (2004)]:

(a) C^1 vector fields on a closed 3-manifold with robust transitive nonwandering sets are Anosov.

(b) Every singularity of a robust attractor of f on a closed 3-manifold is Lorenz-like for f.

(c) Let Λ be a robust attractor of a C^1 vector field f. Then Λ is either hyperbolic or singular hyperbolic.

Now the following theorem was proved in [Arroyo (2007)]:

Theorem 97 *Let Λ be a transitive attractor of a C^2-flow such that: The linear Poincaré flow on the set of regular points has a dominated splitting with contracting direction, all singularities are Lorenz-like and all periodic orbits are hyperbolic, then Λ is singular hyperbolic.*

Next, an important result obtained from Theorem 97 and proved in [Arroyo (2007)]:

If $\Lambda_0 \subset \Lambda$ is a compact invariant set that $\Lambda_0 \cap Sing(f) = \emptyset$, then Λ_0 is hyperbolic of saddle type.

Theorem 97 implies the volume expansion condition along a center unstable bundle over points of the attractor Λ. Achieving the proof of Perturbations of singular hyperbolic attractors for three-dimensional flows is an important topic. Indeed, in [Morales (2004)] some results about the perturbations of singular hyperbolic attractors for three-dimensional flows are given:

- Any attractor obtained from such perturbations contains a singularity with some sufficient conditions to be stably non-isolated in the nonwandering set.

- There is an upper bound for the number of attractors.

- Every three-dimensional flow C^r close to one exhibiting a singular hyperbolic attractor has a singularity non-isolated in the non-wandering set[1].

The main result of Morales, given in [Morales (2004)], that discusses perturbations of singular hyperbolic attractors for three-dimensional flows is given by:

Theorem 98 *For every singular hyperbolic attractor of f there is a neighborhood U such that every attractor in U of every flow C^r close to f is singular.*

Some applications of Theorem 98 for the study of the perturbations of singular hyperbolic attractors in dimension three are listed below [Morales (2004)]:

(a) For every singular hyperbolic attractor of f there is a neighborhood U and $n \in \mathbb{N}^*$ such that every flow C^r close to f has at most n attractors in U.

(b) A singularity σ of a flow f is isolated in the non-wandering set if the set $\Omega(f) \setminus \{\sigma\}$ is not closed in M.

(c) Every flow in $C^r(M)$ which is C^r close to one exhibiting singular hyperbolic attractors has a singularity non- isolated in the non-wandering set.

(d) Every flow in a residual subset of $C^r(M)$ exhibiting a non-trivial singular attractor has a singularity non-isolated in the non-wandering set.

To discuss the existence of singularities stably non-isolated in the non-wandering set, we need to establish the following definition [de Melo & Palis (1982)]:

Definition 84 *(a) If O is a closed orbit of f, then we denote by $O(g)$ the continuation of O for g C^r close to f .*

(b) A hyperbolic singularity σ of f is C^r stably non-isolated in the non-wandering set if $\Omega(g) \setminus \{\sigma(g)\}$ is not closed, $\forall g$ C^r close to f .

An example of a singularity stably non-isolated in the non-wandering set is the geometric Lorenz model. Sufficient condition for a singularity to be stably non-isolated in the non-wandering set was given in the following corollary [Morales & Pacifico (1999)]:

If σ is the unique singularity of a singular hyperbolic attractor of a flow in $C^r(M)$, then σ is C^r stably non- isolated in the non-wandering set.

Here, we collect some important remarks about the above results:

1. The existence of singularities non-isolated in the non-wandering set is an essential obstruction for hyperbolicity [Newhouse (1974)].

---

[1]Recall that C^1 robust attractors with singularities for three-dimensional C^1 flows are singular hyperbolic [Morales, *et al.* (1998)] but the inverse is not true, i.e., singular hyperbolic attractors for three-dimensional C^1 flows are not still C^1 robust [Morales & Pujals (1997)].

2. Every flow close to one exhibiting a geometric Lorenz attractor has a singularity non-isolated in the non-wandering set.

3. Generally, every flow C^r close to one exhibiting a C^r robust singular attractor has a singularity non-isolated in the non-wandering set.

8.11   Consequences of uniform hyperbolicity

The property of uniform hyperbolicity has important topological consequences on the dynamics, the geometry and the statistical properties of the invariant set under consideration, i.e, symbolic dynamics, entropy, invariant foliations, fractal dimensions, physical measures and equilibrium states [Morales, *et al.*, (1999), Carballo, *et al.*, (2000), Morales & Pacifico (2001), Morales & Pacifico (2003), Morales, *et al.* (2004), Morales (2004), Bautista & Morales (2006)]. Some dynamical consequences of singular hyperbolicity are summarized in the following proposition proved in [Morales, *et al.* (2004)]:

Proposition 8.4 *Let Λ be a singular hyperbolic compact set of a C^1 vector field f. Then*

(a) any invariant compact set $\Gamma \subset \Lambda$ without singularities is a hyperbolic set.

(b) A singular hyperbolic attractor Λ of a C^1 vector field f has a uniform positive Lyapunov exponent at every orbit.

(c) For f in a residual (set containing a dense G_δ) subset of a C^1 vector field f, each robust transitive set with singularities is the closure of the stable or unstable manifold of one of its hyperbolic periodic points.

Because, the sensitivity to initial data is a characterization of chaotic systems introduced in Chapter 1, we need the following definition: Let $S(\mathbb{R})$ be the set of surjective increasing continuous functions $h : \mathbb{R} \to \mathbb{R}$:

Definition 85 *(a) (Expansiveness): The flow is sensitive to initial data if there is $\delta > 0$ such that, for any $x \in M$ and any neighborhood N of x, there is $y \in N$ and $t \geq 0$ such that $dist\,(f_t(x), f_t(y)) > \delta$.*

(b) The flow f is expansive if for every $\varepsilon > 0$ there is $\delta > 0$ such that, for any $h \in S(\mathbb{R})$, if $dist\,(f_t(x), f_{h(t)}(y)) \leq \delta$, for all $t \in \mathbb{R}$, then

$$f_{h(t_0)}(y) \in f_{[t_0 - \varepsilon, t_0 + \varepsilon]}(x) = \{f_t(x), t_0 - \varepsilon \leq t \leq t_0 + \varepsilon\}$$

for some $t_0 \in \mathbb{R}$.

(c) The invariant compact set Λ is expansive if the restriction of f_t to Λ is an expansive flow.

This notion is called K^*-*expansiveness* and it was proposed by Komuro in [Komuro (1984)] where he proved that a geometric Lorenz attractor is expansive in this sense. The result given in [Araujo, *et al.* (2009)] about the chaoticity of Singular-hyperbolic attractors is a generalization of this fact to any singular-hyperbolic attractor.

The class of singular-hyperbolic attractors (who are Axiom A attractors for flows) are expansive and have physical measures which are u-Gibbs states. This class of systems is given by:

1. The flow defined by the standard Lorenz equations.

2. The geometric Lorenz flows.

3. The attractors appearing in the unfolding of certain resonant double homoclinic loops.

4. The unfolding of certain singular cycles.

5. Some geometrical models which are singular-hyperbolic but of a different topological type from the geometric Lorenz models.

Hence, the following results have been proved in [Araujo, *et al.* (2009)]:

Theorem 99 *(a) Let Λ be a singular-hyperbolic attractor of a C^1 vector field f, hence, Λ is expansive.*

(b) Let Λ be a singular-hyperbolic attractor. Then Λ supports a unique physical probability measure μ which is ergodic, hyperbolic and its ergodic basin covers a full Lebesgue measure subset of the topological basin of attraction, i.e., $B(\mu) = W^s(\Lambda)$, m mod0.

A singular-hyperbolic attractor of a 3-flow is sensitive to initial data.

The application of the above analysis to the standard Lorenz system and the geometric Lorenz flows gives the following results [Araujo, *et al.* (2009)]:

(a) A geometric Lorenz flow is expansive and has a unique physical invariant probability measure whose basin covers Lebesgue almost every point of the topological basin of attraction. Moreover this measure is a u-Gibbs state and satisfies the Entropy Formula.

(b) The flow defined by the Lorenz equations is expansive and has a unique physical invariant probability measure whose basin covers Lebesgue at almost every point of the topological basin of attraction. Moreover, this measure is a u-Gibbs state and satisfies the Entropy Formula.

8.11.1 Classification of singular-hyperbolic attracting sets

Classification methods for hyperbolic attractors can be found in [Williams (1970-1979), Ghrist and Zhirov (2004)]. In [Morales (2007)], some examples of hyperbolic attractors of singular-hyperbolic attracting sets were constructed.

The basic idea is that these systems accumulate their singularities in a pathological way, explained using matrices with entries in $\{0,1\}$, or have no Lorenz-
like singularities. This type of construction implies the use of *Dehn surgery*
[Goodman (1983)], *templates* [Ghrist, et al., (1997)] and geometric models.
For such a construction, one needs the following notations: Let $\mathcal{A}$ be the set
of (2×2)-matrices with entries in $\{0,1\}$, i.e.,

$$\mathcal{A} = \left\{ \left(\begin{array}{cc} a_{11} & a_{12} \\ a_{21} & a_{22} \end{array} \right) : a_{ij} \in \{0,1\}, \text{ for all } i,j = 1,2 \right\} \tag{8.116}$$

and let $F : \mathcal{A} \rightarrow \mathcal{A}$ and $C : \mathcal{A} \rightarrow \mathcal{A}$ be the bijections which interchange the
rows (resp. columns) of $A \in \mathcal{A}$, i.e.,

$$F \left(\begin{array}{cc} a_{11} & a_{12} \\ a_{21} & a_{22} \end{array} \right) = \left(\begin{array}{cc} a_{21} & a_{22} \\ a_{11} & a_{12} \end{array} \right) \text{ and } C \left(\begin{array}{cc} a_{11} & a_{12} \\ a_{21} & a_{22} \end{array} \right) = \left(\begin{array}{cc} a_{12} & a_{11} \\ a_{22} & a_{21} \end{array} \right) \tag{8.117}$$

Let $\langle F, G \rangle$ be the subgroup generated by $\{F, G\}$ in the group of bijections
of $\mathcal{A}$. Then there is an action $\langle F, G \rangle \times \mathcal{A} \rightarrow \mathcal{A}$, where each matrix $A \in \mathcal{A}$
was identified with its own orbit for the purpose of simplicity. Let Λ be a
singular-hyperbolic set of f and $\sigma \in \Lambda$ be a Lorenz-like singularity of f. This
singularity is equipped with a stable manifold $W^s(\sigma)$, an unstable manifold
$W^u(\sigma)$ and a center-unstable manifold $W^{cu}(\sigma)$. Also, there is a projection
$\pi : V_\sigma \rightarrow W^{cu}(\sigma)$ defined in a neighborhood of V_σ via the strong stable
foliation of Λ. Finally, let $Cl(B)$ be the closure of a subset B:

Definition 86 *Let Λ be a singular-hyperbolic set of f. If Λ is a Lorenz-like
singularity of f, then we define:*

$$\left\{ \begin{array}{c} M(\sigma) = \left(\begin{array}{cc} a_{11} & a_{12} \\ a_{21} & a_{22} \end{array} \right) \\ \\ \mathcal{M}(\sigma) = \{M(\sigma) : \sigma \in \Lambda \text{ is a Lorenz-like singularity}\} \end{array} \right. \tag{8.118}$$

where

$$a_{ij} = \left\{ \begin{array}{l} 0 \text{ if } \sigma \in Cl\left(\pi\left(\Lambda \cap V_\sigma\right) \cap s_{ij}\right) \\ \\ 1 \text{ if } \sigma \notin Cl\left(\pi\left(\Lambda \cap V_\sigma\right) \cap s_{ij}\right) \end{array} \right. \tag{8.119}$$

The $\mathcal{M}(\Lambda)$ is not defined if Λ has no Lorenz-like singularities. This situation includes the case of singular-hyperbolic attracting sets obtained below.
The set $\mathcal{M}(\Lambda)$ is well defined if Λ is a singular-hyperbolic attractor where all
the singularities are Lorenz-like [Morales, et al., (1999)]. Generally, $M(\sigma)$ does
not depend on the chosen center-unstable manifold $W^{cu}(\sigma)$. The following result was proved in [Morales (2007)] using Dehn surgery in the complement of
the Figure-eight knot as in [Goodman (1983)]:

Theorem 100 *There is a C^∞ vector field on a closed 3-manifold whose
non-wandering set is contained in the union of a repelling singularity and
a singular-hyperbolic attracting set without Lorenz-like singularities.*

Some examples of singular-hyperbolic sets in the sense of Definition 86 are given as follows:

1. If Λ is the closure of a generic homoclinic loop associated to a Lorenz-like singularity, then

$$\mathcal{M}(\Lambda) = \left\{ \begin{pmatrix} 0 & 0 \\ 0 & 0 \end{pmatrix} \right\}. \tag{8.120}$$

2. If Λ is the Cherry-like singular-hyperbolic set, then Λ is a not attracting singular-hyperbolic set

$$\mathcal{M}(\Lambda) = \left\{ \begin{pmatrix} 0 & 1 \\ 0 & 1 \end{pmatrix} \right\} \tag{8.121}$$

3. If L is the geometric Lorenz attractor [Guckenheimer & Williams (1979)] and it is analogue to the examples given in [Morales (2000), Morales & Pujals (1997)], then

$$\mathcal{M}(\Lambda) = \left\{ \begin{pmatrix} 1 & 1 \\ 0 & 0 \end{pmatrix} \right\} \tag{8.122}$$

4. If Λ is a singular-hyperbolic attractor, then

$$\mathcal{M}(\Lambda) \cap \left\{ \begin{pmatrix} 0 & 0 \\ 0 & 0 \end{pmatrix}, \begin{pmatrix} 1 & 0 \\ 0 & 0 \end{pmatrix}, \begin{pmatrix} 0 & 1 \\ 0 & 1 \end{pmatrix} \right\} = \emptyset \tag{8.123}$$

This is a result of the transitivity of Λ and the fact that $W^u(\sigma) \subset \Lambda$ is an attracting set.

5. Bautista and Morales give an example of singular-hyperbolic attractors in [Bautista & Morales (2006)] with *more than one singularity* and satisfying the following condition:

$$\begin{pmatrix} 1 & 1 \\ 1 & 1 \end{pmatrix} \in \mathcal{M}(\Lambda) \tag{8.124}$$

This example is a non-trivial, transitive, isolated singular-hyperbolic set which is not a homoclinic class because it has no periodic orbits.

6. Bautista proves in his thesis [Bautista (2005)] that if Λ is a singular-hyperbolic attractor *with only one singularity*, then

$$M(\sigma) \neq \begin{pmatrix} 1 & 0 \\ 0 & 1 \end{pmatrix} \tag{8.125}$$

and it is not known if there is a singular-hyperbolic attractor Λ with only one singularity, such that

$$M(\sigma) = \begin{pmatrix} 1 & 1 \\ 0 & 1 \end{pmatrix} \tag{8.126}$$

The example in item 5 above, permits us to introduce the following conjecture given in [Morales (1996)]:

Conjecture 101 *A singular-hyperbolic attractor of a C^1 vector field, defined in a closed 3-manifold, is a homoclinic class.*

A partial answer for Conjecture 102 can be found in [Morales (2007)]:

Proposition 8.5 *Let Λ be a singular-hyperbolic attractor with dense periodic orbits and only one singularity of a C^1 vector field f defined in a closed 3-manifold. If*

$$M(\sigma) \in \left\{ \begin{pmatrix} 1 & 1 \\ 0 & 0 \end{pmatrix}, \begin{pmatrix} 1 & 1 \\ 0 & 1 \end{pmatrix} \right\} \tag{8.127}$$

and there is $q \in W^u(\sigma)$ such that its omega-limit set is a periodic orbit, then it is a homoclinic class of f.

The item 6 above implies that the Bautista's result, proved in [Bautista (2005)], is false for singular-hyperbolic attractors with more than one singularity. This is shown in Theorem 102 below and proved in [Morales (2007)]. This result gives a positive answer to the question of which matrices belong to $\mathcal{M}(\Lambda)$ for some singular-hyperbolic attractor:

Theorem 102 *For every*

$$A \in \left\{ \begin{pmatrix} 1 & 1 \\ 1 & 1 \end{pmatrix}, \begin{pmatrix} 1 & 1 \\ 0 & 1 \end{pmatrix}, \begin{pmatrix} 1 & 0 \\ 0 & 1 \end{pmatrix} \right\} \tag{8.128}$$

there is a C^∞ vector field in the manifold M exhibiting a singular-hyperbolic attractor Λ such that $A \in \mathcal{M}(\Lambda)$. If

$$A = \begin{pmatrix} 1 & 1 \\ 1 & 1 \end{pmatrix} \tag{8.129}$$

then Λ can be chosen with only one singularity.

Theorem 102 was proved using the fact that it exhibits templates and geometric models. On one hand, Theorem 4.1 in [Bonatti, *et al.* (2005)] implies that the fact that the unstable manifold of a periodic orbit contained in a singular-hyperbolic attractor intersects the stable manifold of every singularity in the attractor is false because it implies that all singular-hyperbolic attractors satisfy

$$\mathcal{M}(\Lambda) \in \left\{ \begin{pmatrix} 1 & 1 \\ 0 & 0 \end{pmatrix}, \begin{pmatrix} 1 & 1 \\ 0 & 1 \end{pmatrix}, \begin{pmatrix} 1 & 1 \\ 1 & 1 \end{pmatrix} \right\} \tag{8.130}$$

which contradicts Theorem 102 since $\begin{pmatrix} 1 & 0 \\ 0 & 1 \end{pmatrix}$ does not belong to the right-hand side set of (8.130). The results given in [Morales & Pacifico (2001)], however, are correct. On the other hand, Proposition 9.25-(1) in [Bonatti, et al. (2005)] claims that a singularity attached to a singular-hyperbolic set must be Lorenz-like. This result is false because the unique singularity of the singular-hyperbolic attracting set in Theorem 102 is attached but not Lorenz-like. Similar counter-examples that exhibit Lorenz-like singularities can be found in [Bautista & Morales (2005), Bautista and Morales (2006)].

Generally, proving hyperbolicity is a hard problem that has many factors involved in its definition. It is useful to use the notion of *structural stability*, since hyperbolicity implies structural stability. In the next section, we give an example of some structurally stable forms of 3-D quadratic mappings.

8.12　Structural stability for 3-D quadratic mappings

In this section, we introduce the concept of structural stability for 3-D quadratic mappings. Notice that the structural stability is considered as a criterion for robustness of invariant sets of dynamical systems. Some important properties of the notion of structural stability, along with its conditions, are given and discussed in order to apply them to the general 3-D quadratic map given by:

$$f : \begin{cases} x' = a_0 + a_1 x + a_2 y + a_3 z + a_4 x^2 + a_5 y^2 + a_6 z^2 + a_7 xy + a_8 xz + a_9 yz \\ \\ y' = b_0 + b_1 x + b_2 y + b_3 z + b_4 x^2 + b_5 y^2 + b_6 z^2 + b_7 xy + b_8 xz + b_9 yz \\ \\ z' = c_0 + c_1 x + c_2 y + c_3 z + c_4 x^2 + c_5 y^2 + c_6 z^2 + c_7 xy + c_8 xz + c_9 yz \end{cases}$$
$$(8.131)$$

where $(a_i, b_i, c_i)_{0 \leq i \leq 9} \in \mathbb{R}^{30}$ are the bifurcation parameters.

8.12.1　The concept of structural stability

The concept of structural stability was introduced by Andronov and Pontryagin in 1937. This notion plays an important role in the development of the theory of dynamical systems. The conditions for structural stability of high-dimensional systems were formulated by Smale in [Smale (1967)] as follows: A system must satisfy both Axiom A and the strong transversality condition. Mathematically, let $C^r(\mathbb{R}^n, \mathbb{R}^n)$ denote the space of C^r vector fields of $\mathbb{R}^n$ into $\mathbb{R}^n$. Let $Diff^r(\mathbb{R}^n, \mathbb{R}^n)$ be the subset of $C^r(\mathbb{R}^n, \mathbb{R}^n)$ consisting of the C^r diffeomorphisms.

Definition 87 *(a) Two elements of $C^r(\mathbb{R}^n, \mathbb{R}^n)$ are C^r ε-close ($k \leq r$), or just C^k close, if they, along with their first k derivatives, are within ε as measured in some norm.*

(b) A dynamical system (vector field or map) is structurally stable if nearby systems have the same qualitative dynamics.

The term *nearby systems*, in Definition 87, can be translated in terms of C^k conjugate for maps, and C^k equivalence for vector fields. This is related to the fact that $\mathbb{R}^n$ is unbounded. To avoid this problem, assume that the maps under consideration act on compact, boundaryless, n-dimensional, differentiable manifolds M, rather than all of $\mathbb{R}^n$. This assumption induces the so-called C^k *topology* described in [Palis and de Melo (1982)] or in [Hirsch (1976)]:

Definition 88 *The C^k topology is the topology induced on $C^r(M, M)$ by the measure of distance between two elements of $C^r(M, M)$.*

Now it is possible to formally define the notion of structural stability as follows:

Definition 89 *(Structural stability) Consider a map $f \in Diff^r(M, M)$ (resp. a C^r vector field in $C^r(M, M)$); then f is structurally stable if there exists a neighborhood N of f in the C^k topology such that f is C^0 conjugate (resp. C^0 equivalent) to every map (resp. vector field) in N.*

As a result, structural stability implies a common and typical or *generic* property of a dynamical system to a dense set of dynamical systems in $C^r(M, M)$. Notice that this case is still not true in the case of the set of rationales, which are dense in $\mathbb{R}$ and also its complement. Topologically, this common property is true when some conditions about residual sets hold: Indeed, let X be a topological space, and let U be a subset of X.

Definition 90 *(Residual set) (a) The subset U is called a residual set if it contains the intersection of a countable number of sets, each of which are open and dense in X.*
(b) If every residual set in X is itself dense in X, then X is called a Baire space.

Now we give the definition of a generic property using the notion of a residual set given in Definition 90 as follows:

Definition 91 *(Generic property) A property of a map (resp. vector field) is said to be C^k generic if the set of maps (resp. vector fields) possessing that property contains a residual subset in the C^k topology.*

8.12.2 Conditions for structural stability

First of all, hyperbolic fixed points, periodic orbits, and the transversal intersection of the stable and unstable manifolds of hyperbolic fixed points and periodic orbits are structurally stable and generic. This fact was proved in

[Palis and de Melo (1992)]. Also, it was shown that structurally stable systems are generic. Indeed, for two-dimensional vector fields on compact manifolds, one has the following result due to Peixoto [Peixoto (1962)]:

Theorem 103 *(Peixoto's Theorem) A C^r vector field on a compact boundaryless two-dimensional manifold M is structurally stable if and only if:*

(i) The number of fixed points and periodic orbits is finite, and each is hyperbolic.

(ii) There are no orbits connecting saddle points.

(iii) The non-wandering set consists of fixed points and periodic orbits.

Moreover, if M is orientable, then the set of such vector fields is open and dense in $C^r(M, M)$ (note: this condition is stronger than generic).

Theorem 103 is useful in practice since it gives precise conditions, under which the dynamics of a vector field on a compact boundaryless two-manifold are structurally stable. However, this result cannot be generalized to n-dimensional diffeomorphisms ($n \geq 2$) or n-dimensional vector fields ($n \geq 3$) due to the presence of Smale horseshoes [Smale (1966)]. Generally, for periodic orbits and fixed points, structural stability can be tested in terms of the eigenvalues of the linearized system. But this approach fails for homoclinic and quasiperiodic orbits since the nearby orbit structure may be exceedingly complicated and defy any local description.

Definition 89 can be reformulated in terms of C^k topology as follows:

Definition 92 *A diffeomorphism f is C^r structurally stable if, for any C^r small perturbation g of f, there is a homeomorphism h of the phase space such that $(h \circ f)(x) = (g \circ h)(x)$ for all points x in the phase space.*

Peixoto's theorem for diffeomorphisms on the circle $\mathbb{S}^1$ can be stated as follows:

Theorem 104 *(Peixoto) A diffeomorphism $f \in Diff^1(\mathbb{S}^1)$ is structurally stable if and only if its non-wandering set $\Omega(f)$ consists of a finite number of fixed points or periodic orbits.*

Generally, it was conjectured that hyperbolicity is equivalent to structural stability in the C^k topology [Anosov (1967), Palis and Smale (1968), Palis (1969), Robbin (1971), de Melo (1973), Robinson (1974)]. The converse was completed in [Mañé (1988)] for diffeomorphisms and in [Hayashi (1997)] for flows, using the results given in Liao [Mañé (1982), Liao (1980-1983), Sannami (1983)]. From these considerations, we have the following conjecture:

Conjecture 105 *(Stability conjecture), A system is C^r structurally stable if and only if it is hyperbolic and all the stable and unstable manifolds associated with the orbits in the limit set are transversal.*

Let us consider a small perturbation g given by

$$g : \begin{cases} x' = d_0 + d_1 x + d_2 y + d_3 z + d_4 x^2 + d_5 y^2 + d_6 z^2 + d_7 xy + d_8 xz + d_9 yz \\ y' = e_0 + e_1 x + e_2 y + e_3 z + e_4 x^2 + e_5 y^2 + e_6 z^2 + e_7 xy + e_8 xz + e_9 yz \\ z' = f_0 + f_1 x + f_2 y + f_3 z + f_4 x^2 + f_5 y^2 + f_6 z^2 + f_7 xy + f_8 xz + f_9 yz \end{cases}$$
$$(8.132)$$

of the mapping f given by (8.131). Here, small perturbation means that the bifurcation parameters $(d_i, e_i, f_i)_{0 \leq i \leq 9} \in \mathbb{R}^{30}$ of systems (8.132) are very close to the bifurcation parameters $(a_i, b_i, c_i)_{0 \leq i \leq 9} \in \mathbb{R}^{30}$ of system (8.131). The linear transformation h is defined by:

$$h(x, y, z) = \begin{pmatrix} h_{11} & h_{12} & h_{13} \\ h_{21} & h_{22} & h_{23} \\ h_{31} & h_{32} & h_{33} \end{pmatrix} \begin{pmatrix} x \\ y \\ z \end{pmatrix} \qquad (8.133)$$

with the condition of invertibility given by

$$d = (h_{22} h_{33} - h_{23} h_{32}) h_{11} + (h_{31} h_{23} - h_{21} h_{33}) h_{12} + (h_{21} h_{32} - h_{22} h_{31}) h_{13} \neq 0.$$
$$(8.134)$$

On one hand, we say that the maps f and g, given by (8.131) and (8.132), respectively, are affinely conjugate if there exists an affine transformation h such that

$$g \circ h(x, y, z) = h \circ f(x, y, z), \text{ for all } (x, y, z) \in \mathbb{R}^3 \qquad (8.135)$$

On the other hand, if such a transformation h exists to satisfy (8.135), then there is an equivalence relation and the set of some 3-D quadratic maps of the form (8.131) is divided into classes of topologically conjugate maps. This implies that f and g have identical topological properties, in particular, they have the same number of fixed and periodic points of the same stability types. If f and g are invertible, then the order of the points is preserved, and if the maps are non-invertible, then the order of points is also preserved but the map h maps forward orbits of f onto the corresponding forward orbits of g.

It is well known that any matrix over $\mathbb{C}$ is similar to an upper triangular matrix, which is the Jordan normal form. Finding this form is related to the knowing of the minimal polynomial. For 3×3 matrices, we have 6 cases of Jordan normal forms [Kaye & Wilson (1998)]:

$$
\begin{cases}
J_1 = \begin{pmatrix} h_{11} & 0 & 0 \\ 0 & h_{22} & 0 \\ 0 & 0 & h_{33} \end{pmatrix}, J_2 = \begin{pmatrix} h_{11} & 0 & 0 \\ 0 & h_{22} & 0 \\ 0 & 0 & h_{22} \end{pmatrix} \\[16pt]
J_3 = \begin{pmatrix} h_{11} & 0 & 0 \\ 0 & h_{22} & 1 \\ 0 & 0 & h_{22} \end{pmatrix}, J_4 = \begin{pmatrix} h_{11} & 0 & 0 \\ 0 & h_{11} & 0 \\ 0 & 0 & h_{11} \end{pmatrix} \\[16pt]
J_5 = \begin{pmatrix} h_{11} & 1 & 0 \\ 0 & h_{11} & 0 \\ 0 & 0 & h_{11} \end{pmatrix}, J_6 = \begin{pmatrix} h_{11} & 1 & 0 \\ 0 & h_{11} & 1 \\ 0 & 0 & h_{11} \end{pmatrix}
\end{cases}
\tag{8.136}
$$

By using these forms and relation (8.135) we find that the values of the bifurcation parameters $(d_i, e_i, f_i)_{0 \le i \le 9} \in \mathbb{R}^{30}$ of systems (8.132) are in terms of the bifurcation parameters $(a_i, b_i, c_i)_{0 \le i \le 9} \in \mathbb{R}^{30}$ of system (8.131). We give details only for the first case of the Jordan normal form J_1.

8.12.3   The Jordan normal form J_1

For this case, we have $h_{11} h_{22} h_{33} \neq 0$ and

$$
(g \circ h)(x, y, z) = \begin{pmatrix}
\begin{array}{c} d_4 h_{11}^2 x^2 + d_7 h_{11} h_{22} xy + d_8 h_{11} h_{33} xz + d_1 h_{11} x \\ + d_5 h_{22}^2 y^2 + d_9 h_{22} h_{33} yz + d_2 h_{22} y \\ + d_6 h_{33}^2 z^2 + d_3 h_{33} z + d_0 \end{array} \\[20pt]
\begin{array}{c} e_4 h_{11}^2 x^2 + e_7 h_{11} h_{22} xy + e_8 h_{11} h_{33} xz + e_1 h_{11} x \\ + e_5 h_{22}^2 y^2 + e_9 h_{22} h_{33} yz + e_2 h_{22} y \\ + e_6 h_{33}^2 z^2 + e_3 h_{33} z + e_0 \end{array} \\[20pt]
\begin{array}{c} f_4 h_{11}^2 x^2 + f_7 h_{11} h_{22} xy + f_8 h_{11} h_{33} xz + f_1 h_{11} x \\ + f_5 h_{22}^2 y^2 + f_9 h_{22} h_{33} yz + f_2 h_{22} y \\ + f_6 h_{33}^2 z^2 + f_3 h_{33} z + f_0 \end{array}
\end{pmatrix}
\tag{8.137}
$$

and

$$
(h \circ f)(x, y, z) = \begin{pmatrix}
\begin{array}{c} a_4 h_{11} x^2 + a_7 h_{11} xy + a_8 h_{11} xz + a_1 h_{11} x + a_5 h_{11} y^2 \\ + a_9 h_{11} yz + a_2 h_{11} y + a_6 h_{11} z^2 + a_3 h_{11} z + a_0 h_{11} \end{array} \\[20pt]
\begin{array}{c} b_4 h_{22} x^2 + b_7 h_{22} xy + b_8 h_{22} xz + b_1 h_{22} x + b_5 h_{22} y^2 \\ + b_9 h_{22} yz + b_2 h_{22} y + b_6 h_{22} z^2 + b_3 h_{22} z + b_0 h_{22} \end{array} \\[20pt]
\begin{array}{c} c_4 h_{33} x^2 + c_7 h_{33} xy + c_8 h_{33} xz + c_1 h_{33} x + c_5 h_{33} y^2 \\ + c_9 h_{33} yz + c_2 h_{33} y + c_6 h_{33} z^2 + c_3 h_{33} z + c_0 h_{33} \end{array}
\end{pmatrix}
\tag{8.138}
$$

By (8.135) we get the equality between the components and get all coefficients zero as follows:

For the first component we get

$$\left\{ \begin{array}{l} d_4 h_{11}^2 - a_4 h_{11} = 0 \\[2mm] d_7 h_{11} h_{22} - a_7 h_{11} = 0 \\[2mm] d_8 h_{11} h_{33} - a_8 h_{11} = 0 \\[2mm] d_1 h_{11} - a_1 h_{11} = 0 \\[2mm] d_5 h_{22}^2 - a_5 h_{11} = 0 \end{array} \right.$$

$$\left\{ \begin{array}{l} d_9 h_{22} h_{33} - a_9 h_{11} = 0 \\[2mm] d_2 h_{22} - a_2 h_{11} = 0 \\[2mm] d_6 h_{33}^2 - a_6 h_{11} = 0 \\[2mm] d_3 h_{33} - a_3 h_{11} = 0 \\[2mm] d_0 - a_0 h_{11} = 0 \end{array} \right. \tag{8.139}$$

For the second component we get

$$\left\{ \begin{array}{l} e_4 h_{11}^2 - b_4 h_{22} = 0 \\[2mm] e_7 h_{11} h_{22} - b_7 h_{22} = 0 \\[2mm] e_8 h_{11} h_{33} - b_8 h_{22} = 0 \\[2mm] e_1 h_{11} - b_1 h_{22} = 0 \\[2mm] e_5 h_{22}^2 - b_5 h_{22} = 0 \end{array} \right.$$

$$\left\{ \begin{array}{l} e_9 h_{22} h_{33} - b_9 h_{22} = 0 \\[2mm] e_2 h_{22} - b_2 h_{22} = 0 \\[2mm] e_6 h_{33}^2 - b_6 h_{22} = 0 \\[2mm] e_3 h_{33} - b_3 h_{22} = 0 \\[2mm] e_0 - b_0 h_{22} = 0 \end{array} \right. \tag{8.140}$$

For the third component we get

$$
\begin{cases}
f_4 h_{11}^2 - c_4 h_{33} = 0 \\[2mm]
f_7 h_{11} h_{22} - c_7 h_{33} = 0 \\[2mm]
f_8 h_{11} h_{33} - c_8 h_{33} = 0 \\[2mm]
f_1 h_{11} - c_1 h_{33} = 0 \\[2mm]
f_5 h_{22}^2 - c_5 h_{33} = 0
\end{cases}
$$

$$
\begin{cases}
f_9 h_{22} h_{33} - c_9 h_{33} = 0 \\[2mm]
f_2 h_{22} - c_2 h_{33} = 0 \\[2mm]
f_6 h_{33}^2 - c_6 h_{33} = 0 \\[2mm]
f_3 h_{33} - c_3 h_{33} = 0 \\[2mm]
f_0 - c_0 h_{33} = 0
\end{cases} \tag{8.141}
$$

Solving with respect to the bifurcation parameters $(d_i, e_i, f_i)_{0 \le i \le 9} \in \mathbb{R}^{30}$, we get

$$
\begin{cases}
d_0 = a_0 h_{11}, d_1 = a_1, d_2 = \frac{a_2 h_{11}}{h_{22}} \\[3mm]
d_3 = \frac{a_3 h_{11}}{h_{33}}, d_4 = \frac{a_4}{h_{11}}, d_5 = \frac{a_5 h_{11}}{h_{22}^2} \\[3mm]
d_6 = \frac{a_6 h_{11}}{h_{33}^2}, d_7 = \frac{a_7}{h_{22}}, d_8 = \frac{a_8}{h_{33}}, d_9 = \frac{a_9 h_{11}}{h_{22} h_{33}}
\end{cases} \tag{8.142}
$$

and

$$
\begin{cases}
e_0 = b_0 h_{22}, e_1 = \frac{b_1 h_{22}}{h_{11}}, e_2 = b_2, e_3 = \frac{b_3 h_{22}}{h_{33}} \\[3mm]
e_4 = \frac{b_4 h_{22}}{h_{11}^2}, e_5 = \frac{b_5}{h_{22}}, e_6 = \frac{b_6 h_{22}}{h_{33}^2} \\[3mm]
e_7 = \frac{b_7}{h_{11}}, e_8 = \frac{b_8 h_{22}}{h_{11} h_{33}}, e_9 = \frac{b_9}{h_{33}}
\end{cases} \tag{8.143}
$$

and

$$
\begin{cases}
f_0 = c_0 h_{33}, f_1 = \frac{c_1 h_{33}}{h_{11}}, f_2 = \frac{c_2 h_{33}}{h_{22}}, f_3 = c_3 \\[3mm]
f_4 = \frac{c_4 h_{33}}{h_{11}^2}, f_5 = \frac{c_5 h_{33}}{h_{22}^2}, f_6 = \frac{c_6}{h_{33}} \\[3mm]
f_7 = \frac{c_7 h_{33}}{h_{11} h_{22}}, f_8 = \frac{c_8}{h_{11}}, f_9 = \frac{c_9}{h_{22}}
\end{cases} \tag{8.144}
$$

The other cases are as follows:

8.12.4 The Jordan normal form J_2

$$h\left(x,y,z\right) = \begin{pmatrix} h_{11} & 0 & 0 \\ 0 & h_{22} & 0 \\ 0 & 0 & h_{22} \end{pmatrix} \begin{pmatrix} x \\ y \\ z \end{pmatrix}, h_{11}h_{22} \neq 0 \qquad (8.145)$$

$$\begin{cases}
d_0 = a_0 h_{11}, d_1 = a_1, d_2 = \frac{a_2 h_{11}}{h_{22}}, d_3 = \frac{a_3 h_{11}}{h_{22}}, d_4 = \frac{a_4}{h_{11}} \\[2mm]
d_5 = \frac{a_5 h_{11}}{h_{22}^2}, d_6 = \frac{a_6 h_{11}}{h_{22}^2}, d_7 = \frac{a_7}{h_{22}}, d_8 = \frac{a_8}{h_{22}}, d_9 = \frac{a_9 h_{11}}{h_{22}^2} \\[2mm]
e_0 = b_0 h_{22}, e_1 = \frac{b_1 h_{22}}{h_{11}}, e_2 = b_2, e_3 = b_3, e_4 = \frac{b_4 h_{22}}{h_{11}^2} \\[2mm]
e_5 = \frac{b_5}{h_{22}}, e_6 = \frac{b_6}{h_{22}}, e_7 = \frac{b_7}{h_{11}}, e_8 = \frac{b_8}{h_{11}}, e_9 = \frac{b_9}{h_{22}} \\[2mm]
f_0 = c_0 h_{22}, f_1 = \frac{c_1 h_{22}}{h_{11}}, f_2 = c_2, f_3 = c_3, f_4 = \frac{c_4 h_{22}}{h_{11}^2} \\[2mm]
f_5 = \frac{c_5 h_{22}}{h_{22}^2}, f_6 = \frac{c_6}{h_{22}}, f_7 = \frac{c_7}{h_{11}}, f_8 = \frac{c_8}{h_{11}}, f_9 = \frac{c_9}{h_{22}}
\end{cases} \qquad (8.146)$$

8.12.5 The Jordan normal form J_3

$$h\left(x,y,z\right) = \begin{pmatrix} h_{11} & 0 & 0 \\ 0 & h_{22} & 1 \\ 0 & 0 & h_{22} \end{pmatrix} \begin{pmatrix} x \\ y \\ z \end{pmatrix}, h_{11}h_{22} \neq 0 \qquad (8.147)$$

$$\begin{cases}
d_0 = a_0 h_{11}, d_1 = a_1, d_2 = \frac{a_2 h_{11}}{h_{22}}, d_3 = -h_{11}\frac{a_2 - a_3 h_{22}}{h_{22}^2} \\[2mm]
d_4 = \frac{a_4}{h_{11}}, d_5 = \frac{a_5 h_{11}}{h_{22}^2}, d_6 = h_{11}\frac{a_5 + a_6 h_{22}^2 - a_9 h_{22}}{h_{22}^4} \\[2mm]
d_7 = \frac{a_7}{h_{22}}, d_8 = -\frac{a_7 - a_8 h_{22}}{h_{22}^2}, d_9 = -h_{11}\frac{2a_5 - a_9 h_{22}}{h_{22}^3} \\[2mm]
e_0 = c_0 + b_0 h_{22}, e_1 = \frac{c_1 + b_1 h_{22}}{h_{11}}, e_2 = \frac{c_2 + b_2 h_{22}}{h_{22}} \\[2mm]
e_3 = -\frac{c_2 - b_3 h_{22}^2 + b_2 h_{22} - c_3 h_{22}}{h_{22}^2}, e_4 = \frac{c_4 + b_4 h_{22}}{h_{11}^2} \\[2mm]
e_5 = \frac{b_5 h_{22} + c_5}{h_{22}^2}, e_6 = \frac{c_5 + b_6 h_{22}^3 + c_6 h_{22}^2 - b_9 h_{22}^2 + b_5 h_{22} - c_9 h_{22}}{h_{22}^4} \\[2mm]
e_7 = \frac{c_7 + b_7 h_{22}}{h_{11} h_{22}}, e_8 = -\frac{c_7 - b_8 h_{22}^2 + b_7 h_{22} - c_8 h_{22}}{h_{11} h_{22}^2}
\end{cases}$$

$$\begin{cases}
e_9 = -\frac{2c_5 - b_9 h_{22}^2 + 2b_5 h_{22} - c_9 h_{22}}{h_{22}^3}, f_0 = c_0 h_{22}, f_1 = \frac{c_1 h_{22}}{h_{11}} \\[2mm]
f_2 = c_2, f_3 = -\frac{c_2 - c_3 h_{22}}{h_{22}}, f_4 = \frac{c_4 h_{22}}{h_{11}^2} \\[2mm]
f_5 = \frac{c_5}{h_{22}}, f_6 = \frac{c_5 + c_6 h_{22}^2 - c_9 h_{22}}{h_{22}^3}, f_7 = \frac{c_7}{h_{11}} \\[2mm]
f_8 = -\frac{c_7 - c_8 h_{22}}{h_{11} h_{22}}, f_9 = -\frac{2c_5 - c_9 h_{22}}{h_{22}^2}
\end{cases} \qquad (8.148)$$

8.12.6   The Jordan normal form J_4

$$h(x,y,z) = \begin{pmatrix} h_{11} & 0 & 0 \\ 0 & h_{11} & 0 \\ 0 & 0 & h_{11} \end{pmatrix} \begin{pmatrix} x \\ y \\ z \end{pmatrix}, h_{11} \neq 0 \qquad (8.149)$$

$$\begin{cases} d_0 = a_0 h_{11}, d_1 = a_1, d_2 = a_2, d_3 = a_3, d_4 = \frac{a_4}{h_{11}}, d_5 = \frac{a_5}{h_{11}} \\[2mm] d_6 = \frac{a_6}{h_{11}}, d_7 = \frac{a_7}{h_{11}}, d_8 = \frac{a_8}{h_{11}}, d_9 = \frac{a_9}{h_{11}} \\[2mm] e_0 = b_0 h_{11}, e_1 = b_1, e_2 = b_2, e_3 = b_3, e_4 = \frac{b_4}{h_{11}}, e_5 = \frac{b_5}{h_{11}} \\[2mm] e_6 = \frac{b_6}{h_{11}}, e_7 = \frac{b_7}{h_{11}}, e_8 = \frac{b_8}{h_{11}}, e_9 = \frac{b_9}{h_{11}} \\[2mm] f_0 = c_0 h_{11}, f_1 = c_1, f_2 = c_2, f_3 = c_3, f_4 = \frac{c_4}{h_{11}}, f_5 = \frac{c_5}{h_{11}} \\[2mm] f_6 = \frac{c_6}{h_{11}}, f_7 = \frac{c_7}{h_{11}}, f_8 = \frac{c_8}{h_{11}}, f_9 = \frac{c_9}{h_{11}} \end{cases} \qquad (8.150)$$

8.12.7   The Jordan normal form J_5

$$h(x,y,z) = \begin{pmatrix} h_{11} & 1 & 0 \\ 0 & h_{11} & 0 \\ 0 & 0 & h_{11} \end{pmatrix} \begin{pmatrix} x \\ y \\ z \end{pmatrix}, h_{11} \neq 0 \qquad (8.151)$$

$$\begin{cases} d_0 = b_0 + a_0 h_{11}, d_1 = \frac{b_1 + a_1 h_{11}}{h_{11}}, d_2 = -\frac{b_1 - a_2 h_{11}^2 + a_1 h_{11} - b_2 h_{11}}{h_{11}^2} \\[2mm] d_3 = \frac{b_3 + a_3 h_{11}}{h_{11}}, d_4 = \frac{b_4 + a_4 h_{11}}{h_{11}^2}, d_5 = \frac{b_4 + a_5 h_{11}^3 - a_7 h_{11}^2 + b_5 h_{11}^2 + a_4 h_{11} - b_7 h_{11}}{h_{11}^4} \\[2mm] d_6 = \frac{b_6 + a_6 h_{11}}{h_{11}^2}, d_7 = -\frac{2b_4 - a_7 h_{11}^2 + 2a_4 h_{11} - b_7 h_{11}}{h_{11}^3} \\[2mm] d_8 = \frac{b_8 + a_8 h_{11}}{h_{11}^2}, d_9 = -\frac{b_8 - a_9 h_{11}^2 + a_8 h_{11} - b_9 h_{11}}{h_{11}^3} \end{cases}$$

$$\begin{cases} e_0 = b_0 h_{11}, e_1 = b_1, e_2 = -\frac{b_1 - b_2 h_{11}}{h_{11}}, e_3 = b_3, e_4 = \frac{b_4}{h_{11}} \\[2mm] e_5 = \frac{b_4 + b_5 h_{11}^2 - b_7 h_{11}}{h_{11}^3}, e_6 = \frac{b_6}{h_{11}}, e_7 = -\frac{2b_4 - b_7 h_{11}}{h_{11}^2}, e_8 = \frac{b_8}{h_{11}}, e_9 = -\frac{b_8 - b_9 h_{11}}{h_{11}^2} \\[2mm] f_0 = c_0 h_{11}, f_1 = c_1, f_2 = -\frac{c_1 - c_2 h_{11}}{h_{11}}, f_3 = c_3, f_4 = \frac{c_4}{h_{11}} \\[2mm] f_5 = \frac{c_4 + c_5 h_{11}^2 - c_7 h_{11}}{h_{11}^3}, f_6 = \frac{c_6}{h_{11}}, f_7 = -\frac{2c_4 - c_7 h_{11}}{h_{11}^2}, f_8 = \frac{c_8}{h_{11}}, f_9 = -\frac{c_8 - c_9 h_{11}}{h_{11}^2} \end{cases}$$

$$\qquad (8.152)$$

8.12.8 The Jordan normal form J_6

$$h\left(x,y,z\right) = \begin{pmatrix} h_{11} & 1 & 0 \\ 0 & h_{11} & 1 \\ 0 & 0 & h_{11} \end{pmatrix} \begin{pmatrix} x \\ y \\ z \end{pmatrix}, h_{11} \neq 0 \qquad (8.153)$$

$$\begin{cases} d_0 = b_0 + a_0 h_{11}, d_1 = \frac{b_1 + a_1 h_{11}}{h_{11}}, d_2 = -\frac{b_1 - a_2 h_{11}^2 + a_1 h_{11} - b_2 h_{11}}{h_{11}^2} \\[2mm] d_3 = \frac{b_1 - a_2 h_{11}^2 + a_3 h_{11}^3 + b_3 h_{11}^2 + a_1 h_{11} - b_2 h_{11}}{h_{11}^3} \\[2mm] d_4 = \frac{b_4 + a_4 h_{11}}{h_{11}^2}, d_5 = \frac{b_4 + a_5 h_{11}^3 - a_7 h_{11}^2 + b_5 h_{11}^2 + a_4 h_{11} - b_7 h_{11}}{h_{11}^4} \\[2mm] d_6 = \frac{b_4 + a_5 h_{11}^3 - a_7 h_{11}^2 + b_5 h_{11}^2 + a_6 h_{11}^5 + a_8 h_{11}^3 + b_6 h_{11}^4 + b_8 h_{11}^2 - a_9 h_{11}^4 - b_9 h_{11}^3 + a_4 h_{11} - b_7 h_{11}}{h_{11}^6} \\[2mm] d_7 = -\frac{2b_4 - a_7 h_{11}^2 + 2a_4 h_{11} - b_7 h_{11}}{h_{11}^3}, d_8 = \frac{2b_4 - a_7 h_{11}^2 + a_8 h_{11}^3 + b_8 h_{11}^2 + 2a_4 h_{11} - b_7 h_{11}}{h_{11}^4} \\[2mm] d_9 = -\frac{2b_4 + 2a_5 h_{11}^3 - 2a_7 h_{11}^2 + 2b_5 h_{11}^2 + a_8 h_{11}^3 + b_8 h_{11}^2 - a_9 h_{11}^4 - b_9 h_{11}^3 + 2a_4 h_{11} - 2b_7 h_{11}}{h_{11}^5} \end{cases}$$

$$(8.154)$$

$$\begin{cases} e_0 = c_0 + b_0 h_{11}, e_1 = \frac{c_1 + b_1 h_{11}}{h_{11}}, e_2 = -\frac{c_1 - b_2 h_{11}^2 + b_1 h_{11} - c_2 h_{11}}{h_{11}^2} \\[2mm] e_3 = \frac{c_1 - b_2 h_{11}^2 + b_3 h_{11}^3 + c_3 h_{11}^2 + b_1 h_{11} - c_2 h_{11}}{h_{11}^3}, e_4 = \frac{c_4 + b_4 h_{11}}{h_{11}^2} \\[2mm] e_5 = \frac{c_4 + b_5 h_{11}^3 - b_7 h_{11}^2 + c_5 h_{11}^2 + b_4 h_{11} - c_7 h_{11}}{h_{11}^4} \\[2mm] e_6 = \frac{c_4 + b_5 h_{11}^3 - b_7 h_{11}^2 + c_5 h_{11}^2 + b_6 h_{11}^5 + b_8 h_{11}^3 + c_6 h_{11}^4 + c_8 h_{11}^2 - b_9 h_{11}^4 - c_9 h_{11}^3 + b_4 h_{11} - c_7 h_{11}}{h_{11}^6} \\[2mm] e_7 = -\frac{2c_4 - b_7 h_{11}^2 + 2b_4 h_{11} - c_7 h_{11}}{h_{11}^3}, e_8 = \frac{2c_4 - b_7 h_{11}^2 + b_8 h_{11}^3 + c_8 h_{11}^2 + 2b_4 h_{11} - c_7 h_{11}}{h_{11}^4} \\[2mm] e_9 = -\frac{2c_4 + 2b_5 h_{11}^3 - 2b_7 h_{11}^2 + 2c_5 h_{11}^2 + b_8 h_{11}^3 + c_8 h_{11}^2 - b_9 h_{11}^4 - c_9 h_{11}^3 + 2b_4 h_{11} - 2c_7 h_{11}}{h_{11}^5} \end{cases}$$

$$(8.155)$$

$$\begin{cases} f_0 = c_0 h_{11}, f_1 = c_1, f_2 = -\frac{c_1 - c_2 h_{11}}{h_{11}}, f_3 = \frac{c_1 + c_3 h_{11}^2 - c_2 h_{11}}{h_{11}^2}, f_4 = \frac{c_4}{h_{11}} \\[2mm] f_5 = \frac{c_4 + c_5 h_{11}^2 - c_7 h_{11}}{h_{11}^3}, f_6 = \frac{c_4 + c_5 h_{11}^2 + c_6 h_{11}^4 + c_8 h_{11}^2 - c_9 h_{11}^3 - c_7 h_{11}}{h_{11}^5}, f_7 = -\frac{2c_4 - c_7 h_{11}}{h_{11}^2} \\[2mm] f_8 = \frac{2c_4 + c_8 h_{11}^2 - c_7 h_{11}}{h_{11}^3}, f_9 = -\frac{2c_4 + 2c_5 h_{11}^2 + c_8 h_{11}^2 - c_9 h_{11}^3 - 2c_7 h_{11}}{h_{11}^4} \end{cases}$$

$$(8.156)$$

8.13   Construction of globally asymptotically stable partial differential systems

An example of partial differential systems converging to a unique steady state that is asymptotically stable is the spatially SVIR model of infectious disease epidemics with immigration of individuals, presented in [Abdelmalek & Bendoukha (2016)]:

$$
\begin{cases}
\partial_t u - d_1 \Delta u = \Lambda_1 - u f(w) - (\mu + \alpha) u = f_1(u, v, w) \quad \text{in } \mathbb{R}^+ \times \Omega, \\[2mm]
\partial_t v - d_2 \Delta v = \Lambda_2 + \alpha u - v g(w) - (\mu + \beta) v = f_2(u, v, w) \quad \text{in } \mathbb{R}^+ \times \Omega, \\[2mm]
\partial_t w - d_3 \Delta w = \Lambda_3 + u f(w) + v g(w) - (\mu + \gamma + \delta) w = f_3(u, v, w), \\
\qquad\qquad\qquad\qquad\qquad\qquad\qquad\qquad \text{in } \mathbb{R}^+ \times \Omega, \\[2mm]
\partial_t R - d_4 \Delta R = \Lambda_4 + \beta v + \delta w - \mu R = f_4(v, w, R), \quad \text{in } \mathbb{R}^+ \times \Omega,
\end{cases}
$$

$$(8.157)$$

Where Ω is an open bounded subset of $\mathbb{R}^n$ with piecewise smooth boundary $\partial\Omega$. The initial conditions are

$$
\begin{cases}
u_0(x) = u(x, 0) \\[2mm]
v_0(x) = v(x, 0) \\[2mm]
w_0(x) = w(x, 0) \\[2mm]
R_0(x) = R(x, 0), \quad \text{in } \Omega,
\end{cases}
$$

$$(8.158)$$

where $u_0(x), v_0(x), w_0(x), R_0(x) \in C^2(\Omega) \cap C^0(\overline{\Omega})$, and homogoneous Neumann boundary conditions:

$$
\frac{\partial u}{\partial \nu} = \frac{\partial v}{\partial \nu} = \frac{\partial w}{\partial \nu} = \frac{\partial R}{\partial \nu} = 0 \quad \text{on } \mathbb{R}^+ \times \partial\Omega, \tag{8.159}
$$

with ν as the unit outer normal to $\partial\Omega$. Assume also that the initial conditions $u_0(x), v_0(x), w_0(x), R_0(x) \in \mathbb{R}_{\geq 0}$. In equations (8.157) the positive functions $u(x, t), v(x, t), w(x, t), R(x, t) \geq 0$ represent the population distributions of four classes of people: Suceptible, vaccinated, infectious, and recovered, respectively. Since the recovered class R does not have an impact on the remaining classes, it will be omitted in the sequel. The parameters $\Lambda_i > 0$ denote the growth of the different classes of individuals whether through birth or immigration and migration. The parameter α denotes the rate at which the suceptible population is vaccinated. The per capita death rate for the former is denoted by γ, whereas the latter is denoted by $\mu > 0$. Since, in reality, it takes a while for the vaccinated individual to develop full immunity, the parameter β has been introduced here, indicating an average duration $\frac{1}{\beta}$. The parameter δ is introduced in order to allow for some of the infected individuals to recover on their own after a duration $\frac{1}{\delta}$. Assume that $\alpha, \beta, \gamma, \delta \geq 0$. The parameters $d_i \geq 0$ represent the diffusivity constants modelling the movement of a certain class as a result of its distribution. The functions $f(w)$ and $g(w)$ are known as the incidence functions and allow for a non-linear relation between the first three classes of individuals. Assuming that the incidence functions satisfy the following conditions for all $w \geq 0$:

(H1) $f(w), g(w) \geq 0$ with equality if and only if $w = 0$,

(H2) $f'(w), g'(w) \geq 0$,

(H3) $f''(w), g''(w) \leq 0$,

(H4) $g(w) \leq f(w)$.

In addition, note that for $(u, v, w) \in \mathbb{R}^3_{\geq 0}$, we have

$$
\begin{cases}
f_1(0, v, w) = \Lambda_1 \geq 0 \\[2mm]
f_2(u, 0, w) = \Lambda_2 + u f(w) \geq 0 \\[2mm]
f_3(u, v, 0) = \Lambda_3 + \beta v \geq 0
\end{cases}
$$

Hence, the function $(f_1, f_2, f_2)^T$ is essentially non-negative and the non-negative octant $\mathbb{R}^3_{\geq 0}$ is an invariant set.

Before presenting the convergence of system (8.157) to its equilibrium point, we need some helpful results:

Lemme 8.6 *[McCluskey & van den Driessche (2004)]: Let M be a 3×3 real matrix. If $tr(M)$, $\det(M)$, and $\det(M^{[2]})$ are all negative, then all of the eigenvalues of M have negative real parts, where*

$$
\begin{cases}
M = \begin{pmatrix} a_{11} & a_{12} & a_{13} \\ a_{21} & a_{22} & a_{23} \\ a_{31} & a_{32} & a_{33} \end{pmatrix} \\[6mm]
M^{[2]} = \begin{pmatrix} a_{11} + a_{22} & a_{23} & -a_{13} \\ a_{32} & a_{11} + a_{33} & -a_{12} \\ -a_{31} & a_{21} & a_{22} + a_{33} \end{pmatrix}
\end{cases}
\tag{8.160}
$$

Proof 44 *Let λ_j, $j = 1, 2, 3$ be the eigenvalues of M with $\Re(\lambda_1) \leq \Re(\lambda_2) \leq \Re(\lambda_3)$. The fact that $\det(M) < 0$ implies that $\lambda_1 \lambda_2 \lambda_3 < 0$. Thus, either $\Re(\lambda_j) < 0$ for $j = 1, 2, 3$ or $\Re(\lambda_1) < 0 \leq \Re(\lambda_2) \leq \Re(\lambda_3)$. Suppose that the second set of inequalities holds. Since $tr(M) < 0$, then $\lambda_1 + \lambda_2 + \lambda_3 < 0$,* which implies that $\Re(\lambda_1 + \lambda_2) < 0$ and $\Re(\lambda_1 + \lambda_3) < 0$. The eigenvalues of $M^{[2]}$ are $\lambda_i + \lambda_j$, $1 \leq i < j \leq 3$, so

$$sgn(\det(M^{[2]})) = sgn(\Re(\lambda_1 + \lambda_2)\Re(\lambda_1 + \lambda_3)\Re(\lambda_2 + \lambda_3))$$

$$= sgn(\Re(\lambda_2 + \lambda_3)).$$

Hence, $\det(M^{[2]}) < 0$ that $\Re(\lambda_2 + \lambda_3) < 0$. Thus, it cannot be that $\Re(\lambda_1) < 0 \leq \Re(\lambda_2) \leq \Re(\lambda_3)$, therefore $\Re(\lambda_j) < 0$ for $j = 1, 2, 3$.

Lemme 8.7 *[Sigdel & McCluskey (2014)]: We have* $f'(w) \leq \frac{f(w)}{w}$ *and* $g'(w) \leq \frac{g(w)}{w}$ *for all $w > 0$.*

Proof 45 *Let $w > 0$. Since $f(w)$ is continuous on $[0, w]$ and differentiable on $(0, w)$, the mean value theorem implies that there exists $c \in (0, w)$ such that $f'(c) \leq \frac{f(w) - f(0)}{w - 0}$. By (**H1**), we have $f'(c) = \frac{f(w)}{w}$. From (**H3**), f' is monotone decreasing. Thus, $f'(w) \leq f'(c) = \frac{f(w)}{w}$. The same can be said about g.*

Lemme 8.8 *[Henshaw & McCluskey (2015)]: Suppose the incidence functions f and g satisfy the criteria in (**H1**)–(**H4**). It follows that if $w > 0$, then*

$$\begin{cases} L\left(\frac{f(w)}{f(w^*)}\right) \leq L\left(\frac{w}{w^*}\right) \\ \\ L\left(\frac{g(w)}{g(w^*)}\right) \leq L\left(\frac{w}{w^*}\right) \end{cases} \tag{8.161}$$

Proof 46 *We prove the property (8.161). Let $w \geq w^*$ and $m(w) = \frac{f(w)}{w}$. Then we have*

$$m'(w) = \frac{f'(w)w - f(w)}{w^2} \leq \frac{f(w) - f(w)}{w^2} = 0$$

Thus, m is decreasing, and then $m(w) \leq m(w^*)$, i.e.,

$$\frac{f(w)}{w} \leq \frac{f(w^*)}{w^*}$$

and so

$$\frac{f(w)}{f(w^*)} \leq \frac{w}{w^*}$$

Since f is increasing, we have

$$1 \leq \frac{f(w)}{f(w^*)} \leq \frac{w}{w^*}$$

Since $L(x) = 1 - \frac{1}{x}$. *Hence,* L *is increasing for* $x > 1$, *and*

$$L\left(\frac{f(w)}{f(w^*)}\right) \leq L(\frac{w}{w^*})$$

In the absence of diffusion, the system (8.157) is reduced to

$$\begin{cases} \partial_t u = \Lambda_1 - uf(w) - (\mu + \alpha)u \\ \\ \partial_t v = \Lambda_2 + \alpha u - vg(w) - (\mu + \beta)v \\ \\ \partial_t w = \Lambda_3 + uf(w) + vg(w) - (\mu + \gamma + \delta)w \end{cases} \quad (8.162)$$

Define

$$\Lambda = \Lambda_1 + \Lambda_2 + \Lambda_3$$

and for any $\epsilon \geq 0$,

$$D_\epsilon = \left\{(u, v, w) : u, v, w > \epsilon \text{ and } u + v + w \leq \frac{\Lambda}{\mu}\right\}. \quad (8.163)$$

It was shown in [Henshaw & McCluskey (2015)] that system (8.162) has the positively invariant non-negative octant $\mathbb{R}^3_{\geq 0}$ *and that there exists a number* $\epsilon > 0$ *such that* D_ϵ *is non-empty, attracting and positively invariant. Also, they showed that the system (8.162) has a unique equilibrium in the attraction region* $(u^*, v^*, w^*) \in D_\epsilon$. *This equilibrium is the solution of the system*

$$\begin{cases} \Lambda_1 = u^* f(w^*) + (\mu + \alpha)u^* \\ \\ \Lambda_2 = -\alpha u^* + v^* g(w^*) + (\mu + \beta)v^* \\ \\ (\mu + \gamma + \delta) = \frac{\Lambda_3 + u^* f(w^*) + vg(w^*)}{w^*} \end{cases} \quad (8.164)$$

The Jacobian and its second additive compound are given by:

$$J = \begin{pmatrix} -H_0 - \mu - \alpha & 0 & -F_2 \\ \alpha & -H_1 - \mu - \beta & -G_2 \\ H_0 & H_1 & -H_2 \end{pmatrix}$$

and

$$J^{[2]} = \begin{pmatrix} -H_0 - H_1 - \alpha - \beta - 2\mu & -G_2 & F_2 \\ H_1 & -H_0 - H_2 - \mu - \alpha & 0 \\ -H_0 & \alpha & -H_1 - H_2 - \mu - \beta \end{pmatrix},$$

respectively, where

$$\begin{cases} F_2 = u^* f'^*) \geq 0, \quad G_2 = v^* g'^*) \geq 0 \\ \\ H_0 = f(w^*) \geq 0, \quad H_1 = g(w^*) \geq 0 \\ \\ H_2 = (\mu + \gamma + \delta) - u^* f'^*) - v^* g'^*) \geq 0 \end{cases} \quad (8.165)$$

The terms F_2, G_2, H_0, H_1 *are positive.*

The local stability is studied by the sign of the determinant of the Jacobian $\det(J)$, its trace $tr(J)$, and the determinant of its second additive compound $\det(J^{[2]})$. We have

$$\det J = -\alpha F_2 H_1 - (H_0 + \mu + \alpha)[(H_1 + \mu + \beta)H_2 + G_2 H_1] \quad (8.166)$$

$$-F_2(H_1 + \mu + \beta)H_0$$

$$tr J = -(H_0 + H_1 + H_2 + \alpha + \beta + 2\mu) \quad (8.167)$$

$$\det J^{[2]} = F_2[\alpha H_1 - H_0(H_0 + H_2 + \mu + \alpha)] - G_2 H_1(H_1 + H_2 + \mu + \beta)$$

$$-(H_0 + H_1 + \alpha + \beta + 2\mu)(H_0 + H_2 + \mu + \alpha)(H_1 + H_2 + \mu + \beta)$$

Clearly, $\det J < 0$ and $tr J < 0$. However, for $\det J^{[2]}$, we need to determine the sign of the term $\alpha H_1 - H_0(H_0 + H_2 + \mu + \alpha)$. Using condition (**H4**), we have $H_1 \leq H_0$, leading to $\alpha H_1 - H_0(H_0 + H_2 + \mu + \alpha)$.

In the presence of diffusion, the steady state solution of system (8.157) satisfies:

$$\begin{cases} d_1 \Delta u + \Lambda_1 - u^* f(w^*) - (\mu + \alpha)u^* = 0, \\ \\ d_2 \Delta v + \Lambda_2 + \alpha u^* - v^* g(w^*) - (\mu + \beta)v^* = 0, \quad (8.168) \\ \\ d_3 \Delta w + \Lambda_3 + u^* f(w^*) + v^* g(w^*) - (\mu + \gamma + \delta)w^* = 0. \end{cases}$$

if the homogeneous Neumann boundary condition $\frac{\partial u}{\partial \nu} = \frac{\partial v}{\partial \nu} = \frac{\partial w}{\partial \nu} = 0$ are satisfied for all $x \in \partial \Omega$.

Let $0 = \lambda_0 < \lambda_1 \leq \lambda_2 \leq \ldots$ be the sequence of eigenvalues for the elliptic operator $(-\Delta)$ subject to the homogeneous Neumann boundary condition on Ω, where each λ_i has multiplicity $m_i \geq 1$. Let $\Phi_{ij}, 1 \leq j \leq m_i$, (recall that $\Phi_0 = const$ and $\lambda_i \to \infty$ at $i \to \infty$) be the normalized eigenfunctions corresponding to λ_i. That is, Φ_{ij} and λ_i satisfy $-\Delta \Phi_{ij} = \lambda_i \Phi_{ij}$ in Ω, with $\frac{\partial \Phi_{ij}}{\partial \nu} = 0$ in $\partial \Omega$, and $\int_\Omega \Phi_{ij}^2(x)dx = 1$.

The following result was proved in [Abdelmalek & Bendoukha (2016)]:

Theorem 106 *The constant steady state (u^*, v^*, w^*) is asymptotically stable.*

Proof 47 *Define the linearizing operator*

$$\mathcal{L} = \begin{pmatrix} -d_1 \Delta - (H_0 + \mu + \alpha) & 0 & -F_2 \\ \alpha & -d_2 \Delta - (H_1 + \mu + \beta) & -G_2 \\ H_0 & H_1 & -d_3 \Delta - H_2 \end{pmatrix}.$$

The asymptotic stability of the steady state solution (u^*, v^*, w^*) can be determined by examining the eigenvalues of the operator $\mathcal{L}$, i.e., the solution is asymptotically stable if all the eigenvalues of $\mathcal{L}$ have negative real parts. Assuming that $(\phi(x), \psi(x), \Upsilon(x))$ is an eigenfunction of $\mathcal{L}$ corresponding to an eigenvalue ξ. We have

$$\mathcal{L}(\phi(x), \psi(x), \Upsilon(x))^t = \xi(\phi(x), \psi(x), \Upsilon(x))^t,$$

leading to

$$(\mathcal{L}-\xi I)\begin{pmatrix} \phi \\ \psi \\ \Upsilon \end{pmatrix} = \begin{pmatrix} 0 \\ 0 \\ 0 \end{pmatrix}.$$

This can be rearranged to the form

$$\sum_{0 \leq i \leq \infty, 1 \leq j \leq m_i} (A_i - \xi I)\begin{pmatrix} a_{ij} \\ b_{ij} \\ c_{ij} \end{pmatrix} \Phi_{ij} = \begin{pmatrix} 0 \\ 0 \\ 0 \end{pmatrix},$$

where

$$\begin{cases} \phi = \sum_{0 \leq i \leq \infty, 1 \leq j \leq m_i} a_{ij} \Phi_{ij} \\ \psi = \sum_{0 \leq i \leq \infty, 1 \leq j \leq m_i} b_{ij} \Phi_{ij} \\ \Upsilon = \sum_{0 \leq i \leq \infty, 1 \leq j \leq m_i} c_{ij} \Phi_{ij}, \end{cases}$$

and

$$A_i = \begin{pmatrix} -d_1\lambda_i - (H_0 + \mu + \alpha) & 0 & -F_2 \\ \alpha & -d_2\lambda_i - (H_1 + \mu + \beta) & -G_2 \\ H_0 & H_1 & -d_3\lambda_i - H_2 \end{pmatrix}.$$

Hence, the stability of the steady state is reduced to examining the eigenvalues of the matrices A_i. Indeed, the real parts of every eigenvalue is negative if the trace and determinant of A_i and the determinant of its second additive compound $A_i^{[2]}$ are all negative. The trace of A_i is given by

$$\mathrm{tr}\, A_i = -(d_1 + d_2 + d_3)\lambda_i + \mathrm{tr}\, J < 0$$

for all $i \geq 0$ since $\mathrm{tr}\, J < 0$ by (8.166). The determinant of A_i is given by:

$$\det A_i = -d_1 d_2 d_3 \lambda_i^3 - B_A \lambda_i^2 - C_A \lambda_i + \det J, \qquad (8.169)$$

where

$$\begin{cases} B_A = H_2 d_1 d_2 + (H_0 + \mu + \alpha)d_2 d_3 + (H_1 + \mu + \beta)d_1 d_3 > 0, \\ C_A = (G_2 H_1 + (H_1 + \mu + \beta)H_2)d_1 + ((H_0 + \alpha + \mu)H_2 + F_2 H_0)d_2 \\ \qquad + (\beta + \mu + H_1)(\alpha + \mu + H_0)d_3 > 0. \end{cases}$$

Clearly, $\det A_i$ *for all* $i \geq 0$ *since* $\det J < 0$. *On the other hand, the matrix* $A_i^{[2]}$ *(the second additive compound of* A_i *) is given by*

$$A_i^{[2]} = \begin{pmatrix} -(d_1 + d_2)\lambda_i - A & -G_2 & F_2 \\ H_1 & -(d_1 + d_3)\lambda_i - B & 0 \\ -H_0 & \alpha & -(d_2 + d_3)\lambda_i - C \end{pmatrix},$$
$$(8.170)$$

where

$$\begin{cases} A = H_0 + H_1 + 2\mu + \alpha + \beta > 0 \\[2mm] B = H_0 + H_2 + \mu + \alpha > 0 \\[2mm] C = H_1 + H_2 + \mu + \beta > 0 \end{cases}$$

Therefore,

$$\det A_i^{[2]} = -(d_2 + d_3)(d_1 + d_3)(d_1 + d_2)\lambda_i^3 - B_{A^{[2]}}\lambda_i^2 - C_{A^{[2]}}\lambda_i + \det J^{[2]}, \quad (8.171)$$

with

$$\begin{cases} B_{A^{[2]}} = (B + C)d_1 d_2 + (A + B + C)d_2 d_3 + (A + B + C)d_1 d_3 \\ \qquad\quad + A d_3^2 + B d_2^2 + C d_1^2 \\[3mm] C_{A^{[2]}} = (AC + BC + F_2 H_0)d_1 + (AB + BC + G_2 H_1)d_2 \\ \qquad\quad + (AB + F_2 H_0 + AC + G_2 H_1)d_3. \end{cases}$$

it is clear that $B_{A^{[2]}}, C_{A^{[2]}} > 0$, *and since* $\det J^{[2]} < 0$, *then* $\det A_i^{[2]} < 0$ *for all* $i \geq 0$. *Finally, the steady state solution is locally asymptotically stable.*

In the presence of diffusion, it was proved in [Abdelmalek & Bendoukha (2016)] that every solution of the system (8.157)-(8.159) with a positive initial value that is different from the equilibrium point will converge to the equilibrium point. Define

$$L(x) = x - 1 - \ln(x) \qquad (8.172)$$

for $x > 0$.

Theorem 107 *Let*

$$V(t) = \int_\Omega [u^* L(\frac{u}{u^*}) + u_2^* L(\frac{v}{v^*}) + u_3^* L(\frac{w}{w^*})]dx$$

Then, $V(t)$ *is non-negative and is strictly minimized at the unique equilibrium* (u^*, v^*, w^*), *i.e., it is a valid Lyapunov functional. Hence,* (u^*, v^*, w^*) *is globally asymptotically stable.*

Proof 48 *The steady state solution (u^*, v^*, w^*) is globally asymptotically stable if $V(t)$ is a Lyapunov functional. Firstly, we differentiate $V(t)$ with respect to time in order to get*

$$\frac{dV}{dt} = \int_\Omega \left[(1 - \frac{u^*}{u})\frac{du}{dt} + (1 - \frac{v^*}{v})\frac{dv}{dt} + (1 - \frac{w^*}{w})\frac{dw}{dt}\right] dx.$$

Substituting the time derivatives with their values from (8.157) yields

$$\frac{dV}{dt} = \int_\Omega (1 - \frac{u^*}{u})[d_1\Delta u + \Lambda_1 - uf(w) - (\mu + \alpha)u]dx$$

$$+ \int_\Omega (1 - \frac{v^*}{v})[d_2\Delta v + \Lambda_2 + \alpha u - vg(w) - (\mu + \beta)v]dx$$

$$+ \int_\Omega (1 - \frac{w^*}{w})[d_3\Delta w + \Lambda_3 + uf(w) + vg(w) - (\mu + \gamma + \delta)w]dx$$

$$= I + J.$$

The first part is

$$I = I_1 + I_2 + I_3, \tag{8.173}$$

where

$$\begin{cases} I_1 = \int_\Omega d_1(1 - \frac{u^*}{u})\Delta u \, dx, \\[2mm] I_2 = \int_\Omega d_2(1 - \frac{v^*}{v})\Delta v \, dx \\[2mm] I_3 = \int_\Omega d_3(1 - \frac{w^*}{w})\Delta w \, dx \end{cases}$$

The second part of the derivative is given by:

$$J = \int_\Omega (1 - \frac{u^*}{u})[\Lambda_1 - uf(w) - (\mu + \alpha)u]dx$$

$$+ \int_\Omega (1 - \frac{v^*}{v})[\Lambda_2 + \alpha u - vg(w) - (\mu + \beta)v]dx$$

$$+ \int_\Omega (1 - \frac{w^*}{w})[\Lambda_3 + uf(w) + vg(w) - (\mu + \gamma + \delta)w]dx$$

By using the Green formula and assuming the Neumann boundary conditions in (8.159), we obtain

$$
\begin{cases}
I_1 = \int_\Omega d_1(1 - \frac{u^*}{u})\Delta u dx = -d_1 \int_\Omega \nabla(1 - \frac{u^*}{u})\nabla u dx = -d_1 \int_\Omega \frac{u^*}{u^2}|\nabla u|^2 dx \\[2ex]
I_2 = \int_\Omega d_2(1 - \frac{v^*}{v})\Delta v dx \\
\qquad = -d_2 \int_\Omega \frac{v^*}{v^2}|\nabla v|^2 dx \\[2ex]
I_3 = \int_\Omega d_3(1 - \frac{w^*}{w})\Delta w dx \\
\qquad = -d_3 \int_\Omega \frac{w^*}{w^2}|\nabla w|^2 dx
\end{cases}
$$

Thus

$$
I = -\int_\Omega \left[d_1 \frac{u^*}{u^2}|\nabla u|^2 + d_2 \frac{v^*}{v^2}|\nabla v|^2 + d_3 \frac{w^*}{w^2}|\nabla w|^2 \right] dx < 0
$$

The second part of the derivative J can be simplified by replacing Λ_1, Λ_2, and $(\mu + \gamma + \delta)$ with their values and rearranging to the form

$$
J = \int_\Omega (1 - \frac{u^*}{u})[u^* f(w^*) + (\mu + \alpha)u^* - uf(w) - (\mu + \alpha)u] dx
$$

$$
+ \int_\Omega (1 - \frac{v^*}{v})[v^* g(w^*) + (\mu + \beta)v^* - \alpha u^* + \alpha u - vg(w) - (\mu + \beta)v] dx
$$

$$
+ \int_\Omega (1 - \frac{w^*}{w})\left(\Lambda_3 + uf(w) + vg(w) - \frac{\Lambda_3 + u^* f(w^*) + v^* g(w^*)}{w^*} w \right) dx
$$

i.e.,

$$
J = \int_\Omega (1 - \frac{u^*}{u})\left[u^* f(w^*)\left(1 - \frac{uf(w)}{u^* f(w^*)}\right) + (\mu + \alpha)u^*(1 - \frac{u}{u^*}) \right] dx
$$

$$
+ \int_\Omega (1 - \frac{v^*}{v})\left[v^* g(w^*)\left(1 - \frac{vg(w)}{v^* g(w^*)}\right) + (\mu + \beta)v^*(1 - \frac{v}{v^*}) + \alpha u^*(\frac{u}{u^*} - 1) \right] dx
$$

$$
+ \int_\Omega (1 - \frac{w^*}{w})\left[\Lambda_3(1 - \frac{w}{w^*}) + u^* f(w^*)\left(\frac{uf(w)}{u^* f(w^*)} - \frac{w}{w^*}\right) \right.
$$

$$
\left. + v^* g(w^*)\left(\frac{vg(w)}{v^* g(w^*)} - \frac{w}{w^*}\right) \right] dx.
$$

Further simplification yields

$$J = \int_\Omega (\mu + \alpha) u^* (1 - \frac{u}{u^*})(1 - \frac{u^*}{u})$$

$$+\Lambda_3 (1 - \frac{w}{w^*})(1 - \frac{w^*}{w})$$

$$+ u^* f(w^*) \Big[\Big(\frac{uf(w)}{u^* f(w^*)}$$

$$- \frac{w}{w^*} \Big)(1 - \frac{w^*}{w}) + (1 - \frac{u^*}{u}) \Big(1$$

$$- \frac{uf(w)}{u^* f(w^*)} \Big) \Big] + v^* g(w^*) \Big[\Big(1$$

$$- \frac{vg(w)}{v^* g(w^*)} \Big)(1 - \frac{v^*}{v}) + \Big(\frac{vg(w)}{v^* g(w^*)}$$

$$- \frac{w}{w^*} \Big)(1 - \frac{w^*}{w}) \Big]$$

$$+(\mu + \beta) v^* (1 - \frac{v}{v^*})(1 - \frac{v^*}{v}) + \alpha u^* (\frac{u}{u^*} - 1)(1 - \frac{v^*}{v}) dx$$

Now, to show that J is negative, we observe the following equalities

$$\begin{cases}
L(\frac{u}{u^*}) + L(\frac{u^*}{u}) = -(1 - \frac{u}{u^*})(1 - \frac{u^*}{u}) \\[2mm]
\quad - L(\frac{u^*}{u}) + L(\frac{f(w)}{f(w^*)}) - L(\frac{w}{w^*}) - L(\frac{uf(w)w^*}{u^* f(w^*)w}) \\[2mm]
= \Big(\frac{uf(w)}{u^* f(w^*)} - \frac{w}{w^*} \Big)(1 - \frac{w^*}{w}) + (1 - \frac{u^*}{u})\Big(1 - \frac{uf(w)}{u^* f(w^*)}\Big) \\[2mm]
-L(\frac{w}{w^*}) - L(\frac{vg(w)w^*}{v^* g(w^*)w}) - L(\frac{v^*}{v}) + L(\frac{g(w)}{g(w^*)}) \\[2mm]
= \Big(1 - \frac{vg(w)}{v^* g(w^*)}\Big)(1 - \frac{v^*}{v}) + \Big(\frac{vg(w)}{v^* g(w^*)} - \frac{w}{w^*} \Big)(1 - \frac{w^*}{w}) \\[2mm]
L(\frac{u}{u^*}) - L(\frac{uv^*}{u^*v}) + L(\frac{v^*}{v}) = (\frac{u}{u^*} - 1)(1 - \frac{v^*}{v})
\end{cases}$$

Substituting these in (8.174) leads to

$$J = -\int_\Omega (\mu + \alpha)u^*\left[L(\frac{u}{u^*}) + L(\frac{u^*}{u})\right]dx - \int_\Omega \Lambda_3 \frac{(w - w^*)^2}{ww^*}dx$$

$$-\int_\Omega u^* f(w^*)\left[L(\frac{u^*}{u}) - L(\frac{f(w)}{f(w^*)}) + L(\frac{w}{w^*}) + L\left(\frac{uf(w)w^*}{u^* f(w^*)w}\right)\right]dx$$

$$-\int_\Omega v^* g(w^*)\left[L(\frac{w}{w^*}) + L\left(\frac{vg(w)w^*}{v^* g(w^*)w}\right) + L(\frac{v^*}{v}) - L\left(\frac{g(w)}{g(w^*)}\right)\right]dx$$

$$- (\mu + \beta)\int_\Omega v^*\left[L(\frac{v}{v^*}) + L(\frac{v^*}{v})\right]dx$$

$$+ \alpha\int_\Omega u^*\left[L(\frac{u}{u^*}) - L(\frac{uv^*}{u^* v}) + L(\frac{v^*}{v})\right]dx$$

By using Lemma 8.6 we get the following inequality:

$$J \leq -\int_\Omega (\mu + \alpha)u^*\left[L(\frac{u}{u^*}) + L(\frac{u^*}{u})\right]dx - \int_\Omega \Lambda_3 \frac{(w - w^*)^2}{ww^*}dx$$

$$-\int_\Omega u^* f(w^*)\left[L(\frac{u^*}{u}) + L\left(\frac{uf(w)w^*}{u^* f(w^*)w}\right)\right]dx$$

$$-\int_\Omega v^* g(w^*)\left[L\left(\frac{vg(w)w^*}{v^* g(w^*)w}\right)\right]dx$$

$$- (\mu + \beta)\int_\Omega v^* L(\frac{v}{v^*})dx - \alpha\int_\Omega u^* L(\frac{uv^*}{u^* v})dx.$$

We have $J \leq 0$, then $\frac{dV}{dt} \leq 0$; $\frac{dV}{dt} = 0$ only at the steady state (u^*, v^*, w^*). By Lyapunov's direct method, the steady state solution (u^*, v^*, w^*) is globally asymptotically stable.

8.14 Construction of globally stable system of delayed differential equations

As an example of a globally stable system of delayed differential equations, let us consider the model proposed in [Jinliang & Tian (2013)]:

$$\begin{cases} x'(t) = \lambda - d_1 x(t) - \frac{\beta x(t) y(t)}{x(t) + y(t)} \\\\ y'(t) = \frac{\beta e^{-d_1 \tau} x(t-\tau) y(t-\tau)}{x(t-\tau) + y(t-\tau)} - d_2 y(t) - a y(t) z(t) \\\\ z'(t) = p y(t) z(t) - d_3 z(t) \end{cases} \qquad (8.175)$$

The system (8.175) is a delayed HBV infection model with CTL immune response. Here $z(t)$ is the density of CTLs. The infected cells $y(t)$ are removed at a rate ayz by the CTL immune response and the virus-specific CTL cells proliferate at a rate pyz by contact with the infected cells, and die at a rate $d_3 z$. Here x, y and z are numbers of uninfected (susceptible) liver cells, infected liver cells and free virions, respectively.

Let $\mathcal{C}$ be the Banach space of continuous real-valued functions $\mathcal{C} = C([-\tau, 0], \mathbb{R}^3)$ with the sup-norm

$$\|\varphi\| = \max \left\{ \sup_{-\tau \le \theta \le 0} |\varphi_1(\theta)|, \sup_{-\tau \le \theta \le 0} |\varphi_2(\theta)|, \sup_{-\tau \le \theta \le 0} |\varphi_3(\theta)| \right\} \qquad (8.176)$$

for $\varphi = (\varphi_1, \varphi_2, \varphi_3) \in \mathcal{C}$. Further, the non-negative cone of $\mathcal{C}$ is defined as $\mathcal{C}^+ = C([-\tau, 0], \mathbb{R}^3_+)$. The initial conditions of system (8.175) at $t = 0$ are given as $x(\theta) = \varphi_1(\theta)$, $y(\theta) = \varphi_2(\theta)$, $z(\theta) = \varphi_3(\theta)$, $\theta \in [-\tau, 0]$, where

$$\varphi = (\varphi_1, \varphi_2, \varphi_3) \in \mathcal{C}^+, \quad \varphi(0) > 0. \qquad (8.177)$$

Before presenting the proof that system (8.175) is globally stable, we need the following result that establishes the positivity and boundedness of solutions for system (8.175) with initial conditions (8.177):

Theorem 108 *Under the preceding initial conditions (8.177), $x(t), y(t)$ and $z(t)$ are all non-negative and bounded for all t at which the solution exists.*

Proof 49 *By the existence and uniqueness theorem [Kuang (1993)] of delayed differential equations, there exists a $t_0 > 0$ such that there exists a solution $(x(t), y(t), z(t))$ of system (8.175) for $0 < t < t_0$. Assume that there exists a solution to system (8.175) for $0 < t < t_1$ for a positive t_1. In fact, $x(t)$ is positive for all $t \ge 0$. Assume the contrary and let $t_1 > 0$ be the first time such that $x(t_1) = 0$. If $x(t)$ loses its non-negativity, there would have to be $x'(t_1) \le 0$, by the first equation of system (8.175) and this is impossible regarding the equation for $x(t)$ in system (8.175). Hence, $x(t) > 0$ for $t > 0$ as long as $x(t)$ exists.*

By the second equation of system (8.175), we have

$$y(t) = y(0) \exp \left(-d_2 t - a \int_0^t z(\theta) d\theta \right)$$

$$+ \int_0^t \frac{\beta e^{-d\tau} x(\theta - \tau) y(\theta - \tau)}{x(\theta - \tau) + y(\theta - \tau)} e^{d_2(\theta - t)} \exp \left(-a \int_\theta^t z(\sigma) d\sigma \right) d\theta$$

then $y(t) > 0$ for $t > 0$. The third equation of system (8.175) implies that $z(t) = z(0)\exp[(py - d_3)t]$. Hence $z(t) \geq 0$ for $0 \leq t < t_1$. Let

$$G(t) = e^{-d_1\tau}x(t) + y(t + \tau) + \frac{a}{p}z(t).$$

Adding all the equations of (8.175) we obtain

$$G'^{-d_1\tau} - d_1 e^{-d_1\tau}x(t) - d_2 y(t + \tau) - \frac{d_3}{a}z(t)$$
$$\leq \lambda e^{-d_1\tau} - dG(t),$$

where $d = \min\{d_1, d_2, d_3\}$. Then $G(t) \leq M_1$ for some $M_1 > 0$ for sufficiently large t. We can take as $M_1 = \frac{2\lambda e^{-d\tau}}{d}$, which implies that $G(t)$ is ultimately bounded, and so are $x(t), y(t)$ and $z(t)$.

In system (8.175) there always exists an infection-free equilibrium $E_0 = (x_0, 0, 0)$, where $x_0 = \frac{\lambda}{d_1}$, which represents the state that the viruses are absent. The basic reproduction number of system (8.175) is given by

$$\Re_0 = \frac{\beta e^{-d\tau}}{d_2}$$

If $\Re_0 \leq 1$, an infection-free equilibrium E_0 is the unique equilibrium, corresponding to the extinction of free viruses. If $\Re_0 > 1$, in addition to E_0, there exists a CTL-inactivated infection equilibrium $E_1(x_1, y_1, 0)$, where

$$\begin{cases} x_1 = \frac{\lambda}{d_1 + d_2 e^{d_1\tau}(\Re_0 - 1)} \\[2mm] y_1 = \frac{\lambda(\Re_0 - 1)}{d_1 + d_2 e^{d_1\tau}(\Re_0 - 1)} \end{cases}$$

which represents the state that the viruses are present, whereas the CTLs are absent. Define a CTL immune-response reproduction number by

$$\Re_1 = \frac{d_1 + d_2 e^{d_1\tau}(\Re_0 - 1)}{p\lambda}(py_1 - d_3) + 1.$$

Given $\Re_1 > 1$, then system (8.175) has an CTL-activated infection equilibrium $E_2(x_2, y_2, z_2)$, where

$$\begin{cases} x_2 = \frac{(\lambda p - d_1 d_3 - \beta d_3) + \sqrt{(\lambda p - d_1 d_3 - \beta d_3)^2 + 4d_1 d_3 \lambda p}}{2d_1 p} \\[3mm] y_2 = \frac{d_3}{p} \\[3mm] z_2 = \frac{\beta p e^{-d_1\tau}x_2}{ax_2 p + d_3} - \frac{d_2}{a} \end{cases}$$

The endemic equilibrium represents the state that both the viruses and CTL response are present. Let

$$g(x) = x - 1 - \ln x$$

Note that $g : \mathbb{R}_+ \to \mathbb{R}_+$ has strict global minimum $g(1) = 0$. The following result was proved in [Wang & Tian (2013)]:

Theorem 109 *If $\Re_0 \le 1$, then the disease free equilibrium E_0 is globally asymptotically stable.*

Proof 50 *Define a Lyapunov functional:*

$$L_0 = e^{-d_1 \tau}\left[x - x_0 - \int_{x_0}^{x} \frac{x_0(\theta + y)}{\theta(x_0 + y)} d\theta\right] + y + \frac{a}{p} z + \beta e^{-d_1 \tau} \int_{t-\tau}^{t} \frac{x(\theta)y(\theta)}{x(\theta) + y(\theta)} d\theta \tag{8.178}$$

The time derivative of L_0 along the solution of (8.175) is given by:

$$\frac{dL_0}{dt}\bigg|_{(8.175)}$$

$$= e^{-d_1 \tau}\left[1 - \frac{x_0(x + y)}{x(x_0 + y)}\right] x' + y' + \frac{a}{p} z' + \frac{\beta e^{-d_1 \tau} xy}{x + y} - \frac{\beta e^{-d_1 \tau} x(t - \tau)y(t - \tau)}{x(t - \tau) + y(t - \tau)}$$

$$= e^{-d_1 \tau}\left[1 - \frac{x_0(x + y)}{x(x_0 + y)}\right]\left[\lambda - d_1 x - \frac{\beta xy}{x + y}\right] + \frac{\beta e^{-d_1 \tau} x(t - \tau)y(t - \tau)}{x(t - \tau) + y(t - \tau)}$$

$$- d_2 y - ayz + \frac{a}{p}(pyz - d_3 z) + \frac{\beta e^{-d_1 \tau} xy}{x + y} - \frac{\beta e^{-d_1 \tau} x(t - \tau)y(t - \tau)}{x(t - \tau) + y(t - \tau)}$$

$$= e^{-d_1 \tau} d_1 (x_0 - x)\left[1 - \frac{x_0(x + y)}{x(x_0 + y)}\right] + \frac{\beta e^{-d_1 \tau} x_0 y}{x_0 + y} - d_2 y - \frac{ad_3 z}{p}$$

$$= e^{-d_1 \tau} d_1 (x_0 - x)\left(1 - \frac{x_0(x + y)}{x(x_0 + y)}\right) + \frac{d_2 x_0 y(\Re_0 - 1)}{x_0 + y} - \frac{d_2 y^2}{x_0 + y} - \frac{ad_3 z}{p},$$

where

$$e^{-d_1 \tau} d_1 (x_0 - x)\left(1 - \frac{x_0(x + y)}{x(x_0 + y)}\right) = -\frac{e^{-d_1 \tau} d_1 y(x_0 - x)^2}{x(x_0 + y)} \le 0.$$

Thus, $\Re_0 \le 1$ implies that $dL_0/dt \le 0$ for all $x > 0$, $y \ge 0$, $z \ge 0$, and $\frac{dL_0}{dt} = 0$ holds if and only if $x = x_0, y = 0$, and $z(t) = 0$ for $\Re_0 \le 1$. Hence, the largest invariant set in $\{(x_t, y_t, v_t, z_t) | \frac{dL_0}{dt} = 0\}$ is E_0. The classical Lyapunov-LaSalle invariance principle [Kuang (1993)], we get that E_0 is globally asymptotically stable when $\Re_0 \le 1$. The following result was proved in [Wang & Tian (2013)]:

Theorem 110 *If $\Re_1 \leq 1 < \Re_0$, then CTL-inactivated infection equilibrium E_1 is globally asymptotically stable.*

Proof 51 *Define a Lyapunov functional:*

$$
L_1 = x - x_1 - \int_{x_1}^{x} \frac{x_1(\theta + y_1)}{\theta(x_1 + y_1)} d\theta + e^{d_1 \tau} y_1 g\left(\frac{y}{y_1}\right) + \frac{ae^{d_1 \tau}}{p} z
$$

$$
+ e^{d_1 \tau} d_2 y_1 \int_{t-\tau}^{t} g\left(\frac{\beta x(\theta) y(\theta)}{e^{d_1 \tau} d_2 y_1 (x(\theta) + y(\theta))}\right) d\theta.
$$

The time derivative of L_1 along the solution of (8.175) is given by:

$$
\frac{dL_1}{dt}\bigg|_{(8.175)} = [1 - \frac{x_1(x + y_1)}{x(x_1 + y_1)}] x' + e^{d_1 \tau} \left(1 - \frac{y_1}{y}\right) y' + \frac{ae^{d_1 \tau}}{p} z' \tag{8.179}
$$

$$
+ \frac{\beta xy}{x + y} - \frac{\beta x(t - \tau) y(t - \tau)}{x(t - \tau) + y(t - \tau)} - e^{d_1 \tau} d_2 y_1 \ln \frac{\beta xy}{e^{d_1 \tau} d_2 y_1 (x + y)}
$$

$$
+ e^{d_1 \tau} d_2 y_1 \ln \frac{\beta x(t - \tau) y(t - \tau)}{e^{d_1 \tau} d_2 y_1 (x(t - \tau) + y(t - \tau))}
$$

$$
= [1 - \frac{x_1(x + y_1)}{x(x_1 + y_1)}] \left[d_1(x_1 - x) + \left(\frac{\beta x_1 y_1}{x_1 + y_1} - \frac{\beta xy}{x + y}\right) \right]
$$

$$
+ [\frac{\beta x(t - \tau) y(t - \tau)}{x(t - \tau) + y(t - \tau)} - e^{d_1 \tau} (d_2 y(t) - ay(t) z(t))] \left(1 - \frac{y_1}{y}\right)
$$

$$
+ \frac{ae^{d_1 \tau}}{p} (py - d_3) z + \frac{\beta xy}{x + y} - \frac{\beta x(t - \tau) y(t - \tau)}{x(t - \tau) + y(t - \tau)}
$$

$$
- e^{d_1 \tau} d_2 y_1 \ln \frac{\beta xy}{e^{d_1 \tau} d_2 y_1 (x + y)}
$$

$$
+ e^{d_1 \tau} d_2 y_1 \ln \frac{\beta x(t - \tau) y(t - \tau)}{e^{d_1 \tau} d_2 y_1 (x(t - \tau) + y(t - \tau))}
$$

Here we used that

$$
\lambda = d_1 x_1 + \frac{\beta x_1 y_1}{x_1 + y_1}, \quad d_2 y_1 = \frac{\beta e^{-d_1 \tau x_1 y_1}}{x_1 + y_1}. \tag{8.180}
$$

Combining the (8.179) and (8.180) we obtain

$$
\Re_1 = \frac{d_1 + d_2 e^{d_1 \tau} (\Re_0 - 1)}{p\lambda} (py_1 - d_3) + 1
$$

which implies that $\Re_1 \leq 1$ is equivalent to $\frac{py_1}{d_3} \leq 1$. The term $\frac{py_1}{d_3}$ is the immune reproductive number, which expresses the average number of activated

CTLs generated from one CTL during its lifetime $\frac{1}{d_3}$ through the stimulation of the infected cells y_1. The immune is activated in the case where $\Re_1 > 1$. Hence, $\frac{dL_1}{dt}$ is always non-positive under the condition $\Re_1 \le 1 < \Re_0$, and it can be verified that $\frac{dL_1}{dt} = 0$ if and only if $x = x_1$ and $\frac{x+y}{x+y_1} = \frac{x_1(x+y_1)}{x(x_1+y_1)} = \frac{(x_1+y_1)x(t-\tau)y(t-\tau)}{x_1y(x(t-\tau)+y(t-\tau))} = 1$. Using the first two equations of system (8.175), we have $0 = x'(t) = \lambda - d_1x_1 - \frac{\beta x_1 y(t)}{x_1 + y(t)}$,

$0 = y'(t) = \frac{\beta e^{-d_1\tau} x_1 y(t-\tau)}{x_1 + y(t-\tau)} - d_2 y(t) - ay(t)z(t)$. This gives $y = y_1, z = 0$ and the global asymptotic stability of E_1 follows from the LaSalle's invariant principle.

Theorem 111 *If $\Re_1 > 1$, then CTL-activated infection equilibrium E_2 is globally asymptotically stable; i.e., E_2 is globally asymptotically stable whenever it exists.*

Proof 52 *Define a Lyapunov functional*

$$
L_2 = x(t) - x_2 - \int_{x_2}^{x(t)} \frac{x_2(\theta + y_2)}{\theta(x_2 + y_2)} d\theta + e^{d_1\tau} y_2 g\left(\frac{y(t)}{y_2}\right) + \frac{ae^{d_1\tau}}{p} z_2 g\left(\frac{z(t)}{z_2}\right)
$$

$$
+ e^{d_1\tau}(d_2 y_2 + a y_2 z_2) \int_{t-\tau}^{t} g\left(\frac{\beta x(\theta)y(\theta)}{e^{d_1\tau}(d_2 y_2 + a y_2 z_2)(x(\theta) + y(\theta))}\right) d\theta
$$

The time derivative of L_2 along the solution of (8.175) is given by

$$
\left.\frac{dL_2}{dt}\right|_{(8.175)} = \left[1 - \frac{x_2(x+y_2)}{x(x_2+y_2)}\right]\left[d_1(x_2 - x) + \left(\frac{\beta x_2 y_2}{x_2 + y_2} - \frac{\beta x y}{x + y}\right)\right] \tag{8.181}
$$

$$
+ \left[\frac{\beta x(t-\tau)y(t-\tau)}{x(t-\tau)+y(t-\tau)} - e^{d_1\tau}(d_2 y(t) - ay(t)z(t))\right]\left(1 - \frac{y_2}{y}\right)
$$

$$
+ \frac{ae^{d_1\tau}}{p}(pyz - d_3 z)\left(1 - \frac{z_2}{z}\right) + \frac{\beta x y}{x+y} - \frac{\beta x(t-\tau)y(t-\tau)}{x(t-\tau)+y(t-\tau)}
$$

$$
- \frac{\beta x_2 y_2}{x_2 + y_2} \ln \frac{\beta x y}{\frac{\beta x_2 y_2}{x_2+y_2}(x+y)} + \frac{\beta x_2 y_2}{x_2 + y_2} \ln \frac{\beta x(t-\tau)y(t-\tau)}{\frac{\beta x_2 y_2}{x_2+y_2}(x(t-\tau)+y(t-\tau))}
$$

Here we used that

$$
\begin{cases}
\lambda = d_1 x_2 + \frac{\beta x_2 y_2}{x_2+y_2} \\[2mm]
d_2 y_2 + a y_2 z_2 = \frac{\beta e^{-d_1\tau} x_2 y_2}{x_2+y_2} \\[2mm]
p y_2 = d_3.
\end{cases} \tag{8.182}
$$

Combining (8.181) and (8.182) we obtain

$$
\begin{aligned}
\frac{dL_2}{dt}\Big|_{(8.175)} &= d_1(x_2 - x)\left[1 - \frac{x_2(x + y_2)}{x(x_2 + y_2)}\right] + \frac{\beta x_2 y_2}{x_2 + y_2}\left[1 - \frac{x_2(x + y_2)}{x(x_2 + y_2)} + \frac{y(x + y_2)}{y_2(x + y)} - \frac{y}{y_2}\right] \\
&\quad + \frac{\beta x_2 y_2}{x_2 + y_2} - \frac{y_2}{y}\frac{x(t - \tau)y(t - \tau)}{x(t - \tau) + y(t - \tau)} \\
&\quad - \frac{\beta x_2 y_2}{x_2 + y_2}\ln\frac{\beta xy}{\frac{\beta x_2 y_2}{x_2 + y_2}(x + y)} + \frac{\beta x_2 y_2}{x_2 + y_2}\ln\frac{\beta x(t - \tau)y(t - \tau)}{\frac{\beta x_2 y_2}{x_2 + y_2}(x(t - \tau) + y(t - \tau))} \\
&= d_1(x_2 - x)\left[1 - \frac{x_2(x + y_2)}{x(x_2 + y_2)}\right] + \frac{\beta x_2 y_2}{x_2 + y_2}\left[\left(1 - \frac{y(x + y_2)}{y_2(x + y)}\right)\left(\frac{x + y}{x + y_2} - 1\right)\right. \\
&\quad - \left(\frac{x + y}{x + y_2} - 1 - \ln\frac{x + y}{x + y_2}\right) - \left(\frac{x_2(x + y_2)}{x(x_2 + y_2)} - 1 - \ln\frac{x_2(x + y_2)}{x(x_2 + y_2)}\right) \\
&\quad \left. - \ln\frac{x + y}{x + y_2} - \ln\frac{x_2(x + y_2)}{x(x_2 + y_2)}\right] \\
&\quad - \frac{\beta x_2 y_2}{x_2 + y_2}\left[\frac{(x_2 + y_2)x(t - \tau)y(t - \tau)}{x_2 y(x(t - \tau) + y(t - \tau))} - 1 - \ln\frac{(x_2 + y_2)x(t - \tau)y(t - \tau)}{x_2 y(x(t - \tau) + y(t - \tau))}\right] \\
&\quad - \frac{\beta x_2 y_2}{x_2 + y_2}\left[\ln\frac{(x_2 + y_2)x(t - \tau)y(t - \tau)}{x_2 y(x(t - \tau) + y(t - \tau))} + \ln\frac{\beta xy}{\frac{\beta x_2 y_2}{x_2 + y_2}(x + y)}\right. \\
&\quad \left. - \ln\frac{\beta x(t - \tau)y(t - \tau)}{\frac{\beta x_2 y_2}{x_2 + y_2}(x(t - \tau) + y(t - \tau))}\right] \\
&= d_1(x_2 - x)\left[1 - \frac{x_2(x + y_2)}{x(x_2 + y_2)}\right] + \frac{\beta x_2 y_2}{x_2 + y_2}\left(1 - \frac{y(x + y_2)}{y_2(x + y)}\right)\left(\frac{x + y}{x + y_2} - 1\right) \\
&\quad - \frac{\beta x_2 y_2}{x_2 + y_2}g\left(\frac{x + y}{x + y_2}\right) - \frac{\beta x_2 y_2}{x_2 + y_2}g\left(\frac{x_2(x + y_2)}{x(x_2 + y_2)}\right) \\
&\quad - \frac{\beta x_2 y_2}{x_2 + y_2}g\left(\frac{(x_2 + y_2)x(t - \tau)y(t - \tau)}{x_2 y(x(t - \tau) + y(t - \tau))}\right)
\end{aligned}
$$

Following the proof of Theorem 110, we get $\frac{dL_2}{dt} < 0$ for all $x > 0, y > 0$ and $z > 0$, and $\frac{dL_2}{dt} = 0$ if and only if $x = x_2$ and $y = y_2, z = z_2$. The largest invariant set in $\{(x_t, y_t, z_t) \mid \frac{dL_2}{dt} = 0\}$ is E_2. The Lyapunov-LaSalle invariance principle implies that the equilibrium $E_2(x_1, y_2, z_2)$ is globally asymptotically stable.

Finally, the dynamics of system (8.175) are described by the following behaviors:

(a) When $\Re_0 \leq 1$, the infection-free equilibrium is globally asymptotically stable. This means that the viruses are cleared and immune is not active.

(b) When $\Re_1 \leq 1 < \Re_0$, the CTL-inactivated infection equilibrium exists and is globally asymptotically stable. This means that CTLs immune response would not be activated and viral infection becomes chronic.

(c) When $\Re_1 > 1$, the CTL-activated infection equilibrium exists and is globally asymptotically stable, in this case the infection causes a persistent CTLs immune response.

8.15 Stabilization by the Jurdjevic-Quinn method

The Jurdjevic-Quinn method (or the speed gradient method) is a method of stabilization and it used in several domains for improving stability performances. This method is described by the following steps:

1. Let a non-linear (affine) system given by:

$$x' = f(x) + \sum_{i=1}^{m} u_i g_i(x) = f(x) + G(x)u \qquad (8.183)$$

where $x \in \mathbb{R}^n$, $u = (u_1, \ldots, u_m) \in \mathbb{R}^m$. The vector fields $f, g_1, \ldots, g_m$ are at least continuous, and $f(0) = 0$. Assume that when the input is disconnected, the system (8.183) has a stable (but not asymptotically stable) equilibrium position.

2. If a (weak) Lyapunov function $V(x)$ for the (unforced) system exists and some other technical assumptions are fulfilled, then the system (8.183) can be asymptotically stabilized at the equilibrium by a feedback law whose construction involves $\nabla V(x)$.

Notice that affine systems of the form (8.183) represent a natural generalization of the well-known linear systems given by:

$$x' = Ax + Bu \qquad (8.184)$$

In order to find the feedback law, let us consider a mechanical system representing a non-linear elastic force $x'' = -f(x) + u$ (with $f(x)x > 0$ for $x \neq 0$). We take

$$V(x, x') = \frac{(x')^2}{2} + \int f(x) \, dx$$

as a Lyapunov function. Hence, asymptotic stabilization can be achieved by *proportional derivative* control, i.e., adding friction to the system, and this is a particular case of feedback depending on the gradient of $V(x, x')$.

The basic assumption of the Jurdjevic-Quinn method is that the unforced system is stable at the origin, and that a smooth, weak Lyapunov function $V(x)$ is known. Now, define the feedback u given by

$$u = k(x) = -\frac{\gamma}{2}(\nabla V(x)G(x)) \qquad (8.185)$$

where $\gamma > 0$.

The second assumption of the Jurdjevic-Quinn method is that the vector fields appearing in (8.183) are C^∞. Notice that the Lie bracket operator associated to an (ordered) pair f_0, f_1 of vector fields is the vector field defined by:

$$[f_0, f_1] = Df_1 \cdot f_0 - Df_0 \cdot f_1$$

(here, Df_i denotes the Jacobian matrix of f_i, $i = 0, 1$). The "ad" operator is iteratively defined by:

$$\begin{cases} ad_{f_0}^1 f_1 = [f_0, f_1] \\ \\ ad_{f_0}^{k+1} f_1 = [f_0, ad_{f_0}^k f_1] \end{cases}$$

Theorem 112 *Assume that a weak Lyapunov function $V(x)$ of class C^∞ for the unforced system associated to (8.184) is known. Assume also that for each $x \neq 0$ in some neighborhood of the origin we have $\nabla V(x) \neq 0$ and $\dim \left\{ f(x), ad_f^k g_i(x), i = 1, 2, k = 0, 1, 2 \right\} = n$. Then, for any $\gamma > 0$, the system is stabilized by the feedback (8.185).*

The proof of this theorem relies on LaSalle's invariance principle.

In many situations, there is a need to limit the class of admissible inputs. For this purpose, assume that the vector fields $f, g_1, \ldots, g_m$ are locally Lipschitz continuous, so that uniqueness of solutions is guaranteed for any admissible input, but not under continuous feedback. An *admissible input* is any piecewise continuous (right-continuous), locally bounded function $u(t) : [0, +\infty) \to \mathbb{R}^m$. Assume that (8.183) can be asymptotically stabilized by a feedback law of the form (8.185).

Definition 93 *A function $h \colon \mathbb{R}^n \mapsto [0, \infty)$ is called radially unbounded if for every $M \in \mathbb{R}$ there exists $R \in \mathbb{R}$ such that for all $x \in \mathbb{R}^n$, if $\|x\| > R$ then $h(x) > M$.*

Hence, an optimization problem can be associated to this stabilization problem:

8.15.1 The minimization problem

Let $h(x)$ be a continuous, positive definite and radially unbounded function. We associate to (8.183) the following cost functional defined by:

$$J(x_0, u(\cdot)) = \frac{1}{2} \int_0^{+\infty} \left(h(\varphi(t)) + \frac{\|u(t)\|^2}{\gamma} \right) dt \qquad (8.186)$$

where $\varphi(t) = \varphi(t; x_0, u(\cdot))$. Let x_0 be the initial state. The minimization problem (8.186) is solvable if there exists an admissible input $u_{x_0}^*(t)$ such that

$$J(x_0, u_{x_0}^*(\cdot)) \leq J(x_0, u(\cdot))$$

for any other admissible input $u(t)$ and the value function is defined by

$$V(x_0) = \inf_u J(x_0, u(\cdot))$$

Here, $V(x_0)$ is a minimum if and only if the minimization problem is solvable for x_0.

In fact, the solvability of the optimization problem (8.186) is equivalent to the following statement:

Theorem 113 *The first order partial differential equation (of the Hamilton-Jacobi type)*

$$\nabla U(x)f(x) - \frac{\gamma}{2}\left(\nabla U(x)G(x)\right)^2 = -\frac{h(x)}{2} \tag{8.187}$$

has a solution $U(x)$ which is radially unbounded, positive definite and of class C^1.

Example 26 *As an example of Theorem 113 is the linear, $h(x) = 2\|x\|^2$ and $\gamma = \frac{1}{2}$. In this case, the Hamilton Jacobi equation reduces to the algebraic Riccati equation:*

$$PA + AP - PBBP = -I \tag{R}$$

where I is the identity matrix of $\mathbb{R}^n$ and the unknown P is symmetric and positive definite.

8.15.2 The inverse optimization problem

Theorem 114 *(A) (Inverse optimization problem): Assume that there exists a radially unbounded, positive definite, C^1 function $V(x)$ and a positive number γ such that (8.183) is asymptotically stabilizable by the continuous feedback (8.185). Assume also that the closed-loop system admits $V(x)$ as a strict Lyapunov function and the derivative of $V(x)$ with respect to the closed loop system is radially unbounded. Define*

$$h(x) = -2\nabla V(x)f(x) + \gamma\|\nabla V(x)G(x)\|^2$$

Then, the optimization problem (8.186) has a solution given in feedback form by:

$$u = k(x) = -\gamma(\nabla V(x)G(x)) \tag{8.188}$$

for each x_0 and the value function coincides with $V(x_0)$.

(B) Assume that there exists a continuous, positive definite, radially unbounded function $h(x)$ and a positive number γ such that the minimization problem (8.186) is solvable for each initial state x_0. Moreover, assume that the value function $V(x_0)$ is radially unbounded and of class C^1. Then, system (8.183) is asymptotically stabilizable by the following continuous feedback

$$u = k(x) = -\alpha(\nabla V(x)G(x)) \tag{8.189}$$

for any $\alpha \geq \frac{\gamma}{2}$, and the value function V represents a strict Lyapunov function for the closed loop system.

8.15.3 Input-to-state stability

The input-to-state stability relates the behavior of the output to the size of the external input. Let us consider the system

$$x' = f(x, u) \tag{8.190}$$

Recall that $\beta \in \mathcal{LK}$ means that $\beta : [0, +\infty) \times [0, +\infty) \to \mathbb{R}$ is decreasing to zero with respect to the first variable and of class $\mathcal{K}_0$ with respect to the second one.

Definition 94 *We say that the system (8.190) possesses the input-to-state stability (iISS property) if there exist two maps $\beta \in \mathcal{LK}$, $\gamma \in \mathcal{K}_0$ such that, for each initial state x_0, each admissible input $u : [0, +\infty) \to \mathbb{R}^m$, each solution $\varphi(\cdot) \in S_{x_0, u(\cdot)}$ and each $t \geq 0$*

$$\|\varphi(t)\| \leq \beta(t, \|x_0\|) + \gamma(\|u\|_\infty) .$$

If the system (8.190) is ISS and $u \equiv 0$, then (8.190) is a globally asymptotically stable system. In fact, ISS systems are special cases of dissipative systems.

There exists a Lyapunov-like characterization of the ISS property as follows:

Theorem 115 *For the system (8.190), the following statements are equivalent:*

(i) The system possesses the ISS property

(ii) There exists a positive definite, radially unbounded C^∞ function $V : \mathbb{R}^n \to \mathbb{R}$ and two functions $\rho, \chi \in \mathcal{K}_0^\infty$ such that

$$\nabla V(x) \cdot f(x, u) < -\chi(\|x\|)$$

for all $x \in \mathbb{R}^n$ $(x \neq 0)$ and $u \in \mathbb{R}^m$, provided that $\|x\| \geq \rho(\|u\|)$.

(iii) There exists a positive definite, radially unbounded C^∞ function $V : \mathbb{R}^n \to \mathbb{R}$ and two functions $\omega, \alpha \in \mathcal{K}_0^\infty$ such that

$$\nabla V(x) \cdot f(x, u) \leq \omega(\|u\|) - \alpha(\|x\|)$$

for all $x \in \mathbb{R}^n$ and $u \in \mathbb{R}^m$.

We have the following definition of the *IS-stabilizable systems*:

Definition 95 *A systems is said to be IS-stabilizable if the ISS property can be recovered by applying a suitable feedback law of the form:*

$$u = k(x) + \tilde{u}$$

Thus, we have the following result concerns the affine system (8.183):

Theorem 116 *Every globally asymptotically stable (or continuously globally asymptotically stabilizable) affine system of the form (8.184) is IS-stabilizable.*

It is necessary to complete the description of the system by associating with system (8.190) an observation function $c(x) : \mathbb{R}^n \to \mathbb{R}^p$ such that:

$$\begin{cases} x' = f(x, u) \\ \\ y = c(x) \end{cases} \qquad (8.191)$$

The variable y is called the *output* and represents the available information about the evolution of the system. Let $w : \mathbb{R}^p \times \mathbb{R}^m \to \mathbb{R}$ be a given function called the supply rate, and consider the following three dissipation inequalities:

(D1) (Intrinsic version): For each admissible input $u(\cdot)$, each $\varphi \in \mathcal{S}_{0,u(\cdot)}$ and for each $t \geq 0$

$$\int_0^t w(c(\varphi(s)), u(s)) \, ds \geq 0$$

(note the initialization at $x_0 = 0$).

(D2) (Integral version): There exists a positive semi-definite function $S(x)$ called the *storage function*, such that for each admissible input $u(\cdot)$, each initial state x_0, each $\varphi \in \mathcal{S}_{x_0,u(\cdot)}$ and for each $t \geq 0$, we have the following inequality:

$$S(\varphi(t)) \leq S(x_0) + \int_0^t w(c(\varphi(s)), u(s)) \, ds \ .$$

(D3) (Differential version): There exists a positive semi-definite function $S(x) \in C^1$ such that for each $x \in \mathbb{R}^n, u \in \mathbb{R}^m$ we have

$$\nabla S(x) f(x, u) \leq w(c(x), u) \ .$$

Inequalities **(D1)**, **(D2)**, **(D3)** are used to define dissipative systems. Notice that **(D3)** implies **(D2)** implies **(D1)**. In fact, the implication **(D1)** $\implies$ **(D2)** requires a complete controllability assumption and the implication **(D2)** $\implies$ **(D3)** requires the existence of at least one storage function of class C^1.

Special forms of the supply rate w give several notions of *external stability* as follows:

1. *passivity*, for $w = yu$.

2. *finite L_2-gain*, for $w = k^2 \|u\|^2 - \|y\|^2$, where k is some real constant.

The second property implies the following estimation

$$\int_0^t \|y(s)\|^2 \, ds \leq k^2 \int_0^t \|u(s)\|^2 \, ds$$

Also, we have the following result:

Lemme 8.9 *The ISS property can be interpreted as an extension of the finite L_2-gain property.*

Proof 53 *According to Theorem 115(iii), ISS systems are dissipative in the sense of **(D3)**, with $c(x) = $ Identity and the supply rate $w(x, u) = \omega(\|u\|) - \alpha(\|x\|)$. Thus, for zero initialization, the following estimation holds*

$$\int_0^t \alpha(\|\varphi(s)\|)\, ds \leq k^2 \int_0^t \omega(\|u(s)\|)\, ds \tag{8.192}$$

At the same time, one can set $c(x) = \sqrt{\alpha(\|x\|)}$ and the integrand on the left-hand side becomes $\|y\|^2$.

Now, if (8.191) possesses the finite L_2-gain property and a suitable observability condition, then the unforced part of the system is asymptotically stable at the origin. In particular, when $c(x)$ is positive definite, then the observability condition is automatically satisfied. On the other hand, assume that (8.183) is smoothly stabilizable, then, by using an appropriate feedback the system can be rendered ISS, as shown by Theorem 116 and an estimation of the form (8.192)

Let $k^2 = \frac{1}{2\gamma}$, then one has the following result:

Theorem 117 *(A) Assume that there exists a positive semi-definite function $\Phi(x) \in C^1$ which solves the equation (of the Hamilton-Jacobi type)*

$$\nabla\Phi(x)f(x) + \frac{\gamma}{2}\|\nabla\Phi(x)G(x)\|^2 = -\|c(x)\|^2 \tag{8.193}$$

for each $x \in \mathbb{R}^n$. Then, the affine system (8.192) has a finite L_2-gain.

(B) Associated with the affine system (8.183) we consider the optimization problem (8.186), where h is positive definite and continuous. Assume that the problem is solvable for each x_0, and that the value function $V(x)$ is C^1. Then, by applying the feedback

$$u = k(x, \tilde{u}) = -\gamma(\nabla V(x)G(x)) + \tilde{u}$$

and choosing the observation function $c(x) = \sqrt{\frac{h(x)}{2}}$, the system (8.192) has a finite L_2-gain.

(C) If the affine system (8.183) is stabilizable by the following damping feedback

$$u = k(x) = -\frac{\gamma}{2}(\nabla V(x)G(x))$$

where $V(x)$ can be taken as a strict Lyapunov function for the closed loop system, then the doubled feedback $u = 2k(x) + \tilde{u}$ gives rise to a system with finite L_2-gain.

Notice that Theorem 117(A) is invertible under some restrictive assumptions.

In the next section, we will present some results regarding the damping control: We know that if a linear system is stabilizable by means of a continuous feedback, then it is also stabilizable by means of a feedback in damping form $u = -\alpha BPx$, where $\alpha \geq \frac{1}{2}$ and P is a solution of (8.R). This fact has an analogue for the non-linear case (8.183), as shown in the following result:

Theorem 118 *Consider the affine system (8.183) and assume that*

$$\|f(x)\| \leq A\|x\|^2 + C \quad and \quad \|G(x)\| \leq D$$

for some positive constants A, C, D. Assume further that (8.183) admits a stabilizer $u = k(x)$ such that:

(i) $k(x)$ is of class C^1 and $k(0) = 0$,

(ii) $k(x)$ guarantees sufficiently fast decay: More precisely, we require that each solution of the closed loop system is square integrable, i.e.,

$$\int_0^{+\infty} \|\varphi_{k(\cdot)}(t; x_0)\|^2 \, dt < +\infty \tag{8.194}$$

for each $x_0 \in \mathbb{R}^n$.

Then, there exists a map $V(x)$ such that the feedback law (8.185) is a global stabilizer for the systems, i.e., the system also admits a damping control.

Chapter 9

Transformation of Dynamical Systems to Hyperjerky Motions

9.1   Introduction

This chapter is devoted to presenting some relevant results about transforming some dynamical systems to jerky and hyperjerk forms. We begin in Sec. 9.2, by describing the transformation of 3-D dynamical systems to jerk form. Sec. 9.3 deals with the transformation of 3-D dynamical systems to rational and cubic jerks forms in a case where the inverse transformation has some singularities. In Sec. 9.4, we present in detail the method used to transform a 4-D dynamical system to hyperjerk form and the expression of the transformation is given in Sec. 9.4.1. Some examples of 4-D hyperjerky dynamics are presented in Sec. 9.4.2. Sec. 9.5 discusses some numerical examples of crackle and top dynamics.

9.2   Transformation of 3-D dynamical systems to jerk form

A jerky dynamical system is a differential equation of the form $x''' = J(x, x', x'')$ with an explicit time dependence [Eichhorn, et al., 1998, Linz, 1998]. The first three derivatives (x', x'', x''') are called *velocity*, *acceleration*, and *jerk*, respectively. This class of systems in one scalar variable or, mechanically interpreted, jerky dynamics exhibit both regular and irregular or chaotic dynamical behavior [Sprott (1994-1997(a-b)]. In particular, a wide class of 3-D vector fields with polynomial and non-polynomial non-linearities possess this property. Thus, they can be considered as the natural generalization of oscillator dynamics, hence, new methods of studying chaos are possible.

Jerky dynamics can be found in several non-mechanical areas of physics, one example of which is the chaotic third-order differential equation representing an oscillator model of thermal convection given in [Moore & Spiegel (1966)].

The following result was proved in [Eichhorn, et al., (1998)]:

Theorem 119 *Any dynamical system of the functional form:*

$$\begin{cases} x' = c_1 + b_{11}x + b_{12}y + b_{13}z + n_1\,(x,y,z) \\[2mm] y' = c_2 + b_{21}x + b_{22}y + b_{23}z + n_2\,(x,y,z) \\[2mm] z' = c_3 + b_{31}x + b_{32}y + b_{33}z + n_3\,(x,y,z) \end{cases} \tag{9.1}$$

with n_i $(i = 1,2,3)$ being non-linear functions of the indicated arguments can be reduced to a jerky dynamics $x^{(3)} = J\,(x, x', x'')$, if the conditions

$$b_{12}n_2\,(x,y,z) + b_{13}n_3\,(x,y,z) = f\,(x, b_{12}y + b_{13}z) \tag{9.2}$$

with f being an arbitrary function of the indicated arguments and

$$b_{12}^2 b_{23} + b_{13}^2 b_{32} + b_{12}b_{13}\,(b_{33} - b_{22}) \neq 0 \tag{9.3}$$

hold.

The condition (9.2) implies that the non-linearity $n_2(x,y,z)$ is solely a function of x and y, and $n_3(x,y,z)$ can be an arbitrary function of x. Hence, system (9.1) takes the following form:

$$\begin{cases} x' = c_1 + b_{11}x + b_{12}y + n_1\,(x) \\[2mm] y' = c_2 + b_{21}x + b_{22}y + b_{23}z + n_2\,(x,y) \\[2mm] z' = c_3 + b_{31}x + b_{32}y + b_{33}z + n_3\,(x,y,z) \end{cases} \tag{9.4}$$

Also, system (9.4) can be obtained from system (9.1) by using the invertible transformation given by

$$\eta = y + \frac{b_{13}}{b_{12}}z \tag{9.5}$$

because the function $f\,(x, b_{12}y + b_{13}z)$ corresponds to the non-linearity $n_2(x, \eta)$ in the new η equation. The transformation T between (9.1) and the jerky dynamics $x^{(3)} = J\,(x, x', x'')$ is given by:

$$\begin{cases} T_1\,(x,y,z) = x \\[2mm] T_2\,(x,y,z) = c_1 + \mathbf{b}^1.\,(x,y,z) + n_1\,(x) \\[2mm] T_3\,(x,y,z) = \mathbf{c.b}^1 + \mathbf{b}^1.\mathbf{b}_1 x + \mathbf{b}^1.\mathbf{b}_2 xy + \mathbf{b}^1.\mathbf{b}_3 xz + \mathbf{b}^1.\,(n_1, n_2, n_3)\,(x,y,z) \\[2mm] \qquad + \left[c_1 + \mathbf{b}^1.\,(x,y,z) + n_1\,(x)\right]\partial_x n_1\,(x) \end{cases} \tag{9.6}$$

where

$$\mathbf{b}^i = (b_{i1}, b_{i2}, b_{i3})^T, i = 1,2,3 \tag{9.7}$$

and

$$\mathbf{c} = (c_1, c_2, c_3)^T \qquad (9.8)$$

and

$$\mathbf{b}_j = (b_{1j}, b_{2j}, b_{3j})^T, j = 1, 2, 3 \qquad (9.9)$$

as the column vectors of the matrix B_{ij}. The dot denotes the scalar product. The inverse transformation is given by:

$$
\left\{
\begin{aligned}
& T_1^{-1}(u_1, u_2, u_3) = x \\[2mm]
& T_2^{-1}(u_1, u_2, u_3) = \frac{c.\mathbf{b}^1 b_{13} - c_1 \mathbf{b}^1.\mathbf{b}_3 + (b_{12} A_{32} - b_{13} A_{22})x + \left[\mathbf{b}^1.\mathbf{b}_3 + b_{13}\partial_x n_1(x)\right]v}{\det M} \\[2mm]
& \qquad + \frac{-b_{13}a - (b_{12}b_{23} + b_{13}b_{33})n_1(x) + b_{13}f(x, v - r(x))}{\det M} \\[2mm]
& T_3^{-1}(u_1, u_2, u_3) = \frac{c_1 \mathbf{b}^1.\mathbf{b}_2 - c.\mathbf{b}^1.b_{12} + (b_{12} A_{33} - b_{13} A_{23})x - \left[\mathbf{b}^1.\mathbf{b}_2 + b_{12}\partial_x n_1(x)\right]v}{\det M} \\[2mm]
& \qquad + \frac{b_{12}a + (b_{12}b_{22} + b_{13}b_{32})n_1(x) - b_{12}f(x, v - r(x))}{\det M}
\end{aligned}
\right.
$$

$$(9.10)$$

where A_{ij} denotes the adjunct or the cofactor to the element b_{ij} of the matrix **B** and

$$
\left\{
\begin{aligned}
& M = \begin{pmatrix} b_{12} & b_{13} \\ \mathbf{b}^1.\mathbf{b}_2 & \mathbf{b}^1.\mathbf{b}_3 \end{pmatrix} \\[2mm]
& r(x) = c_1 + b_{11}x + n_1(x) \\[2mm]
& s(x, v) = \mathbf{c}.\mathbf{b}^1 + \mathbf{b}^1.\mathbf{b}_1 x + v\partial_x n_1(x) + b_{11}n_1(x) + f(x, v - r(x))
\end{aligned}
\right.
$$

$$(9.11)$$

The general form of the jerky dynamics corresponding to the dynamical system (9.1) is given by:

$$J(u_1, u_2, u_3) = g(x, v) a + h(x, v) v + k(u_1, u_2, u_3) \qquad (9.12)$$

where

$$
\left\{
\begin{aligned}
& g(x, v) = \mathbf{tr}\mathbf{B} + \partial_x n_1(x) + f'(x, v - r(x)) \\[2mm]
& h(x, v) = -\mathbf{tr}A - (b_{22} + b_{33})\partial_x n_1(x) + \partial_x f(x, v - r(x)) \\[2mm]
& \qquad + v\partial_x^2 n_1(x) + [b_{11} + \partial_x n_1(x)] f'(x, v - r(x)) \\[2mm]
& k(u_1, u_2, u_3) = \mathbf{c}.\mathbf{A}_1 + \mathbf{b}^1 \mathbf{A}^1 x + \\[2mm]
& \mathbf{A}_1.(n_1, n_2, n_3)\left(x, T_2^{-1}(u_1, u_2, u_3), T_3^{-1}(u_1, u_2, u_3)\right)
\end{aligned}
\right.
$$

where $\partial_x f$ is the derivative with respect to x only if the first argument of f and f' is the derivative with respect to the second argument of f and **tr** is the trace of a matrix. Here, the row and column vectors of A are $\mathbf{A}^i$ and $\mathbf{A}_j$ respectively.

The application of the above analysis to minimal chaotic flows gives two types of systems: (1) transformable systems and (2) non-transformable systems. For the first case, the system

$$
\begin{cases}
x' = -y \\
y' = x + z \\
z' = xz + \alpha y^2
\end{cases}
\tag{9.14}
$$

is transformed to the jerky dynamics:

$$
x''' = xx'' - x' - \alpha \left(x'\right)^2 + x^2
\tag{9.15}
$$

via the transformation given by:

$$
\begin{cases}
x' = -y \\
x'' = -x - z
\end{cases}
\tag{9.16}
$$

which has the inverse transformation given by:

$$
\begin{cases}
y = -x' \\
z = -x - x''
\end{cases}
\tag{9.17}
$$

For the second case, the system

$$
\begin{cases}
x' = y \\
y' = -x + yz \\
z' = \alpha - y^2
\end{cases}
\tag{9.18}
$$

cannot be transformed into jerky dynamics.

Examples of chaotic jerky dynamics includes the system given by:

$$
x''' = -1.8x'' - 2x + xx' - 1
\tag{9.19}
$$

The chaotic attractor of system (9.19) is given in Fig. 9.1.

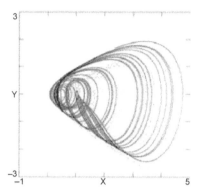

FIGURE 9.1: The chaotic attractor of system (9.19).

9.3   Transformation of 3-D dynamical systems to rational and cubic jerks forms

The inverse transformation T^{-1} exists if the procedure that transforms the 3-D dynamical system (9.1) to the jerk form (9.12) is without any singularities. In some cases [Malasoma (2000-2002)], it is possible to describe a 3-D chaotic system with five terms and a single quadratic non-linearity whose jerk representation contains rational functions with four terms:

$$\begin{cases} x' = z \\ \\ y' = -\alpha y + z \\ \\ z' = -x + xy \end{cases} \tag{9.20}$$

which can be written in jerk form as follows:

$$x''' = \alpha x - \alpha x'' + x x' - \frac{x' x''}{x} \tag{9.21}$$

The system (9.20) is chaotic over most of the narrow range $10.2849 < \alpha < 10.3716$ with an attractor as shown in Fig. 9.2 for $\alpha = 10.3$. The second example that contains cubic non-linearities [Sprott (1997(a))] is the Moore and Spiegel system (1966) which is considered as a model for the irregular variability in the luminosity of stars:

$$x''' = x'' + 9x' - x^2 x' - 5x \tag{9.22}$$

with an attractor as shown in Fig. 9.3.

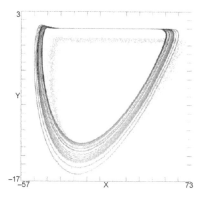

FIGURE 9.2: Chaotic attractor for the Malasoma rational jerk system (9.20) with $\alpha = 10.3$.

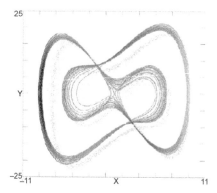

FIGURE 9.3: Chaotic attractor for the Moore-Spiegel system (9.22)

The simplest cubic case [Malasoma (2000-2002)] is given by:

$$x''' = -ax'' + x\left(x'\right)^2 - x \tag{9.23}$$

which is chaotic for $a = 2.03$ with an attractor as shown in Fig. 9.4. The Lorenz and the Rôssler systems can also be written in jerk form [Linz (1997)]. The Lorenz system (1.8) can be written as:

$$x''' + \left(1 + a + b - \frac{x'}{x}\right)x'' + \left[\frac{b\left(1 + a + x^2\right)x - (1+a)\,x'}{x}\right]x'$$

$$-ba\left(r - 1 - x^2\right)x = 0$$

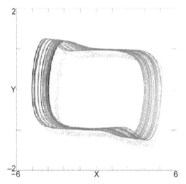

FIGURE 9.4: Chaotic attractor for the cubic jerk system (9.23).

and the Rôssler system given by:

$$\begin{cases} x' = -y - z \\ y' = x + ay \\ z' = b + z(x - c) \end{cases} \quad (9.24)$$

has the chaotic attractor shown in Fig. 9.5 and can be written as [Innocenti, et al., (2008)]:

$$y''' + (c - a)y'' + (1 - ac)y' + cy - b - (y' - ay)(y'' - ay' + y) = 0 \quad (9.25)$$

The Rôssler system (9.24) can also be written as a jerk function of the x variable [Linz (1997)] or the z variable [Lainscsek, et al., (2003)].

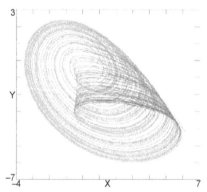

FIGURE 9.5: The Rôssler chaotic attractor from system (9.24) for $a = 0.5, b = 1$ and $c = 3$.

9.4 Transformation of 4-D dynamical systems to hyperjerk form

Generally, a is a dynamical system governed by an nth-order ordinary differential equation with $n > 3$ describing the time evolution of a single scalar variable of the following form:

$$x^{(4)} = H\left(x, x', x'', x''', ...\right) \qquad (9.26)$$

The fourth derivative $x^{(n)} = x''''$ has been called a *snap*, with successive derivatives *crackle* and *pop*, and there is no universally accepted name for the higher derivatives.

Some simple forms of 4th and 5th order, displaying chaos and hyperchaos, were presented in [Chlouverakis & Sprott (2006)]. One example of the situation when $n = 4$ is given by:

$$x^{(4)} + 6x'' = 1 - x^2 \qquad (9.27)$$

and it is shown in Fig. 9.6. System (9.27) is equivalent to the following 4-D quadratic dynamical system:

$$\begin{cases} x' = y \\ \\ y' = z \\ \\ z' = u \\ \\ u' = -6z + 1 - x^2 \end{cases} \qquad (9.28)$$

The connection between externally driven non-linear oscillators and specific uni- and bi-directionally coupled systems of two autonomous oscillators was

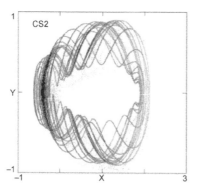

FIGURE 9.6: A chaotic attractor for the Chlouverakis snap system (9.27).

discussed in [Linz (2008)] along with a reinterpretation of simple chaotic forms of hyperjerk systems with some criteria that exclude chaotic behavior in some classes of these hyperjerk systems. However, some 4-D quadratic continuous-time autonomous systems cannot be reduced to hyperjerk dynamics, or they possess a functionally complicated hyperjerk form.

Theorem 120 *Any sinusoidally driven oscillator determined by:*

$$x'' + \Omega(x, x') = A\sin(\omega t + \phi) \tag{9.29}$$

with (i) an oscillator function $\Omega(x, x')$ being an arbitrary function of x and x' that possesses a second time-derivative $\Omega''(x, x')$ that is at least Lipschitz continuous and (ii) A, ω, ϕ being arbitrary real parameters can be transformed to a hyperjerky system of the form (9.26).

Proof 54 *First of all, any sinusoidal oscillation $A\sin(\omega t + \phi)$ can be obtained by a linear undamped oscillator equation, $h'' + \omega h^2 = 0$ with appropriately chosen initial conditions, $h(0) = A\sin(\phi)$, and $h'(0) = \omega A\cos(\phi)$. Hence, any additively and sinusoidally driven oscillator can be interpreted as a system of two oscillator equations where the linear oscillator is unidirectionally coupled to the other. Define the variables $y(t) = x'(t)$ and $g(t) = h'(t)$. Then, the differential equation for the driven oscillator can be rewritten in form of an autonomous dynamical system in the four variables x, y, h, g as follows:*

$$x' = y \tag{9.30}$$
$$y' = -\Omega(x, y) + h \tag{9.31}$$
$$h' = g \tag{9.32}$$
$$g' = -\omega^2 h \tag{9.33}$$

Let us introduce two simple sequential transformations in order to reduce it to the normal form of a hyperjerky dynamics given by (9.26). The first transformation is given by:

$$z = -\Omega(x, y) + h \text{ with } z' = -\Omega'(x, y) + g \tag{9.34}$$

Transformation (9.34) allows us to replace the equation for h' with an equation for z' where the dependencies on h are eliminated by the dependencies on g using the third equation of (9.33), $h' = g$. The second transformation is given by:

$$w = -\Omega'(x, y) + g \text{ with } w' = -\Omega''(x, y) + g' \tag{9.35}$$

The fact that

$$g' = -\omega^2 h = -\omega^2 [\Omega(x, y) + z] \tag{9.36}$$

implies that we can replace the equation in g' with an equation in w'. Hence, the system (9.30)-(9.33) can be written in the transformed variables x, y, z, w as follows:

$$x' = y \tag{9.37}$$

$$y' = z \tag{9.38}$$

$$z' = w \tag{9.39}$$

$$w' = -\Omega''(x, y) - \omega^2 [\Omega(x, y) + z] \tag{9.40}$$

and is it equivalent to a hyperjerky system (9.26) after identifying $y = x', z = x''$ and $w = x'''$, i.e.,

$$x^{(4)} = H(x, x', x'', x''') = -\Omega''(x, x') - \omega^2 \Omega(x, x') - \omega^2 x''$$

Another important result is given as follows [Linz (2008)]:

Theorem 121 *Any system of coupled oscillators determined by:*

$$\begin{cases} x'' + \Omega(x, x') = h \\ h'' + Q(h, h') = \sigma(x, x') \end{cases} \tag{9.41}$$

with oscillator functions $\Omega(x, x')$ and $Q(h, h')$ being arbitrary functions of their arguments and with the second time-derivative $\Omega(x, x'), \Omega''(x, x')$, $Q(h, h')$ and $\sigma(x, x')$ being at least Lipschitz continuous can be reduced to a hyperjerky system of the form (9.26).

Also, it was shown in [Linz (2008)] that the two specific functional forms of a hyperjerky system cannot be chaotic, hence, the following result:

Theorem 122 *Consider the hyperjerky system given by:*

$$x'''' + ax''' + bx'' = \Sigma(x, x') \tag{9.42}$$

with a and b being time-independent coefficients. This system cannot exhibit chaotic behavior if

(i) $\Sigma(x, x')$ is an arbitrary function of x and x' that is either positive semidefinite, $\Sigma(x, x') \geq 0$, or negative semidefinite $-\Sigma(x, x') \geq 0$, for all times t or, equivalently, all x and x'.

(ii) $\Sigma(x, x')$ has the form $\Sigma(x, x') = -cx' + \gamma(x)$ with c being a time-independent coefficient and $\gamma(x)$ being an arbitrary function of x that is either positive semidefinite, $\gamma(x) \geq 0$, or negative semidefinite $-\gamma(x) \geq 0$, for all times t or all x.

Proof 55 *The proof of Theorem 122 is a generalization of the strategy used in [Linz (1998-2000), Eichhorn, et al., (1998)] in order to obtain no-chaos criteria for jerky dynamics.*

(i) Integrating Eq. (9.42) once with respect to time t yields

$$x''' + ax'' + bx' =^t \Sigma\left(x\left(\tau\right), x'\left(\tau\right)\right) d\tau \qquad (9.43)$$

The fact that the left hand side of Eq. (9.43) is linear in x, implies that system (9.42) cannot be chaotic if $\Sigma\left(x, x'\right) = 0$ and it converges only to fixed point solutions, or periodic solutions or diverging solutions. Now, if the right hand side $\Sigma\left(x, x'\right)$ is either positive or negative semidefinite for all t or x, x', the integrand of the right hand side of (9.43) does not change its sign with time t or x, x'. Thus, $^t\Sigma\left(x\left(\tau\right), x'\left(\tau\right)\right) d\tau$ can only approach zero or a non-zero constant or diverge to infinity in the limit $t \to \infty$. Hence, there is no bounded oscillation around zero in the long-time limit. This result and the linear dynamics of the left hand side of Eq. (9.43) imply that no chaotic behavior can occur in system (9.42).

(ii) The same method, yields for the specific form of $\Sigma\left(x, x'\right)$

$$x''' + ax'' + bx' + cx =^t \gamma\left(x\left(\tau\right)\right) d\tau \qquad (9.44)$$

The left hand side of Eq. (9.44) is linear in x and cannot be chaotic if $\gamma\left(x\right) = 0$. If $\gamma\left(x\right)$ is either positive or negative semidefinite for all t or x, x' then the integrand of the right hand side of (9.44) cannot change its sign with time t or x, x'. Again, $^t\gamma\left(x\left(\tau\right)\right) d\tau$ can only approach zero or a non-zero constant or diverge to infinity. Thus, no bounded oscillations around zero of the right hand side of Eq. (9.44) can occur in the long-time limit. Hence, chaotic behavior is excluded.

In [Zeraoulia & Sprott (2013(a))] the possible cases for equivalence between four-dimensional autonomous dynamical systems and hyperjerk dynamics are investigated. In fact, a wide class of four-dimensional vector fields possess this property. Indeed, let us consider a 4-D system of the form:

$$\begin{cases} x' = c_1 + b_{11}x + b_{12}y + b_{13}z + b_{14}u + n_1\left(x\right) \\[2mm] y' = c_2 + b_{21}x + b_{22}y + b_{23}z + b_{24}u + n_2\left(x, y, z, u\right) \\[2mm] z' = c_3 + b_{31}x + b_{32}y + b_{33}z + b_{34}u + n_3\left(x, y, z, u\right) \\[2mm] u' = c_4 + b_{41}x + b_{42}y + b_{43}z + b_{44}u + n_4\left(x, y, z, u\right) \end{cases} \qquad (9.45)$$

where the functions $n_1\left(x\right)$ and $\left(n_i\left(x, y, z, u\right)\right)_{2 \leq i \leq 4}$ are assumed to be at least of class C^2 in their corresponding arguments. We set

$$\begin{cases} n_{11}(x) = \frac{\partial n_i(x)}{\partial x} \\\\ n_{111} = \frac{\partial^2 n_1(x)}{\partial x^2} \\\\ n_{i1}(x,y,z,u) = \frac{\partial n_i(x,y,z,u)}{\partial x} \\\\ n_{i2}(x,y,z,u) = \frac{\partial n_i(x,y,z,u)}{\partial y} \\\\ n_{i3}(x,y,z,u) = \frac{\partial n_i(x,y,z,u)}{\partial z} \\\\ n_{i4}(x,y,z,u) = \frac{\partial n_i(x,y,z,u)}{\partial u}, i = 2,3,4. \end{cases}$$

In this section, we derive only the hyperjerk form of system (9.45) with respect to x. The same logic applies for y, z, and u. Assume that $b_{12} \neq 0$. Then from the first equation of (9.45), we have

$$y(x,x',z,u) = -\frac{1}{b_{12}}\left(-x' + c_1 + n_1(x) + b_{11}x + b_{13}z + b_{14}u\right)$$

Thus,

$$y'(x,x',x'',z,u) = f_1 z + f_2 x + f_3 u + f_4 + f_5,$$

where

$$\begin{cases} f_1 = \frac{b_{13}^2 b_{32} - b_{12}b_{13}b_{33} - b_{12}b_{14}b_{43} + b_{13}b_{14}b_{42}}{b_{12}^2} \\\\ f_2 = \frac{b_{11}b_{13}b_{32} - b_{12}b_{13}b_{31} + b_{11}b_{14}b_{42} - b_{12}b_{14}b_{41}}{b_{12}^2} \\\\ f_3 = \frac{b_{14}^2 b_{42} - b_{12}b_{13}b_{34} + b_{13}b_{14}b_{32} - b_{12}b_{14}b_{44}}{b_{12}^2} \\\\ f_4(x,x',z,u) = -\frac{b_{13}\left(c_3 + n_3((x,y,z,u)) - \frac{b_{32}(c_1 + n_1(x))}{b_{12}}\right)}{b_{12}} \\\\ \qquad\qquad\qquad - \frac{b_{14}\left(c_4 + n_4(x,y,z,u) - \frac{b_{42}(c_1 + n_1(x))}{b_{12}}\right)}{b_{12}} \\\\ f_5(x,x',x'') = -\frac{b_{11}b_{12} + b_{13}b_{32} + b_{14}b_{42} + b_{12}n_{11}(x)}{b_{12}^2}x' + \frac{1}{b_{12}}x'' \end{cases} \qquad (9.46)$$

By equating the formula for $y' = y'(x,x',x'',z,u)$ with the second equation of (9.45), we obtain

$$\left(f_1 - b_{23} + \frac{1}{b_{12}}b_{13}b_{22}\right)z = n_2 - f_4 - f_6 \qquad (9.47)$$

where

$$f_6(x,x',x'') = \frac{f_2 b_{12} + b_{11}b_{22} - b_{12}b_{21}}{b_{12}}x - \frac{b_{22}}{b_{12}}x' - c_2 + \frac{b_{22}c_1}{b_{12}} + \frac{b_{22}}{b_{12}}n_1(x) + f_5 \qquad (9.48)$$

The second member of (9.47) depends on z in the part $n_2 - f_4$, so if we assume that $n_2(x, y, z, u) - f_4(x, x', z, u) = g_1(x, u)$, i.e., this part does not depend on the variable z, with g_1 being an arbitrary function of the indicated arguments, then

$$z(x, x', x'') = f_7 f_6(x, x', x'') \tag{9.49}$$

where f_7 is given by:

$$f_7 = \frac{b_{12}^2}{b_{12}^2 b_{23} - b_{13}^2 b_{32} - b_{12} b_{13} b_{22} + b_{12} b_{13} b_{33} + b_{12} b_{14} b_{43} - b_{13} b_{14} b_{42}} \tag{9.50}$$

with the condition

$$b_{12}^2 b_{23} - b_{13}^2 b_{32} - b_{12} b_{13} b_{22} + b_{12} b_{13} b_{33} + b_{12} b_{14} b_{43} - b_{13} b_{14} b_{42} \neq 0 \tag{9.51}$$

The function g_1 does not depend on y since y itself depends on z. Hence the condition $n_2 - f_4 = g_1(x, u)$ is equivalent to

$$(b_{12} n_2 + b_{13} n_3 + b_{14} n_4)(x, y, z, u) = b_{12} g_1(x, u) - b_{12} f_8(x) \tag{9.52}$$

where

$$f_8(x) = \frac{\left(b_{13}\left(c_3 - \frac{b_{32}(c_1 + n_1(x))}{b_{12}} \right) + b_{14}\left(c_4 - \frac{b_{42}(c_1 + n_1(x))}{b_{12}} \right) \right)}{b_{12}} \tag{9.53}$$

From (9.49), we have $z' = f_7\left(\frac{1}{b_{12}} x''' + f_{10}(x) x'' + f_9(x, x') x' \right)$, where

$$\begin{cases} f_9(x, x') = \frac{b_{12}(f_2 - b_{21}) + b_{22} b_{11} + b_{22} n_{11}(x) - n_{111}(x) x'}{b_{12}} \\[2mm] f_{10}(x) = -\frac{b_{22}}{b_{12}} - \frac{(b_{11} b_{12} + b_{13} b_{32} + b_{14} b_{42} + b_{12} n_{11}(x))}{b_{12}^2} \end{cases} \tag{9.54}$$

By equating the formula for z' with the third equation of (9.45), we obtain

$$-\frac{b_{12} b_{34} - b_{14} b_{32}}{b_{12}} u = -f_{11} - f_{12} - \frac{f_7}{b_{12}} x''' \tag{9.55}$$

where

$$\begin{cases} f_{11}(x, x', x'') = \frac{b_{11} b_{32} - b_{12} b_{31}}{b_{12}} x + \frac{-b_{32} + f_7 f_9(x, x') b_{12}}{b_{12}} x' + f_7 f_{10}(x) x'' \\[2mm] f_{12}(x, x', x'', u) = -n_3(x, y, z, u) - c_3 + \frac{b_{32}(c_1 + n_1(x)) + f_6 f_7 b_{13}}{b_{12}} - f_6 f_7 b_{33} \end{cases} \tag{9.56}$$

The second member of (9.55) depends on u in the part $-f_{12}$, so if we assume that $-f_{12} = g_2(x, z)$, i.e., this part does not depend on the variable u, with g_2 being an arbitrary function of the indicated arguments, then

$$u = \frac{f_7 x''' + b_{12} f_{11}(x, x', x'') + b_{12} f_{12}(x, x', x'')}{b_{12} b_{34} - b_{14} b_{32}} \tag{9.57}$$

with the condition

$$b_{12}b_{34} - b_{14}b_{32} \neq 0 \qquad (9.58)$$

Thus, the condition $-f_{12} = g_2(x, z)$ is equivalent to

$$-n_3(x, y, z, u) - c_3 + \frac{b_{32}(c_1 + n_1(x) + f_6 f_7 b_{13})}{b_{12}} - f_6 f_7 b_{33} = g_2(x, z) \quad (9.59)$$

From (9.57), we have $u' = f_{13}x^{(4)} + f_{14}$, where

$$
\begin{cases}
f_{13} = \frac{f_7}{b_{12}b_{34} - b_{14}b_{32}} \\[2mm]
f_{14}(x, x', x'', x''') = \frac{b_{12}f'_{11}(x, x', x'', x''')}{b_{12}b_{34} - b_{14}b_{32}} + \frac{b_{12}f'_{12}(x, x', x'', x''')}{b_{12}b_{34} - b_{14}b_{32}}
\end{cases}
\qquad (9.60)
$$

because $\frac{df_{11}(x, x', x'')}{dt} = f'_{11}(x, x', x'', x''')$, and since $f_{12} = -g_2(x, z)$, then $\frac{f_{12}(x, y, z, u)}{dt} = f'_{12}(x, x', x'', x''')$. Again, by equating the formula for u' with the fourth equation of (9.45), we obtain

$$x^{(4)} = \frac{f_{15}(x, x', x'', x''') + f_{16}(x, x', x'', x''')}{f_{13}} = H(x, x', x'', x''') \qquad (9.61)$$

where

$$
\begin{cases}
f_{15}(x, x', x'', x''', u) = c_4 + b_{41}x + b_{43}f_7 f_6 + \rho_0 \\[3mm]
f_{16}(x, x', x'', x''', u) = b_{42}\left(-\frac{1}{b_{12}}(-x' + c_1 + n_1(x) + b_{11}x + \rho_1)\right) \\[3mm]
\rho_0(x, x', x'', x''', u) = b_{44}\left(\frac{f_7 x''' + b_{12}f_{11} + b_{12}f_{12}}{b_{12}b_{34} - b_{14}b_{32}}\right) + n_4 - f_{14} \\[3mm]
\rho_1(x, x', x'', x''', u) = b_{13}f_7 f_6 + b_{14}\left(\frac{f_7 x''' + b_{12}f_{11} + b_{12}f_{12}}{b_{12}b_{34} - b_{14}b_{32}}\right)
\end{cases}
$$

$$(9.62)$$

At this stage, the form (9.61)-(9.62) is the corresponding hyperjerk form of the 4-D dynamical system (9.45).

9.4.1 The expression of the transformation between (9.45) and (9.61)-(9.62)

In this section, a rigorous expression for the transformation between the 4-D dynamical system (9.45) and the hyperjerk form (9.61)-(9.62) is described. Indeed, the above procedure defines an invertible transformation $T = T(x, x', x'', x''')$ if the following conditions hold:

$$
\begin{cases}
b_{12} \neq 0 \\[2mm]
b_{12}^2 b_{23} - b_{13}^2 b_{32} - b_{12} b_{13} b_{22} + b_{12} b_{13} b_{33} + b_{12} b_{14} b_{43} - b_{13} b_{14} b_{42} \neq 0 \\[2mm]
b_{12} b_{34} - b_{14} b_{32} \neq 0 \\[2mm]
n_2 - f_4 = g_1(x, u) \\[2mm]
-n_3 - c_3 + \frac{1}{b_{12}} b_{32}(c_1 + n_1(x) + f_6 f_7 b_{13}) - f_6 f_7 b_{33} = g_2(x, z)
\end{cases}
$$
$$(9.63)$$

where g_1 and g_2 are arbitrary functions of the indicated arguments. The transformation $T(x, x', x'', x''')$ is defined by:

$$
\begin{cases}
T_1(x, y, z, u) = x \\[2mm]
T_2(x, y, z, u) = x' = c_1 + b_{11} x + b_{12} y + n_1(x) \\[2mm]
T_3(x, y, z, u) = x'' = f_{17}(x, y) + f_{18}(x, z, u) + f_{19}(x, y, z, u) \\[2mm]
T_4(x, y, z, u) = x''' = \Psi_1 + \Psi_2 \\[2mm]
\Psi_1 = f_{20} x^3 + f_{21} x^2 y + f_{22} x^2 + f_{23} xy + f_{24} xz + f_{25} xu \\[2mm]
\Psi_2 = f_{26} x + f_{27} y^2 + f_{28} yz + f_{29} yu + f_{30} y + f_{31} z + f_{32} u + f_{33}
\end{cases}
$$
$$(9.64)$$

and its inverse is defined by:

$$
\begin{cases}
T_1^{-1}(x, x', x'', x''') = x \\[2mm]
T_4^{-1}(x, x', x'', x''') = u = \dfrac{f_7 x''' + b_{12} f_{11}(x, x', x'') + b_{12} f_{12}(x, x', x'')}{b_{12} b_{34} - b_{14} b_{32}} \\[2mm]
T_3^{-1}(x, x', x'', x''') = z = f_7 f_6(x, x', x'') \\[2mm]
T_2^{-1}(x, x', x'', x''') = y = -\dfrac{1}{b_{12}}(-x' + c_1 + n_1(x) + b_{11} x + b_{13} z + b_{14} u)
\end{cases}
$$
$$(9.65)$$

where

$$
\begin{cases}
f_{17} = \left(b_{11}^2 + n_{11}(x) b_{11} + b_{12}b_{21}\right) x + \left(b_{11}b_{12} + b_{12}b_{22} + b_{12}n_{11}(x)\right) y \\[2mm]
f_{18} = \left(b_{11}b_{13} + b_{12}b_{23} + b_{13}n_{11}(x)\right) z + \left(b_{11}b_{14} + b_{12}b_{24} + b_{14}n_{11}(x)\right) u \\[2mm]
f_{19} = \left(b_{11}\left(c_1 + n_1(x)\right) + b_{12}\left(c_2 + n_2(x,y,z,u)\right) + n_{11}(x)\left(c_1 + n_1(x)\right)\right) \\[2mm]
\qquad f_{20} = b_{11}^2 n_{111}(x),\, f_{21} = b_{11}b_{12}n_{111}(x) \\[2mm]
\qquad f_{22} = \xi_1 + \xi_2,\, f_{23} = \xi_3 + \xi_4 \\[2mm]
f_{24} = b_{12}b_{23}n_{11}(x) + b_{13}b_{33}n_{11}(x) + b_{14}b_{43}n_{11}(x) + b_{11}b_{13}n_{111}(x)
\end{cases}
\tag{9.66}
$$

and

$$
\begin{cases}
f_{25} = b_{12}b_{24}n_{11} + b_{13}b_{34}n_{11} + b_{14}b_{44}n_{11} + b_{11}b_{14}n_{111} \\[2mm]
\qquad f_{26} = \xi_5 + \xi_6 + \xi_7 + \xi_8 + \xi_9 \\[2mm]
f_{27} = b_{12}^2 n_{111},\, f_{28} = b_{12}b_{13}n_{111},\, f_{29} = b_{12}b_{14}n_{111} \\[2mm]
f_{30} = \xi_{10} + \xi_{11} + \xi_{12},\, f_{31} = \xi_{13} + \xi_{14},\, f_{32} = \xi_{15} + \xi_{16} \\[2mm]
\qquad f_{33} = \xi_{17} + \xi_{18} + \xi_{19} + \xi_{20}
\end{cases}
\tag{9.67}
$$

where

$$
\begin{cases}
\xi_1 = b_{11}\left(n_{11}^2 + b_{11}n_{11} + c_1 n_{111} + n_1 n_{111}\right)(x) \\[2mm]
\xi_2 = b_{11}n_{111}(x)\left(c_1 + n_1(x)\right) + \left(b_{12}b_{21} + b_{13}b_{31} + b_{14}b_{41}\right) n_{11}(x) \\[2mm]
\xi_3 = b_{12}\left(n_{11}^2(x) + b_{11}n_{11}(x) + \left(c_1 + n_1(x)\right) n_{111}(x)\right) \\[2mm]
\xi_4 = \left(b_{12}b_{22} + b_{13}b_{32} + b_{14}b_{42}\right) n_{11}(x) + b_{11}b_{12}n_{111}(x) \\[2mm]
\xi_5 = \left(c_1 + n_1(x)\right)\left(n_{11}^2(x) + b_{11}n_{11}(x) + \left(c_1 + n_1(x)\right) n_{111}(x)\right) \\[2mm]
\xi_6 = b_{11}\left(b_{11}^2 + \left(b_{11} + c_1 + n_1\right) n_{11}(x) + b_{12}b_{21} + b_{12}n_{21}(x,y,z,u)\right) \\[2mm]
\xi_7 = b_{21}\left(b_{11}b_{12} + b_{12}b_{22} + b_{12}n_{22}\right) + b_{31}\left(b_{11}b_{13} + b_{12}b_{23} + b_{12}n_{23}\right) \\[2mm]
\xi_8 = b_{41}\left(b_{11}b_{14} + b_{12}b_{24} + b_{12}n_{24}\right) + b_{12}n_{11}(x)\left(c_2 + n_2\right) \\[2mm]
\xi_9 = b_{13}n_{11}(x)\left(c_3 + n_3(x,y,z,u)\right) + b_{14}n_{11}\left(c_4 + n_4(x,y,z,u)\right) \\[2mm]
\xi_{10} = b_{12}\left(b_{11}^2 + n_{11}b_{11} + c_1 n_{11} + n_1 n_{11} + b_{12}b_{21} + b_{12}n_{21}\right)
\end{cases}
\tag{9.68}
$$

and

$$
\left\{
\begin{array}{l}
\xi_{11} = b_{22} \left(b_{11}b_{12} + b_{12}b_{22} + b_{12}n_{22} \right) + b_{32} \left(b_{11}b_{13} + b_{12}b_{23} + b_{12}n_{23} \right) \\[2mm]
\xi_{12} = b_{42} \left(b_{11}b_{14} + b_{12}b_{24} + b_{12}n_{24} \right) + b_{12}n_{111} \left(c_1 + n_1 \right) \\[2mm]
\xi_{13} = b_{23} \left(b_{11}b_{12} + b_{12}b_{22} + b_{12}n_{22} \right) + b_{33} \left(b_{11}b_{13} + b_{12}b_{23} + b_{12}n_{23} \right) \\[2mm]
\xi_{14} = b_{43} \left(b_{11}b_{14} + b_{12}b_{24} + b_{12}n_{24} \right) + b_{13}n_{111} \left(c_1 + n_1 \right) \\[2mm]
\xi_{15} = b_{24} \left(b_{11}b_{12} + b_{12}b_{22} + b_{12}n_{22} \right) + b_{34} \left(b_{11}b_{13} + b_{12}b_{23} + b_{12}n_{23} \right) \\[2mm]
\xi_{16} = b_{44} \left(b_{11}b_{14} + b_{12}b_{24} + b_{12}n_{24} \right) + b_{14}n_{111} \left(c_1 + n_1 \right) \\[2mm]
\xi_{17} = \left(c_1 + n_1 \right) \left(b_{11}^2 + n_{11}b_{11} + c_1 n_{11} + n_1 n_{11} + b_{12}b_{21} + b_{12}n_{21} \right) \\[2mm]
\xi_{18} = \left(c_2 + n_2 \left(x, y, z, u \right) \right) \left(b_{11}b_{12} + b_{12}b_{22} + b_{12}n_{22} \left(x, y, z, u \right) \right) \\[2mm]
\xi_{19} = \left(c_4 + n_4 \left(x, y, z, u \right) \right) \left(b_{11}b_{14} + b_{12}b_{24} + b_{12}n_{24} \left(x, y, z, u \right) \right) \\[2mm]
\xi_{20} = \left(c_3 + n_3 \left(x, y, z, u \right) \right) \left(b_{11}b_{13} + b_{12}b_{23} + b_{12}n_{23} \left(x, y, z, u \right) \right)
\end{array}
\right.
$$

$$(9.69)$$

Note that the inverse transformation exits because the procedure described in Sec. 9.4 transforms the 4-D dynamical system (9.45) to the hyperjerk form in (9.61)-(9.62) without any singularities. Therefore, the inverse procedure defined by (9.65)-(9.69) can give the initial 4-D dynamical system (9.45).

9.4.2 Examples of 4-D hyperjerky dynamics

Many examples of chaotic hyperjerk motion were studied in the literature [Chlouverakis & Sprott (2006)]. Generally, all periodically forced oscillators and some of coupled oscillators are equivalent to a snap form, as demonstrated in [Linz (2008)]. This includes the frictionless forced pendulum and the periodically forced undamped oscillator with a cubic restoring force. The same type of system with damping was studied in [Sprott (1997(a))]. The simplest chaotic hyperjerk system is studied in [Chlouverakis & Sprott (2006)]. These include snap systems with one non-linear function and the simplest dissipative chaotic case with a single quadratic non-linearity given by (9.27).

Using the above method, we can prove that any dynamical system of the functional form:

$$\begin{cases} x' = c_1 + b_{11}x + b_{12}y + n_1\left(x\right) \\[2mm] y' = c_2 + b_{21}x + b_{22}y + b_{23}z + n_2\left(x,y\right) \\[2mm] z' = c_3 + b_{31}x + b_{32}y + b_{33}z + b_{34}u + n_3\left(x,y,z\right) \\[2mm] u' = c_4 + b_{41}x + b_{42}y + b_{43}z + b_{44}u + n_4\left(x,y,z,u\right) \end{cases} \tag{9.70}$$

can be reduced to a hyperjerk dynamic form $x'''' = H\left(x,x',x'',x'''\right)$ if the following conditions hold:

$$b_{12} \neq 0, b_{23} \neq 0, b_{34} \neq 0 \tag{9.71}$$

That is, $b_{13} = b_{14} = b_{24} = 0$ in system (9.45). If n_2 and n_3 do not depend on z and u, respectively, then there is no need to consider the functions g_1 and g_2 defined above. In this case, we have

$$y\left(x,x'\right) = -\frac{1}{b_{12}}\left(-x' + c_1 + n_1\left(x\right) + b_{11}x\right)$$

Thus

$$\begin{cases} y'\left(x,x',x''\right) = f_5 = -\frac{b_{11}b_{12}+b_{12}n_{11}(x)}{b_{12}^2}x' + \frac{1}{b_{12}}x'' \\[2mm] f_1 = 0, f_2 = 0, f_3 = 0, f_4 = 0 \\[2mm] z\left(x,x',x''\right) = f_7 f_6 \end{cases}$$

where

$$\begin{cases} f_6 = \left(\frac{b_{11}b_{22}-b_{12}b_{21}}{b_{12}}\right)x - \frac{b_{22}}{b_{12}}x' - c_2 + \frac{b_{22}c_1}{b_{12}} + \frac{b_{22}}{b_{12}}n_1\left(x\right) + f_5 \\[2mm] f_7 = \frac{1}{b_{23}} \\[2mm] f_8\left(x\right) = 0 \end{cases}$$

Hence

$$z' = f_7\left(\frac{1}{b_{12}}x''' + f_{10}\left(x\right)x'' + f_9 x'\right)$$

where

$$\begin{cases} f_9 = \frac{b_{12}(f_2-b_{21})+b_{22}b_{11}+b_{22}n_{11}(x)-n_{111}(x)x'}{b_{12}} \\[2mm] f_{10}\left(x\right) = \left(-\frac{b_{22}}{b_{12}} - \frac{(b_{11}+n_{11}(x))}{b_{12}}\right) \end{cases}$$

By equating the formula for z' with the third equation of (9.70), we obtain

$$u = \frac{f_7 x''' + b_{12}f_{11} + b_{12}f_{12}}{b_{12}b_{34}}$$

where

$$\begin{cases} f_{11} = \frac{b_{11}b_{32}-b_{12}b_{31}}{b_{12}}x + \frac{-b_{32}+f_7f_9(x,x')b_{12}}{b_{12}}x' + f_7f_{10}(x)x'' \\[2mm] f_{12} = -n_3 - c_3 + \frac{b_{32}(c_1+n_1(x))}{b_{12}} - f_6f_7b_{33} \end{cases}$$

Hence, we have

$$u' = f_{13}x^{(4)} + f_{14}$$

where

$$\begin{cases} f_{13} = \frac{f_7}{b_{12}b_{34}} \\[3mm] f_{14} = \frac{f'_{11}}{b_{34}} + \frac{f'_{12}}{b_{34}} \end{cases}$$

Finally, we obtain

$$x^{(4)} = \frac{f_{15} + f_{16}}{f_{13}} = H(x, x', x'', x''')$$

where

$$\begin{cases} f_{15} = c_4 + b_{41}x + b_{43}f_7f_6 + \rho_0 \\[3mm] f_{16} = b_{42}\left(-\frac{1}{b_{12}}(-x' + c_1 + n_1(x) + b_{11}x)\right) \\[3mm] \rho_0 = b_{44}\left(\frac{f_7x''' + b_{12}f_{11} + b_{12}f_{12}}{b_{12}b_{34}}\right) + n_4 - f_{14} \end{cases}$$

In this case, the expression for the transformation $T = T(x, x', x'', x''')$ is given by

$$\begin{cases} T_1(x, y, z, u) = x \\[2mm] T_2(x, y, z, u) = x' = c_1 + b_{11}x + b_{12}y + n_1(x) \\[2mm] T_3(x, y, z, u) = x'' = f_{17}(x, y) + f_{18}(x, z, u) + f_{19}(x, y, z, u) \\[2mm] T_4(x, y, z, u) = x''' = \Psi_1 + \Psi_2 \\[2mm] \Psi_1 = f_{20}x^3 + f_{21}x^2y + f_{22}x^2 + f_{23}xy + f_{24}xz \\[2mm] \Psi_2 = f_{26}x + f_{27}y^2 + f_{30}y + f_{31}z + f_{32}u + f_{33} \end{cases} \qquad (9.72)$$

and its inverse is defined by

$$\begin{cases} T_1^{-1}(x, x', x'', x''') = x \\[3mm] T_4^{-1}(x, x', x'', x''') = u = \frac{f_7x''' + b_{12}f_{11}(x,x',x'') + b_{12}f_{12}(x,x',x'')}{b_{12}b_{34}} \\[3mm] T_3^{-1}(x, x', x'', x''') = z = f_7f_6(x, x', x'') \\[3mm] T_2^{-1}(x, x', x'', x''') = y = -\frac{1}{b_{12}}(-x' + c_1 + n_1(x) + b_{11}x) \end{cases} \qquad (9.73)$$

where

$$
\left\{
\begin{aligned}
f_{17} &= \left(b_{11}^2 + n_{11}\left(x\right)b_{11} + b_{12}b_{21}\right)x + \left(b_{11}b_{12} + b_{12}b_{22} + b_{12}n_{11}\left(x\right)\right)y \\
f_{18} &= b_{12}b_{23}z \\
f_{19} &= \left(b_{11}\left(c_1 + n_1\left(x\right)\right) + b_{12}\left(c_2 + n_2\left(x,y\right)\right) + n_{11}\left(x\right)\left(c_1 + n_1\left(x\right)\right)\right) \\
f_{20} &= b_{11}^2 n_{111}\left(x\right), f_{21} = b_{11}b_{12}n_{111}\left(x\right), f_{22} = \xi_1 + \xi_2 \\
f_{23} &= \xi_3 + \xi_4, f_{24} = b_{12}b_{23}n_{11}\left(x\right), f_{25} = 0, f_{28} = 0, f_{29} = 0 \\
f_{26} &= \xi_5 + \xi_6 + \xi_7 + \xi_8, f_{27} = b_{12}^2 n_{111} \\
f_{30} &= \xi_{10} + \xi_{11} + \xi_{12}, f_{31} = \xi_{13} + \xi_{14}, f_{32} = \xi_{15} + \xi_{16} \\
f_{33} &= \xi_{17} + \xi_{18} + \xi_{19} + \xi_{20}
\end{aligned}
\right.
\tag{9.74}
$$

where

$$
\left\{
\begin{aligned}
\xi_1 &= b_{11}\left(n_{11}^2 + b_{11}n_{11} + c_1 n_{111} + n_1 n_{111}\right)\left(x\right) \\
\xi_2 &= b_{11}n_{111}\left(x\right)\left(c_1 + n_1\left(x\right)\right) + \left(b_{12}b_{21}\right)n_{11}\left(x\right) \\
\xi_3 &= b_{12}\left(n_{11}^2\left(x\right) + b_{11}n_{11}\left(x\right) + \left(c_1 + n_1\left(x\right)\right)n_{111}\left(x\right)\right) \\
\xi_4 &= \left(b_{12}b_{22}\right)n_{11}\left(x\right) + b_{11}b_{12}n_{111}\left(x\right) \\
\xi_5 &= \left(c_1 + n_1\left(x\right)\right)\left(n_{11}^2\left(x\right) + b_{11}n_{11}\left(x\right) + \left(c_1 + n_1\left(x\right)\right)n_{111}\left(x\right)\right) \\
\xi_6 &= b_{11}\left(b_{11}^2 + \left(b_{11} + c_1 + n_1\right)n_{11}\left(x\right) + b_{12}b_{21} + b_{12}n_{21}\left(x,y\right)\right) \\
\xi_7 &= b_{21}\left(b_{11}b_{12} + b_{12}b_{22} + b_{12}n_{22}\left(x,y\right)\right) + b_{31}\left(b_{12}b_{23} + b_{12}n_{23}\left(x,y\right)\right) \\
\xi_8 &= b_{41}\left(b_{12}n_{24}\left(x,y\right)\right) + b_{12}n_{11}\left(x\right)\left(c_2 + n_2\left(x,y\right)\right) \\
\xi_9 &= 0 \\
\xi_{10} &= b_{12}\left(b_{11}^2 + n_{11}b_{11} + c_1 n_{11} + n_1 n_{11} + b_{12}b_{21} + b_{12}n_{21}\right)
\end{aligned}
\right.
\tag{9.75}
$$

and

$$
\left\{
\begin{aligned}
\xi_{11} &= b_{22}\left(b_{11}b_{12} + b_{12}b_{22} + b_{12}n_{22}\right) + b_{32}\left(b_{12}b_{23} + b_{12}n_{23}\right) \\
\xi_{12} &= b_{42}\left(b_{12}n_{24}\right) + b_{12}n_{111}\left(c_1 + n_1\right) \\
\xi_{13} &= b_{23}\left(b_{11}b_{12} + b_{12}b_{22} + b_{12}n_{22}\right) + b_{33}\left(b_{12}b_{23} + b_{12}n_{23}\right) \\
\xi_{14} &= b_{43}\left(b_{12}n_{24}\right) \\
\xi_{15} &= b_{34}\left(b_{12}b_{23} + b_{12}n_{23}\right) \\
\xi_{16} &= b_{44}b_{12}n_{24} \\
\xi_{17} &= \left(c_1 + n_1\right)\left(b_{11}^2 + n_{11}b_{11} + c_1 n_{11} + n_1 n_{11} + b_{12}b_{21} + b_{12}n_{21}\right) \\
\xi_{18} &= \left(c_2 + n_2\left(x, y\right)\right)\left(b_{11}b_{12} + b_{12}b_{22} + b_{12}n_{22}\left(x, y\right)\right) \\
\xi_{19} &= \left(c_4 + n_4\left(x, y, z, u\right)\right)\left(b_{12}n_{24}\left(x, y\right)\right) \\
\xi_{20} &= \left(c_3 + n_3\left(x, y, z\right)\right)\left(b_{12}b_{23} + b_{12}n_{23}\left(x, y\right)\right)
\end{aligned}
\right. \tag{9.76}
$$

Now, if $n_1\left(x\right) = 0$, then the expression for the transformation $T = T\left(x, x', x'', x'''\right)$ is given by

$$
\left\{
\begin{aligned}
T_1\left(x, y, z, u\right) &= x \\
T_2\left(x, y, z, u\right) &= x' = c_1 + b_{11}x + b_{12}y \\
T_3\left(x, y, z, u\right) &= x'' = f_{17}\left(x, y\right) + f_{18}\left(z\right) + f_{19}\left(x, y\right) \\
T_4\left(x, y, z, u\right) &= x''' = f_{26}x + f_{30}y + f_{31}z + f_{32}u + f_{33}\left(x, y, z, u\right)
\end{aligned}
\right. \tag{9.77}
$$

and its inverse is defined by

$$
\left\{
\begin{aligned}
T_1^{-1}\left(x, x', x'', x'''\right) &= x \\
T_4^{-1}\left(x, x', x'', x'''\right) &= u = \frac{f_7 x''' + b_{12}f_{11}\left(x, x', x''\right) + b_{12}f_{12}\left(x, x', x''\right)}{b_{12}b_{34}} \\
T_3^{-1}\left(x, x', x'', x'''\right) &= z = f_7 f_6\left(x, x', x''\right) \\
T_2^{-1}\left(x, x', x'', x'''\right) &= y = -\frac{1}{b_{12}}\left(-x' + c_1 + n_1\left(x\right) + b_{11}x\right)
\end{aligned}
\right. \tag{9.78}
$$

where

$$
\begin{cases}
f_{17} = \left(b_{11}^2 + b_{12}b_{21}\right) x + \left(b_{11}b_{12} + b_{12}b_{22}\right) y \\[2mm]
f_{18} = b_{12}b_{23}z, \; f_{19} = \left(b_{11}\left(c_1\right) + b_{12}\left(c_2 + n_2\left(x, y\right)\right)\right) \\[2mm]
f_{20} = 0, \, f_{21} = 0, \, f_{22} = 0, \, f_{23} = 0, \, f_{24} = 0, \, f_{25} = 0, \, f_{28} = 0, \, f_{29} = 0, \, f_{27} = 0 \\[2mm]
f_{26} = \xi_6 + \xi_7 + \xi_8, \, f_{30} = \xi_{10} + \xi_{11} + \xi_{12}, \, f_{31} = \xi_{13} + \xi_{14} \\[2mm]
f_{32} = \xi_{15} + \xi_{16}, \, f_{33} = \xi_{17} + \xi_{18} + \xi_{19} + \xi_{20}
\end{cases}
\tag{9.79}
$$

where

$$
\begin{cases}
\xi_1 = 0, \xi_2 = 0, \xi_3 = 0, \xi_4 = 0, \xi_5 = 0, \xi_9 = 0 \\[2mm]
\xi_6 = b_{11}\left(b_{11}^2 + b_{12}b_{21} + b_{12}n_{21}\left(x, y\right)\right) \\[2mm]
\xi_7 = b_{21}\left(b_{11}b_{12} + b_{12}b_{22} + b_{12}n_{22}\left(x, y\right)\right) + b_{31}\left(b_{12}b_{23} + b_{12}n_{23}\left(x, y\right)\right) \\[2mm]
\xi_8 = b_{41}\left(b_{12}n_{24}\left(x, y\right)\right), \, \xi_{10} = b_{12}\left(b_{11}^2 + b_{12}b_{21} + b_{12}n_{21}\left(x, y\right)\right) \\[2mm]
\xi_{11} = b_{22}\left(b_{11}b_{12} + b_{12}b_{22} + b_{12}n_{22}\left(x, y\right)\right) + b_{32}\left(b_{12}b_{23} + b_{12}n_{23}\left(x, y\right)\right) \\[2mm]
\xi_{12} = b_{42}\left(b_{12}n_{24}\right) \\[2mm]
\xi_{13} = b_{23}\left(b_{11}b_{12} + b_{12}b_{22} + b_{12}n_{22}\left(x, y\right)\right) + b_{33}\left(b_{12}b_{23} + b_{12}n_{23}\left(x, y\right)\right) \\[2mm]
\xi_{14} = b_{43}\left(b_{12}n_{24}\right)
\end{cases}
\tag{9.80}
$$

and

$$
\begin{cases}
\xi_{15} = b_{34}\left(b_{12}b_{23} + b_{12}n_{23}\left(x, y\right)\right) \\[2mm]
\xi_{16} = b_{44}b_{12}n_{24}\left(x, y\right) \\[2mm]
\xi_{17} = c_1\left(b_{11}^2 + b_{12}b_{21} + b_{12}n_{21}\left(x, y\right)\right) \\[2mm]
\xi_{18} = \left(c_2 + n_2\left(x, y\right)\right)\left(b_{11}b_{12} + b_{12}b_{22} + b_{12}n_{22}\left(x, y\right)\right) \\[2mm]
\xi_{19} = b_{12}n_{24}\left(x, y\right)\left(c_4 + n_4\left(x, y, z, u\right)\right) \\[2mm]
\xi_{20} = \left(c_3 + n_3\left(x, y, z\right)\right)\left(b_{12}b_{23} + b_{12}n_{23}\left(x, y\right)\right)
\end{cases}
\tag{9.81}
$$

The above analysis gives the sufficient conditions for the equivalence of a 4-D quadratic system with a hyperjerk system. Hence, the general 4-D quadratic system has the following form:

$$\begin{cases} x' = c_1 + b_{11}x + b_{12}y + n_1\left(x\right) \\ \\ y' = c_2 + b_{21}x + b_{22}y + b_{23}z + n_2\left(x, y, z, u\right) \\ \\ z' = c_3 + b_{31}x + b_{32}y + b_{33}z + b_{34}u + n_3\left(x, y, z, u\right) \\ \\ u' = c_4 + b_{41}x + b_{42}y + b_{43}z + b_{44}u + n_4\left(x, y, z, u\right) \end{cases} \tag{9.82}$$

where

$$\begin{cases} n_1\left(x\right) = a_4x^2 \\ \\ n_2 = b_4x^2 + b_5y^2 + b_6z^2 + b_7u^2 + b_8xy + b_9xz + b_{10}yz \\ \qquad + b_{11}xu + b_{12}zu + b_{13}yu \\ \\ n_3 = e_4x^2 + e_5y^2 + e_6z^2 + e_7u^2 + e_8xy + e_9xz + e_{10}yz \\ \qquad + e_{11}xu + e_{12}zu + e_{13}yu \\ \\ n_4 = d_4x^2 + d_5y^2 + d_6z^2 + d_7u^2 + d_8xy + d_9xz + d_{10}yz \\ \qquad + d_{11}xu + d_{12}zu + d_{13}yu \end{cases} \tag{9.83}$$

Now, any 4-D quadratic system of the form:

$$\begin{cases} x' = c_1 + b_{11}x + b_{12}y + a_4x^2 \\ \\ y' = c_2 + b_{21}x + b_{22}y + b_{23}z + b_4x^2 + b_5y^2 + b_8xy \\ \\ z' = c_3 + b_{31}x + b_{32}y + b_{33}z + b_{34}u + e_4x^2 + e_5y^2 + e_6z^2 \\ \qquad + e_8xy + e_9xz + e_{10}yz \\ \\ u' = c_4 + b_{41}x + b_{42}y + b_{43}z + b_{44}u + n_4\left(x, y, z, u\right) \end{cases} \tag{9.84}$$

can be reduced to a hyperjerk dynamic form $x'''' = H\left(x, x', x'', x'''\right)$ if the conditions (9.71) hold.

Notice that the form (9.82) contains 20 non-linearities, and the best known hyperchaotic 4-D quadratic system is the Rössler system that contains one non-linearity xz and is given by:

$$\begin{cases} x' = -y - z \\ \\ y' = x + ay + u \\ \\ z' = b + xz \\ \\ u' = cu - dz, \end{cases} \tag{9.85}$$

However, system (9.85) is not equivalent to any hyperjerk system due to the presence of singularities.

In [Chlouverakis & Sprott (2006)], the algebraically simplest hyperchaotic snap system was studied, and it is given by:

$$x'''' + x^4 x''' + A x'' + x' + x = 0 \tag{9.86}$$

This system displays hyperchaos when $A = 3.6$ with Lyapunov exponents $(0.1310, 0.0358, 0, -1.2550)$ and it is shown in Fig. 9.7.

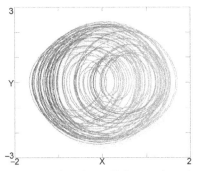

FIGURE 9.7: An attractor for the 4-D hyperchaotic snap system (9.86).

9.5   Examples of crackle and top dynamics

One example of the situation when $n = 5$ is given by:

$$x^{(5)} + x^{(4)} + 4x''' + x'' + x' = 0.2 \left(x^2 - 1 \right) \tag{9.87}$$

and it is shown in Fig. 9.8.

Because system (9.87) is a hyperjerk with a fifth derivative, it is called a *crackle system*. System (9.87) is equivalent to the following 5-D quadratic

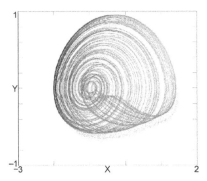

FIGURE 9.8: A chaotic attractor for the 5-D crackle system (9.87).

dynamical system:

$$\begin{cases} x' = y \\ y' = z \\ z' = u \\ u' = v \\ v' = -v - 4u - z - y + 0.2\left(x^2 - 1\right) \end{cases} \tag{9.88}$$

One example of the situation when $n = 6$ is given by:

$$x^{(6)} + x^{(5)} + 2.6x^{(4)} + 2x''' + x'' + 0.5x' = 0.1\left(x^2 - 1\right) \tag{9.89}$$

and it is shown in Fig. 9.9. Because system (9.89) is a hyperjerk with a six derivative, it is called a *top system*. System (9.89) is equivalent to the following 6-D quadratic dynamical system:

$$\begin{cases} x' = y \\ y' = z \\ z' = u \\ u' = v \\ v' = w \\ w' = -w - 2.6v - 2u - z - 0.5y + 0.1\left(x^2 - 1\right) \end{cases} \tag{9.90}$$

In the end of this chapter, we notice that finding similar examples with derivatives higher than the sixth is a very hard and challenging problem.

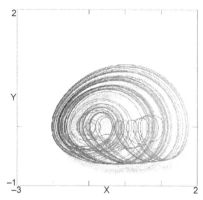

FIGURE 9.9: Attractor for the 6-D pop system in system (9.89).

References

Abdeljawad, T., On conformable fractional calculus, J. Comput. Appl. Math. 279, 57–66 (2015).

Adler, R. and Weiss, B., Similarity of automorphisms of the torus. Memoirs of the American Mathematical Society. 98. American Mathematical Society, Providence, R.I. (1970).

Afraimovich, V. S., Bykov, V. V. and Shilnikov, L. P., On the appearance and structure of Lorenz attractor. DAN SSSR. 234, 336–339 (1977).

Afraimovich, V. S., Bykov, V. V. and Shilnikov, L. P., On structurally unstable attracting limit set of the type of Lorenz attractor. Trans. Moscow. Math. Soc. 44, 153-216 (1982).

Afraimovich, V. S. and Shilnikov, L. P., Strange attractors and quasiattractors in Non-linear Dynamics and Turbulence eds. by Barenblatt, G. I., Iooss, G. and Joseph, D. D., Pitman, NY, 1–28 (1983).

Afraimovich, V. S., Bykov, V. V. and Shilnikov, L. P., On the structurally unstable attracting limit sets of Lorenz attractor type. Tran. Moscow. Math. Soc. 2, 153–215 (1983).

Afraimovich, V. S. and Shilnikov, L. P., Strange attractors and quasi-attractors. In Dynamics and Turbulence. New York Pitman (1983).

Afraimovich, V. S., Chernov, N. I. and Sataev, E. A., Statistical properties of two-dimensional generalized hyperbolic attractors. Chaos. 5(1), 238–252 (1995).

Aharonov, D., Devaney, R. L. and Elias, U., The dynamics of a piecewise linear map and its smooth approximation. Int. J. of Bif. and Chaos. 7(2), 351–372 (1997).

Albers, D. J. and Sprott, J. C., Structural stability and hyperbolicity violation in high-dimensional dynamical systems. Non-linearity. 19(8), 1801–1849 (2006).

Albers, D. J., Sprott, J. C., and Dechert, W. D., Routes to chaos in neural networks with random weights. Int. J. Bifurcation and Chaos. 8, 1463–1478 (1998).

Albers, D. J., Crutchfield, J. P. and Sprott, J. C., Persistent chaos in high dimensions. Phys. Rev. E. 74, 057201 (2006).

Alexander, J. C., York, J. A., You, Z. P. and Kan, I., Riddled basins. Int. Bifurc. Chaos. 2, 795–813 (1992).

Alvarez-Llamoza, Cosenzab, M. G. and Poncec, G. A., Critical behavior of the Lyapunov exponent in type-III intermittency. Chaos, Solitons & Fractals. 36(1), 150–156 (2008).

Anatoliy, A. Martynyuk, M. Stamova, Fractional-like derivative of Lyapunov-type functions and applications to stability analysis of motion. Electronic Journal of Differential Equations (2018) (62).

Andrecut, M. and Ali, M. K., On the occurrence of robust chaos in a smooth system. Modern Physics Letters B. 15(12-13), 391–395 (2001).

Anishchenko, V., Complex Oscillations in Simple Systems. Nauka, Moscow (1990).

Anishchenko, V., Dynamical Chaos Models and Experiments. World Scientific, Singapore (1995).

Anosov, D. V., Geodesic flows on closed Riemannian manifolds of negative curvature. Proc. Steklov Math. Inst. 90, 1–235 (1967).

Amit, D. J., Modeling Brain Function, The World of Attractor Neural Networks. Cambridge University Press (1989).

Araujo, V., Pacifico, M. J., Pujals, E. R. and Viana, M., Singular-hyperbolic attractors are chaotic. Trans. Amer. Math. Soc. 361, 2431–2485 (2009).

Arroyo, A., Singular hyperbolicity for transitive attractors with singular points of 3-dimensional C^2-flows. Bull. Braz. Math. Soc, New Series. 38(3), 455–465 (2007).

Arroyo, A. and Pujals, E., Dynamical properties of singular hyperbolic attractors. Discrete an Continuous Dynamical Systems. 19(1), 67–87 (2007).

Ashwin, P., Synchronization from Chaos. Nature. 422, 384–385 (2003).

Avrutin, V. and Schanz. M., Border-collision period-doubling scenario. Phys. Rev. E. 70(3), 026222 (2004).

Aziz Alaoui, M. A., Multi-Fold in a Lozi-type map. Preprint (2000).

Aziz Alaoui, M. A., Robert. C. and Celso Grebogi. C., Dynamics of a Hénon-Lozi-type map. Chaos. Solitons & Fractals. 12, 2323–2341 (2001).

Baillieul, J., Brockett, R. W. and Washburn, R. B. Chaotic motion in nonlinear feedback systems. IEEE Trans. Auto. Control. 27, 990–997 (1980).

Bamon, R., Labarca, R., Mane, R. and Pacifico, M. J. The explosion of singular cycles. Publ. Math. IHES. 78, 207–232 (1993).

Bautista, S., Morales, C. and Pacífico, M. J., There are singular hyperbolic flows without spectral decomposition. Preprint IMPA. Série A278 (2004).

Bautista, S., Sobre Conjuntos Hiperblicos´Singulares, In Portuguese. Thesis, Universidade Federal do Rio de Janeiro (2005).

Bautista, S., Morales, C. and Pacifico, M. J. Intersecting invariant manifolds on singular-hyperbolic sets. Preprint (2005).

Bautista, S. and Morales, C., Existence of periodic orbits for singular-hyperbolic sets. Moscow Mathematical Journal. 6(2), 265–297 (2006).

Banergee, S., York, J. A. and Grebogi, C., Robust chaos. Phys. Rev. Lettres. 80(14), 3049–3052 (1998).

Banerjee, S. and Chakrabarty, K., Non-linear modeling and bifurcations in the boost converter. IEEE Trans. Power Electron. 13, 252–260 (1998).

Banerjee, S. and Grebogi, C., Border collision bifurcations in two-dimensional piecewise smooth maps. Phy. Rev. E. 59(4), 4052–4061 (1999).

Banerjee, S., Kastha, D., Das, S., Vivek, G. and Grebogi, C., Robust chaos-the theoretical formulation and experimental evidence. ISCAS (5), 293–296 (1999).

Banerjee, S. and Verghese, G. C., Non-linear Phenomena in Power Electronics, Attractors Bifurcations, Chaos, and Non-linear Control. IEEE Press, New York, USA, (2001).

Banerjee, J. P. and Banerjee, S., Border collision bifurcations in one-dimensional discontinuous switching maps. Inter. J. Bifur. Chaos. 13, 3341–3352 (2003).

Banerjee, S., Parui, S. and Gupta, A., Dynamical effects of missed switching in current-mode controlled dc-dc converters. IEEE Trans. Circuits & Systems–II 51, 649–654 (2004).

Banks, J. and Dragan, V., Smale's Horseshoe map via ternary numbers. SIAM Review. 36(2), 265–271 (1994).

Barreto, E., Hunt, B., Grebogi, C., and Yorke, J. A., From high dimensional chaos to stable periodic orbits—The structure of parameter space. Phys. Rev. Lett. 78, 4561 (1997).

Barton. S. A., Two-dimensional movement controlled by a chaotic neural network. Automatica 31(8), 1149–1155 (1995).

Bauer, M. and Martienssen, W., Quasi-periodicity route to chaos in neural networks. Europhys. Lett. 10, 427–431 (1989).

Benettin, G., Galgani, L., Giorgilli, A. and Strelcyn, J.M., Lyapunov characteristic exponents for smooth dynamical systems and for Hamiltonian systems; a method for computing all of them, Part 2: Numerical applications. Meccanica, 15(9), 21–30 (1980).

Béguin, F. and Bonatti, C., Flots de Smale en dimension 3, présentations finies de voisinages invariants d'ensembles selles (French. English, French summary), (Smale flows in dimension 3 finite presentations of invariant neighborhoods of saddle sets). Topology. 41(1), 119–162 (2002).

Benedicks, M. and Carleson, L., The dynamics of the Hénon maps. Ann. Math. 133, 73–169 (1991).

Berry, R. M. and Nowak, M. A., Defective escape mutants of HIV, J. Theor. Biol. 171, 387–395 (1994).

Bessa, M., The Lyapunov exponents of generic zero divergence 3-dimensional vector fields. Ergod. Th. Dynam. Syst. 27(5), 1445–1472 (2007).

Boccaletti, S., Kurths, J., Valladares, D.L., Osipov, G. and Zhou, C., The synchronization of chaotic systems. Physics Reports 366, 1–101 (2002).

Bonatti, C., Diaz, L. J. and Pujals, E. R., A C^1-generic dichotomy for diffeomorphisms, Weak forms of hyperbolicity or infinitely many sinks or sources. Annals of Mathematics. 158, 355–418 (2003).

Bonatti, C., Diaz, L. and Viana, M., Dynamics beyond uniform hyperbolicity. A Global Geometric and Probabilistic Perspective. Encyclopaedia Mat. Sci. 102, Mathematical Physics, III (BerlinSpringer-Verlag) (2005).

Bonetto, F., Falco, P. and Giuliani, A., Analyticity of the SRB measure of a lattice of coupled Anosov diffeomorphisms of the torus. J. Math. Phys. 45, 3282–3300 (2004).

Bornholdt. S and Schuster, H. G., Handbook of Graphs and Networks. From the Genome to the Internet, Wiley-VCH, Weinheim (2003).

Boukhalfa, E. and Laskri, Y., Quasi-controlling of chaotic discrete dynamical systems. Global Journal of Pure and Applied Mathematics. 12(5), 4317–4323 (2016).

Bowen, R., Topological entropy and Axiom. A. Proc. Sympos. Pure Math., Vol. XIV, Berkeley, Calif. 23–41 (1968). Amer. Math. Soc., Providence, R.I. (1970).

Bowen, R., Periodic points and measures for axiom A diffeomorphisms. Trans. Amer. Math. Soc. 15(4), 377–397 (1971).

Bunimovich, L. A. and Sinai, Y., In Non-linear Waves, edited by Gaponov-Grekhov, A. V., Nauka, Moscow (1980). P. 212 (in Russian).

Campbell, D. K., Galeeva, R., Tresser, C. and Uherka, D. J., Piecewise linear models for the quasiperiodic transition to chaos. Chaos. 6(2), 121–154 (1996).

Caroppo, D., Mannarelli, M., Nardulli, G. and Stramaglia, S., Chaos in neural networks with a non-monotonic transfer function. Physical Review E. 60, 2186–2192 (1999).

Carpenter, G. A. and Grossberg, S. (Eds.). Pattern Recognition by Self-Organizing Neural Networks, MIT Press, Cambridge, MA (1991).

Cao, Y. and Kiriki, S., The basin of the strange attractors of some Hénon maps. Chaos, Solitons & Fractals. 11(5), 729–734 (2000).

Cao, Y., A note about Milnor attractor and riddled basin. Chaos, Solitons and Fractals. 19, 759–764 (2004).

Cao, H. and Wang, C., Effect of step size on bifurcations and chaos of a map-based BVP oscillator," Inter. J. Bifurcation & Chaos. 20(6), 1789–1795 (2010).

Castillo-Chavez, C., Cooke, K., Hunag, W. and Levin, S. A., Results on the Dynamics for Models for the Sexual Transmission of the Human Jrnmunodeficiency Virus. Appl. Math. Lett. 2(4), 327–331 (1989(a)).

Castillo-Chavez, C., Cooke, K., Huang, W. and Levin, S. A., The Role of long periods of infectiousness in the dynamics of acquired immunodeficiency syndrome (AIDS). In, Castillo-Chavez C., Levin, S. A., Shoemaker C. A., (eds.) Mathematical Approaches to Problems in Resource Management and Epidemiology. Lecture Notes in Biomathematics, vol **81**. Springer, Berlin, Heidelberg (1989(b)).

Castillo-Chavez, C., Cooke, K., Huang, W. and Levin, S. A., On the Role of Long Incubation Periods in the Dynamics of Acquired Immunodeficiency Syndrome (AIDS), Part 1. Single Population Models. J. Math. Biol. 27(4), 373–398 (1989(c)).

Castillo-Chavez, C., Cooke, K. L., Huang, W. and Levin, S. A., On the Role of Long Incubation Periods in the Dynamics of Acquired Immunodeficiency Syndrome (AIDS). Part 2, Multiple Group Models. In, Castillo-Chavez C. (eds.) Mathematical and Statistical Approaches to AIDS Epidemiology. Lecture Notes in Biomathematics, vol **83**. Springer, Berlin, Heidelberg (1989(d)).

Cessac, B., Doyon, B., Quoy, M. and Samuelides, M., Mean-field equations, bifurcation map and route to chaos in discrete time neural networks. Physica D. 74, 24–44 (1994).

Chang, A., Bienfang, J. C., Hall, G. M., Gardner, J. R. and Gauthiera, D. J., Stabilizing unstable steady states using extended timedelay autosynchronization. Chaos. 8(4), 782–790 (1998).

Charles, M., Gray, P. K., Andreas, K. E. and Wolf, S., Oscillatory responses in cat visual cortex exhibit inter-columnar synchronization which reflects global stimulus properties. Nature. 338, 334–337 (1989).

Charlotte, Y. F. H. and Bingo, W. K. L., Initiation of HIV therapy, Inter. J. Bifur & Chaos. 20(4), 1279–1292 (2010).

Chen, G. and Lai, D., Anticontrol of chaos via feedback. In Proc. of IEEE Confence on Decision and Control, San Diego, CA, 367–372 (1997).

Chen, G. and Lai, D., Feedback anticontrol of chaos. Int. J. Bifur. Chaos. 8, 1585–1590 (1998).

Chen, G., Hsu, S. and Zhou, J., Snap-back repellers as a cause of chaotic vibration of the wave equation with a an der Pol boundary condition and energy injection at the middle of the span. J. Math. Phys. 39(12), 6459–6489 (1998).

Chen, G. and Lai, D., Making a discrete dynamical system chaotic, feedback control of lyapunov exponents for discrete-time dynamical system. IEEE Trans. Circ. Syst. I. 44, 250–253 (1997).

Chen, Y.G., Urban chaos and perplexing dynamics of urbanization. Lett. Spat. Resour. Sci. 2(2-3), 85–95 (2009).

Chen, Y.G., A new model of urban population density indicating latent fractal structure. International Journal of Urban Sustainable Development, 1, 1-2, 89–110 (2010).

Chlouverakis, K. E. and Sprott, J. C., Chaotic hyperjerk systems. Chaos, Solitons and Fractals. 28, 739–746 (2006).

Christy, J., Branched surfaces and attractors. I. Dynamic branched surfaces. Trans. Amer. Math. Soc. 336, 759–784 (1993).

Christine, A. S. and Walter, J. F., How brains make chaos in order to make sense of the world. Behavioral and Brain Sciences. 10 161–195 (1987).

Christiansen, F. and Rugh, H. H., Computing Lyapunov spectra with continuous Gram–Schmidt orthonormalization. Non-linearity. 10(5), 1063–1073 (1997).

Chua, L. O. and Lin, T., Chaos in digital filters. IEEE Trans. Circuits Syst. 35, 648–658 (1988).

Chua, L. O., Komuro, M. and Matsumoto, T., The double scroll family. IEEE Trans. Circuits Syst. CAS-33(11), 1073–1118 (1986).

Chua, L. O. and Tichonicky, I., 1-D map for the double scroll. IEEE Trans. Circuits Syst.-IFund. Th. Appl. 38(3), 233–243 (1991).

Cooper, R., Winter, A. L., Crow, H. J. and Grey, W., Electroencephalogr. Clin. Neurophysiol. 18, 217 (1965).

Courchamp, F., Pontier, D., Langlais, M. and Artois, M., Population dynamics of feline immunodeficiency virus within cat populations. J. Theor. Biol. 175, 553–560 (1995).

Craig, I. K., Xia, X. and Venter, J. W., Introducing HIV/AIDS education into the electrical engineering curriculum at the university of Pretoria. IEEE Trans. Educ. 47, 65–73 (2004).

Crovisier, S., Birth of homoclinic intersectionsa model for the central dynamics of partially hyperbolic systems. Arxiv preprint math.DS/0605387 (2006).

Cohen, S. D. and Hindmarsh, A. C., CVODE, A stiff/non-stiff ODE solver in C. Comput. Phys. 10, 138–141 (1996).

Crovisier, S., Partial hyperbolicity far from homoclinic bifurcations. Preprint (2008).

Cvitanovic, P., Gunaratne, G. and Procaccia. I., Topological and metric properties of Hénon-type strange attractors. Phys. Rev. A. 38, 1503–1520 (1988).

Dafilis, M. P., Liley, D. T. J. and Cadusch, P. J., Robust chaos in model of the electroencephalogram—Implications for brain dynamics. Chaos. 11(3), 474–478 (2001).

de Carvalho, A. and Hall, T., How to prune a horseshoe. Non-linearity. 15, R19–68 (2002).

de Melo, W., Structural stability of diffeomorphisms on two-manifolds. Invent. Math. 21, 233–246 (1973).

de Melo, W. and Palis, J., Geometric theory of dynamical systems. An introduction. Springer-Verlag, New York–Berlin (1982).

de Oliveira, K. A., Vannucci, A. and da Silva, E. C., Using artificial neural networks to forecast chaotic time series. Phys. A. 284(1), 393–404 (2000).

de Souza, F. M. C., Modeling the dynamics of HIV-1 and CD4 and CD8 lymphocytes. IEEE Engin. Med. Biol. 18, 21–24 (1999).

di Bernardo, M., Feigin, M. I., Hogan, S. J. and Homer, M. E., Local analysis of C-bifurcations in n-dimensional piecewise smooth dynamical systems. Chaos, Solitons & Fractals. 10(11), 1881–1908 (1999).

di Bernardo, M., Kowalczyk, P. and Nordmark, A. B., Bifurcations of dynamical systems with sliding, derivation of normal form mappings. Physica. D. 170, 175–205 (2002).

di Bernardo, M., Kowalczyk, P. and Nordmark, A. B., Sliding bifurcations, a novel mechanism for the sudden onset of chaos in dry-friction oscillators. Int. J. Bifurc. Chaos. 13, 2935–2948 (2003).

Djellit, I. and Boukemara, I., Dynamics of a three parameters family of piecewise maps. Facta universitatis. Nis. Ser. Elec. Energ. 20(1), 85–92 (2007).

Dixit, N. M. and Perelson, A. S., Complex patterns of viral load decay under antiretroviral therapy, Influence of pharmacokinetics and intracellular delay. J. Theor. Biol. 226, 95–109 (2004).

Doering, C. I., Persistently transitive vector fields on three-dimensional manifolds. Dynamical Systems and Bifurcation Theory, Pitman Research Notes in Mathematics Series. 160, 59–89 (1987).

Dogaru, R., Murgan, A. T., Ortmann, S. and Glesner, M., Searching for robust chaos in discrete time neural networks using weight space exploration. 1996 Proceedings ICNN'96 Washington D.C. 2–6 June (1996).

Chen, G. and Dong, X., From chaos to order: Methodologies, Perspectives and Applications. World Scientific, Singapore (1998).

Eckhorn, R., Bauer, R., Jordan, W., Brosch, M., Kruse, W., Munk, M. and Reitboeck, H. J., Coherent oscillations. Biol. Cybern. 60, 121–130 (1988).

Eichhorn, R., Linz, J. and Hanggi, P., Transformations of non-linear dynamical systems to jerky motion and its application to minimal chaotic flows. Phys. Rev. 58(6), 7151–7164 (1998).

Eckmann, J. P. and Ruelle, D., Ergodic theory of chaos and strange attractors. Rev. Mod. Phys. 57, 617–656 (1985).

Elsgolts, L. E. and Norkin, S. B., Application of Differential Equations with Deviating Arguments, Volume **105**, Academic Press, (1973).

Essunger, P. and Perelson, A. S., Modeling HIV infection of CD4+ T-cell subpopulations. J. Theor. Biol. 170, 367–391 (1994).

Ermentrout, G. B., XPPAUT, http//www.pitt.edu/;phase.

Farmer, J. D., Ott, E. and Yorke, J. A., The dimension of chaotic attractors. Physica D. 7, 153–180 (1983).

Feigin, M. I., Doubling of the oscillation period with C-bifurcations in piecewise continuous systems. Prikladnaya Matematika i Mechanika. 34, 861–869 (1970).

Feigin, M. I., Forced oscillations in systems with discontinuous non-linearities (Moscow, Nauka) (1994) (in Russian).

Feely, O., Chua, L. O., Non-linear dynamics of a class of analog-to-digital converters. Inter. J. Bifur. Chaos. 22, 325–340 (1992).

Feely, O. and Chua, L. O., The effect of integrator leak in modulation. IEEE Trans. Circuits Syst. 38, 1293–1305 (1991).

Filter, R. A., Xia, X. and Gray, C. M., Dynamic HIV/AIDS parameter estimation with application to a vaccine readiness study in Southern Africa. IEEE Trans. Biomed. Engin. 52, 784–791 (2005).

Fischer, K. H. and Hertz, J. A. *Spin Glasses* (Cambridge University Press, Cambridge) (1991).

Fishman, M. A. and Perelson, A. S., Th1/Th2 cross regulation, J. Theor. Biol. 170, 25–56 (1994).

Freeman, W. J., Societies of Brains. Lawrence Erlbaum Associates, Mahwah (1995).

Fujisaka, H., Statistical dynamics generated by fluctuations of local Lyapunov exponents. Prog. Theor. Phys. 70, 1264–1275 (1983).

Fujisaka, H. and Sato, C., Computing the number, location and stability of fixed points of Poincaré maps. Circuits and Systems I, Fundamental Theory and Applications, IEEE Transactions on (see also Circuits and Systems IRegular Papers, IEEE Transactions on). 44(4), 303–311 (1997).

Funahashi, K. and Nakamura, Y., Approximation of dynamical systems by continuous time recurrent neural networks. Neural Networks. 6, 801–806 (1993).

Galias, Z., Existence and uniqueness of low-period cycles and estimation of topological entropy for the Hénon map. In Proc. Int. Symposium on Nonlinear Theory and its Applications, NOLTA'98. Crans-Montana, 1, 187–190 (1998).

Gasri, A., Different schemes of coexistence of full state hybrid function projective synchronization and inverse full state hybrid function projective synchronization. Non-linear Dynamics and Systems Theory. 18 (2), 154–169 (2018).

Gavrilov, N. K. and Shilnikov, L. P., On three dimensional dynamical system close to systems with a structuraly satble homoclinic curve. Math. USSR. Sb. 19, 139–156 (1973).

Ghrist, R. W., Holmes, P. J. and Sullivan, M. C., Knots and Links in Three-Dimensional Flows, Lecture Notes in Mathematics, **1654**, Berlin. Springer-Verlag (1997).

Ghrist, R. W. and Zhirov, A. Yu., Combinatorics of one-dimensional hyperbolic attractors of diffeomorphisms of surfaces, (Russian summary) Trudy Matematicheskogo Instituta Imeni V. A. Steklova. Rossi kaya Akademiya Nauk, 244 Din. Sist. i Smezhnye Vopr. Geom., 143–215 (2004).

Gilchrist, M. A., Coombs, D. and Perelson, A. S., Optimizing within-host viral fitness, Infected cell lifespan and virion production rate. J. Theor. Biol. 229, 281–288 (2004).

Goodman, S., Dehn Surgery on Anosov Flows, Geometric Dynamics (Rio de Janeiro, (1981) 300–307 Lecture Notes in Mathematics, 1007, Berlin Springer (1983).

Grebogi, C., Ott, E. and Yorke, J. A., Basin Boundary Metamorphoses, Changes in Accessible Boundary Orbits. Physica. D. 24, 243 (1987).

Guckenheimer, J. and Williams, R. F., Structural Stability of Lorenz Attractors. Publ. Math. IHES. 50, 307–320 (1979).

Hale, J. K. and Waltman, P., Persistence in infinite-dimensional systems. SIAM J. Math. Anal. 20, 388–395 (1989).

Hammel, S. M., Jones, C. K. R. T. and Moloney, J. V., Global dynamical behavior of the optical field in a ring cavity. J. Opt. Soc. Am. B. 2, 552–564 (1985).

Hammond, B. J., Quantitative study of the control of HIV-1 gene expression. J. Theor. Biol. 163, 199–221 (1993).

Han, P., Perturbed basins of attraction. Mathematische Annalen. 337(1), 1–13 (2007).

Hassouneh, M. A. and Abed, E. H., Feedback control of border collision bifurcations in piecewise smooth systems. ISR Technical research report, TR (2002)-26 (2002).

Hassouneh, M. A., Abed, E. H. and Banerjee, S., Feedback control of border collision bifurcations in two-dimensional discrete-time systems. ISR Technical Research Report (2002).

Hayashi, S., Diffeomorphisms in $C^1(M)$ satisfy Axiom A. Ergod. Th. and Dynam. Sys. 12, 233–253 (1992).

Hayashi, S., Connecting invariant manifolds and the solution of the C^1 stability and 2126-stability conjectures for flows. Ann. of Math. 145, 81–137 (1997).

Haykin, S. Neural Networks, A Comprehensive Foundation (2nd edn.) Prentice Hall, Upper Saddle River, NJ (1999).

Hegger, R., Kantz, H. and Schreiber, T., Practical implementation of nonlinear time series methods, The TISEAN package. Chaos. 9, 413–424 (1999).

Hempel, J., 3-Manifolds. Annals of Mathematics Studies. 86 (2004) (Paperback) AMS Chelsea Publishing.

Hénon, M., A two dimensional mapping with a strange attractor. Commun. Math. Phys. 50, 69–77 (1976).

Henshaw, S. and McCluskey, C. C., Global stability of a vaccination model with immigration. Electron. J. of Diff. Eqs., 2015 (92), 1–10 (2015).

Hirsch, M. and Pugh, C., Stable manifolds and hyperbolic sets. Proc. Sympos. Pure Math., Vol. XIV, Berkeley, Calif. 133–163 (1968), Amer. Math. Soc., Providence, R.I. (1970).

Hirsch, M. and Pugh, C., Stable manifolds and hyperbolic sets. Proc. of Symposium in Pure Math., Amer. Math. Soc. 14, 133–165 (1970).

Hirsch, M., Palis, J., Pugh, C. and Shub, M., Neighborhoods of hyperbolic sets. Invent. Math, 9, 121–134 (1969)/(1970).

Hirsh, M. and Smale, S., Differential Equations, and Linear Algebra. Dynamical Systems. New York, Academic Press (1974).

Hirsch, M., Differentiable Topology. Graduate Texts in Mathematics. 33 Springer-Verlag (1976).

Hirsh, M., Pigh, C. and Shub, M., Invariant manifolds. Lecture Notes in Math. Springer-Verlag (1977).

Hirsch, M., Smale, S. and Devaney, R. L., Differential Equations Dynamical Systems and an Introduction to Chaos. (2004), Elsevier.

Hornik, K., Stinchocombe, M. and White, H., Universal approximation of an unknown mapping and its derivatives using multilayer feedforward networks. Neural Networks. 3, 551–560 (1990).

Hraba, T. and Dolezal, J., A Mathematical Model and $CD4^+$ Lymphocyte Dynamics in HIV Infection. Emerging Infectious Diseases. 2(4), 299–305 (1996).

Hu, H. and Young, L. S., Non-existence of SBR measure for some diffeomorphisms that are almost Anosov. Ergodic Theory Dynamical Systems. 15, 67–76 (1995).

Hurewicz, W. and Wallman, H., Dimension Theory. Princeton University Press, Princeton, NJ (1984).

Kanou, N. and Horio, Y., Switched-current chaotic neural network with chaotic simulated annealing. In Proceedings ICNN'95, Perth, Australia, 6, 3146–3149 (1995).

Kapitaniak, T., Maistrenko, Y. and Grebogi, C., Bubbling and riddling of higher-dimensional attractors. Chaos, Solitons and Fractals. 17(1), 61-66 (2003).

Kaplan, J. and Yorke, J. A., Chaotic behavior of multidimensional difference equations. In Functional Differential Equations and Approximation of Fixed Points (H. O. Peitgen and H. O. Walther, eds.) Springer, New York (1987).

Katok, A. and Hasselblatt, B., Introduction to the Modern Theory of Dynamical Systems, Cambridge University Press (1995).

Kaye, R. and Wilson, R., Linear Algebra, OUP (1998).

Khalil, H. K., Non-linear systems. Prentice Hall, New Jersey (1996).

Khalil, R., Al Horani, M., Yousef, A. and Sababheh, M., A new definition of fractional derivative. J. Comput. Appl. Math. 264, 65–70 (2014).

Khan, A. M., Mar, D. J. and Westervelt, R. M., Spatial measurements near the instability threshold in ultrapure G_e. Phys. Rev. B. 45, 8342–8347 (1992).

Kiriki, S., Li, M. C. and Soma, T., Coexistence of invariant sets with and without SRB measures in Hénon family. Non-linearity. 23(9), 2253 (2010).

Kennedy, M. P. and Kolumban, G., Introduction to the special issue on non-coherent chaotic communications. IEEE Trans. Circuits Syst., I, Fundam. Theory Appl. 47, 1661 (2000).

Kennedy, J. and York, J. A., Topological horseshoes. Trans. Amer. Math. Soc. 353, 2513–2530 (2001).

Ko, J. H., Kim, W. H. and Chung, C. C., Optimized structured treatment interruption for HIV therapy and its performance analysis on controllability. IEEE Trans. Biomed. Engin. 53, 380–386 (2006).

Kollar, L. E., Stepan, G. and Turi, J., Dynamics of piecewise linear discontinuous maps. Inter. J. Bifur. Chaos. 14, 2341–2351 (2004).

Komuro, M., Expansive properties of Lorenz attractors. In The theory of dynamical systems and its applications to non-linear problems, 4–26. World Sci. Publishing, Kyoto (1984).

Kouyos, R. D., Gordon, S. N., Staprans, S. I., Silvestri, G. and Regoes, R. R., Similar impact of CD8$^+$ T cell responses on early virus dynamics during SIV infections of rhesus macaques and sooty mangabeys. PLoS Comput. Biol. 6(8), e1000901 (2010).

Kuang, Y., Delay Differential Equations with Applications in Population Dynamics. Boston, Academics Press (1993).

Keener, J. P. and Glass, L., Global bifurcations of a periodically forced non-linear oscillator. Journal of Mathematical Biology, 21(2), 175–190 (1984).

Krstik, M., Kanellakopoulos, I. and Koketovid, P. V., Non-linear and adaptive control design. J. Wiley, New York (1995).

Krichman, M., Sontag, E. D. and Wang, Y., Input-output-state stability. SIAM. J. Control. 39, 1874–1928 (2001).

Kuznetsov, Y. A., Elements of Applied Bifurcation Theory. Springer, 3rd edition (2004).

Kwon, O. J., Park, J. D. and Hoyun, L., Stabilizing chaos to fixed points by continuous proportional feedback. Journal of Korean Physical Society. 38(6), 635–641 (2001).

Iggidr, A. and Bensoubaya, M., Stability of discrete-time systems, New criteria and applications to control problems, Inria, RR3003 (1996).

Innocenti, G., Genesio, R. and Ghilardi, C., Oscillations and chaos in simple quadratic systems. Chaos. 18, 1917–1937 (2008).

Itik, M., Salamci, M. U. and Banks, S. P., Optimal control of drug therapy in cancer treatment. Non-lin. Anal. Th. Meth. Appl. 71, e1473–e1486 (2009).

Iwasa, Y., Michor, F. and Nowak, M. A., Virus evolution within patients increases pathogenicity. J. Theor. Biol. 232, 17–26 (2005).

Iwami, S., Benjamin, P. H., Catherine, A. A., Morita, S., Tada, T., Sato, K., Igarashi, T. and Miura, T., Quantification system for the viral dynamics of a highly pathogenic simian/human immunodeficiency virus based on an *in vitro* experiment and a mathematical model. Retrovirology. 9, 18 (2012).

Jafelice, R. M., Barros, L. C., Bassanezi, R. C. and Gomide, F., Fuzzy set-based model to compute the life expectancy of HIV infected populations," IEEE Ann. Meeting of the Fuzzy Information Processing. 27–30 (2004), NAFIPS. 1, 314–318 (2004).

Jafarizadeh, M. A. and Behnia, S., Hierarchy of Chaotic maps with an invariant measure and their compositions. J. Non-linear. Math. Phy. 9(1), 26–41 (2002).

Jarvenpaa, E. and Jarvenpaa, M., On the de nition of SRB measures for coupled map lattices. Comm. Math. Phys. 220, 109–143 (2001).

Jiang, M., SRB measures for lattice dynamical systems. J. Statistical Physics. 111(3-4), 863–902 (2003).

Jin, X., Bauer, D. E., Tuttleton, S. E., Lewin, S., Gettie, A., *et al.*, Dramatic rise in plasma viremia after $CD8^+$ T cell depletion in simian immunodeficiency virus-infected macaques. J. Exp. Med. 189, 991–998 (1999).

Jinliang, W. and Xinxin, T., Global stability of a delay differential equation of hepatitis B virus infection with immune response. EJDE-2013/94.

Jing-ling, S., Hua-Wei, Y., Jian-Hua, D. and Hong-Jun, Z., Riddled basin of laser cooled-Ions in a Paul trap. Chinese. Phys. Lett. 13 81–84 (1996).

Labarca, R. and Pacifico, M. J., Stability of singular horsshoes. Toplogy. 25(3), 337–352 (1986).

Lai, D. and Chen, G., Making a discrete dynamical system chaotic, Theorical results and numerical simulations. Int. J. Bifur. Chaos. 13(11), 3437–3442 (2003).

Lainscsek, C., Lettellier, C. and Gorodnitsky, I., Global modeling of the Rôssler system from the z-variable. Phys. Lett. A. 314, 409–427 (2003).

Lakshmikantham, V., Leela, S. and Martynyuk, A., Stability Analysis of Nonlinear Systems. Marcel Dekker, New York (1989).

Lakshmikantham, V., Leela, S. and Vasundhara Devi, J., Theory of Fractional Dynamic Systems, Cambridge Scientific Publisher, Cambridge (2009).

LaSalle, J. P., The Stability and Control of Discret Processes. Springer-Verlag, New York (1986).

Laughton, S. N. and Coolen, A. C. C., Quasi-periodicity and bifurcation phenomena in using spin neural networks with asymmetric interactions. J. Phys. A. 27, 8011–8028 (1994).

Lee, E. and Markus, L., Foundation of Optimal Control Theory. John Wiley, New York (1967).

Li, T. Y. and Yorke, J. A., Period three implies chaos. Amer. Math. Monthly. 82, 985–992 (1975).

Li, C. P. and Chen, G., An improved version of the Marotto theorem. Chaos, Solitons & Fractals. 18, 69–77 (2003).

Li, C. P. and Chen, G., Erratum to An improved version of the Marotto theorem (Chaos, Solitons & Fractals. 18, 69–77 (2003)), Chaos. Solitons & Fractals, 20, 655 (2004).

Li, Z., Park, J. B., Joo, Y. H., Choi, Y. H. and Chen, G. Anticontrol of chaos for discrete TS fuzzy systems. IEEE Trans. Circ. Syst.-I. 49, 249–253 (2002(a)).

Li, X., Chen, G., Chen, Z. and Yuan, Z., Chaotifying linear Elman networks. IEEE Trans. Neural Networks. 13, 1193–1199 (2002(b)).

Li, Z., Park, J. B., Chen, G. and Joo, Y. H. Generating chaos via feedback control from a stable TS fuzzy system through a sinusoidal non-linearity. Int. J. Bifur. Chaos. 12, 2283–2291 (2002(c)).

Li, Y., Merrill, J. D., Mooney, K., Song, L., Wang, X., Guo, C. J. et al., Morphine enhances HIV infection of neonatal macrophages. Pediatr. Resn. 54(2), 282–8 (2003).

Li, T. Y. and Yorke, J. A., Periodic three implies chaos. Am. Math. Monthly. 82, 985–989 (1975).

Li, C., On super-chaotifying discrete dynamical systems. Chaos, Solitons and Fractals. 21, 855–861 (2004).

Li, C. and Chen, G., Estimating the Lyapunov exponents of discrete systems. Chaos. 14(2), 343–346 (2004).

Lian, K. Y., Chiang, T. S., Chiu, C. S. and Liu, P., Synthesis of fuzzy model-based designs to synchronization and secure communications for chaotic systems. IEEE. Trans. Syst. Man. Cybern. B. Cybern. 31(1) 66–83 (2001).

Liberzon, D., Switching in systems and control. Boston, Birkhauser (2003).

Liley, D. T. J., Cadusch, P. J. and Wright, J. J., A continuum theory of electrocortical activity. Neurocomputing. 26-27, 795–800 (1999).

Lin, W., Ruan, J. and Zhou, W., On the mathematical clarification of the snapback repeller in high-dimensional systems and chaos in a discrete neural network model. Int. J. Bifur. Chaos. 12(5), 1129–1139 (2002).

Linh, N. M. and Phat, V. N., Exponential stability of non-linear time-varying differential equations and applications. Electronic Journal of Differential Equations. 2001(34), 1–13 (2001).

Lipsitch, M. and Nowak, M. A., The evolution of virulence in sexually transmitted HIV/AIDS. J. Theor. Biol. 174, 427–440 (1995).

Linz, S. J., Non-linear dynamical models and jerky motion. Am. J. Phys. 65, 523-526 (1997).

Linz, S. J., Newtonian jerky dynamics, Some general properties. Am J. Phys. 66, 1109–14 (1998).

Linz, S. J., No-chaos criteria for certain jerky dynamics. Phys. Lett. A. 275, 204–210 (2000).

Linz, S. J., On hyperjerky systems. Chaos, Solitons and Fractals. 37 741–747 (2008).

Liu, S., Zhang, Y. and Liu, Y., Construction of a class of chaotic maps and Gaussian primitive property. Dynam. Contin. Discre. and Impulsive Syst. Computers & Mathematics with Applications. 45(6–9), 1213–1219 (2003).

Lorenz, E. N., Deterministic Non-periodic Flow. J. Atmos. Sci. 20, 130–141 (1963).

Lozi, R., Un attracteur etrange (?) du type attracteur de Hénon. J. Phys. (Paris) **39** Colloq. C5, 9–10 (1978).

Lû, J., Zhou, T., Chen, G. and Yang, X., Generating chaos with a switching piecewise-linear controller. Chaos, 12(2), 344–349 (2002).

Mahla, A. I. and Badan Palhares, A. G., Chua's circuit with discontinuous non-linearity. J. Circuit, System and Computer. 3(1), 231–237 (1993).

Malasoma, J. M., What is the simplest dissipative chaotic jerk equation which is parity invariant? Phys. Lett. A. 264, 383–389 (2000).

Malasoma, J. M., A new class of minimal chaotic flows. Phys. Lett. A. 305, 52–58 (2002).

Marotto, F. R., Snap-back repellers imply chaos in $\mathbb{R}^n$. J. Math. Anal. Appl. 3, 199–223 (1978).

Mascagni, M., In Algorithms for Parallel Processing (Heath, M. T, Ranade, A. and Schreiber, R. S., eds.) Springer-Verlag, New York (1999).

McShane, E. J., Integration. Princeton Univ. Press, New Jersey (1947).

McCluskey, C. C. and van den Driessche, P., Global analysis of two tuberculosis models. J. Dynam. Diff. Eqs. 16(1), 139-166, (2004).

Matsuoka, Y., Saito, T. and Torikai, H., Complicated superstable periodic behavior in piecewise constant circuits with impulsive excitation. Proc. of 2005 International Symposium on Non-linear Theory and its Applications (NOLTA05, Bruges, Belgium), 513–516 (2005).

Matsuoka, Y., Torikai, H. and Saito, T., Chaotic superstable periodic orbits in non-linear circuits. Proc. of 2006 International Conference on Signals and Electronic Systems (ICSES'06, Lodz,Poland), 67–70 (2006(a)).

Matsuoka, Y., Saito, T. and Torikai, H., Complicated superstable behavior in a piecewise constant circuit with impulsive switching, Conference: Circuits and Systems, 2006. ISCAS 2006. Proceedings. 2006 IEEE International Symposium on. (2006(b)).

Matsuoka, Y. and Saito, T., Analysis of co-existence phenomena of superstable periodic orbit and chaos in a non-autonomous piecewise constant circuit. Proc. of 2006 International Symposium on Non-linear Theory and its Applications, (NOLTA'06, Bologna, Italy), 1035-1038 (2006(c)).

Matsuoka, Y. and Saito, T., Superstable phenomena of 1-D map with a trapping window and its application. Proc. of 2007 International Symposium on Non-linear Theory and its Applications, (NOLTA'07, Vancouver, Canada), 55–58 (2007).

Milnor, J., On the concept of Attractor. Commun. Math. Phys. 99, 177–195 (1985(a)).

Milnor, J., On the concept of attractor, correction and remarks. Commun. Math. Phys. 102, 517–519 (1985(b)).

Misiurewicz, M. and Szewc, B., Existence of a homoclinic point for the Hénon map. Comm. Math. Phys. 75(3), 285–291 (1980).

Molgedey, L., Schuchhardt, J. and Schuster, H. G., Suppressing chaos in neural networks by noise. Phys. Rev. Lett. 69, 3717–3719 (1992).

Moore, D. W. and E. Spiegel, A., A thermally excited non-linear oscillator, Astrophys. J. 143, 871–887 (1966).

Morales, C. and Pujals, E., Singular strange attractors on the boundary of Morse-Smale systems. Ann. Sci. école Norm. Sup. 30, 693–717 (1997).

Morales, C., M. Pacifico, J. and Pujals, E., On C^1 robust singular transitive sets for three-dimensional flows. C. R. Acad. Sci. Paris, Série I. 326, 81–86 (1998).

Morales, C., Pacifico, M. J. and Pujals, E. R., Singular hyperbolic systems. Proc. Amer. Math. Soc. 127(11), 3393–3401 (1999).

Morales, C. A. and Pacifico, M. J., Attractors and singularities robustly accumulated by periodic orbits. International Conference on Differential Equations. 1, 2 (Berlin, (1999)) World Sci. Publishing, 64–67 (1999).

Morales, C., Pacifico, M. J. and Pujals, E., Strange attractors across the boundary of hyperbolic systems. Comm. Math. Phys. 211(3), 527–558 (2000).

Morales, C. and Pacifico, M. J., Mixing attractors for 3-flows. Non-linearity. 14, 359–378 (2001).

Morales, C. A., Singular-hyperbolic sets and topological dimension. Dynamical Systems. 18(2), 181–189 (2003).

Morales, C. and Pacifico, M. J., A dichotomy for three-dimensional vector fields. Ergodic Theory Dynam. Systems. 23, 1575–1600 (2003(a)).

Morales, C. and Pacifico, M. J., Transitivity and homoclinic classes for singular-hyperbolic systems. Preprint Série A, 208 (2003(b)).

Morales, C., The explosion of singular hyperbolic attractors. Ergod. Th. Dynam. Syst. 24(2), 577–592 (2004).

Morales, C. A., Pacifico, M. J. and Pujalls, E. R., Robust transitive singular sets for 3-flows are partially hyperbolic attractors or repellers. Annals of Mathematics. 160(2), 375–432 (2004).

Morales, C. A. and Pacifico, M. J., Sufficient conditions for robustness of attractors. Pacific. J. Mathematics. 216(2), 327–342 (2004).

Morales, C., A note on periodic orbits for singular-hyperbolic flows. Discrete Contin. Dyn. Syst. 11(2-3), 615–619 (2004).

Morales, C. A., Pacifico, M. J. and San Martin, B., Expanding Lorenz attractors through resonant double homoclinic loops. SIAM. J Math. Anal. 36(6), 1836–1861 (2005).

Morales, C., Poincaré-Hopf index and singular-hyperbolic sets on 3-balls. Preprint (2006).

Morales, C. A., Pacifico, M. J. P. and San Martin, B., Contracting Lorenz attractors through resonant double homoclinic loops. SIAM. J. Mathematical Analysis. 38, 309–332 (2006).

Morales, C., Singular-hyperbolic attractors with handlebody basins. J. Dynamical and Control Systems. 13(1), 15–24 (2007).

Morales, C., Topological dimension of singular-hyperbolic attractors. Preprint published at IMPA (2008).

Morales, C., Poincaré-Hopf index and partial hyperbolicity. Ann. Fac. Sci. Toulouse Math. XVII(1), 193–206 (2008).

Morita, M., Associative memory with non-monotonic dynamics. Neural Networks. 6, 115–126 (1993).

Nijmeijer, H., Mareels, I. M. Y., An observer looks at synchronization. IEEE Transactions on Circuits and Systems I. 44(10), 874–890 (1997).

Newhouse, S., Diffeomorphisms with infinitely many sinks. Topology. 13, 9–18 (1974).

Nowak, M. A., May, R. M. and Sigmund, K., Immune responses against multiple epitopes. J. Theor. Biol. 175, 352–353 (1995).

Nowak, M. A., Bonhoeffer, S., Shaw, G. M. and May, R. M., Anti-viral drug treatment, Dynamics of resistance in free virus and infected cell populations. J. Theor. Biol. 184, 203–217 (1997).

Nusse, E. H. and Yorke, J. A., Border-collision bifurcations including period two to period threefor piecewise smooth systems. Physica D. 57, 39–57 (1992).

Nusse, H. E. and Tedeschini-Lalli, L., Wild Hyperbolic Sets, Yet no chance for the coexistence of infinitely many KLUS-simple newhouse attracting sets. Commun. Math. Phys. 144, 429–442 (1992).

Nusse, H. E. and Yorke, J. A., Border-collision bifurcations for piecewise smooth one-dimensional maps. Inter. J. Bifur. Chaos. 5, 189–207 (1995).

Nusse, E. H. and Yorke, J. A., Border-collision bifurcations including period two to period threefor piecewise smooth systems. Physica D. 57, 39–57 (1992).

Nusse, H. E. and Yorke, J. A., Basins of Attraction. Science. 27(1), 1376–1380 (1996).

Nunez, P. L., Electric Fields of the Brain. Oxford University Press, New York (1981).

Nunez, P. L., Toward a quantitative description of large-scale neocortical dynamic function and EEG. Behav. Brain Sci. 23(3), 371–437 (2000).

Ogorzalek, M. J., Complex behavior in digital filters. Int. J. Bifurcation Chaos. 2 (1), 11–29 (1992).

Ohnishi, M. and Inaba, N., A singular bifurcation into instant chaos in piecewise-linear circuit. IEEE Transactions on Circuits and Systems I Communications and Computer Sciences. 41(6), 433–442 (1994).

Ottino, J. M., The kinematics of mixingstretching, chaos, and transport. Cambridge Cambridge University Press (1989).

Ottino, J. M., Muzzion, F. J., Tjahjadi, M., Franjione, J. G., Jana, S. C. and Kusch, H. A., Chaos, symmetry, and self-similarity, exploring order and disorder in mixing processes. Science. 257, 754–760 (1992).

Paar, V. and Pavin, N., Intermingled fractal arnold tongues. Phy. Rev. E. 57(2), 1544–1549 (1998).

Pacifico, M. J., Pujals, E. R. and Viana, M., Sensitiveness and SRB measure for singular hyperbolic attractors. Preprint (2002).

Palis, J., On Morse–Smale dynamical systems. Topology. 8, 385–405 (1969).

Palis, J., On the structure of hyperbolic points in Banach spaces. Anais. Acad. Bras. Ciecias. 40 (1968).

Palis, J. and Smale, S., Structural stability theorems, in Global Analysis, Berkeley (1968) in Proc. Sympos. Pure Math., vol. XIV, Amer. Math. Soc. 223–232 (1970).

Palis, J. and Takens, F., Hyperbolicity and the creation of homoclinic orbits. Annals of Mathematics. 125, 337–374 (1987).

Palis, J. and Takens, F., Hyperbolicity and sensitive chaotic dynamic at homoclinic bifurcation. Cambridge University Press (1993).

Palis, J. and Viana, M., High dimension diffeomorphisms displaying infinitely sinks. Ann. Math. 140, 1–71 (1994).

Palis, J., A global view of dynamics and a conjecture on the denseness of finitude of attractors. Asterisque. 261, 339–351 (2000).

Palis, J., A global perspective for non-conservative dynamics. Ann. I. H. Poincaré - AN. 22, 485–507 (2005).

Palis, J., Open questions leading to a global perspective in dynamics. Non-linearity. 21, T37–T43 (2008).

Papaschinopoulos, G. and Schinas, C. J., Invariant boundedness and persistence of non-autonomous difference equations of rational form. Comm. Appl. Non-linear Anal. 6, 71–88 (1999).

Park, J. H. and Kwon, O. M., Chaos, novel criterion for delayed feedback control of time-delay chaotic systems. Solitons & Fractals. 23(2), 495–501 (2005).

Pasemann, F., A simple chaotic neuron. Physica D. 104, 205–211 (1997).

Papaschinopoulos, G. and Schinas, C. J., Invariant boundedness and persistence of non-autonomous difference equations of rational form. Comm. Appl. Non-linear Anal. 6, 71–88 (1999).

Parmananda, P. Stabilization of unstable steady states and periodic orbits in an electrochemical system using delayed feedback control. Physical Review E. 59, 52–66 (1999).

Parui, S. and Banerjee, S., Border collision bifurcations at the change of state-space dimension. Chaos. 12, 1054–1069 (2002).

Pham, Q. and Slotine, J. J. E., Stable concurrent synchronization in dynamics system networks. MIT-NSL Report (2005).

Pecora, L. M. and Carroll, T. L., Synchroniasation in chaotic systems. Phys. Rev. Lett. 64, 821–824 (1990).

Peixoto, M., Structural stability on two-dimensional manifolds. Topology. 1, 101–120 (1962).

Perelson, A. S. and Ribeiro, R. M., Modeling the within-host dynamics of HIV infection. BMC. Biol. 11, 96 (2013).

Pesin, Y., Dynamical systems with generalized hyperbolic attractors, hyperbolic, ergodic and topological properties. Ergodic Theory Dynam. Systems. 12, 123–151 (1992).

Poincaré, H., Sur le problème des trois corps et les equations de la dynamique. Acta. Math. 13, 1–270 (1890).

Pomeau, Y. and Manneville, P., Intermittent transition to turbulence in dissipative dynamical systems. Commun. Math. Phys. 74, 189–97 (1980).

Pospíšil, M., Pospiílova Škripkova, L., Sturm's theorems for conformable fractional differential equation. Math. Commun. 21, 273–281 (2016).

Potapov, A. and Ali, M. K., Robust chaos in neural networks. Phy. Lett. A. 277(6), 310–322 (2000).

Priel, A. and Kanter, I., Robust chaos generation by a perceptron. Europhys. Lett. 51(2), 230–236 (2000).

Przytycki, F., Construction of invariant sets for Anosov diffeomorphisms and hyperbolic attractors. Studia Math. 68, 199–213 (1980).

Pujals, E. R. and Sambarino, M., On homoclinic tangencies, hyperbolicity, creation of homoclinic orbits and variation of entropy. Non-linearity. 13, 921–926 (2000).

Pujals, E. R., Tangent bundles dynamics and its consequences. ICM (2002). III, 1–3 (2002).

Pujals, E. R., On the density of hyperbolicity and homoclinic bifurcations for 3D-diffeomorphisms in attracting regions. Discrete and Continuous Dynamical Systems. 16(1), 179–226 (2006).

Pujals, E. R., Robert, L. and Shub, M., Expanding maps of the circle rerevisited, Positive Lyapunov exponents in a rich family. Ergod. Th. Dynam. Syst. 26, 1931–1937 (2006).

Pujals, E. and Sambarino, M., Integrability on codimension one dominated splitting. Bull. Braz. Math. Soc. 38, 1–19 (2007).

Pujals, E. R., Density of hyperbolicity and homoclinic bifurcations for topologically hyperbolic sets. Discrete and Continuous Dynamical System. 20(2), 337–408 (2008).

Ram, P. S. and Connell, C. M., Global stability for an SEI model of infectious disease with immigration. Applied Math. and Comp. 243, 684–689 (2014).

Regoes, R. R., Antia, R., Garber, D. A., Silvestri, G., Feinberg, M. B. *et al.*, Roles of target cells and virus-specific cellular immunity in primary simian immunodeficiency virus infection. J. Virol. 78, 4866–4875 (2004).

Robert, B. and Robert, C., Border collision bifurcations in a one-dimensional piecewise smooth map for a PWM currentprogrammed H-bridge inverter. Int. J. Control. 75(16-17), 1356–1367 (2002).

Robbin, J., A structural stability theorem. Ann. of Math. 94, 447–493 (1971).

Robinson, C., Structural stability of vector fields. Ann. of Math. 99, 154–175 (1974); Errata in Robinson, C., Ann. of Math. 101, 368 (1975).

Robinson, C., Dynamical systems stability, symbolic dynamics, and chaos. CRC Press (2004).

Rôssler, O. E., An equation for continuous chaos. Phys. Lett. A. 57, 397–98 (1976).

Rôssler, O. E., Continuous chaos-Four prototype equations. Ann. N.Y. Acad. Sci. 31, 376–392 (1979).

Routroy, B., Dutta, P. S. and Banerjee, S., Border collision bifurcations in n-dimensional piecewise linear discontinuous maps. National Conference on Non-linear Systems and Dynamics (2006).

Ruoting, Y., Yiguang, H., Huashu, Q. and Chen, G., Anticontrol of chaos for dynamic systems in p-normal form, A homogeneity-based approach. Chaos, Solitons and Fractals, 25, 687–697 (2005).

Sacker, R. J. and Sell, G. R., Lifting properties in skew product flows with applications to differential equations. Memoir. Amer. Math. Soc. No **190**(1977).

Salem Abdelmalek, Samir Bendoukha, Global asymptotic stability of a diffusive SVIR epidemic model with immigration of individuals, EJDE-2016/284.

Sanchez-Salas, F. J., Sinai-Ruelle-Bowen measures for piecewise hyperbolic transformations. Divulgaciones Matematicas. 9(1), 35–54 (2001).

Sannami, A., The stability theorems for discrete dynamical systems on two-dimensional manifolds. Nagoya Math. J. 90(1983) 1–55.

Sauer, T., Yorke, J. and Casdagli, M., Embedology. J. Stat. Phys. 65, 579–616 (1991).

Schmitz, J. E., Kuroda, M. J., Santra, S., Sasseville, V. G., Simon, M.A. et al., Control of viremia in simian immunodeficiency virus infection by CD8$^+$ lymphocytes. Science. 283, 857–860 (1999).

Schwartz, E. J., Biggs, K. R. H., Bailes, C. et al., HIV dynamics with immune responses, perspectives from mathematical modeling. Curr. Clin. Micro Rpt. 3, 216 (2016).

Shimada, I. and Nagashima, T., A numerical approach to ergodic problem of dissipative dynamical systems. Prog. Theor. Phys. 61, 1605–1616 (1979).

Silva, P., Shilnikov theorem; a tutorial, IEEE transactions. Circuits Syst. I. 40, 675–682 (1993).

Sinai, Y., Dynamical systems with elastic collisions. Russian Math Surveys. 25, 141–92 (1970).

Sinai, Y., Gibbs measures in ergodic theory. Uspehi. Mat. Nauk, 27(4), 21–64 (1972). English translation. Russian. Math. Surveys. 27(4), 21–69 (1972).

Sinai, Y., Gibbs measure in ergodic theory. Russian Math. Surveys. 27, 21–69 (1972).

Sinai, Y., Stochasticity of Dynamical Systems. In Non-linear Waves, edited by Gaponov-Grekhov, A. V., Nauka, Moscow, 192 (1979) (in Russian).

Sivasamy, R., Compound synchronization of different chaotic systems based on active backstepping control. Annual Review of Chaos Theory, Bifurcations and Dynamical Systems. 7, 68–80 (2017).

Sharkovskii, A. N., Coexistence of cycles of a continuous map of a line into itself. Ukranian Mat. Z. 16, 61–71 (1964).

Sharkovsky, A. N. and Chua, L. O., Chaos in some 1–D discontinuous maps that appear in the analysis of electrical circuits. IEEE Trans. Circuits & Systems–I. 40, 722–731 (1993).

Shilnikov, L. P., A case of the existence of a countable number of periodic motions. Sov. Math. Docklady. 6, 163–166 (1965) (translated by S. Puckette).

Shilnikov, L. P., A contribution of the problem of the structure of an extended neighborhood of rough equilibrium state of saddle-focus type. Math. U.S.S.R. Shornik. 10, 91–102 (1970) (translated by F. A. Cezus).

Shilnikov, L. P., The bifurcation theory and quasi-hyperbiloc attractors. Uspehi Mat. Nauk. 36, 240–241 (1981).

Shilnikov, L. P., Bifurcations and chaos in the Shimizu-Marioka system (In Russian) in Methods and qualitative theory of differential equations, Gorky State University, 180–193 (1986). (English translation in Selecta Mathematica Sovietica, 10, 105–117 (1991)).

Shilnikov, L. P., Strange attractors and dynamical models. J. Circuits Syst. Comput. 3(1), 1–10 (1993).

Shuai, J. W., Ji, A. N., Chen, Z. X., Liu, R. T., Wu, B. X., Chaos and hyperchaos in the non-monotonic neuronal model, unpublished.

Shub, M., Topological Transitive Diffeomorphism on T^4. Lecture Notes in Math. 206 Springer-Verlag, New York (1971).

Shub, M., Global stability of dynamical systems. Springer-Verlag, 1987.

Smale, S., Diffeomorphisms with many periodic points, In Differential and Combinatorial Topology, A Symposium in Honor of Marston Morse, 63–70 (1965), S. S. Cairns (ed). Princeton University Press, Princeton, NJ.

Smale, S., Diffeomorphisms with many periodic points, in Differential and Combinatorial Topology". A Symp. In Honor of Marston Morse, 63–80 (1965), Princeton Univ. Press, Princeton, N.J.

Smale, S., Differentiable dynamical systems. Bull. Amer. Math. l Soc, 73, 747–817 (1967).

Smale, S., The Ω-stability theorem. Proc. Sympos. Pure. Math., Vol. XIV, Berkeley, Calif., 289–298 (1968), Amer. Math. Soc., Providence, R.I. 4.

Sompolinsky, H., Cristanti, A. and Sommers, H. J., Chaos in random neural networks. Phys. Rev. Lett. 61, 259–260 (1988).

Sprott, J. C., Some simple chaotic flows. Phys. Rev. E. 50, R647–R650 (1994).

Sprott, J. C., Some simple chaotic jerk functions. Am. J. Phys. 65, 537–543 (1997(a)).

Sprott, J. C., Simplest dissipative chaotic flow. Phys. Lett. A. 228, 271–274 (1997(b)).

Sprott, J. C., Chaos and Time-Series Analysis. Oxford University Press (2003).

Sprott, J. C., Chaotic dynamics on large networks. Chaos. 18, 023135 (2008).

Stamova, I. M. and Stamov, G. T., Functional and Impulsive Differential Equations of Fractional Order, Qualitative Analysis and Applications, CRC Press, Taylor and Francis Group, Boca Raton (2017).

Steriade, M., McCormick, D. A. and Sejnowski, T. J., Thalamocortical oscillations in the sleeping and aroused brain. Science. 262, 679–685 (1993).

Strafford, M. A., Corey, L., Cao, Y., Daar, E. S., Ho, D. D. and Perelson, A. S., Modeling plasma virus concentration during primary HIV infection. J. Theor. Biol. 203, 285–301 (2000).

Stofer, D. and Palmer, K. J., Rigorous verification of chaotic behaviour of maps using validated shadowing. Non-linearity. 12(6), 1683–1698 (1999).

Sun, Y. J., Hsieh, J. G. and Hsieh, Y. C., Exponential stability criteria for uncertain retarded systems with multiple delays. J. Math. Anal. Appl. 20, 430–446 (1996).

Sun, Y. J., Solution bounds of generalized Lorenz chaotic systems. Chaos, Solitons & Fractals. 40(2), 691–696 (2009).

Taixiang, S. and Hongjian, X., On the basin of attraction of the two cycle of the difference equation. J. Difference Equations and Applications. 13(10), 945–952 (2007).

Tan, Z., Hepburn, B. S., Tucker, C. and Ali, M. K., Pattern recognition using chaotic neural networks. Discrete Dynamics in Nature and Society. 2, 243–247 (1998).

Thieme, H. R. and Castillo-Chavez, C., On the Role of Variable Infectivity in the dynamics of the Human Immunodeficiency Virus epidemic. In, Castillo-Chavez C. (eds) Mathematical and Statistical Approaches to AIDS Epidemiology. Lecture Notes in Biomathematics, vol **83**. Springer, Berlin, Heidelberg (1989).

Tang, K. S., Man, K. F., Zhong, G. Q. and Chen, G., Generating chaos via $x|x|$, IEEE Trans. Circuits Syst., I, Fundam. Theory Appl. 48, 636 (2001).

Tigan, G., Opriş, D., Analysis of a $3D$ chaotic system, Chaos. Solitons & Fractals. 36(5), 1315–1319 (2008).

Tirozzi, B. and Tsokysk, M., Chaos in highly diluted neural networks. Europhys. Lett. 14, 727–732 (1991).

Tsuji, R. and Ido. S., Computation of Poincar e map of chaotic torus magnetic eld line using parallel computation of data table and its interpolation. Parallel Computing in Electrical Engineering, PARELEC apos; **02**, 386–390 (2002).

Tufillaro, N., Abbott, T. and Reilly, J., An Experimental Approach to Non-linear Dynamics and Chaos. Addison-Wesley, Redwood City, CA (1992).

Tucker, W., A Rigorous ODE Solver and Smale's 14th Problem. Found. Comput. Math. 2, 153–117 (2002(a)).

Tucker, W., Computing accurate Poincar e maps. Physica. D, 171(3), 127–137 (2002(b)).

Tunc, C., The boundedness to non-linear differential equations of fourth order with delay. Non-linear Studies. 17(1), 47–56 (2010(a)).

Tunc, C., Bound of solutions to third-order non-linear differential equations with bounded delay. Journal of the Franklin Institute. 347, 415–425 (2010(b)).

Umberger, D. K. and Farmer, J. D., Fat fractals on the energy surface. Phys. Rev. Lett. 55(7), 661–664 (1985).

Vaidyanathan, S., Global chaos synchronization of a novel 3-D chaotic system with two quadratic non-linearities via active and adaptive control. In, Azar, A. T, Vaidyanathan, S (eds) Advances in Chaos Theory and Intelligent Control, Studies in computational intelligence, Springer-Verlag, Germany (2016(a)).

Vaidyanathan, S. and Azar, A. T., Qualitatife study and adaptive control of a novel 4-D hyperchaotic system with three quadratic non-linearities. In Azar A. T., Vaidyanathan, S. (eds.) Advances in Chaos Theory and Intelligent Control, Studies in computational intelligence, Springer-Verlag, Germany (2016(b)).

Vaidyanathan, S., A novel 3-D conservative Jerk chaotic system with two quadratic non-linearities and its adaptive control. In, Azar, A. T., Vaidyanathan, S. (eds.) Advances in Chaos Theory and Intelligent Control, Studies in computational intelligence, Springer-Verlag, Germany (2016(c)).

Vaidyanathan, S., Volos, C. K. and Pham, V. T., Hyperchaos, control, synchronization and circuit simulation of a novel 4-D hyperchaotic system with three quadratic non-linearities. In Azar, A. T., Vaidyanathan, S. (eds.) Advances in Chaos Theory and Intelligent Control, Studies in computational intelligence, Springer-Verlag, Germany (2016(d)).

Vaidya, N. K., Ribeiro, R. M., Perelson, A. S. and Kumar, A., Modeling the effects of Morphine on Simian Immunodeficiency Virus dynamics. PLoS Comput Biol, 12(9), e1005127 (2016).

Wang, X., Chen, G. and Yu, X., Anticontrol of chaos in continuous-time systems via time-delayed feedback. Chaos. 10, 771 (2000).

Wang, X. F. and Chen, G. On feedback anticontrol of discrete chaos. Int. J. Bifur. Chaos. 9, 1435–1441 (1999).

Wang, X. and Chen, G., Chaotifying a stable LTI system by tiny feedback control, IEEE Trans. Circuits Syst., I, Fundam. Theory Appl. 47, 410 (2000(a)).

Wang, X. F. and Chen, G. Chaotifying a stable map via smooth small amplitude high-frequency feedback control. Int. J. Circ. Theory Appl. 28, 305–312 (2000(b)).

Wang, X., Chen, G. and Yu, X., Anticontrol of chaos in continuous-time systems via time-delayed feedback. Chaos. 10, 771 (2000).

Weihong, H., Theory of adaptive adjustmen. Discrete Dynamics in Nature and Society. 5, 247–263 (2001).

Weihong, H., Stabilizing high-order discrete dynamical systems by a lagged adaptive adjustment mechanism. Discrete Dynamics in Nature and Society. 7(1), 59–68 (2002).

Wein, L. M., D'Amato, R. M. and Perelson, A. S., Mathematical analysis of antiretroviral therapy aimed at HIV-1 eradication or maintenance of low viral loads. J. Theor. Biol. 192, 81–98 (1988).

Wen-Xin, Q. and Chen, G., On the boundedness of solutions of the Chen system. Journal of Mathematical Analysis and Applications. 329(1), 445–451 (2007).

Widrow, B. and Lehr, M. A., 30 years of Adaptive Neural Networks, Perceptron, Madaline, and Backpropagation. Proc. IEEE. 78(9), 1415–1442 (1990).

Williams, F. R., Classification of one-dimensional attractors. Proc. Symp. in Pure Math. 361–393 (1970).

Williams, R. F., Expanding attractors. IHES Publication, Math. 43, 169–203 (1974).

Williams, R. F., The Structure of Lorenz Attractors. Publ. Math. IHES. 50, 321–347 (1979).

Weisstein, E., Homoclinic tangle. In Smale Horseshoe. Homoclinic Point. Eric Weisstein's World of Mathematics., http//mathworld.wolfram.com (2002).

Wodarz, D., Lloyd, A. L., Jansen, V. A. A. and Nowak, M. A., Dynamics of macrophage and T cell infection by HIV. J. Theor. Biol. 196, 101–113 (1999).

Wodarz, D., Modeling T cell responses to antigenic challenge. J. Pharmacokinet Phar-macodyn. 41, 415–29 (2014).

Wolf, D. M., Varghese, M. and Sanders, S. R., Bifurcation of power electronic circuits. J. Franklin. Inst. 331B(6), 957–999 (1994).

Wolf, C., Generalized physical and SRB measures for hyperbolic diffeomorphisms. J. Statistical Physics, 122(6), 1111–1138 (2006).

Xavier, J. C. and Rech, P. C., Regular and chaotic dynamics of the Lorenz–Stenflo system. International Journal of Bifurcation and Chaos. 20(1), 145–152 (2010).

Xia, X. and Moog, C. H., Identifiability of non-linear systems with application to HIV/AIDS models. IEEE Trans. Autom. Contr. 48, 330–336 (2003).

Yang, X. S. and Tang, Y., Horseshoes in piecewise continuous maps. Chaos, Solitons and Fractals. 19, 841–845 (2004).

Yoshizawa, T., Stability Theory by Lyapunov Second Method. The Math. Soci. of Japan, Tokyo (1966).

Yuan, G. H., Shipboard crane control, simulated data generation and border-collision bifurcations. Ph. D. dissertation, Univ. of Maryland, College Park, USA (1997).

Yuan, G. H., Banerjee, S., Ott, E. and Yorke, J. A., Border collision bifurcations in the buck converter. IEEE Trans. Circuits & Systems–I 45, 707–716 (1998).

Zhang, H. Z. and Chen, G., Single-input multi-output state-feedback chaotification of general discrete systems. Int. J. Bifur. Chaos. 14(9), 3317–3323 (2004).

Zhang, G., Chen, G., Chen, T. and Lin, Y., Analysis of a type of non-smooth dynamical systems. Chaos, Solitons and Fractals. 30, 1153–1164 (2006).

Zeraoulia Elhadj and Hamri Nasr-eddine, A new chaotic attractor from modified Chen equation. Far East Journal of Applied Mathematics. 18(2), 185–198 (2005(a)).

Zeraoulia Elhadj, A new chaotic attractor from 2-D discrete mapping via border-collision period doubling scenario, Discrete dynamics in nature and society, Volume 2005, 235–238 (2005(b)).

Zeraoulia Elhadj, A new switching piecewise-linear chaotic attractor with two equilibria of saddle-focus type. Non-linear Phenomena in Complex System. 9(4), 399–402 (2006).

Zeraoulia Elhadj, Analysis of a new chaotic system with three quadratic non-linearities. Dynamics of Continuous, Discrete and Impulsive Systems. 14, 603–613 (2007(a)).

Zeraoulia Elhadj, On the dynamics of a n-D piecewise linear map. Electronic Journal of Theoretical Physics. 4(14), 1–7 (2007(b)).

Zeraoulia Elhadj, A new 3-D piecewise linear system for chaos generation. Radioengeneering. 16(2), 40–43 (2007(c)).

Zeraoulia Elhadj, Dynamical analysis of a 3-D chaotic system with only two quadratic non-linearities. Journal of Systems Science and Complexity. 21, 67–75 (2008(a)).

Zeraoulia Elhadj and Sprott, J. C., A two-dimensional discrete mapping with C^∞-multifold chaotic attractors. Electronic Journal of Theorical Physics. 5(17), 111–124 (2008(b)).

Zeraoulia Elhadj, Analysis of a new three-dimensional quadratic chaotic system. Radioengeneering. 17(1), 9–13 (2008(c)).

Zeraoulia Elhadj and Sprott, J. C., On the robustness of chaos in dynamical systems, Theories and applications. Front. Phys. China. 3, 195–204 (2008(d)).

Zeraoulia Elhadj, On the occurrence of chaos via different routes to chaos, period doubling and border-collision bifurcations. Translated from Sovremennaya Matematika i Ee Prilozheniya (Contemporary Mathematics and Its Applications), Vol. **61**, Optimal Control, (2008(e)).

Zeraoulia Elhadj and Sprott, J. C., The discrete hyperchaotic double scroll. International Journal of Bifurcations & Chaos. 19(3), 1023–1027 (2009(a)).

Zeraoulia Elhadj and Sprott, J. C., Classification of three-dimensional quadratic diffeomorphisms with constant Jacobian. Front. Phys. China. 4(1) 111–121 (2009(b)).

Zeraoulia Elhadj and Sprott, J. C., Some explicit formulas of Lyapunov exponents for 3D quadratic mappings. Front. Phys. China. 4(4), 549–555 (2009(c)).

Zeraoulia Elhadj and Sprott, J. C., A new simple 2-D piesewise lineare map. Journal of Systems Science and Complexity. 23(2), 379–389 (2010(a)).

Zeraoulia Elhadj and Sprott, J. C., Chaotifying 2-D linear maps via a piecewise linear controller function. Non-linear Oscillations. 13(3), 352–360 (2010(b)).

Zeraoulia Elhadj, On the occurrence of chaos via different routes to chaos, period doubling and border-collision bifurcations. Journal of Mathematical Sciences. 161(2), 194–199 (2009(c)).

Zeraoulia Elhadj, An example of superstable quadratic mapping of the space. Facta Universitatis, SER., Elec. & Energ. 22(3), 373–378 (2009(d)).

Zeraoulia Elhadj and Sprott, J. C., About the boundedness of 3D continuous time quadratic systems. Non-linear Oscillations. 13(4), 550–557 (2010(a)).

Zeraoulia Elhadj, A familly of superstable n-D mappings. International Journal of Artificial Life Research. 1(1), 72–77 (2010(b)).

Zehrour, O. and Zeraoulia Elhadj, Ellipsoidal Chaos. Non-linear Studies. 19(1), 71–77 (2010(c)).

Zeraoulia Elhadj and Sprott, J. C., 2-D quadratic maps and 3-D ODE systems. A Rigorous Approach, World Scientific Series on Non-linear Science Series A, no **73** (2010(d)).

Zeraoulia Elhadj, Robust chaos in piecewise non-smooth map of the plane. Journal of Mathematical Sciences. 177(3), 366–372 (2011(a)).

Zeraoulia Elhadj and Sprott, J. C., On the dynamics of a new 2-D rational mapping. International Journal of Bifurcations & Chaos. 21(1), 155–160 (2011(b)).

Zeraoulia Elhadj, The discrete butterfly. Non-linear Studies. 18(3), 529–539 (2011(c)).

Zeraoulia Elhadj and Sprott, J. C., A universal non-linear control law for the synchronization of arbitrary 4-D continuous-time quadratic systems. Electronic Journal of Theoretical Physics. 8(25), 267–272 (2011(d)).

Zeraoulia Elhadj and Sprott, J. C., Robustification of chaos in 2-D maps. Advances in Complex Systems. 14(6), 817-827 (2011(e)).

Zeraoulia Elhadj, Generating fully bounded chaotic attractors. International Journal of Artificial Life Research. 2(3), 36–42 (2011(f)).

Zeraoulia Elhadj and Sprott, J. C., Robust chaos and its applications. World Scientific Series on Non-linear Science Series A, no **79**, (2011(g)).

Zeraoulia Elhadj, Sprott, J.C., (Editors), Frontiers in the study of chaotic dynamical systems with open problems. World Scientific Series on Non-linear Science Series B, Vol. **16** (2011(h)).

Zeraoulia Elhadj, Models and Applications of Chaos Theory in Modern Sciences, Science Publishers, (An Imprint of Edenbridge Ltd., British Isles), (2011(m)).

Zeraoulia Elhadj and Sprott, J. C., A unified piecewise smooth chaotic mapping that contains the Hénon and the Lozi systems. Annual Review of Chaos Theory, Bifurcations and Dynamical Systems. 1, 50-60 (2012(a)).

Zeraoulia Elhadj and Sprott, J. C., Boundedness of certain forms of jerky dynamics. Qualitative Theory of Dynamical Systems. 11(2), 199–213 (2012(b)).

Zeraoulia Elhadj, Hyperbolicity and its consequences regarding robustness and density, In Advanced Topics in Non-linear Chaotic Systems (Z. Elhadj, ed.) LAP LAMBERT Academic Publishing, 76–126 (2012(c)).

Zeraoulia Elhadj and Sprott, J. C., An example of a fully bounded chaotic sea that surrounds an infinite set of nested invariant tori. Palestine Journal of Mathematics 1, 71–73 (2012(d)).

Zeraoulia Elhadj and Sprott, J. C., Hyperbolification of dynamical systems, The case of continuous-time systems. Journal of Experimental and Theoretical Physics 115(2), 356–360 (2012(e)).

Zeraoulia Elhadj, Sprott, J. C., Transformation of 4-D dynamical systems to hyperjerk form. Palestine Journal of Mathematics. 2(1), 38–45 (2013(a)).

Zeraoulia Elhadj, Lozi Mappings, Theory and Applications. Science Publishers (2013(b)).

Zgliczynski, P., Computer assisted proof of the horseshoe dynamics in the Hénon map. Random & Computational Dynamics. 5, 1–17 (1997).

Zhang, W., Agarwal, R. P., Construction of mappings with attracting cycles. Computers and Mathematics with Applications. 45, 1213–1219 (2003).

Zhirov, A. Yu., Hyperbolic attractors of diffeomorphisms of oriented surfaces. I. Coding, classification, and coverings. Mat. Sb, 1856, 3–50 (1994(a)), English transl. Russian Acad. Sci. Sb. Math. 82, 135–174 (1995).

Zhirov, A. Yu., Hyperbolic attractors of diffeomorphisms of oriented surfaces. II. Enumeration and application to pseudo-Anosov diffeomorphisms. Mat. Sb. 1859, 29–80 (1994(b)), English transl. Russian Acad. Sci. Sb. Math. 83, 23–65 (1996).

Zhirov, A. Yu., Hyperbolic attractors of diffeomorphisms of oriented surfaces. III. A classification algorithm. Mat. Sb. 1862 (1995) 69–82, English transl. Sb. Math. 186 (1995) 221–244.

Zhou, X. and Zhang, X., Control of discrete-chaotic dynamic system based on improved relativity. TENCON'02. Proceedings. 2002 IEEE Region 10 Conference on Computers Communication, Control and Power Engineering, 28–31 Oct. 2002, 3, 1393–1396 (2002).

Index

1-D singular map, 83
2-D Hénon-like mapping with unknown bounded functions, 233
2-D piecewise smooth system, 86
3-D cancer model, 181

fragility of chaos, 103
presence of diffusion, 316

activation function, 101
AIDS, 48
algebraically simplest hyperchaotic snap system, 359
amount of HIV, 58
Anosov diffeomorphism, 31
antibodies, 49
antibody escape rates, 49
artificial neural network, 114
asymptotically stable, 244

Bandcount adding scenario, 13
Basin boundaries, 14
basin of attraction, 13
bewildering complexity, 135
border, 79
border collision bifurcations, 80
boundary crisis, 95
bounded derivative, 193
boundedness of 3-D quadratic continuous-time autonomous systems, 197
butterfly, 10

cannot exhibit chaotic behavior, 345
Cantor-like set, 5
Cardan method's, 175
CD4 lymphocyte population, 70

CD4 T cells, 47
CD4? lymphocyte dynamics, 56
CD8 lymphocyte population, 70
CD8? T cell, 49
chaos control problem, 181
chaos in neural networks, 97
chaotic attractor, 131
chaotic jerky dynamics, 339
chaotic orbit, 3
chaotification, 124
Chen system, 144
Cherry-like singular-hyperbolic set, 300
Chua attractor, 12
classical urban models, 71
co-existence of the full-state hybrid function projective synchronization, 155
coexistence phenomenon, 10
coexisting attractors, 74
Collet-Eckmann maps, 28
comparison theorem, 253
compound synchronization, 143
contained in the sphere, 225
control matrix, 157
controlling homoclinic chaotic attractor, 182
Converse of Sarkovskii's Theorem, 37
crackle system, 359
critical boundary, 85
critical point, 231
critically robust singular-hyperbolic, 18
CTL escape, 49
cubic non-linearities, 340

dangerous bifurcation, 94

degree of chaos, 99

delayed differential equations, 322

derivative, 199

descriptor map, 98

Diluted circulant networks, 114

Direct Lyapunov's method for a fractional-like system, 264

discontinuities, 204

discontinuity, 85

discontinuous piecewise linear switching map, 129

discrete butterfly, 123

Discrete hyperchaotic double scroll map, 117

disease-free equilibrium, 55

disease-free population, 55

disease-free state, 51

distributed delay model, 49

dynamics of the Human Immunodeficiency Virus (HIV), 67

effects of morphine, 65

ellipsoid, 200

elliptic operator, 316

endemic equilibrium, 53

endemic state, 51

enumeration of attractors, 16

existence of a fixed point, 127

existence of a solution, 203

exponential stability, 251

exponentially stabilizable, 252

expression for the transformation, 349

externally driven non-linear oscillators, 343

feedforward neural network, 114

four-dimensional ellipsoidal surface, 227

four-dimensional vector fields, 346

fractional-like derivative, 261

fractional-like integral, 262

fundamental matrix, 244

gain parameter, 97

generalized Hénon map, 169

generalized Lyapunov-like function, 253

Generalized Poincaré map, 3

generic structure, 100

geometric Lorenz attractor, 300

global superstability of the 3-D quadratic map, 280

globally superstable 1-D quadratic mappings, 275

Hénon map, 13, 233

Hausdorff dimension, 25

Hessian matrix, 231

heteroclinic orbit, 40

HIV-1 viral load, 70

homoclinic bifurcation , 39

homoclinic orbit, 40

homoclinic point, 6

homoclinic tangency, 6

homogeneous Neumann boundary condition, 316

Human immunodeficiency virus, 46

hyper-chaotification method, 193

hyperbolic attractor, 287

hyperbolic repeller, 287

hyperbolic saddle focus, 40

hyperbolic theory of dynamical systems, 285

hyperchaotic attractor, 131

hyperchaotic model, 224

hyperjerk system, 343

hyperjerky dynamics, 220

immune system, 47

infection-free equilibrium, 324

influx-constraining function, 57

input-to-state stability, 332

instant chaos, 95

interconnection matrix, 101

Intermittent chaos, 83

invariant manifold, 6

inverse full-state hybrid function projective synchronization, 155

Inverse optimization problem, 331

inverse period doubling bifurcation, 85

invertible transformation, 207
IS-stabilizable, 332

jerk function, 342
jerky dynamic systems, 205
jerky dynamical system, 336
jerky dynamics, 206
Jordan normal forms, 305
Jurdjevic-Quinn method, 329

Lû system, 144
Lasalle's invariance principle theorem, 259
Lebesgue measure, 23
Li-Chen-Marotto Theorem, 37
Lie bracket operator, 329
limiting system, 51
Lorenz system, 10
Lorenz-Haken system, 224
Lorenz-like singularity, 19, 300
Lorenz-Stenflo system, 224
lower left fractional-like derivative, 263
lower right fractional-like derivative, 263
Lyapunov exponent, 27
Lyapunov function, 199
Lyapunov functional, 318
Lyapunov stability, 243
Lyapunov's second or direct method, 245
Lyapunov-like function, 253

Marotto theorem, 37
maximal solution, 253
maximum dissipation, 74
maximum Lyapunov exponent, 179
maximum point, 200
maximum value, 200
minimal chaotic attractors, 215
minimal chaotic flows, 339
minus infinity Lyapunov exponent, 271
model of electroencephalogram, 110
Modeling of HIV infection dynamics, 48

Models of intracellular eclipse, 49
Moore and Spiegel system, 340
multi-fold strange attractors, 238
multivariate function analysis, 225

narrow range, 340
natural marriage, 48
neural processes, 97
non-isolated fixed point, 134
non-linear differential equations with bounded delay, 203
non-linear elements, 102
non-linear feedback, 173
non-negative semi-definite Lyapunov function, 267
non-smooth Lyapunov functions, 251
Non-trivial hyperbolic attractor, 287
non-trivial periodic solution, 247
non-wandering set, 293
normal form, 79
number of cytotoxic T cells, 58

one wing, 179
one-dimensional piecewise smooth system, 78
optimization problem, 330

parameterized family, 201
perceptron, 114
period 3 theorem, 34
period two to period three, 134
persistence, 16
pincers map, 105
Poincaré map, 1
polynomial function, 213
pre-existence, 71

quadratic jerky equation, 207
quadratic map, 34
quasi-synchronization, 169
quasi-synchronization errors, 170
quasiattractors, 75
quasiquadratic maps, 29

Rôssler system, 342
radially unbounded, 330

rational functions, 340
reproductive number, 51
retrovirus, 47
Riccati equation, 331
Riddled basins, 14
rigorous mathematical proof of chaos, 128
Robust chaos, 11
robust chaos, 83
robust singular transitive set, 294
robust transitive, 17
robustification, 186
rotational system, 225
route to chaos, 74
Runge-Kutta algorithm, 179
rural-urban interaction models, 73

S-unimodal maps, 29
saddle point, 38
Sarkovskii's ordering, 36
Sarkovskii's Theorem, 36
saturated modulus, 100
scaling behavior, 86
Schwarzian derivative, 84
second additive compound, 315
sensitive dependence on initial conditions, 10
sensitivity to initial data, 297
sexually active homosexual population, 50
sexually active individuals, 54
sexually active subpopulations, 51
Shilnikov theorems, 38
sigmoid neurons, 102
simplest chaotic hyperjerk system, 352
simplest cubic case, 341
singular Axiom A vector fields, 293
singular hyperbolic attractor, 296
singular-hyperbolic attractors, 30
singularities, 204
sinusoidally driven oscillator, 344
Smale horseshoe, 3
Snap-back repeller, 37

spatially SVIR model of infectious disease epidemics, 312
SRB (Sinai-Ruelle-Bowen) measure, 23
stability of the chaotic attractor, 95
stability performances, 329
stabilization problems, 251
stabilizing dynamical system, 267
Stable period-2 orbit, 134
stable period-3 orbit, 134
standard models, 48
statistical signature of intermittency, 83
steady state solution, 316
strange attractors, 10
structural stability for 3-D quadratic mappings, 302
structured uncertainties, 181
superstable, 271
suppressing chaotic behaviors, 173
switching law, 129
switching piecewise-linear system, 174
symbolic dynamics, 7

tangent bifurcation, 85
The first bifurcation theorem, 34
The second bifurcation theorem, 34
thermal convection, 336
time-independent Hamiltonian systems, 248
time-varying differential equations, 251
timing, 49
top system, 360
topological conjugacy problem, 16
topological entropy, 25
total viral load, 60
transverse homoclinic point , 39
triggers, 49
two wings, 179
Type-I intermittency, 84
Type-II intermittency , 84
Type-III intermittency, 84

ultimate bound, 224

Unbounded orbits, 134
uniformly exponentially stable, 254
uniformly hyperbolic, 289
unimodal map, 28, 84
universal control law, 153
universal synchronization approach,
        151
unstable equilibrium, 250
upper and lower bounds, 30
upper bound, 198
upper left fractional-like derivative,
        263
upper right fractional-like derivative,
        263

vaccine strategies, 49
viral dynamics, 64

weak attractor, 15
Weierstrass Intermediate Value Theo-
        rem, 33
weight-space exploration, 98

zero Lyapunov exponent, 134